MW01156541

Texts in Applied Mathematics **6**

Editors
J.E. Marsden
L. Sirovich
M. Golubitsky
W. Jäger

Advisor
G. Iooss
P. Holmes

Springer
New York
Berlin
Heidelberg
Barcelona
Budapest
Hong Kong
London
Milan
Paris
Singapore
Tokyo

Texts in Applied Mathematics

1. *Sirovich:* Introduction to Applied Mathematics.
2. *Wiggins:* Introduction to Applied Nonlinear Dynamical Systems and Chaos.
3. *Hale/Koçak:* Dynamics and Bifurcations.
4. *Chorin/Marsden:* A Mathematical Introduction to Fluid Mechanics, 3rd ed.
5. *Hubbard/West:* Differential Equations: A Dynamical Systems Approach: Ordinary Differential Equations.
6. *Sontag:* Mathematical Control Theory: Deterministic Finite Dimensional Systems, 2nd ed.
7. *Perko:* Differential Equations and Dynamical Systems, 2nd ed.
8. *Seaborn:* Hypergeometric Functions and Their Applications.
9. *Pipkin:* A Course on Integral Equations.
10. *Hoppensteadt/Peskin:* Mathematics in Medicine and the Life Sciences.
11. *Braun:* Differential Equations and Their Applications, 4th ed.
12. *Stoer/Bulirsch:* Introduction to Numerical Analysis, 2nd ed.
13. *Renardy/Rogers:* A First Graduate Course in Partial Differential Equations.
14. *Banks:* Growth and Diffusion Phenomena: Mathematical Frameworks and Applications.
15. *Brenner/Scott:* The Mathematical Theory of Finite Element Methods.
16. *Van de Velde:* Concurrent Scientific Computing.
17. *Marsden/Ratiu:* Introduction to Mechanics and Symmetry.
18. *Hubbard/West:* Differential Equations: A Dynamical Systems Approach: Higher-Dimensional Systems.
19. *Kaplan/Glass*: Understanding Nonlinear Dynamics.
20. *Holmes:* Introduction to Perturbation Methods.
21. *Curtain/Zwart:* An Introduction to Infinite-Dimensional Linear Systems Theory.
22. *Thomas:* Numerical Partial Differential Equations: Finite Difference Methods.
23. *Taylor:* Partial Differential Equations: Basic Theory.
24. *Merkin:* Introduction to the Theory of Stability.
25. *Naber:* Topology, Geometry, and Gauge Fields: Foundations.
26. *Polderman/Willems:* Introduction to Mathematical Systems Theory: A Behavioral Approach.
27. *Reddy:* Introductory Functional Analysis: with Applications to Boundary-Value Problems and Finite Elements.
28. *Gustafson/Wilcox:* Analytical and Computational Methods of Advanced Engineering Mathematics.
29. *Tveito/Winther:* Introduction to Partial Differential Equations: A Computational Approach.
30. *Gasquet/Witomski:* Fourier Analysis and Applications: Filtering, Numerical Computation, Wavelet.
31. *Bremaud:* Markov Chains: Gibbs Fields and Monte Carlo.
32. *Durran:* Numerical Methods for Wave Equations in Geophysical Fluid Dynamics.

Eduardo D. Sontag

Mathematical Control Theory

Deterministic Finite Dimensional Systems

Second Edition

With 26 Illustrations

Springer

Eduardo D. Sontag
Department of Mathematics
Rutgers University
New Brunswick, NJ 08903
USA

Series Editors

J.E. Marsden
Control and Dynamical Systems 107-81
California Institute of Technology
Pasadena, CA 91125
USA

L. Sirovich
Division of Applied Mathematics
Brown University
Providence, RI 02912
USA

M. Golubitsky
Department of Mathematics
University of Houston
Houston, TX 77204-3476
USA

W. Jäger
Department of Applied Mathematics
Universität Heidelberg
Im Neuenheimer Feld 294
69120 Heidelberg
Germany

Mathematics Subject Classification (1991): 93xx, 68Dxx, 49Exx, 34Dxx

Library of Congress Cataloging-in-Publication Data
Sontag, Eduardo D.
Mathematical control theory : deterministic finite dimensional systems / Eduardo D. Sontag. — 2nd ed.
p. cm. — (Texts in applied mathematics ; 6)
Includes bibliographical references and index.
ISBN 0-387-98489-5 (hardcover : alk. paper)
1. Control theory. 2. System analysis. I. Title. II. Series.
QA402.3.S683 1998
515′.64—dc21 98-13182

Printed on acid-free paper.

© 1998 Springer-Verlag New York, Inc.
All rights reserved. This work may not be translated or copied in whole or in part without the written permission of the publisher (Springer-Verlag New York, Inc., 175 Fifth Avenue, New York, NY 10010, USA), except for brief excerpts in connection with reviews or scholarly analysis. Use in connection with any form of information storage and retrieval, electronic adaptation, computer software, or by similar or dissimilar methodology now known or hereafter developed is forbidden.
The use of general descriptive names, trade names, trademarks, etc., in this publication, even if the former are not especially identified, is not to be taken as a sign that such names, as understood by the Trade Marks and Merchandise Marks Act, may accordingly be used freely by anyone.

Production managed by Terry Kornak; manufacturing supervised by Thomas King.
Camera-ready copy prepared from the author's LaTex files.
Printed and bound by Edwards Brothers, Inc., Ann Arbor, MI.
Printed in the United States of America.

9 8 7 6 5 4 3 2 1

ISBN 0-387-98489-5 Springer-Verlag New York Berlin Heidelberg SPIN 10670158

Series Preface

Mathematics is playing an ever more important role in the physical and biological sciences, provoking a blurring of boundaries between scientific disciplines and a resurgence of interest in the modern as well as the classical techniques of applied mathematics. This renewal of interest, both in research and teaching, has led to the establishment of the series Texts in Applied Mathematics (TAM).

The development of new courses is a natural consequence of a high level of excitement on the research frontier as newer techniques, such as numerical and symbolic computer systems, dynamical systems, and chaos, mix with and reinforce the traditional methods of applied mathematics. Thus, the purpose of this textbook series is to meet the current and future needs of these advances and to encourage the teaching of new courses.

TAM will publish textbooks suitable for use in advanced undergraduate and beginning graduate courses, and will complement the Applied Mathematics Sciences (AMS) series, which will focus on advanced textbooks and research-level monographs.

Preface to the Second Edition

The most significant differences between this edition and the first are as follows:

- Additional chapters and sections have been written, dealing with:
  - nonlinear controllability via Lie-algebraic methods,
  - variational and numerical approaches to nonlinear control, including a brief introduction to the Calculus of Variations and the Minimum Principle,
  - time-optimal control of linear systems,
  - feedback linearization (single-input case),
  - nonlinear optimal feedback,
  - controllability of recurrent nets, and
  - controllability of linear systems with bounded controls.

- The discussion on nonlinear stabilization has been expanded, introducing the basic ideas of control-Lyapunov functions, backstepping, and damping control.

- The chapter on dynamic programming and linear-quadratic problems has been substantially edited, so that the material on linear systems can be read in a fully independent manner from the nonlinear preliminaries.

- A fairly large number of errors and typos have been corrected.

- A list of symbols has been added.

I would like to strongly encourage readers to send me suggestions and comments by e-mail (*sontag@control.rutgers.edu*), and also to visit the following Web site:

http://www.math.rutgers.edu/~sontag/

where I expect to post updates, additional material and references, links, and errata.

The current contents of the text far exceed what can be done in a year, if all material is covered and complete proofs are given in lectures. However, there are several ways to structure a year-long course, or two such courses, based on parts of the book. For example, one may cover only the linear theory, skipping the optional sections as well as the chapters on nonlinear controllability and multiplier (variational) methods. A separate course, fairly independent, could cover the more advanced nonlinear material. Ultimately, the topics should reflect student and faculty background and interests, and I'll be happy to discuss syllabi with potential instructors.

I wish to thank all those colleagues, students, and readers who have sent me suggestions and comments, including in particular Brian Ingalls, Gerardo Lafferriere, Michael Malisoff, and Konrad Reif. A most special acknowledgment goes to Jose Luis Mancilla Aguilar and Sarah Koskie, who pointed out a very large number of typos and errors, and proposed appropriate corrections. Of course, many mistakes surely remain, and I am solely responsible for them. I also reiterate my acknowledgment and thanks for the continued support from the Air Force Office of Scientific Research, and to my family for their infinite patience.

Eduardo D. Sontag

Piscataway, NJ
May, 1998.

Preface to the First Edition

This textbook introduces the basic concepts and results of mathematical control and system theory. Based on courses that I have taught during the last 15 years, it presents its subject in a self-contained and elementary fashion. It is geared primarily to an audience consisting of mathematically mature advanced undergraduate or beginning graduate students. In addition, it can be used by engineering students interested in a rigorous, proof-oriented systems course that goes beyond the classical frequency-domain material and more applied courses.

The minimal mathematical background that is required of the reader is a working knowledge of linear algebra and differential equations. Elements of the theory of functions of a real variable, as well as elementary notions from other areas of mathematics, are used at various points and are reviewed in the appendixes. However, the book was written in such a manner that readers not comfortable with these techniques should still be able to understand all concepts and to follow most arguments, at the cost of skipping a few technical details and making some simplifying assumptions in a few places —such as dealing only with piecewise continuous functions where arbitrary measurable functions are allowed. In this dual mode, I have used the book in courses at Rutgers University with mixed audiences consisting of mathematics, computer science, electrical engineering, and mechanical and aerospace engineering students. Depending on the detail covered in class, it can be used for a one-, two-, or three-semester course. By omitting the chapter on optimal control and the proofs of the results in the appendixes, a one-year course can be structured with no difficulty.

The book covers what constitutes the common core of control theory: The algebraic theory of linear systems, including controllability, observability, feedback equivalence, and minimality; stability via Lyapunov, as well as input/output methods; ideas of optimal control; observers and dynamic feedback; parameterization of stabilizing controllers (in the scalar case only); and some very basic facts about frequency domain such as the Nyquist criterion. Kalman filtering is introduced briefly through a deterministic version of "optimal observation;" this avoids having to develop the theory of stochastic processes and represents a natural application of optimal control and observer techniques. In general, no stochastic or infinite dimensional results are covered, nor is a detailed treatment given of nonlinear differential-geometric control; for these more advanced areas, there are many suitable texts.

The introductory chapter describes the main contents of the book in an intuitive and informal manner, and the reader would be well advised to read this in detail. I suggest spending at least a week of the course covering this material. Sections marked with asterisks can be left out without loss of continuity. I have only omitted the proofs of what are labeled "Lemma/Exercises," with the intention that these proofs *must* be worked out by the reader in detail. Typically, Lemma/Exercises ask the reader to prove that some elementary property holds, or to prove a difference equation analogue of a differential equation result just shown. Only a trivial to moderate effort is required, but filling in the details forces one to read the definitions and proofs carefully. "Exercises" are for the most part also quite simple; those that are harder are marked with the symbol $\diamond$.

Control and system theory shares with some other areas of "modern" applied mathematics (such as quantum field theory, geometric mechanics, and computational complexity), the characteristic of employing a broad range of mathematical tools, providing many challenges and possibilities for interactions with established areas of "pure" mathematics. Papers often use techniques from algebraic and differential geometry, functional analysis, Lie algebras, combinatorics, and other areas, in addition to tools from complex variables, linear algebra, and ordinary and partial differential equations. While staying within the bounds of an introductory presentation, I have tried to provide pointers toward some of these exciting directions of research, particularly through the remarks at the ends of chapters and through references to the literature. (At a couple of points I include further details, such as when I discuss Lie group actions and families of systems, or degree theory and nonlinear stabilization, or ideal theory and finite-experiment observability, but these are restricted to small sections that can be skipped with no loss of continuity.)

This book should be useful too as a research reference source, since I have included complete proofs of various technical results often used in papers but for which precise citations are hard to find. I have also compiled a rather long bibliography covering extensions of the results given and other areas not treated, as well as a detailed index facilitating access to these.

Although there are hundreds of books dealing with various aspects of control and system theory, including several extremely good ones, this text is unique in its emphasis on foundational aspects. I know of no other book that, while covering the range of topics treated here and written in a standard theorem/proof style, also develops the necessary techniques from scratch and does not assume any background other than basic mathematics. In fact, much effort was spent trying to find consistent notations and terminology across these various areas. On the other hand, this book does not provide realistic engineering examples, as there are already many books that do this well. (The bibliography is preceded by a discussion of such other texts.)

I made an effort to highlight the distinctions as well as the similarities between continuous- and discrete-time systems, and the process (sampling) that is used in practice for computer control. In this connection, I find it highly probable that future developments in control theory will continue the movement toward a more "computer-oriented" and "digital/logical" view of systems, as opposed to the more classical continuous-time smooth paradigm motivated by analogue devices. Because of this, it is imperative that a certain minimal amount of the 'abstract nonsense' of systems (abstract definitions of systems as actions over general time sets, and so forth) be covered: It is impossible to pose, much less solve, systems design problems unless one first understands what a system is. This is analogous to understanding the definition of function as a set of ordered pairs, instead of just thinking of functions as only those that can be expressed by explicit formulas.

A few words about notation and numbering conventions. Except for theorems, which are numbered consecutively, all other environments (lemmas, definitions, and so forth) are numbered by section, while equations are numbered by chapter. In formal statements, I use roman characters such as x to denote states at a given instant, reserving Greek letters such as ξ for state trajectories, and a similar convention is used for control values versus control functions, and output values versus output functions; however, in informal discussions and examples, I use roman notation for both points and functions, the meaning being clear from the context. The symbol ∎ marks an end of proof, while □ indicates the end of a remark, example, etc., as well as the end of a statement whose proof has been given earlier.

This volume represents one attempt to address two concerns raised in the report "Future Directions in Control Theory: A Mathematical Perspective," which was prepared by an international panel under the auspices of various American funding agencies (National Science Foundation, Air Force Office of Scientific Research, Army Research Office, and Office of Naval Research) and was published in 1988 by the Society for Industrial and Applied Mathematics. Two of the main recommendations of the report were that further efforts be made to achieve conceptual unity in the field, as well as to develop better training for students and faculty in mathematics departments interested in being exposed to the area. Hopefully, this book will be useful in helping achieve these goals.

It is hard to even begin to acknowledge all those who influenced me professionally and from whom I learned so much, starting with Enzo Gentile and other faculty at the Mathematics Department of the University of Buenos Aires. Without doubt, my years at the Center for Mathematical System Theory as a student of Rudolf Kalman were the central part of my education, and the stimulating environment of the center was unequaled in its possibilities for learning. To Professor Kalman and his long-range view of the area, originality of thought, and emphasis on critical thinking, I will always owe major gratitude. Others who spent considerable time at the Center, and from whose knowledge I also benefited immensely at that time and ever since, include Roger Brockett, Samuel Eilenberg, Michel Fliess, Malo Hautus, Michiel Hazewinkel, Hank Hermes, Michel Heymann, Bruce Lee, Alberto Isidori, Ed Kamen, Sanjoy Mitter, Yves Rouchaleau, Yutaka Yamamoto, Jan Willems, and Bostwick Wyman. In latter years I learned much from many people, but I am especially indebted to my Rutgers colleague Héctor Sussmann, who introduced me to the continuous-time nonlinear theory. While writing this book, I received constant and very useful feedback from graduate students at Rutgers, including Francesca Albertini, Yuandan Lin, Wen-Sheng Liu, Guoqing Tang, Yuan Wang, and Yudi Yang. I wish to especially thank Renee Schwarzschild, who continuously provided me with extensive lists of misprints, errors, and comments. Both Pramod Khargonekar and Jack Rugh also gave me useful suggestions at various points.

The continued and generous research support that the Air Force Office of Scientific Research provided to me during most of my career was instrumental in my having the possibility of really understanding much of the material treated in this book as well as related areas. I wish to express my sincere gratitude for AFOSR support of such basic research.

Finally, my special thanks go to my wife Fran and to my children Laura and David, for their patience and understanding during the long time that this project took.

Eduardo D. Sontag

Piscataway, NJ
June, 1990.

Contents

*Can be skipped with no loss of continuity.

*Can be skipped with no loss of continuity.

*Can be skipped with no loss of continuity.

Chapter 1

Introduction

1.1 What Is Mathematical Control Theory?

Mathematical control theory is the area of application-oriented mathematics that deals with the basic principles underlying the analysis and design of control systems. To *control* an object means to influence its behavior so as to achieve a desired goal. In order to implement this influence, engineers build devices that incorporate various mathematical techniques. These devices range from Watt's steam engine governor, designed during the English Industrial Revolution, to the sophisticated microprocessor controllers found in consumer items —such as CD players and automobiles— or in industrial robots and airplane autopilots.

The study of these devices and their interaction with the object being controlled is the subject of this book. While on the one hand one wants to understand the fundamental limitations that mathematics imposes on what is achievable, irrespective of the precise technology being used, it is also true that technology may well influence the type of question to be asked and the choice of mathematical model. An example of this is the use of difference rather than differential equations when one is interested in digital control.

Roughly speaking, there have been two main lines of work in control theory, which sometimes have seemed to proceed in very different directions but which are in fact complementary. One of these is based on the idea that a good model of the object to be controlled is available and that one wants to somehow *optimize* its behavior. For instance, physical principles and engineering specifications can be —and are— used in order to calculate that trajectory of a spacecraft which minimizes total travel time or fuel consumption. The techniques here are closely related to the classical calculus of variations and to other areas of optimization theory; the end result is typically a preprogrammed flight plan. The other main line of work is that based on the constraints imposed by *uncertainty* about the model or about the environment in which the object operates. The central tool here is the use of *feedback* in order to correct for deviations from the desired behavior. For instance, various feedback control systems are used during actual space flight in order to compensate for errors from the precomputed trajectory. Mathematically, stability theory, dynamical systems, and especially the theory of functions of a complex variable, have had a strong influence on this approach. It is widely recognized today that these two broad lines of work deal just with different aspects of the same problems, and we do not make an artificial distinction between them in this book.

Later on we shall give an axiomatic definition of what we mean by a "system" or "machine." Its role will be somewhat analogous to that played in mathematics by the definition of "function" as a set of ordered pairs: not itself the object of study, but a necessary foundation upon which the entire theoretical development will rest. In this Chapter, however, we dispense with precise definitions and will use a very simple physical example in order to give an intuitive presentation of some of the goals, terminology, and methodology of control theory.

The discussion here will be informal and not rigorous, but the reader is encouraged to follow it in detail, since the ideas to be given underlie everything else in the book. Without them, many problems may look artificial. Later, we often refer back to this Chapter for motivation.

1.2 Proportional-Derivative Control

One of the simplest problems in robotics is that of controlling the position of a single-link rotational joint using a motor placed at the pivot. Mathematically, this is just a pendulum to which one can apply a torque as an external force (see Figure 1.1).

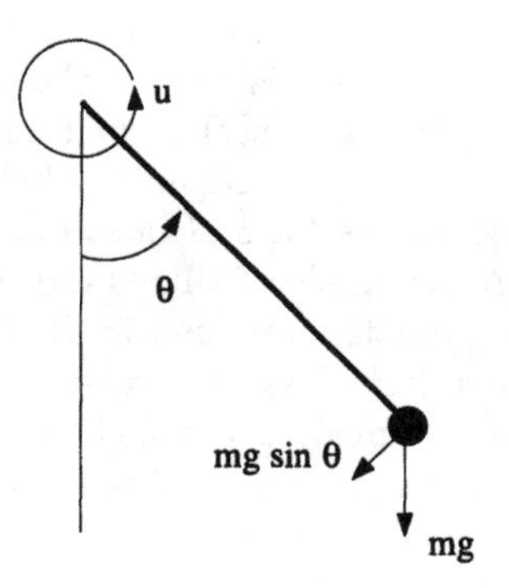

Figure 1.1: *Pendulum.*

We assume that friction is negligible, that all of the mass is concentrated at the end, and that the rod has unit length. From Newton's law for rotating objects, there results, in terms of the variable θ that describes the counterclockwise angle with respect to the vertical, the second-order nonlinear differential equation

$$m\ddot{\theta}(t) + mg\sin\theta(t) = u(t), \tag{1.1}$$

where m is the mass, g the acceleration due to gravity, and $u(t)$ the value of the external torque at time t (counterclockwise being positive). We call $u(\cdot)$ the *input* or *control* function. To avoid having to keep track of constants, let us assume that units of time and distance have been chosen so that $m = g = 1$.

The vertical stationary position ($\theta = \pi, \dot{\theta} = 0$) is an equilibrium when no control is being applied ($u \equiv 0$), but a small deviation from this will result in an unstable motion. Let us assume that our objective is to apply torques as needed to correct for such deviations. For small $\theta - \pi$,

$$\sin\theta = -(\theta - \pi) + o(\theta - \pi)\,.$$

Here we use the standard "little-o" notation: $o(x)$ stands for some function $g(x)$ for which

$$\lim_{x \to 0} \frac{g(x)}{x} = 0\,.$$

Since only small deviations are of interest, we drop the nonlinear part represented by the term $o(\theta - \pi)$. Thus, with $\varphi := \theta - \pi$ as a new variable, we replace equation (1.1) by the linear differential equation

$$\ddot{\varphi}(t) - \varphi(t) = u(t) \tag{1.2}$$

as our object of study. (See Figure 1.2.) Later we will analyze the effect of the ignored nonlinearity.

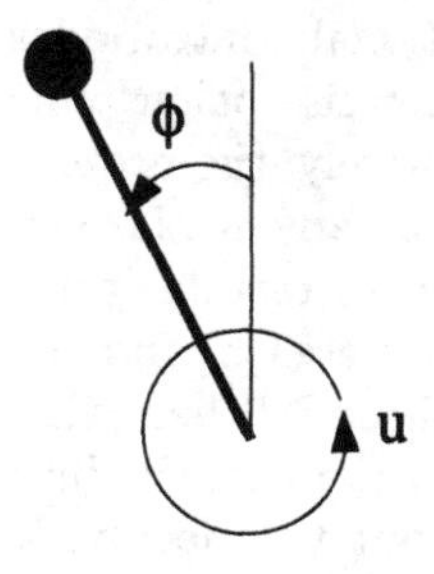

Figure 1.2: *Inverted pendulum.*

Our objective then is to bring φ and $\dot{\varphi}$ to zero, for any small nonzero initial values $\varphi(0), \dot{\varphi}(0)$ in equation (1.2), and preferably to do so as fast as possible, with few oscillations, and without ever letting the angle and velocity become too large. Although this is a highly simplified system, this kind of "servo" problem illustrates what is done in engineering practice. One typically wants to achieve a desired value for certain variables, such as the correct idling speed in an automobile's electronic ignition system or the position of the read/write head in a disk drive controller.

A naive first attempt at solving this control problem would be as follows: If we are to the left of the vertical, that is, if $\varphi = \theta - \pi > 0$, then we wish to move to the right, and therefore, we apply a negative torque. If instead we are to the right, we apply a positive, that is to say counterclockwise, torque. In other words, we apply *proportional feedback*

$$u(t) = -\alpha\varphi(t)\,, \tag{1.3}$$

where α is some positive real number, the *feedback gain.*

Let us analyze the resulting *closed-loop* equation obtained when the value of the control given by (1.3) is substituted into the *open-loop* original equation (1.2), that is

$$\ddot{\varphi}(t) - \varphi(t) + \alpha\varphi(t) = 0\,. \tag{1.4}$$

If $\alpha > 1$, the solutions of this differential equation are all oscillatory, since the roots of the associated characteristic equation

$$z^2 + \alpha - 1 = 0 \tag{1.5}$$

are purely imaginary, $z = \pm i\sqrt{\alpha - 1}$. If instead $\alpha < 1$, then all of the solutions except for those with

$$\dot{\varphi}(0) = -\varphi(0)\sqrt{1-\alpha}$$

diverge to $\pm\infty$. Finally, if $\alpha = 1$, then each set of initial values with $\dot{\varphi}(0) = 0$ is an equilibrium point of the closed-loop system. Therefore, in none of the cases is the system guaranteed to approach the desired configuration.

We have seen that proportional control does not work. We proved this for the linearized model, and an exercise below will show it directly for the original nonlinear equation (1.1). Intuitively, the problem can be understood as follows. Take first the case $\alpha < 1$. For any initial condition for which $\varphi(0)$ is small but positive and $\dot{\varphi}(0) = 0$, there results from equation (1.4) that $\ddot{\varphi}(0) > 0$. Therefore, also $\dot{\varphi}$ and hence φ increase, and the pendulum moves away, rather than toward, the vertical position. When $\alpha > 1$ the problem is more subtle: The torque is being applied in the correct direction to counteract the natural instability of the pendulum, but this feedback helps build too much inertia. In particular, when already close to $\varphi(0) = 0$ but moving at a relatively large speed, the controller (1.3) keeps pushing toward the vertical, and overshoot and eventual oscillation result.

The obvious solution is to keep $\alpha > 1$ but to modify the proportional feedback (1.3) through the addition of a term that acts as a brake, penalizing velocities. In other words, one needs to add damping to the system. We arrive then at a *PD*, or *proportional-derivative* feedback law,

$$u(t) = -\alpha\varphi(t) - \beta\dot{\varphi}(t)\,, \tag{1.6}$$

with $\alpha > 1$ and $\beta > 0$. In practice, implementing such a controller involves measurement of both the angular position and the velocity. If only the former is easily available, then one must estimate the velocity as part of the control algorithm; this will lead later to the idea of *observers*, which are techniques for

reliably performing such an estimation. We assume here that $\dot{\varphi}$ can indeed be measured. Consider then the resulting closed-loop system,

$$\ddot{\varphi}(t) + \beta\dot{\varphi}(t) + (\alpha - 1)\varphi(t) = 0\,. \tag{1.7}$$

The roots of its associated characteristic equation

$$z^2 + \beta z + \alpha - 1 = 0 \tag{1.8}$$

are

$$\frac{-\beta \pm \sqrt{\beta^2 - 4(\alpha - 1)}}{2}\,,$$

both of which have negative real parts. Thus all the solutions of (1.2) converge to zero. The system has been ***stabilized*** under feedback. This convergence may be oscillatory, but if we design the controller in such a way that in addition to the above conditions on α and β it is true that

$$\beta^2 > 4(\alpha - 1)\,, \tag{1.9}$$

then all of the solutions are combinations of decaying exponentials and no oscillation results.

We conclude from the above discussion that through a suitable choice of the gains α and β it is possible to attain the desired behavior, at least for the linearized model. That this same design will still work for the original nonlinear model, and, hence, assuming that this model was accurate, for a real pendulum, is due to what is perhaps the most important fact in control theory —and for that matter in much of mathematics— namely that first-order approximations are sufficient to characterize local behavior. Informally, we have the following *linearization principle*:

Designs based on linearizations work **locally** *for the original system*

The term "local" refers to the fact that satisfactory behavior only can be expected for those initial conditions that are close to the point about which the linearization was made. Of course, as with any "principle," this is not a theorem. It can only become so when precise meanings are assigned to the various terms and proper technical assumptions are made. Indeed, we will invest some effort in this text to isolate cases where this principle may be made rigorous. One of these cases will be that of stabilization, and the theorem there will imply that if we can stabilize the linearized system (1.2) for a certain choice of parameters α, β in the law (1.6), then the same control law does bring initial conditions of (1.1) that start close to $\theta = \pi, \dot{\theta} = 0$ to the vertical equilibrium.

Basically because of the linearization principle, a great deal of the literature in control theory deals exclusively with linear systems. From an engineering point of view, local solutions to control problems are often enough; when they are not, ad hoc methods sometimes may be used in order to "patch" together such local solutions, a procedure called *gain scheduling*. Sometimes, one may even be lucky and find a way to transform the problem of interest into one that is globally linear; we explain this later using again the pendulum as an example. In many other cases, however, a genuinely nonlinear approach is needed, and much research effort during the past few years has been directed toward that goal. In this text, when we develop the basic definitions and results for the linear theory we will always do so with an eye toward extensions to nonlinear, global, results.

An Exercise

As remarked earlier, proportional control (1.3) by itself is inadequate for the original nonlinear model. Using again $\varphi = \theta - \pi$, the closed-loop equation becomes

$$\ddot{\varphi}(t) - \sin\varphi(t) + \alpha\varphi(t) = 0\,. \tag{1.10}$$

The next exercise claims that solutions of this equation typically will not approach zero, no matter how the feedback gain α is picked.

Exercise 1.2.1 Assume that α is any fixed real number, and consider the ("energy") function of two real variables

$$V(x,y) := \cos x - 1 + \frac{1}{2}(\alpha x^2 + y^2)\,. \tag{1.11}$$

Show that $V(\varphi(t),\dot{\varphi}(t))$ is constant along the solutions of (1.10). Using that $V(x,0)$ is an analytic function and therefore that its zero at $x=0$ is isolated, conclude that there are initial conditions of the type $\varphi(0)=\varepsilon, \dot{\varphi}(0)=0$, with ε arbitrarily small, for which the corresponding solution of (1.10) does not satisfy that $\varphi(t)\to 0$ and $\dot{\varphi}(t)\to 0$ as $t\to\infty$. □

1.3 Digital Control

The actual physical implementation of (1.6) need not concern us here, but some remarks are in order. Assuming again that the values $\varphi(t)$ and $\dot{\varphi}(t)$, or equivalently $\theta(t)$ and $\dot{\theta}(t)$, can be measured, it is necessary to take a linear combination of these in order to determine the torque $u(t)$ that the motor must apply. Such combinations are readily carried out by circuits built out of devices called *operational amplifiers*. Alternatively, the damping term can be separately implemented directly through the use of an appropriate device (a "dashpot"), and the torque is then made proportional to $\varphi(t)$.

A more modern alternative, attractive especially for larger systems, is to convert position and velocity to digital form and to use a computer to calculate the necessary controls. Still using the linearized inverted pendulum as an illustration, we now describe some of the mathematical problems that this leads to.

A typical approach to computer control is based on the *sample-and-hold* technique, which can be described as follows. The values $\varphi(t)$ and $\dot{\varphi}(t)$ are measured only at discrete instants or *sampling times*

$$0,\ \delta,\ 2\delta,\ 3\delta,\ldots,k\delta,\ldots$$

The control law is updated by a program at each time $t = k\delta$ on the basis of the sampled values $\varphi(k\delta)$ and $\dot{\varphi}(k\delta)$. The output of this program, a value v_k, is then fed into the system as a control (*held constant* at that value) during the interval $[k\delta, k\delta + \delta]$.

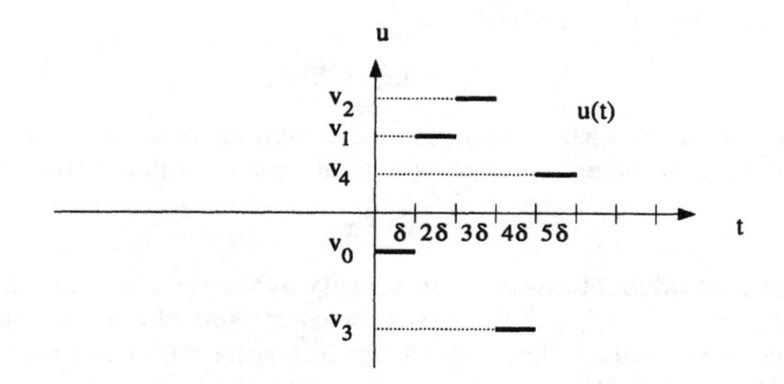

Figure 1.3: *Sampled control.*

For simplicity we assume here that the computation of v_k can be done quickly relative to the length δ of the sampling intervals; otherwise, the model must be modified to account for the extra delay. To calculate the effect of applying the constant control

$$u(t) \equiv v_k \ \text{ if } \ t \in [k\delta, k\delta + \delta] \tag{1.12}$$

we solve the differential equation (1.2) with this function u. By differentiation one can verify that the general solution is, for $t \in [k\delta, k\delta + \delta]$,

$$\varphi(t) = \frac{\varphi(k\delta) + \dot{\varphi}(k\delta) + v_k}{2} e^{t-k\delta} + \frac{\varphi(k\delta) - \dot{\varphi}(k\delta) + v_k}{2} e^{-t+k\delta} - v_k\,, \tag{1.13}$$

so

$$\dot{\varphi}(t) = \frac{\varphi(k\delta) + \dot{\varphi}(k\delta) + v_k}{2} e^{t-k\delta} - \frac{\varphi(k\delta) - \dot{\varphi}(k\delta) + v_k}{2} e^{-t+k\delta}\,. \tag{1.14}$$

Thus, applying the constant control u gives rise to new values for $\varphi(k\delta + \delta)$ and $\dot{\varphi}(k\delta + \delta)$ at the end of the interval via the formula

$$\begin{pmatrix} \varphi(k\delta + \delta) \\ \dot{\varphi}(k\delta + \delta) \end{pmatrix} = A \begin{pmatrix} \varphi(k\delta) \\ \dot{\varphi}(k\delta) \end{pmatrix} + Bv_k\,, \tag{1.15}$$

where

$$A = \begin{pmatrix} \cosh\delta & \sinh\delta \\ \sinh\delta & \cosh\delta \end{pmatrix} \tag{1.16}$$

and

$$B = \begin{pmatrix} \cosh\delta - 1 \\ \sinh\delta \end{pmatrix}. \tag{1.17}$$

In other words, if we let $x_0, x_1, \ldots$ denote the sequence of two dimensional vectors

$$x_k := \begin{pmatrix} \varphi(k\delta) \\ \dot{\varphi}(k\delta) \end{pmatrix},$$

then $\{x_k\}$ satisfies the recursion

$$x_{k+1} = Ax_k + Bv_k\,. \tag{1.18}$$

Assume now that we wish to program our computer to calculate the constant control values v_k to be applied during any interval via a linear transformation

$$v_k := Fx_k \tag{1.19}$$

of the measured values of position and velocity at the start of the interval. Here F is just a row vector (f_1, f_2) that gives the coefficients of a linear combination of these measured values. Formally we are in a situation analogous to the PD control (1.6), except that we now assume that the measurements are being made only at discrete times and that a constant control will be applied on each interval. Substituting (1.19) into the difference equation (1.18), there results the new difference equation

$$x_{k+1} = (A + BF)x_k\,. \tag{1.20}$$

Since for any k

$$x_{k+2} = (A + BF)^2 x_k\,, \tag{1.21}$$

it follows that, if one finds gains f_1 and f_2 with the property that the matrix $A + BF$ is nilpotent, that is,

$$(A + BF)^2 = 0\,, \tag{1.22}$$

then we would have a controller with the property that after two sampling steps necessarily $x_{k+2} = 0$. That is, both φ and $\dot{\varphi}$ vanish after these two steps, and the system remains at rest after that. This is the objective that we wanted to achieve all along. We now show that this choice of gains is always possible. Consider the characteristic polynomial

$$\begin{aligned} \det(zI - A - BF) &= z^2 + (-2\cosh\delta - f_2\sinh\delta - f_1\cosh\delta + f_1)z \\ &\quad - f_1\cosh\delta + 1 + f_1 + f_2\sinh\delta\,. \end{aligned} \tag{1.23}$$

It follows from the Cayley-Hamilton Theorem that condition (1.22) will hold provided that this polynomial reduces to just z^2. So we need to solve for the

f_i the system of equations resulting from setting the coefficient of z and the constant term to zero. This gives

$$f_1 = -1/2\frac{2\cosh\delta - 1}{\cosh\delta - 1} \quad \text{and} \quad f_2 = -1/2\frac{2\cosh\delta + 1}{\sinh\delta}\,. \tag{1.24}$$

We conclude that it is always possible to find a matrix F as desired. In other words, using sampled control we have been able to achieve stabilization of the system. Moreover, this stability is of a very strong type, in that, at least theoretically, it is possible to bring the position and velocity *exactly* to zero in finite time, rather than only asymptotically as with a continuous-time controller. This strong type of stabilization is often called *deadbeat control*; its possibility (together with ease of implementation and maintenance, and reliability) constitutes a major advantage of digital techniques.

1.4 Feedback Versus Precomputed Control

Note that the first solution that we provided to the pendulum control problem was in the form of a *feedback* law (1.6), where $u(t)$ could be calculated in terms of the current position and velocity, which are "fed back" after suitable weighings. This is in contrast to "open-loop" design, where the expression of the entire control function $u(\cdot)$ is given in terms of the initial conditions $\varphi(0), \dot{\varphi}(0)$, and one applies this function $u(\cdot)$ blindly thereafter, with no further observation of positions and velocities. In real systems there will be random perturbations that are not accounted for in the mathematical model. While a feedback law will tend to correct automatically for these, a precomputed control takes no account of them. This can be illustrated by the following simple examples.

Assume that we are only interested in the problem of controlling (1.2) when starting from the initial position $\varphi(0) = 1$ and velocity $\dot{\varphi}(0) = -2$. Some trial and error gives us that the control function

$$u(t) = 3e^{-2t} \tag{1.25}$$

is adequate for this purpose, since the solution when applying this forcing term is

$$\varphi(t) = e^{-2t}\,.$$

It is certainly true that $\varphi(t)$ and its derivative approach zero, actually rather quickly. So (1.25) solves the original problem. If we made any mistakes in estimating the initial velocity, however, the control (1.25) is no longer very useful:

Exercise 1.4.1 Show that if the differential equation (1.2) is again solved with the right-hand side equal to (1.25) but now using instead the initial conditions

$$\varphi(0) = 1,\ \dot{\varphi}(0) = -2 + \varepsilon\,,$$

where ε is any positive number (no matter how small), then the solution satisfies

$$\lim_{t \to +\infty} \varphi(t) = +\infty\,.$$

If $\varepsilon < 0$, show that then the limit is $-\infty$. □

A variation of this is as follows. Suppose that we measured correctly the initial conditions but that a momentary power surge affects the motor controlling the pendulum. To model this, we assume that the differential equation is now

$$\ddot{\varphi}(t) - \varphi(t) = u(t) + d(t), \tag{1.26}$$

and that the disturbance $d(\cdot)$ is the function

$$d(t) = \begin{cases} \varepsilon & \text{if } t \in [1,2] \\ 0 & \text{otherwise.} \end{cases}$$

Here ε is some positive real number.

Exercise 1.4.2 Show that the solution of (1.26) with initial conditions $\varphi(0) = 1, \dot{\varphi}(0) = -2$, and u chosen according to (1.25) diverges, but that the solution of

$$\ddot{\varphi}(t) - \varphi(t) = -\alpha\varphi(t) - \beta\dot{\varphi}(t) + d(t) \tag{1.27}$$

still satisfies the condition that φ and $\dot{\varphi}$ approach zero. □

One can prove in fact that the solution of equation (1.27) approaches zero even if $d(\cdot)$ is an arbitrary decaying function; this is an easy consequence of results on the *input/output stability* of linear systems.

Not only is it more robust to errors, but the feedback solution is also in this case simpler than the open-loop one, in that the explicit form of the control as a function of time need not be calculated. Of course, the cost of implementing the feedback controller is that the position and velocity must be continuously monitored.

There are various manners in which to make the advantages of feedback mathematically precise. One may include in the model some characterization of the uncertainty, for instance, by means of specifying a probability law for a disturbance input such as the above $d(\cdot)$. In any case, one can always pose a control problem directly as one of finding feedback solutions, and we shall often do so.

The second solution (1.12)-(1.19) that we gave to the pendulum problem, via digital control, is in a sense a combination of feedback and precomputed control. But in terms of the sampled model (1.18), which ignores the behavior of the system in between sampling times, digital control can be thought of as a purely feedback law. For the times of interest, (1.19) expresses the control in terms of the "current" state variables.

1.5 State-Space and Spectrum Assignment

We now have seen two fundamentally different ways in which to control the linearized inverted pendulum (1.2). They both involve calculating certain gains: α and β in the case of the continuously acting PD controller (1.6), or the entries f_1 and f_2 of F for sampled feedback. In both cases we found that appropriate choices can be made of these coefficients that result, for instance, in decaying exponential behavior if (1.9) holds, or deadbeat control if (1.24) is used.

That such choices are possible is no coincidence. *For very general linear systems, it is possible to obtain essentially arbitrary asymptotic behavior.* One of the basic results in control theory is the *Pole-Shifting Theorem*, also called the "Pole-Assignment" or the "Spectrum Assignment" Theorem, which makes this fact precise.

In order to discuss this Theorem, it is convenient to use the *state-space* formalism. In essence, this means that, instead of studying a high-order differential equation such as (1.1) or (1.2), we replace it by a system of first-order differential equations. For instance, instead of (1.2) we introduce the first-order vector equation

$$\dot{x}(t) = Ax(t) + Bu(t)\,, \tag{1.28}$$

where $x(t)$ is the column vector $(\varphi(t), \dot{\varphi}(t))'$ (prime indicates transpose), and where A and B are the matrices

$$A = \begin{pmatrix} 0 & 1 \\ 1 & 0 \end{pmatrix}, \quad B = \begin{pmatrix} 0 \\ 1 \end{pmatrix}. \tag{1.29}$$

An equation such as (1.28) is an example of what is called a *linear, continuous-time, time-invariant, finite dimensional control system.*

The term "continuous-time" refers to the fact that the time variable t is continuous, in contrast to a system defined by a difference equation such as (1.18). The terminology "time-invariant" is used to indicate that the time t does not appear as an independent variable in the right-hand side of the equation. If the mass of the pendulum were to change in time, for instance, one might model this instead through a time-*varying* equation. "Finite dimensional" refers to the fact that the state $x(t)$ can be characterized completely at each instant by a finite number of parameters (in the above case, two). For an example of a system that is not finite dimensional, consider the problem of controlling the temperature of an object by heating its boundary; the description in that case incorporates a partial differential equation, and the characterization of the state of the system at any given time requires an infinite amount of data (the temperatures at all points).

If instead of the linearized model we had started with the nonlinear differential equation (1.1), then in terms of the same vector x we would obtain a set of first-order equations

$$\dot{x}(t) = f(x(t), u(t))\,, \tag{1.30}$$

where now the right-hand side f is not necessarily a linear function of x and u. This is an example of what we will later call a nonlinear continuous-time system.

Of course, we are interested in systems described by more than just two state variables; in general $x(t)$ will be an n-dimensional vector. Moreover, often there is more than one independent control acting on a system. For instance, in a three-link robotic arm there may be one motor acting at each of its joints (the "shoulder," "elbow," and "wrist"). In such cases the control $u(t)$ at each instant is not a scalar quantity but an m-dimensional vector that indicates the magnitude of the external forces being applied at each input channel. Mathematically, one just lets f in (1.30) be a function of $n + m$ variables. For linear systems (1.28) or (1.18), one allows A to be an $n \times n$ matrix and B to be a rectangular matrix of size $n \times m$.

The transformation from a high-order single equation to a system of first-order equations is exactly the same as that often carried out when establishing existence results for, or studying numerical methods for solving, ordinary differential equations. For any σ, the vector $x(\sigma)$ contains precisely the amount of data needed in order to solve the equation for $\tau > \sigma$, assuming that the control $u(\cdot)$ is specified for $\tau > \sigma$. The *state* $x(\sigma)$ thus summarizes all the information about the past of the system needed in order to understand its future behavior, except for the purely external effects due to the input.

Later we reverse things and *define* systems through the concept of state. We will think of (1.28) not as a differential equation but rather, for each pair of times $\tau > \sigma$, as a rule for obtaining any value $x(\tau)$ from the specifications of $x(\sigma)$ and of the restriction ω of the control $u(\cdot)$ to the interval between σ and τ. We will use the notation

$$x(\tau) = \phi(\tau, \sigma, x, \omega) \tag{1.31}$$

for this rule. We may read the right-hand side as "the state at time τ resulting from starting at time σ in state x and applying the input function ω." Because solutions of differential equations do not exist in general for all pairs of initial and final times $\sigma < \tau$, typically ϕ may be defined only on a subset of the set of possible quadruples $(\tau, \sigma, x, \omega)$.

An advantage of doing things at this level of abstraction is that many objects other than those described by ordinary differential equations also are representable by a rule such as (1.31). For example, discrete-time equations (1.18) also give rise to systems in this sense, since x_{k+i} depends only on

$$x_k, v_k, v_{k+1}, \ldots, v_{k+i-1}$$

for all k and i. The idea of using an abstract transition mapping ϕ originates with the mathematical theory of dynamical systems. The difference with that theory is that here we are interested centrally in understanding the effect of controls.

Returning to the issue of feedback, note that using the matrix formalism (1.28) and denoting by F the vector of gains

$$F = (\,-\alpha \quad -\beta\,)$$

we may write the PD feedback law (1.6) as

$$u(t) = Fx(t)\,. \tag{1.32}$$

When we substitute this into (1.28), we obtain precisely the closed-loop equations (1.7). So the characteristic equation (1.8) of the closed-loop system is the determinant of the matrix

$$\begin{pmatrix} z & -1 \\ \alpha - 1 & z + \beta \end{pmatrix},$$

that is, the characteristic polynomial of $A+BF$. The zeros of the characteristic equation are the eigenvalues of $A + BF$.

We conclude from the above discussion that the problems of finding gains α, β for the PD controller with the properties that all solutions of (1.7) converge to zero, or that all solutions converge with no oscillation, are particular cases of the problem of finding a row vector F so that $A + BF$ has eigenvalues with certain desired properties. This is precisely the same problem that we solved for the discrete-time system (1.18) in order to obtain deadbeat control, where we needed $A+BF$ to have a double eigenvalue at zero. Note that in the second case the matrices A and B are different from the corresponding ones for the first problem. Note also that when dealing with vector-input systems, for which $u(t)$ is not a scalar, the F in equation (1.32) must be a matrix instead of a row vector.

Thus one arrives at the following purely algebraic question suggested by both the PD and the deadbeat digital control problems:

> *Given a pair of real matrices (A, B), where A is square of size $n \times n$ and B is of size $n \times m$, characterize the set of all possible eigenvalues of $A + BF$ as F ranges over the set of all possible real matrices of size $m \times n$.*

The Spectrum Assignment Theorem says in essence that for almost any pair of matrices A and B it is possible to obtain *arbitrary* eigenvalues for $A + BF$ using suitable feedback laws F, subject only to the obvious constraint (since $A + BF$ is a real matrix) that complex eigenvalues must appear in conjugate pairs. "Almost any" means that this will be true for all pairs that describe what we will call *controllable* systems, those systems for which it is possible to steer any state to any other state. We will see later that controllability corresponds to a simple nondegeneracy condition relating A and B; when there is just one control ($m = 1$) the condition is simply that B, seen as a column vector, must be cyclic for A.

This Theorem is most often referred to as the *Pole-Shifting Theorem*, a terminology that is due to the fact that the eigenvalues of $A+BF$ are also the poles of the resolvent operator

$$(zI - A - BF)^{-1},$$

or equivalently, of the rational function $1/\det(zI - A - BF)$; this rational function appears often in classical control design.

As a consequence of the Pole-Shifting Theorem we know that, save for very exceptional cases, given any system

$$\dot{x}(t) = Ax(t) + Bu(t)$$

it will be possible to find a feedback law $u(t) = Fx(t)$ such that all solutions of the closed-loop equation

$$\dot{x}(t) = (A + BF)x(t)$$

decay exponentially. To conclude this, one uses the Pole-Shifting Theorem to obtain a matrix F so that all eigenvalues of $A+BF$ are negative real numbers. Another illustration is the general deadbeat control problem: Given any discrete-time system (1.18), one can in general find an F so that $(A+BF)^n = 0$, simply by assigning zero eigenvalues. Again, a nondegeneracy restriction will be needed, as illustrated by the counterexample $A = I, B = 0$ for which no such F exists.

The Pole-Shifting Theorem is central to linear systems theory and is itself the starting point for more interesting analysis. Once we know that arbitrary sets of eigenvalues can be assigned, it becomes of interest to compare the performance of different such sets. Among those that provide stability, some may be more useful than others because they have desirable transient characteristics such as lack of "overshoot." Also, one may ask what happens when certain entries of F are restricted to vanish, which corresponds to constraints on what can be implemented.

Yet another possibility is to consider a cost criterion to be optimized, as we do when we discuss optimal control theory. For example, it is possible to make trajectories approach the origin arbitrarily fast under the PD controller (1.6), but this requires large gains α, β, which means that the controls $u(t)$ to be applied will have large magnitudes. Thus, there is a trade-off between the cost of controlling and the speed at which this may be achieved.

In a different direction, one may investigate to what extent the dynamics of nonlinear and/or infinite dimensional systems are modifiable by feedback, and this constitutes one of the main areas of current research in control theory.

Classical Design

The next exercise on elementary ordinary differential equations is intended to convey the flavor of some of the simpler performance questions that might be asked when assigning eigenvalues and to mention some standard terminology. Techniques from the theory of functions of a complex variable are often used in this type of study, which is typically the subject of undergraduate engineering courses in the "classical design" or "frequency design" of control systems. We include this discussion as an illustration but will not treat frequency design in this text.

Exercise 1.5.1 Consider again the pendulum linearized about its unstable upper position, given by the equation $\ddot{\varphi}(t) - \varphi(t) = u(t)$, and assume that we use the PD control law $u(t) = -\alpha\varphi(t) - \beta\dot{\varphi}(t)$ to obtain an asymptotically stable closed-loop system $\ddot{\varphi}(t) + b\dot{\varphi}(t) + a\varphi(t) = 0$ (with $a = \alpha - 1 > 0$ and $b = \beta > 0$). Introduce the *natural frequency* $\omega := \sqrt{a}$ and the *damping factor* $\zeta := b/(2\sqrt{a})$, so that the equation now reads

$$\ddot{\varphi}(t) + 2\zeta\omega\dot{\varphi}(t) + \omega^2\varphi(t) = 0\,. \tag{1.33}$$

(A) Prove the following facts:

1. If $\zeta < 1$ (the "underdamped" case), all solutions are decaying oscillations.
2. If $\zeta = 1$ (the "critically damped" case) or if $\zeta > 1$ ("overdamped"), then all solutions for which $\varphi(0) \neq 0$ are such that $\varphi(t) = 0$ for at most one $t > 0$.
3. If $\zeta \geq 1$, then every solution that starts from rest at a displaced position, that is, $\varphi(0) \neq 0$ and $\dot{\varphi}(0) = 0$, approaches zero monotonically. (In this case, we say that there is no "overshoot.")
4. Show rough plots of typical solutions under the three cases $\zeta < 1$, $\zeta = 1$, and $\zeta > 1$.

(B) Consider again the underdamped case. Using rough plots of the solutions of the equation with $\varphi(0) = -1, \dot{\varphi}(0) = 0$ both in the case when ζ is near zero and in the case when ζ is near 1, illustrate the fact that there is a trade-off between the speed of response (how fast one gets to, and stays near, the desired value 0 for φ) and the proportion of overshoot (the maximum of $\varphi(t)$ given the initial magnitude 1). Show that for $\zeta = \sqrt{2}/2$ this overshoot is $e^{-\pi} < 0.05$.

(C) Assume now that the objective is not to stabilize the system (1.2) about $\varphi = \dot{\varphi} = 0$, but to make it assume some other desired value $\varphi = \varphi_d$, still with velocity $\dot{\varphi}$ approaching zero. If we were already at rest at this position $\varphi = \varphi_d$, then the constant control $u \equiv -\varphi_d$ would keep us there. Otherwise, we must add to this a correcting term. We then modify the PD controller to measure instead the deviation from the desired value φ_d:

$$u(t) = -\alpha(\varphi(t) - \varphi_d) - \beta\dot{\varphi}(t) - \varphi_d\,.$$

Show that with this control, and still assuming $\alpha > 1, \beta > 0$, the solutions of the closed-loop system indeed approach $\varphi = \varphi_d, \dot{\varphi} = 0$. One says that the solution has been made to *track* the desired *reference* value φ_d. □

The value $\zeta = \sqrt{2}/2$ appears often in engineering practice, because this results in an overshoot of less than 5%, a figure considered acceptable. If we write θ for the angle between the negative real axis and the vector z, where z is either root of the characteristic equation

$$z^2 + 2\zeta\omega z + \omega^2 = 0\,,$$

then $\zeta = \cos\theta$. Thus, a value $\theta = 45°$ or less is often sought. If in addition one wants the magnitude of solutions to decay at a given exponential rate, say, as $e^{-\lambda t}$, then one looks at placing eigenvalues in a region that forces the desired stability degree, such as that illustrated in Figure 1.4, which exhibits a "damping margin" of 45° and a "stability margin" of λ.

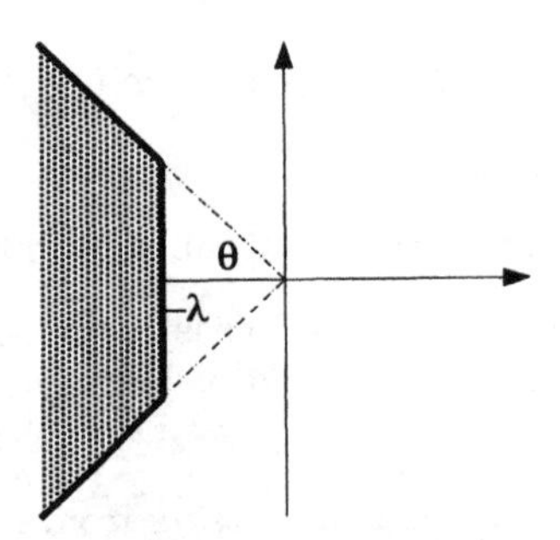

Figure 1.4: *Typical desired eigenvalue locations.*

Because solutions of more general linear equations can be written as sums of those corresponding to complex pairs and real roots, designers often apply these same rules to systems of order higher than two.

1.6 Outputs and Dynamic Feedback

We introduce now another major component of control theory, the idea of measurements and outputs.

In some situations all of the variables needed in order to implement a feedback control are readily available, but other times this is not the case. For instance, the speed of a car can be obtained easily via a speedometer, but the ground speed of an aircraft may be difficult to estimate. To model mathematically such a constraint, one adds to the description of a system the specification of a *measurement* function. This is a function of the states, with scalar or vector values, which summarizes all of the information that is available to the control algorithm.

Returning once more to the inverted pendulum example, let us assume that in designing the PD controller (1.6) we are restricted to measuring directly only the angular position $\varphi(t)$. We model this by adding to the state space equation

(1.28), where A and B are given by (1.29), the mapping

$$C : \mathbb{R}^2 \to \mathbb{R} : \; x \mapsto x_1 \,, \tag{1.34}$$

or equivalently its matrix in the canonical basis, $C = (1, 0)$, that picks the variable φ. We write $y(t)$ for the allowed measurement at time t, that is,

$$y(t) = Cx(t) \,. \tag{1.35}$$

More generally, we define a *linear system with output* by giving triples of matrices (A, B, C), where for some integers n, m, p, A is $n \times n$, B is $n \times m$, and C is $p \times n$. The integer p can be thought of as the number of independent measurements that can be taken at each instant. One also defines nonlinear and discrete-time systems with output in an analogous way, or more abstractly by adding an output map to the axiomatic definition of system via transitions (1.31).

A linear feedback law that depends only on the allowed measurements is then, in terms of this notation, of the form

$$u(t) = KCx(t) \,. \tag{1.36}$$

In other words, F is restricted to factor as $F = KC$, where C is given. As discussed when dealing with proportional-only control, such *static output feedback* is in general inadequate for control purposes. The set of eigenvalues achievable for matrices $A + BKC$ by varying K is severely restricted.

Of course, there is no theoretical reason for restricting attention to static feedback laws. A controller that would incorporate a device for differentiating the observed position $x_1(t) = \varphi(t)$ could then use this derivative.

However, differentiation tends to be an undesirable operation, because it is very sensitive to noise. To illustrate this sensitivity, consider the problem of finding the derivative c of the function

$$\xi(t) = ct \,,$$

where c is an unknown constant. Assume that the data available consists of $\xi(t)$ measured under additive noise,

$$y(t) = \xi(t) + n(t) \,,$$

where $n(t)$ is of the form

$$n(t) = d \sin \omega t \,,$$

and d and ω are unknown constants, ω being large. (This is realistic in that noise effects tend to be of "high frequency" compared to signals, but of course in real applications the noise cannot be modeled by a deterministic signal of constant frequency, and a probabilistic description of $n(t)$ is needed for an accurate analysis.) If we simply differentiate, the result

$$\dot{y}(t) = c + d\omega \cos \omega t$$

can be very far from the desired value c since ω is large. An alternative is to use the fact that the noise $n(t)$ averages to zero over long intervals; thus, we can cancel its effect by integration. That is, if we compute

$$\int_0^t y(\tau)\, d\tau$$

and we then multiply this by $2/t^2$, we obtain

$$c - \frac{2d}{\omega t^2}(1 - \cos \omega t)\,, \tag{1.37}$$

which converges to c as $t \to \infty$ and hence provides an accurate estimate, asymptotically, of the desired derivative.

The topic of *observers* deals with obtaining estimates of all the state variables from the measured ones, using only integration —or, in discrete-time, summations. It is an easy consequence of general theorems on the existence of observers that one can achieve also stabilization of the inverted pendulum, as well as of most other continuous-time linear systems, using a controller that is itself a continuous-time linear system. Such a controller is said to incorporate *integral* or *dynamic* feedback, and it includes a differential equation, driven by the observations, that calculates the necessary estimate. Analogously, for discrete-time systems with partial observations, the theorems result in the existence of observers that are themselves discrete-time linear systems.

In particular, there exist for the pendulum coefficients α, β, μ, ν, so that

$$\dot{z}(t) = \nu z(t) + \mu x_1(t) \tag{1.38}$$

together with a feedback law

$$u(t) = -\alpha x_1(t) - \beta z(t) \tag{1.39}$$

stabilizes the system. The controller solves the differential equation (1.38) with an arbitrary initial condition $z(0)$ and feeds the linear combination (1.39) of the measured variable x_1 and the estimated variable z back into the system. Later we develop in detail a systematic approach to the construction of such dynamic controllers, but in this simple example it is easy to find the necessary parameters by analyzing the equations.

Exercise 1.6.1 Consider the system of three differential equations obtained from (1.28) together with (1.38), where $u(t)$ is given by (1.39). Find numbers α, β, μ, ν, such that all solutions of this system approach zero. □

When observations are subject to noise, the speed at which one can estimate the unmeasured variables —and hence how fast one can control— is limited by the magnitude of this noise. For instance, the convergence of the expression in equation (1.37) is slower when d is large (noise has high intensity) or ω is small (noise has large bandwidth). The resulting trade-offs give rise to problems of *stochastic filtering* or *stochastic state estimation*, and the *Kalman Filter* is a technique for the analysis of such problems.

PID Control

Dynamic control is also useful in order to cancel unwanted disturbances even if all state variables are available for feedback. We illustrate this using again the inverted pendulum model. Assume that an unknown but constant disturbance $d(t) \equiv e$ can act on the system, which is therefore described by

$$\ddot{\varphi}(t) - \varphi(t) = u(t) + e. \tag{1.40}$$

One might still try to use the PD controller (1.6) to stabilize this system independently of e. Since stability should hold in particular when $e = 0$, one should have $\alpha > 1$ and $\beta > 0$ as before. But the PD controller is not sufficient:

Exercise 1.6.2 Show that if $e \neq 0$ and $\alpha > 1$, $\beta > 0$, then no solution of

$$\ddot{\varphi}(t) + \beta\dot{\varphi}(t) + (\alpha - 1)\varphi(t) = e$$

is such that $\varphi(t) \to 0$ as $t \to \infty$. □

Since e is a constant, it can be thought of as the solution of the differential equation

$$\dot{x}_0(t) = 0\,. \tag{1.41}$$

If we think of x_0 as a quantity that cannot be measured, then the previous discussion about observers suggests that there may exist a dynamic controller:

Exercise 1.6.3 Find numbers α, β, μ with the following property: For each $e \in \mathbb{R}$, all of the solutions of the system of equations

$$\begin{aligned}
\dot{x}_0(t) &= x_1(t) \\
\dot{x}_1(t) &= x_2(t) \\
\dot{x}_2(t) &= -\mu x_0(t) + (1 - \alpha)x_1(t) - \beta x_2(t) + e
\end{aligned}$$

converge to $(e/\mu, 0, 0)$. (*Hint:* First solve this problem for the homogeneous system that results if $e = 0$ by finding parameters that make the matrix associated to the equations have all eigenvalues with negative real parts. Then show that the same parameters work in the general case.) □

Thus, using a controller that integrates $x_1(t) = \varphi(t)$ and feeds back the combination

$$u(t) = -\alpha\varphi(t) - \beta\dot{\varphi}(t) - \mu \int_0^t \varphi(\tau)d\tau \tag{1.42}$$

one can ensure that both φ and $\dot{\varphi}$ converge to zero. Another way of thinking of this controller is as follows: If the integral term is not used, the value $\varphi(t)$ approaches a constant *steady-state error*; the effect of the integral is to offset a nonzero error. The controller (1.42) is a *PID*, or *proportional-integral-derivative* feedback, the control mechanism most widely used in linear systems applications.

More generally, the problem of canceling the effect of disturbances that are not necessarily constant gives rise to the study of *disturbance rejection* or *disturbance decoupling* problems. The resulting dynamic controllers typically incorporate an "internal model" of the possible mechanism generating such disturbances, as with the simple case (1.41).

Observability, Duality, Realization, and Identification

A basic system theoretic property that arises naturally when studying observers and dynamic feedback is that of *observability.* Roughly speaking, this means that all information about the state x should in principle be recoverable from knowledge of the measurements y that result; in a sense it is a counterpart to the concept of controllability —mentioned earlier— which has to do with finding controls u that allow one to attain a desired state x. In the case of linear systems, this analogy can be made precise through the idea of *duality,* and it permits obtaining most results for observers as immediate corollaries of those for controllability. This duality extends to a precise correspondence between optimal control problems —in which one studies the trade-off between cost of control and speed of control— and filtering problems —which are based on the trade-off between magnitude of noise and speed of estimation.

Once outputs or measurements are incorporated into the basic definition of a system, one can pose questions that involve the relationship between controls u and observations y. It is then of interest to characterize the class of *input/output (i/o) behaviors* that can be obtained and conversely to study the *realization* problem: "Given an observed i/o behavior, what possible systems could give rise to it?" In other words, if we start with a "black box" model that provides only information about how u affects y, how does one deduce the differential —or, in discrete-time, difference— equation that is responsible for this behavior?

Besides obvious philosophical interest, such questions are closely related to *identification* problems, where one wishes to estimate the i/o behavior itself from partial and possibly noisy data, because state-space descriptions serve to parametrize such possible i/o behaviors. Conversely, from a synthesis viewpoint, realization techniques allow us to compute a state representation, and hence also construct a physical system, that satisfies given i/o specifications.

The main results on the realization problem show that realizations essentially are unique provided that they satisfy certain minimality or irredundancy requirements. We will provide the main theorems of realization theory for linear systems. The underlying properties turn out to be closely related to other system theoretic notions such as controllability and observability.

1.7 Dealing with Nonlinearity

Ultimately, linear techniques are limited by the fact that real systems are more often than not nonlinear. As discussed above, local analysis and control design in general can be carried out satisfactorily using just first-order information, but global analysis requires more powerful methods.

One case in which global control *can* be achieved with ideas from linear system theory is that in which the system can be reduced, after some transformation, to a linear one. For instance, a change of coordinates in the space of the x variable might result in such a simplification. A situation that has appeared many times in applications is that in which a particular *feedback* control linearizes the system. Consider again the pendulum, but now the original nonlinear model.

If we first subtract the effect of the term $\sin\theta$ by using u, we are left with a simple linear system

$$\ddot{\theta}(t) = u(t)\,, \tag{1.43}$$

which can be controlled easily. For instance, the PD feedback law

$$u(t) = -\theta(t) - \dot{\theta}(t) \tag{1.44}$$

stabilizes (1.43). In order to stabilize the original system, we now add the subtracted term. That is, we use

$$u(t) = \sin\theta(t) - \theta(t) - \dot{\theta}(t)\,. \tag{1.45}$$

Passing from the original model to (1.43) can be thought of as the effect of applying the feedback law

$$u(t) = f(x(t)) + u'(t)\,,$$

where $f(x) = \sin x_1$ and u' is a new control; the study of such feedback transformations and the characterization of conditions under which they simplify the system is an active area of research.

The above example illustrates another issue that is important when dealing with the global analysis of nonlinear control problems. This is the fact that the mathematical structure of the state space might impose severe restrictions on what is achievable. The proposed feedback law (1.45) will stabilize

$$\ddot{\theta}(t) + \sin\theta(t) = u(t)$$

provided we can think of θ as taking arbitrary real values. However, in the physical system $\theta + 2\pi$ describes the same state as θ. Thus, the natural space to describe θ is a unit circle. When applying the control u as defined by the formula (1.45), which of the infinitely many possible values of θ should one use? It turns out that everything will behave as desired as long as one chooses u continuously in time —as the pendulum finishes one complete counterclockwise revolution, start using an angle measured in the next 2π-interval. But this choice is not unambiguous in terms of the physical coordinate θ. In other words, (1.45) is not a feedback law when the "correct" state space is used, since it is not a function of states.

The correct state space for this example is in fact the Cartesian product of a circle and the real line, since pairs $(\theta(t), \dot{\theta}(t))$ belong naturally to such a space. It can be proved that in such a space —mathematically the tangent bundle to the unit circle— there cannot exist *any* continuous feedback law that stabilizes the system, because the latter would imply that the space is diffeomorphic (smoothly equivalent) to a Euclidean space. In general, techniques from differential geometry —or, for systems described by polynomial nonlinearities, algebraic geometry— appear when studying global behavior.

When dealing with the problems posed by infinite dimensionality as opposed to nonlinearity, similar considerations force the use of techniques from functional analysis.

1.8 A Brief Historical Background

Control mechanisms are widespread in nature and are used by living organisms in order to maintain essential variables such as body temperature and blood sugar levels at desired setpoints. In engineering too, feedback control has a long history: As far back as the early Romans, one finds water levels in aqueducts being kept constant through the use of various combinations of valves.

Modern developments started during the seventeenth century. The Dutch mathematician and astronomer Christiaan Huygens designed pendulum clocks and in doing so analyzed the problem of speed control; in this work he competed with his contemporary Robert Hooke. The needs of navigation had a strong influence on scientific and technological research at that time, and accurate clocks —to allow determinations of solar time— were in great demand. The attention turned to windmills during the eighteenth century, and speed controls based on Hooke's and Huygens' ideas were built. A central idea here is the use of flyballs: Two balls attached to an axis turn together with the windmill, in such a manner that centrifugal force due to angular velocity causes them to rise; in turn this upward movement is made to affect the positions of the mill's sails. Thus, feedback was implemented by the linkages from the flyballs to the sails.

But it was the Industrial Revolution, and the adaptation by James Watt in 1769 of flyball governors to steam engines, that made control mechanisms very popular; the problem there was to regulate engine speed despite a variable load. Steady-state error could appear, and various inventors introduced variations of the integral feedback idea in order to deal with this problem.

The mathematician and astronomer George Airy was the first to attempt, around 1840, an analysis of the governor and similar devices. By 1868, there were about 75,000 Watt governors in use; that year, the Scottish physicist James Clerk Maxwell published the first complete mathematical treatment of their properties and was able to explain the sometimes erratic behavior that had been observed as well as the effect of integral control. His work gave rise to the first wave of theoretical research in control, and characterizations of stability were independently obtained for linear systems by the mathematicians A. Hurwitz and E.J. Routh. This theory was applied in a variety of different areas, such as the study of ship steering systems.

During the 1930s, researchers at Bell Telephone Laboratories developed the theory of feedback amplifiers, based on assuring stability and appropriate response for electrical circuits. This work, by H. Nyquist, H.W. Bode, and others, constitutes even today the foundation of much of frequency design. Analog computers appeared also around that time and formed the basis for implementing controllers in the chemical and petroleum industries. During the Second World War, these techniques were used in the design of anti-aircraft batteries and fire-control systems, applications that continue to be developed today. The mathematician Norbert Wiener, who developed a theory of estimation for random processes used in these applications, coined at the end of the war the term "cybernetics" to refer to control theory and related areas.

These so-called classical approaches were for the most part limited by their restriction to linear time-invariant systems with scalar inputs and outputs. Only during the 1950s did control theory begin to develop powerful general techniques that allowed treating multivariable, time-varying systems, as well as many nonlinear problems. Contributions by Richard Bellman (dynamic programming) and Rudolf Kalman (filtering, linear/quadratic optimal control, and algebraic analysis) in the United States, and by L. Pontryagin (nonlinear optimal control) in the Soviet Union, formed the basis for a very large research effort during the 1960s, which continues to this day. Present day theoretical research in control theory involves a variety of areas of pure mathematics, as illustrated for instance by the remarks in Section 1.7. Concepts and results from these areas find applications in control theory; conversely, questions about control systems give rise to new open problems in mathematics.

Excellent references for the early historical background are the papers [149], which contain a large number of literature citations, and the book [299]. See also the introductory article [226]. Other references, which in addition contain overviews of current research and open problems, are the reports [275] and [135].

1.9 Some Topics Not Covered

In an area as wide as control theory, it is impossible to cover all topics in a single text, even briefly. For instance, we concentrate exclusively on *deterministic* systems. Incorporating models for uncertainty leads to *stochastic* models, which are the subject of much research activity, when this uncertainty is expressed in probabilistic or statistical terms; references to the literature on stochastic aspects of many of the problems considered are given at various points in the text. Mathematically different but closely related is the area of *robust control*, which deals with the design of control laws that are guaranteed to perform even if the assumed model of the system to be controlled is incorrect —with the allowed deviations quantified in appropriate norms— or under the possibility of imperfect controller design. See the collection of papers [123] for pointers to a large literature. The area of *adaptive control* deals also with the control of partially unknown systems, but differs from robust control in the mathematics employed. Adaptive controllers typically make use of *identification techniques*, which produce estimates of the system parameters for use by controllers; see e.g. [282].

When using computers one should consider the effect of quantization errors on the implementation of control laws, which arise due to limited precision when real-valued signals are translated into fixed-point representations (*A/D* or *analog to digital* conversion); see e.g. [304]. Other questions relate to the interface between higher-level controllers implemented in software and lower-level servos of an analog and physical character; this gives rise to the area of *hybrid systems*; see the volume [15].

The notes at the end of the various chapters give further bibliographical references to many of the other areas that have been omitted.

Chapter 2

Systems

This Chapter introduces concepts and terminology associated with discrete- and continuous-time systems, linearization, and input/output expansions, and establishes some of their elementary properties.

2.1 Basic Definitions

As seen in the previous Chapter, it is of interest to study both discrete-time and continuous-time systems. In the first case, the underlying time set is discrete, and it can be thought of as the set of integers $\mathbb{Z}$; in the second case the time variable is a real number. To treat both simultaneously, we introduce the following notion. In most of our results it will be the case that either $\mathcal{T} = \mathbb{Z}$ or $\mathbb{R}$.

Definition 2.1.1 *A* **time set** $\mathcal{T}$ *is a subgroup of* $(\mathbb{R}, +)$. □

For any such set, $\mathcal{T}_+$ is the set of nonnegative elements $\{t \in \mathcal{T} | t \geq 0\}$. By notational convention, when the time set $\mathcal{T}$ is understood from the context, all intervals are assumed to be restricted to $\mathcal{T}$. Thus, for instance,

$$[a, b) = \{t \in \mathcal{T}, a \leq t < b\}$$

and similarly for open, closed, or infinite intervals.

For each set $\mathcal{U}$ and interval $\mathcal{I}$, the set of all maps from $\mathcal{I}$ into $\mathcal{U}$ is denoted by

$$\mathcal{U}^{\mathcal{I}} = \{\omega \mid \omega : \mathcal{I} \to \mathcal{U}\}. \tag{2.1}$$

If $\mathcal{T} = \mathbb{Z}$ and k is a nonnegative integer, the set $\mathcal{U}^{[0,k)}$ can be identified naturally with the set of all sequences

$$\omega(0), \ldots, \omega(k-1)$$

of length k consisting of elements of $\mathcal{U}$, i.e. the Cartesian product $\mathcal{U}^k$. In the particular case in which $\mathcal{I}$ is an empty interval, the set in (2.1) consists of just one element, which we denote as $\diamond$; this can be thought of as the "empty sequence" of zero length.

Assume that σ, τ, μ, are any three elements of $\mathcal{T}$ that satisfy $\sigma \leq \tau \leq \mu$. If $\omega_1 \in \mathcal{U}^{[\sigma,\tau)}$ and $\omega_2 \in \mathcal{U}^{[\tau,\mu)}$, their *concatenation*, denoted simply as $\omega_1\omega_2$, is the function $\omega \in \mathcal{U}^{[\sigma,\mu)}$ defined by

$$\omega(t) := \begin{cases} \omega_1(t) & \text{if } t \in [\sigma, \tau), \\ \omega_2(t) & \text{if } t \in [\tau, \mu). \end{cases}$$

Note that $\omega\diamond = \diamond\omega = \omega$.

The next definition provides the abstraction of the concept of system that was discussed on page 12.

Definition 2.1.2 *A* **system** *or* **machine** $\Sigma = (\mathcal{T}, \mathcal{X}, \mathcal{U}, \phi)$ *consists of:*

- *A time set* $\mathcal{T}$*;*
- *A nonempty set* $\mathcal{X}$ *called the* **state space** *of* Σ*;*
- *A nonempty set* $\mathcal{U}$ *called the* **control-value** *or* **input-value space** *of* Σ*; and*
- *A map* $\phi : \mathcal{D}_\phi \to \mathcal{X}$ *called the* **transition map** *of* Σ*, which is defined on a subset* $\mathcal{D}_\phi$ *of*

$$\{(\tau, \sigma, x, \omega) \mid \sigma, \tau \in \mathcal{T}, \sigma \leq \tau, x \in \mathcal{X}, \omega \in \mathcal{U}^{[\sigma,\tau)}\}, \tag{2.2}$$

such that the following properties hold:

nontriviality *For each state* $x \in \mathcal{X}$*, there is at least one pair* $\sigma < \tau$ *in* $\mathcal{T}$ *and some* $\omega \in \mathcal{U}^{[\sigma,\tau)}$ *such that* ω **is admissible for** x*, that is, so that* $(\tau, \sigma, x, \omega) \in \mathcal{D}_\phi$*;*

restriction *If* $\omega \in \mathcal{U}^{[\sigma,\mu)}$ *is admissible for* x*, then for each* $\tau \in [\sigma, \mu)$ *the restriction* $\omega_1 := \omega|_{[\sigma,\tau)}$ *of* ω *to the subinterval* $[\sigma, \tau)$ *is also admissible for* x *and the restriction* $\omega_2 := \omega|_{[\tau,\mu)}$ *is admissible for* $\phi(\tau, \sigma, x, \omega_1)$*;*

semigroup *If* σ, τ, μ *are any three elements of* $\mathcal{T}$ *so that* $\sigma < \tau < \mu$*, if* $\omega_1 \in \mathcal{U}^{[\sigma,\tau)}$ *and* $\omega_2 \in \mathcal{U}^{[\tau,\mu)}$*, and if* x *is a state so that*

$$\phi(\tau, \sigma, x, \omega_1) = x_1 \quad and \quad \phi(\mu, \tau, x_1, \omega_2) = x_2\,,$$

then $\omega = \omega_1\omega_2$ *is also admissible for* x *and* $\phi(\mu, \sigma, x, \omega) = x_2$*;*

identity *For each* $\sigma \in \mathcal{T}$ *and each* $x \in \mathcal{X}$*, the empty sequence* $\diamond \in \mathcal{U}^{[\sigma,\sigma)}$ *is admissible for* x *and* $\phi(\sigma, \sigma, x, \diamond) = x$*.* □

As discussed in Chapter 1, Section 1.6, one often needs to model explicitly the fact that limited measurements are available. The following concept is then natural.

Definition 2.1.3 *A system or machine* **with outputs** *is given by a system* Σ *together with*

- *A set* $\mathcal{Y}$ *called the* **measurement-value** *or* **output-value space**; *and*
- *A map* $h : \mathcal{T} \times \mathcal{X} \to \mathcal{Y}$ *called the* **readout** *or* **measurement map**. □

We use the same symbol for the system with outputs $(\mathcal{T}, \mathcal{X}, \mathcal{U}, \phi, \mathcal{Y}, h)$ and its **underlying system** $(\mathcal{T}, \mathcal{X}, \mathcal{U}, \phi)$. Thus Σ could denote either the first or the second of these, depending on the context. Unless otherwise stated, every definition given for systems will apply to systems with outputs by reference to its underlying system. For instance, we define in a later section the notion of a controllable system; a controllable system with outputs will then be, by definition, one whose underlying system is controllable.

Elements of $\mathcal{X}$ are called **states**, elements of $\mathcal{U}$ are **control values** or **input values**, and those of $\mathcal{Y}$ are **output values** or **measurement values**. The functions $\omega \in \mathcal{U}^{[\sigma,\tau)}$ are called **controls** or **inputs**. If ω is like this, with σ and τ in $\mathcal{T}$, one denotes by

$$|\omega| = \tau - \sigma$$

the **length** of ω. If ω is admissible for x, we also say that (x, ω) forms an **admissible pair**.

The definition of system is intended to capture the intuitive notion of a machine that evolves in time according to the transition rules specified by ϕ. At each instant, the state x summarizes all of the information needed in order to know the future evolution of the system. For instance, in the case of the pendulum we took as the state the pair (angular position, angular velocity). We read $\phi(\tau, \sigma, x, \omega)$ as "the state at time τ resulting from starting at time σ in state x and applying the input function ω." The definition allows the possibility of undefined transitions, when the input ω is not admissible for the given state, which correspond for differential equations to the phenomenon of "explosion times" at which the solution ceases to be defined.

A few simplifying conventions are useful. When σ and τ are clear from the context, we write $\phi(\tau, \sigma, x, \omega)$ simply as $\phi(x, \omega)$. Often, a formula such as

$$z = \phi(x, \omega) \tag{2.3}$$

will implicitly mean "ω is admissible for x and $z = \phi(x, \omega)$." When more than one system is being studied, we sometimes add a subscript and write $\mathcal{X}_\Sigma$, $\mathcal{U}_\Sigma$, ϕ_Σ to emphasize that we are considering the set $\mathcal{X}$, $\mathcal{U}$, or the map ϕ, corresponding to the particular system Σ. Sometimes we refer simply to "the system $(\mathcal{X}, \mathcal{U}, \phi)$," or even just $(\mathcal{X}, \phi)$, when the time set $\mathcal{T}$, or $\mathcal{U}$, is clear from the context, and similarly for systems with outputs. A particularly useful notation is the following: We write

$$x \underset{\omega}{\leadsto} z \tag{2.4}$$

instead of equation (2.3). Then the semigroup axiom is simply

$$x \underset{\omega_1}{\leadsto} x_1 \text{ and } x_1 \underset{\omega_2}{\leadsto} x_2 \quad \Rightarrow \quad x \underset{\omega_1\omega_2}{\leadsto} x_2 \tag{2.5}$$

and the identity axiom is

$$x \underset{\diamond}{\leadsto} x \tag{2.6}$$

for all x.

It is also convenient to introduce a notation for the mapping ψ that assigns to each admissible pair (x, ω) the complete state trajectory on $[\sigma, \tau]$ rather than just the final state at time τ. That is,

$$\psi : \mathcal{D}_\phi \to \bigcup_{\sigma \leq \tau} \mathcal{X}^{[\sigma,\tau]},\ (\tau, \sigma, x, \omega) \mapsto \xi\,,$$

where

$$\xi(t) := \phi(t, \sigma, x, \omega|_{[\sigma,t)}) \tag{2.7}$$

for each $t \in [\sigma, \tau]$. Note that the right-hand side of (2.7) is well defined because of the restriction axiom. As with ϕ, we write simply $\psi(x, \omega)$ instead of $\psi(\tau, \sigma, x, \omega)$ when σ and τ are clear from the context.

It is useful to define also admissibility for infinite-length controls.

Definition 2.1.4 *Given a system Σ and a state $x \in \mathcal{X}$, the function $\omega \in \mathcal{U}^{[\sigma,\infty)}$* **is admissible for** *x provided that every restriction $\omega|_{[\sigma,\tau)}$ is admissible for x, for each $\tau > \sigma$.* □

We also use the term "control" to refer to such an infinite length ω, and write $\psi(\infty, \sigma, x, \omega)$ for the function $\xi \in \mathcal{X}^{[\sigma,\infty)}$ which satisfies (2.7) for all $t \in [\sigma, \infty)$. Note that the expression $\phi(\infty, \sigma, x, \omega)$ has *not* been defined; a natural definition, however, would be as

$$\lim_{t\to\infty} \psi(\infty, \sigma, x, \omega)(t)\,,$$

and this is basically what will be done later when dealing with asymptotic notions, for those systems for which $\mathcal{X}$ has a topological structure.

Definition 2.1.5 *A* **trajectory** Γ *for the system Σ on the interval $\mathcal{I} \subseteq \mathcal{T}$ is a pair of functions (ξ, ω), $\xi \in \mathcal{X}^{\mathcal{I}}, \omega \in \mathcal{U}^{\mathcal{I}}$ such that*

$$\xi(\tau) = \phi(\tau, \sigma, \xi(\sigma), \omega|_{[\sigma,\tau)})$$

holds for each pair $\sigma, \tau \in \mathcal{I}, \sigma < \tau$. □

Definition 2.1.6 *A* **path** *of the system Σ on the interval $\mathcal{I} \subseteq \mathcal{T}$ is a function $\xi : \mathcal{I} \to \mathcal{X}$ for which there exists some $\omega \in \mathcal{U}^{\mathcal{I}}$ such that (ξ, ω) is a trajectory on $\mathcal{I}$ for Σ. If ξ is a path and $\mathcal{I}$ has the form $[\sigma, \tau)$ or $[\sigma, \tau]$ (σ finite), then $\xi(\sigma)$ is the* **initial state** *of the path. If $\mathcal{I}$ has the form $(\sigma, \tau]$ or $[\sigma, \tau]$ (τ finite), then $\xi(\tau)$ is the* **terminal state** *of the path.* □

Some Taxonomy

Various general properties serve to classify systems into broad classes, as discussed in the introductory chapter. We introduce next some of these classes. Discrete- and continuous-time systems will be defined in the next section.

Definition 2.1.7 *The system* Σ *is* **complete** *if every input is admissible for every state:*

$$\mathcal{D}_\phi = \{(\tau, \sigma, x, \omega) \mid \sigma \leq \tau, x \in \mathcal{X}, \omega \in \mathcal{U}^{[\sigma,\tau)}\}.$$

More generally, for any family of controls $\mathcal{V} \subseteq \bigcup_{\sigma\leq\tau} \mathcal{U}^{[\sigma,\tau)}$, $\mathcal{V}$**-completeness** *means that* $(\tau, \sigma, x, \omega)$ *is in* $\mathcal{D}_\phi$ *whenever* $\sigma \leq \tau, x \in \mathcal{X}, \omega \in \mathcal{V}$. □

For continuous-time systems (see below) completeness will always mean $\mathcal{V}$-completeness, where $\mathcal{V}$ is the class of all essentially bounded measurable controls.

The nontriviality and restriction axioms in the definition of system are automatically true if the system is complete. In that case, a system is precisely the same as an *action* of $\mathcal{U}^*$ on the set $\mathcal{X}$, where $\mathcal{U}^*$ is the union of the sets $\mathcal{U}^{[\sigma,\tau)}$, thought of as a semigroup with a partially defined binary operation (concatenation). Many of the concepts that we study are direct analogues of those studied for semigroup or group actions; for instance, "controllability" will be the analogue of "transitivity."

Systems defined by classical differential or difference equations become "systems" in our sense when they are reinterpreted as having a one-element control-value set. The closed-loop system that we obtained after designing a feedback law for the pendulum was such a system, as would be the unforced pendulum that results when setting the control to zero.

Definition 2.1.8 *The system* Σ **has no controls** *if* $\mathcal{U}$ *is a one-element set.* □

We also say then that Σ is a *classical dynamical system.* The often-used terminology *autonomous* has the disadvantage of appearing in the theory of differential equations to refer to what we call below time-invariance. For systems with no controls, there is for each σ, τ only one possible element in $\mathcal{U}^{[\sigma,\tau)}$, so $\phi(\tau, \sigma, x, \omega)$ is independent of the last coordinate; we then write simply $\phi(\tau, \sigma, x)$.

The most important general subclass of systems is that of time-invariant systems, for which the structure is independent of time, that is, $\phi(\tau, \sigma, x, \omega)$ depends only on $\tau - \sigma, x, \omega$:

Definition 2.1.9 *The system* Σ *is* **time-invariant** *if for each* $\omega \in \mathcal{U}^{[\sigma,\tau)}$, *each* $x \in \mathcal{X}$, *and each* $\mu \in \mathcal{T}$, *if* ω *is admissible for* x *then the translation*

$$\omega^\mu \in \mathcal{U}^{[\sigma+\mu,\tau+\mu)}, \ \omega^\mu(t) := \omega(t - \mu) \tag{2.8}$$

is admissible for x, *and*

$$\phi(\tau, \sigma, x, \omega) = \phi(\tau + \mu, \sigma + \mu, x, \omega^\mu).$$

For systems with outputs, it is also required that $h(\tau, x)$ *be independent of* τ. □

When dealing with time-invariant systems, it is customary to identify controls with their translations. Thus, for instance, given $\omega_1 \in \mathcal{U}^{[0,\sigma)}$ and $\omega_2 \in \mathcal{U}^{[0,\tau)}$, one thinks of their concatenation $\omega_1\omega_2$ as an element of $\mathcal{U}^{[0,\sigma+\tau)}$, meaning the concatenation $\omega_1\nu$, where ν is the translation of ω_2 to $[\sigma, \sigma+\tau)$. A system that is not necessarily time-invariant is sometimes called, to emphasize that fact, a **time-varying** system.

2.2 I/O Behaviors

Frequently physical systems are specified not by means of their state-space descriptions but rather through their input/output behavior, that is to say, the effect that inputs have on observed outputs. In this section we introduce i/o behaviors and show how these are induced by systems with outputs in the sense of Definition 2.1.3.

Technically, there are at least two possible ways of specifying i/o behavior. The first, the one to be followed here, is to give a rule that tells us what is the output value that results right after an input has been applied. The second possibility is to specify the entire output, as a function of time, that is generated by this input.

Definition 2.2.1 *An* **i/o behavior** $\Lambda = (\mathcal{T}, \mathcal{U}, \mathcal{Y}, \lambda)$ *consists of:*

- *A time set* $\mathcal{T}$*;*
- *A nonempty set* $\mathcal{U}$ *called the* **control-value** *or* **input-value space** *of* Λ*;*
- *A nonempty set* $\mathcal{Y}$ *called the* **output-value** *or* **measurement-value space** *of* Λ*; and*
- *A map*

  $$\lambda : \mathcal{D}_\lambda \to \mathcal{Y}$$

  called the **response map** *of* Λ*, which is defined on a nonempty subset* $\mathcal{D}_\lambda$ *of*

  $$\{(\tau, \sigma, \omega) \mid \sigma, \tau \in \mathcal{T}, \sigma \leq \tau, \omega \in \mathcal{U}^{[\sigma,\tau)}\},$$

such that the following property holds:

restriction *If* (τ, σ, ω) *is in* $\mathcal{D}_\lambda$*, then also* $(\mu, \sigma, \omega|_{[\sigma,\mu)}) \in \mathcal{D}_\lambda$ *for each* $\mu \in [\sigma, \tau]$. □

For each fixed σ, τ, we often denote

$$\lambda^{\sigma,\tau}(\omega) := \lambda(\tau, \sigma, \omega)$$

or simply $\lambda(\omega)$ if σ and τ are clear from the context.

Remark 2.2.2 Given an i/o behavior Λ, one may define also the **i/o map** associated to Λ, as follows: This is the map that gives the entire output function when a control has been applied, that is,

$$\bar{\lambda} : \mathcal{D}_\lambda \to \bigcup_{\sigma \leq \tau} \mathcal{Y}^{[\sigma,\tau]},$$

where

$$\bar{\lambda}(\tau, \sigma, \omega)(t) := \lambda(t, \sigma, \omega|_{[\sigma,t)})$$

for each $t \in [\sigma, \tau]$. Note that the following property holds:

strict causality If (τ, σ, ω) is in $\mathcal{D}_\lambda$, then for each $\mu \in [\sigma, \tau]$,

$$\bar{\lambda}(\mu, \sigma, \omega|_{[\sigma,\mu)}) = \bar{\lambda}(\tau, \sigma, \omega)|_{[\sigma,\mu]}.$$

Note also that the response can be recovered from the i/o map, since

$$\lambda(\tau, \sigma, \omega) = \bar{\lambda}(\tau, \sigma, \omega)(\tau).$$ □

Our interest will be mainly in those behaviors that arise from systems after choosing an initial state.

Definition 2.2.3 *An* **initialized system** *is a pair* (Σ, x^0), *where* x^0 *is a state, the* **initial state**. *Similarly for systems with outputs.* □

Definition 2.2.4 *Let* (Σ, x^0) *be an initialized system. The* **input to state (i/s) behavior of the system** (Σ, x^0) *is the behavior with the same sets* $\mathcal{T}$ *and* $\mathcal{U}$ *as* Σ, *and with output value space* $\mathcal{X}$, *whose response map is defined on the projection*

$$\{(\tau, \sigma, \omega) \mid (\tau, \sigma, x^0, \omega) \in \mathcal{D}_\phi\} \tag{2.9}$$

by the formula

$$\lambda(\tau, \sigma, \omega) := \phi(\tau, \sigma, x^0, \omega).$$

If (Σ, x^0) *is an initialized system with output, then the* **i/o behavior of** (Σ, x^0) *has the same* $\mathcal{T}$, $\mathcal{U}$, *and* $\mathcal{Y}$ *as* Σ, *domain again equal to (2.9), and*

$$\lambda(\tau, \sigma, \omega) := h(\tau, \phi(\tau, \sigma, x^0, \omega)).$$

This map is the **response map of** (Σ, x^0). □

Subscripts such as in Λ_Σ or Λ_{Σ,x^0} may be used to indicate that one is referring to the behavior or the response map associated to a given system.

It is clear from the definitions that both the i/s and the i/o behaviors of a system are indeed i/o behaviors. Observe that the i/o map associated to the i/s response in the sense of Remark 2.2.2 is precisely the same as $\psi(\cdot, \cdot, x^0, \cdot)$ (cf. formula (2.7)). Note also that from the definitions and the identity axiom,

$$\lambda_{\Sigma,x^0}(\sigma, \sigma, \diamond) = h(\sigma, x^0)$$

for all $\sigma \in \mathcal{T}$.

In general there will be infinitely many systems that induce a given i/o behavior, the situation being entirely analogous to that which arises if we take the composition of two maps $\gamma = \alpha\beta$: Even though γ is uniquely defined from α and β, the latter are far from being determined just from the knowledge of γ, unless stringent hypotheses are made as to their form. This converse issue, that of passing from a given i/o behavior to a system with outputs that gives rise to the behavior, and the classification of the possible solutions, is the *realization problem*, to be studied in detail later (Sections 6.5 and 6.8).

The next definition introduces some standard families of behaviors.

Definition 2.2.5 *The i/o behavior Λ is*

complete *if $\mathcal{D}_\lambda = \{(\tau, \sigma, \omega) \mid \sigma, \tau \in \mathcal{T}, \sigma \leq \tau, \omega \in \mathcal{U}^{[\sigma,\tau)}\}$;*

time-invariant *if for each (τ, σ, ω) in $\mathcal{D}_\lambda$ and for each $\mu \in \mathcal{T}$ it holds that $(\tau + \mu, \sigma + \mu, \omega^\mu)$ is again in $\mathcal{D}_\lambda$, where ω^μ is the shift of ω defined in equation (2.8), and*

$$\lambda(\tau + \mu, \sigma + \mu, \omega^\mu) = \lambda(\tau, \sigma, \omega)$$

for each $t \in [\sigma, \tau)$. □

As for systems, one also defines $\mathcal{V}$-completeness of behaviors, with respect to any family of controls (for continuous-time systems, these will be the measurable essentially bounded ones).

Intuitively, complete behaviors are those with the property that the output is defined for every possible input, and time-invariant ones are those for which the same input but applied at a different time will still produce the same output.

Exercise 2.2.6 Prove the following facts, for initialized systems:

1. If Σ is ($\mathcal{V}$-)complete, then Λ_Σ also is.
2. If Σ is time-invariant, then Λ_Σ also is.

Give counterexamples to the converse implications. □

2.3 Discrete-Time

Definition 2.3.1 *A* **discrete-time** *system or i/o behavior is one for which $\mathcal{T} = \mathbb{Z}$.* □

Assume that Σ is a discrete-time system. For each $t \in \mathbb{Z}$, $x \in \mathcal{X}$, and $u \in \mathcal{U}$ for which the right-hand side is defined, let

$$\mathcal{P}(t, x, u) := \phi(t + 1, t, x, \omega), \tag{2.10}$$

where ω is that control of length 1 in $\mathcal{U}^{\{t\}}$ that has the value $\omega(t) = u$. This defines a *next-state* or *transition* mapping

$$\mathcal{P} : \mathcal{E} \to \mathcal{X},$$

where

$$\mathcal{E} = \mathcal{E}(\phi) = \{(t, x, u) \in \mathbb{Z} \times \mathcal{X} \times \mathcal{U} \mid (t+1, t, x, u) \in D_\phi\} .$$

In the particular case of complete systems, $\mathcal{E} = \mathbb{Z} \times \mathcal{X} \times \mathcal{U}$. In general, the following property is an immediate consequence of the nontriviality and restriction axioms:

(*) For each $x \in \mathcal{X}$, there exist some $u \in \mathcal{U}$ and some $t \in \mathbb{Z}$ such that $(t, x, u) \in \mathcal{E}$.

It is possible to recover the system description from the knowledge of $\mathcal{P}$. For instance, if $\omega \in \mathcal{U}^{[t,t+2)}$, then

$$\phi(t+2, t, x, \omega) = \mathcal{P}(t+1, \mathcal{P}(t, x, \omega(t)), \omega(t+1))$$

provided that both applications of $\mathcal{P}$ on the right-hand side are defined, and ϕ is undefined at $(t+2, t, x, \omega)$ otherwise. By induction one can establish the following fact:

Lemma/Exercise 2.3.2 If $\mathcal{E}$ is a subset of $\mathbb{Z} \times \mathcal{X} \times \mathcal{U}$ that satisfies (*) and $\mathcal{P} : \mathcal{E} \to \mathcal{X}$ is any mapping, then there exists a unique discrete-time system $\Sigma = (\mathbb{Z}, \mathcal{X}, \mathcal{U}, \phi)$ such that $\mathcal{E} = \mathcal{E}(\phi)$ and property (2.10) holds on $\mathcal{E}$. □

The triple $(\mathcal{X}, \mathcal{U}, \mathcal{P})$ provides a local-in-time description of the discrete-time system Σ. One often refers to "the discrete-time system $\Sigma = (\mathcal{X}, \mathcal{U}, \mathcal{P})$" when in fact meaning the system induced through Lemma 2.3.2. Alternatively, one may introduce such a system through its associated *evolution equation*

$$x(t+1) = \mathcal{P}(t, x(t), \omega(t)) \ , \ t \in \mathbb{Z}$$

together with, for systems with outputs, the measurement equation

$$y(t) = h(t, x(t)) .$$

We often drop the t-arguments and simply write

$$x^+ = \mathcal{P}(t, x, u), \ y = h(t, x) .$$

Note that if Σ is time-invariant then $\mathcal{P}$ and h are independent of t, and vice versa.

In the special case in which $\mathcal{X}$ and $\mathcal{U}$ are *finite* sets, discrete-time systems are called *automata* in computer science. For systems with outputs, and $\mathcal{Y}$ also finite, the terminology there is *sequential machines.*

In order to develop some intuition with the formalism, we analyze a typical sequential machine problem. Later we concentrate on systems for which all of the sets appearing are infinite —in fact, real or complex vector spaces— but automata and sequential machine theory did provide much of the original motivation for the development of the basic system theory concepts.

Example 2.3.3 In digital data communications, one is interested in reliably transmitting a sequence of bits (zeros and ones) over a noisy channel. In order to detect errors, one extra bit often is added to each block of, say, k bits. This *parity check* bit is made to be a 0 or a 1 as needed so that the resulting sequence of $k+1$ elements has an even (or odd, depending on the conventions) number of 1's. Assume that we are interested in designing a system that will detect errors. That is, after each $k+1$ inputs, the output will be "0" if the parity is even, and "1," indicating that an error occurred, otherwise. While scanning a block, no error should be signaled. For simplicity, we take the case $k = 2$. (A more typical value is $k = 7$.)

The desired i/o behavior can be modeled as follows. We chose an initial time, say, $t = 0$, and assume that the inputs will be presented starting at this time. Thus we chose

$$\mathcal{U} = \mathcal{Y} = \{0, 1\}$$

and the following domain for λ:

$$\mathcal{D}_\lambda = \{(\tau, 0, \omega) \mid \tau \geq 0, \omega \in \mathcal{U}^{[0,\tau)}\}.$$

The response map is $\lambda(\tau, 0, \omega) =$

$$\begin{cases} 1 & \text{if } \omega(\tau-3) + \omega(\tau-2) + \omega(\tau-1) \text{ is odd and 3 divides } \tau > 0, \\ 0 & \text{otherwise.} \end{cases}$$

We formalize the problem as that of finding an initialized discrete-time system (Σ, x^0) with outputs whose i/o behavior matches λ for $\tau \geq 1$, that is,

$$\lambda^{0,\tau}_{\Sigma,x^0} = \lambda^{0,\tau} \tag{2.11}$$

for all $\tau \geq 1$. Here we will just guess a solution. One possibility is as follows: Consider the complete system with

$$\mathcal{X} = \{0, 1\}^3$$

having the following local description:

$$\begin{aligned} x_1(t+1) &= \omega(t) \\ x_2(t+1) &= x_1(t) \\ x_3(t+1) &= x_2(t) \\ y(t) &= \alpha(t).\{[x_1(t) + x_2(t) + x_3(t)] \bmod 2\} \end{aligned} \tag{2.12}$$

and initial state x^0 arbitrary, where

$$\alpha := \begin{cases} 1 & \text{if 3 divides } t, \\ 0 & \text{otherwise.} \end{cases}$$

We are using x_i to denote the ith coordinate of $x \in \mathcal{X}$. It is easy to see that this system has the desired input/output behavior.

The system is time-varying. Alternatively, we could also use the time-invariant system which has

$$\mathcal{X} = \{0,1\}^3 \times \{0,1,2\},$$

and equations

$$\begin{aligned} x_1(t+1) &= \omega(t) \\ x_2(t+1) &= x_1(t) \\ x_3(t+1) &= x_2(t) \\ x_4(t+1) &= x_4(t) + 1 \ (\text{mod}\, 3) \\ y(t) &= \beta(x_4(t)).\{[x_1(t) + x_2(t) + x_3(t)] \bmod 2\} \end{aligned}$$

where $\beta(1) = \beta(2) = 0, \beta(0) = 1$. As initial state we may take for instance $(0,0,0,0)$. One can always include a "clock" in order to pass from a given system to a time-invariant system that behaves as the original one. Mathematically, however, this associated time-invariant system is a different object; for instance, the set $\mathcal{X}$ has changed. Moreover, many desirable properties may be lost in the reformulation; for example, the first system that we gave is linear over the field $GF(2)$ (that is, $\mathbb{Z}_2$ under mod-2 operations) in the sense to be defined below, but the corresponding time-invariant system is not. □

Of course, there are infinitely many different systems Σ that solve the above problem. For instance, we change $\mathcal{X}$ to $\{0,1\}^4$ and add an equation such as

$$x_4(t+1) = 0$$

to the equations (2.12), and this will still produce the desired behavior. Note that the value of x_4 is not in any way observable from the output data.

Exercise 2.3.4 The first system given above has a state set of cardinality 8, while for the second (time-invariant one) this cardinality is 24. Find:

1. A system (and initial state) with state space of just two elements $\{0,1\}$ for which (2.11) still holds.

2. A time-invariant system with a 6-element state set for which this also holds.

3. Prove that it is impossible to solve the problem using a time-invariant system with less than 6 states. □

Example 2.3.3 serves to illustrate the fact that measurement maps are useful not only in modeling constraints imposed by what can be observed in a physical system but also the fact that only certain functions of the states may be of interest in a given problem.

2.4 Linear Discrete-Time Systems

In this section we let $\mathbb{K}$ be any field, such as $\mathbb{R}$, $\mathbb{C}$, or the field of two elements $GF(2)$ that appeared in Example 2.3.3. The terms *vector space* and *linear* will always be interpreted with respect to this fixed field $\mathbb{K}$.

The Cartesian product of any two vector spaces $\mathcal{X}$ and $\mathcal{U}$ is endowed naturally with a vector space structure (the direct sum) under coordinatewise operations:

$$(x, u) + k(z, v) = (x + kz, u + kv) .$$

Similarly, for any $\sigma < \tau$ we may also consider $\mathcal{U}^{[\sigma,\tau)}$ as a vector space, with pointwise operations:

$$(k\omega + \nu)(t) := k\omega(t) + \nu(t) .$$

If a space is finite dimensional and a basis has been chosen, we use column vectors to list the coordinates; if this space is $\mathbb{K}^n$, we always assume that the standard basis has been chosen, and we identify linear maps $\mathbb{K}^m \to \mathbb{K}^n$ with their matrices in the respective standard bases. When $\mathcal{T} = \mathbb{Z}$ and $\mathcal{U} = \mathbb{K}^m$, we identify $\mathcal{U}^{[\sigma,\tau)}$ with $\mathcal{U}^k$, where $k = m(\tau - \sigma)$, and write coordinates of elements in this space by listing those of $\omega(\sigma), \ldots, \omega(\tau - 1)$ in that order.

Definition 2.4.1 *The discrete-time system Σ is* **linear** *(over the field $\mathbb{K}$) if:*

- *It is complete;*
- *$\mathcal{X}$ and $\mathcal{U}$ are vector spaces; and*
- *$\mathcal{P}(t, \cdot, \cdot)$ is linear for each $t \in \mathbb{Z}$.*

The system with outputs Σ is linear if in addition:

- *$\mathcal{Y}$ is a vector space; and*
- *$h(t, \cdot)$ is linear for each $t \in \mathbb{Z}$.*

The system is **finite dimensional** *if both $\mathcal{U}$ and $\mathcal{X}$, as well as $\mathcal{Y}$ for a system with outputs, are finite dimensional; the* **dimension** *of Σ is in that case the dimension of $\mathcal{X}$.* □

Recall that completeness means simply that $\mathcal{P}(t, x, u)$ is defined for every triple (t, x, u). Linearity is equivalent to the existence of linear maps

$$A(t) : \mathcal{X} \to \mathcal{X} \quad \text{and} \quad B(t) : \mathcal{U} \to \mathcal{X}, \quad t \in \mathbb{Z},$$

(namely, $\mathcal{P}(t, \cdot, 0)$ and $\mathcal{P}(t, 0, \cdot)$, respectively) such that

$$\boxed{\mathcal{P}(t, x, u) = A(t)x + B(t)u} \tag{2.13}$$

For systems with outputs also

$$\boxed{h(t,x) = C(t)x} \tag{2.14}$$

for some linear

$$C(t) : \mathcal{X} \to \mathcal{Y}, \quad t \in \mathbb{Z}.$$

The time-invariant case results precisely when all of these maps are independent of t.

For finite dimensional systems, the integers n, m, and p are conventionally used for denoting the dimensions of $\mathcal{X}$, $\mathcal{U}$, and $\mathcal{Y}$, respectively. Thus, in a basis each $A(t)$ is of size $n \times n$, $B(t)$ is $n \times m$, and $C(t)$ is $p \times n$.

One typically specifies discrete-time linear systems simply by giving the families of maps $\{A(t)\}$, $\{B(t)\}$, and $\{C(t)\}$, or, in the finite dimensional case, their matrices in some basis. For instance, equations (2.12) correspond using the standard basis for $GF(2)^3$ to

$$A(t) \equiv \begin{pmatrix} 0 & 0 & 0 \\ 1 & 0 & 0 \\ 0 & 1 & 0 \end{pmatrix}, \; B(t) \equiv \begin{pmatrix} 1 \\ 0 \\ 0 \end{pmatrix}, \; C(t) = \begin{pmatrix} \alpha(t) & \alpha(t) & \alpha(t) \end{pmatrix}$$

and α is the function defined before.

Note that, given any linear difference equation

$$x(t+1) = A(t)x(t),$$

we can think of this as a discrete-time linear system with no controls, simply by letting $\mathcal{U}$ be the trivial vector space $\{0\}$ and $B(t) \equiv 0$.

Linearity of the local-in-time transitions gives linearity of the transition map, by an easy induction:

Lemma/Exercise 2.4.2 Prove that, for any linear discrete-time system and any pair of integers $\sigma < \tau$, $\phi(\tau, \sigma, \cdot, \cdot)$ is linear. □

Linear Discrete-Time Behaviors

We now define linear behaviors. We choose a definition that incorporates the fact that an input identically equal to zero should not affect the system.

Definition 2.4.3 *The discrete-time i/o behavior* Λ *is* **linear** *(over the field* $\mathbb{K}$*) if:*

- *It is complete;*
- $\mathcal{U}$ *and* $\mathcal{Y}$ *are vector spaces;*
- *For every* $\sigma \leq \tau$ *in* $\mathbb{Z}$, $\lambda(\sigma, \sigma, \diamond) = 0$ *and the mapping* $\lambda^{\sigma,\tau}$ *is linear; and*

- *For each $\sigma < \mu < \tau$ in $\mathbb{Z}$ and each input $\omega \in \mathcal{U}^{[\mu,\tau)}$, it holds that*

$$\lambda(\tau, \sigma, \mathbf{0}\omega) = \lambda(\tau, \mu, \omega),$$

where $\mathbf{0}$ is the input identically equal to 0 on $[\sigma, \mu)$. □

A number of elementary facts hold for linear i/o behaviors and about their relation with linear systems initialized at $x^0 = 0$. We leave them as (easy) exercises. All systems and i/o behaviors are assumed to be discrete-time here.

Lemma/Exercise 2.4.4 If Σ is linear, then $\Lambda_{\Sigma,0}$ is linear. □

Lemma/Exercise 2.4.5 If Λ is linear, then there exists a unique family of linear maps

$$\widetilde{\mathcal{A}}_{ij},\ i, j \in \mathbb{Z}, i > j$$

such that for each $\sigma < \tau$

$$\lambda^{\sigma,\tau}(\omega) = \sum_{j=\sigma}^{\tau-1} \widetilde{\mathcal{A}}_{\tau j}\omega(j) \tag{2.15}$$

for all $\omega \in \mathcal{U}^{[\sigma,\tau)}$. □

Lemma/Exercise 2.4.6 If Σ and Λ are linear, then

$$\Lambda_{\Sigma,0} = \Lambda$$

if and only if

$$\widetilde{\mathcal{A}}_{ij} = \begin{cases} C(i)B(j) & \text{if } i = j+1, \\ C(i)A(i-1)A(i-2)\ldots A(j+1)B(j) & \text{otherwise} \end{cases} \tag{2.16}$$

in terms of the linear maps introduced in Lemma 2.4.5. □

Lemma/Exercise 2.4.7 Let Λ be a linear i/o behavior. Consider the family of linear maps $\left\{\widetilde{\mathcal{A}}_{ij}\right\}_{i>j}$ introduced in Lemma 2.4.5. Then, Λ is time-invariant if and only if there exists a sequence of linear maps $\{\mathcal{A}_i, i = 1, 2, 3, \ldots\}$ such that

$$\widetilde{\mathcal{A}}_{ij} = \mathcal{A}_{i-j}$$

for each $i > j$. □

The sequence $\{\mathcal{A}_i\}$ is called the **impulse response** of Λ (or of a system realizing Λ), since its columns result from the application of an "impulse" in one input coordinate, i.e., an input that is equal to zero everywhere except at a single instant, and at that instant the value equals 1. Another common name for the elements of this sequence is *Markov parameters.*

In the time-invariant case, an input $\omega \in \mathcal{U}^{[\sigma,\tau)}$ produces the final output value

$$y(\tau) = \sum_{j=\sigma}^{\tau-1} \mathcal{A}_{\tau-j}\omega(j). \tag{2.17}$$

In other words, the output is the (discrete) *convolution* of the input with the impulse response.

We conclude from Lemmas 2.4.6 and 2.4.7 that a discrete-time linear time-invariant system realizes a discrete-time linear time-invariant behavior precisely when the following factorization condition holds among their associated linear operators:

$$\boxed{\mathcal{A}_i = CA^{i-1}B \text{ for all } i > 0} \tag{2.18}$$

2.5 Smooth Discrete-Time Systems

In this section $\mathbb{K} = \mathbb{R}$ or $\mathbb{C}$. The first part of Appendix B on Jacobians and differentials should be consulted for terminology regarding smoothness and Jacobians.

For any discrete-time system Σ and each $t \in \mathbb{Z}$, let

$$\mathcal{E}_t := \{(x,u) \mid (t,x,u) \in \mathcal{E}\} \subseteq \mathcal{X} \times \mathcal{U}$$

be the domain of $\mathcal{P}(t,\cdot,\cdot)$.

Definition 2.5.1 *Let $k = 0,1,2,\ldots,\infty$. A C^k* **discrete-time system** *(over $\mathbb{K}$) is one that satisfies, for some nonnegative integers n, m:*

1. *$\mathcal{X}$ is an open subset of $\mathbb{K}^n$;*
2. *$\mathcal{U}$ is an open subset of $\mathbb{K}^m$; and*
3. *For each $t \in \mathbb{Z}$, the set $\mathcal{E}_t$ is open and the map $\mathcal{P}(t,\cdot,\cdot)$ is of class C^k there.*

If Σ is a system with outputs, it is required in addition that $\mathcal{Y}$ be an open subset of some Euclidean space $\mathbb{K}^p$, and that $h(t,\cdot)$ be of class C^k. □

When $k = 1$ or ∞, one also says that Σ is *differentiable* or *smooth*, respectively.

We define linearization only for real systems. In the complex case, differentiability is understood in the sense that the real and complex parts must be differentiable, as discussed in the Appendix on differentials, and linearizations could be computed for the real system of twice the dimension associated to any given system over $\mathbb{C}$.

Definition 2.5.2 *Let Σ be a C^1 discrete-time system over $\mathbb{R}$, and assume that $\Gamma = (\bar{\xi}, \bar{\omega})$ is a trajectory for Σ on an interval $\mathcal{I}$. The* **linearization of Σ**

along Γ *is the discrete-time linear system* $\Sigma_*[\Gamma]$ *with local-in-time description* $(\mathbb{R}^n, \mathbb{R}^m, \mathcal{P}_*)$, *where*

$$\mathcal{P}_*(t,\cdot,\cdot) := \begin{cases} \mathcal{P}(t,\cdot,\cdot)_*[\bar{\xi}(t),\bar{\omega}(t)] & \text{if } t \in \mathcal{I}, \\ 0 & \text{if } t \notin \mathcal{I}. \end{cases} \tag{2.19}$$

If Σ *is a system with outputs, then* $\Sigma_*[\Gamma]$ *is the discrete-time linear system with outputs having the above* $\mathcal{P}$, $\mathcal{Y} = \mathbb{R}^p$, *and the readout map*

$$h_*(t,\cdot) := \begin{cases} h(t,\cdot)_*[\bar{\xi}(t)] & \text{if } t \in \mathcal{I}, \\ 0 & \text{if } t \notin \mathcal{I}. \end{cases} \tag{2.20}$$

When $\mathcal{I} = \mathbb{Z}$ *and* $\bar{\xi}$ *and* $\bar{\omega}$ *are both constant,* $\bar{\xi}(t) \equiv x$ *and* $\bar{\omega}(t) \equiv u$, *then* $\Sigma_*[\Gamma]$ *is the* **linearization of** Σ **at** (x, u). □

When Γ is clear from the context, we write simply Σ_*.

Note that a pair (x, u) gives rise to a constant trajectory for Σ, as in the last part of the above definition, if and only if $\mathcal{P}(t, x, u) = x$ holds for all t. We call such a pair $(x, u) \in \mathcal{X} \times \mathcal{U}$ an **equilibrium pair** and the corresponding x an **equilibrium state**. If (x, u) is like this *and* Σ is time-invariant, then Σ_* is again time-invariant.

In terms of the matrices appearing in equations (2.13) and (2.14), the system Σ_* has

$$\begin{aligned} A(t) &= \mathcal{P}_x(t, \bar{\xi}(t), \bar{\omega}(t)) \\ B(t) &= \mathcal{P}_u(t, \bar{\xi}(t), \bar{\omega}(t)) \end{aligned}$$

whenever $t \in \mathcal{I}$ and $A(t) = B(t) = 0$ for $t \notin \mathcal{I}$, and

$$C(t) = h_x(t, \bar{\xi}(t))$$

for $t \in \mathcal{I}$ and $C(t) = 0$ otherwise. (Subscripts indicate partial derivatives.)

As an example take the system Σ:

$$x^+ = (2x + u)^2$$

and its equilibrium pair $(x = 1, u = -1)$. In terms of perturbations of these values

$$x = 1 + \widetilde{x},\ u = -1 + \widetilde{u}\,,$$

the equations are

$$\begin{aligned} \widetilde{x}^+ = x^+ - 1 &= (2x + u)^2 - 1 \\ &= (2[1 + \widetilde{x}] - 1 + \widetilde{u})^2 - 1 \\ &= 4\widetilde{x} + 2\widetilde{u} + 4\widetilde{x}^2 + \widetilde{u}^2 + 4\widetilde{x}\widetilde{u} \\ &= 4\widetilde{x} + 2\widetilde{u} + o(\widetilde{x}, \widetilde{u})\,. \end{aligned}$$

Thus, Σ_* has matrices $A = (4)$ and $B = (2)$. Alternatively, one could of course obtain these matrices by evaluation of

$$\mathcal{P}_x = 4(2x + u), \;\; \mathcal{P}_u = 2(2x + u)$$

at the equilibrium pair. Thus, Σ_* is

$$x^+ = 4x + 2u\,,$$

which is a time-invariant system since the trajectory was a constant one (equilibrium pair). A variation of this example is provided by the same system Σ but along the following trajectory on $\mathcal{I} = [0, +\infty)$:

$$\omega(0) = 1, \;\; \omega(t) \equiv 0 \;\text{ for all } t \geq 1$$

(that is, ω is an impulsive input) and

$$\xi(0) = 0, \;\; \xi(t) = 2^{2^t - 2} \;\text{ for all } t \geq 1\,,$$

which results in the time-varying linearization

$$x^+ = a_t x + b_t u\,,$$

where $a_t = b_t \equiv 0$ for $t < 0$, $a_0 = 4$, $b_0 = 2$, and

$$a_t = 2^{2^t + 1}, \;\; b_t = 2^{2^t}$$

for each $t \geq 1$.

Abusing terminology, one often simply says that (A, B), or (A, B, C) for systems with outputs, is the linearization of the system along the given trajectory. Note that for systems with no controls, that is, classical dynamical systems, the linearization of $x^+ = \mathcal{P}(x)$ along a trajectory is just the linear recursion $x^+ = A(t)x$, with $A(t)$ being the Jacobian of $\mathcal{P}$ computed at each step.

2.6 Continuous-Time

In this Section we define continuous-time systems as those that are described by finite dimensional differential equations with sufficiently regular right-hand sides. More generally, one could also consider systems whose evolution is described by partial or functional differential equations, or systems with a discontinuous right-hand side, but these more general concepts will not be treated in the present volume.

Motivational Discussion

Essentially, we wish to define continuous-time systems as those that are described by differential equations

$$\dot{x} = f(t, x, u)$$

where the right-hand side may depend on t. For a well-defined system to arise, we need the solutions of this equation for arbitrary initial conditions and measurable locally essentially bounded controls to exist at least for small times. Weaker regularity hypotheses can be made, but as far as x and u are concerned, we shall settle essentially on continuity on both x and u and differentiability on x. This still leaves to be settled the issue of what regularity must be assumed in t, when dealing with time-varying systems. In order to decide what is reasonable in that regard, consider one of the main ways in which time-varying systems arise. Typically, there is a trajectory (ζ, ν) that has been given for the above equation, and one wants to compare ζ with a trajectory (ξ, ω) corresponding to a different control. In other words, one is interested in the dynamics of the deviations

$$\xi' := \xi - \zeta, \; \omega' := \omega - \nu$$

from (ζ, ν). Note that, omitting the "t" arguments where clear,

$$\dot{\xi}' = g(t, \xi', \omega'),$$

where

$$g(t, x, u) := f(t, x + \zeta(t), u + \nu(t)) - f(t, \zeta(t), \nu(t)).$$

Even if f is independent of t, the function g will in general depend on t. But this dependence of g on t, assuming that $f = f(x, u)$, is at most through a measurable essentially bounded function —ν or the continuous function ξ— substituted into f. Motivated by this, we will *define* systems asking that the time variation appear in precisely that way (substitution of a measurable function into a continuous one). Differentiable and smooth systems will be defined by substitutions into differentiable or smooth functions. An advantage of doing things in this manner is that deviations such as those above remain in the class, and many results can be proved just for time-invariant systems, with the generalizations to the time-varying case being immediate.

The Definition

One important technical point is the introduction of concepts from Lebesgue integration theory. Appendix C on ordinary differential equations, which should be consulted before covering this section, reviews the basic notions. **However, readers who have not seen measure theory before should still be able to follow the rest of this book.** Most results still will be correct if one substitutes *piecewise continuous* functions instead of "measurable."

In this section, $\mathbb{K} = \mathbb{R}$ or $\mathbb{C}$. At the beginning we wish to allow control values to belong to an arbitrary metric space $\mathcal{U}$; but for most results we later assume that $\mathcal{U}$ is an open subset of a Euclidean space $\mathbb{R}^m$ or $\mathbb{C}^m$. The use of more general metric spaces permits including discrete control-value sets, which appear in optimal control theory as well as in other areas.

Definition 2.6.1 *Let $\mathcal{X}$ be an open subset of $\mathbb{K}^n$ and let $\mathcal{U}$ be a metric space. A* **right-hand side (rhs)** *with respect to $\mathcal{X}$ and $\mathcal{U}$ is a function*

$$f:\ \mathbb{R}\times\mathcal{X}\times\mathcal{U}\to\mathbb{K}^n$$

which can be obtained in the following way: There must exist another metric space $\mathcal{S}$ as well as maps

$$\widetilde{f}:\ \mathcal{S}\times\mathcal{X}\times\mathcal{U}\to\mathbb{K}^n$$

and

$$\pi:\mathbb{R}\to\mathcal{S}$$

so that

$$f(t,x,u)=\widetilde{f}(\pi(t),x,u) \tag{2.21}$$

and the following properties hold:

1. *$\widetilde{f}(s,\cdot,u)$ is of class C^1 for each fixed s,u ;*
2. *both $\widetilde{f}$ and the partial derivative $\widetilde{f}_x$ are continuous on $\mathcal{S}\times\mathcal{X}\times\mathcal{U}$;*
3. *π is a measurable locally essentially bounded function.* □

As discussed above, this definition is somewhat complicated by the need to deal with rather general time variations. For f independent of t, the definition reduces to the requirement that $\mathcal{U}$ be a metric space and

$$f:\mathcal{X}\times\mathcal{U}\to\mathbb{K}^n$$

is so that

$$f(\cdot,u)\ \text{ is of class }\ C^1\ \text{ for each fixed }\ u \tag{2.22}$$

and

$$f\ \text{ and }\ f_x\ \text{ are continuous on }\mathcal{X}\times\mathcal{U}. \tag{2.23}$$

Also, of course, if $\mathcal{U}$ is any open subset of $\mathbb{K}^m$ and

$$f:\mathbb{R}\times\mathcal{X}\times\mathcal{U}\to\mathbb{K}^n$$

is continuously differentiable, then f is a rhs.

Lemma 2.6.2 Any rhs in the sense of Definition 2.6.1 satisfies the following property: For any real numbers $\sigma<\tau$, any measurable essentially bounded $\omega\in\mathcal{U}^{[\sigma,\tau)}$, and any $x^0\in\mathcal{X}$, there is some nonempty subinterval $J\subseteq\mathcal{I}:=[\sigma,\tau]$ open relative to $\mathcal{I}$ and containing σ, and there exists a solution of

$$\begin{aligned}\dot{\xi}(t) &= f(t,\xi(t),\omega(t))\\ \xi(\sigma) &= x^0\end{aligned} \tag{2.24}$$

on J. Furthermore, this solution is maximal and unique: that is, if

$$\zeta:J'\to\mathcal{X}$$

is any other solution of (2.24) defined on a subinterval $J' \subseteq \mathcal{I}$, then necessarily

$$J' \subseteq J \text{ and } \xi = \zeta \text{ on } J'.$$

If $J = \mathcal{I}$, then ω is said to be **admissible** for x^0.

Proof. Consider

$$g(t,x) := f(t,x,\omega(t)) = \widetilde{f}(\pi(t),x,\omega(t))$$

as a function on $\mathcal{I} \times \mathcal{X}$. Note that this is continuous on x, so hypothesis (H2) in Appendix C is satisfied for g. Since the mapping

$$\mathcal{I} \to \mathcal{U} \times \mathcal{S},\ t \mapsto (\omega(t), \pi(t))$$

is measurable, $g(\cdot, x)$ is the composition of a measurable and a continuous function and is therefore measurable in its first coordinate. Thus, (H1) holds too. We claim that the conditions in the existence and uniqueness Theorem 54 (p. 476) in the Appendix hold. Because of Proposition C.3.4, we need to verify that for each compact subset $K \subseteq \mathcal{X}$ and each $x^0 \in \mathcal{X}$ there exist locally bounded measurable maps α and β so that

$$|g(t,x^0)| \leq \beta(t)$$

for all $t \in \mathcal{I}$, and

$$|g_x(t,x)| \leq \alpha(t)$$

for all $x \in K$ and all $t \in \mathcal{I}$.

Since ω is bounded and π is locally bounded, there exist compact sets K_1, K_2 such that $\omega(t) \in K_1$ and $\pi(t) \in K_2$ for almost all $t \in \mathcal{I}$. By continuity of $\widetilde{f}$ and $\widetilde{f}_x$, there are upper bounds M_1, M_2 for the values of $|\widetilde{f}|$ and $|\widetilde{f}_x|$, respectively, on

$$K_2 \times K \times K_1,$$

so the functions $\alpha \equiv M_2$ and $\beta \equiv M_1$ are as desired. ∎

Lemma/Exercise 2.6.3 Let f be a rhs, and consider

$$\mathcal{D} := \{(\tau,\sigma,x,\omega) \mid \sigma < \tau, x \in \mathcal{X}, \omega \in \mathcal{U}^{[\sigma,\tau)} \text{ is admissible for } x^0\}.$$

(Note that, in particular, admissibility means that such ω's are measurable.) On this set define

$$\phi(\tau,\sigma,x,\omega) := \xi(\tau),$$

where $\xi(\tau)$ is the (unique) solution of (2.24) on $[\sigma,\tau]$. Then, $\Sigma_f := (\mathbb{R}, \mathcal{X}, \mathcal{U}, \phi)$ is a system. □

Definition 2.6.4 *A* **continuous-time system** *(over $\mathbb{K} = \mathbb{R}$ or $\mathbb{C}$) is a system of the type Σ_f, where f is a rhs.* □

Note that this system will be time-invariant when f is independent of t. We often refer to Σ_f simply as "the system $\dot{x} = f(t,x,u)$," and call f the "rhs of Σ." The map f is also said to provide a **local in time** description of Σ.

Definition 2.6.5 *A* **continuous-time system with outputs** *is a system Σ with outputs whose underlying system is a continuous-time system, $\mathcal{Y}$ is a metric space, and for which $h : \mathcal{T} \times \mathcal{X} \to \mathcal{Y}$ is continuous.* □

Definition 2.6.6 *Let $k = 1, 2, \ldots, \infty$. A* **continuous-time system of class** $\mathcal{C}^k$ *is one for which $\mathcal{U}$ is an open subset of $\mathbb{K}^m$, for some nonnegative integer m, and for which $\mathcal{S}$ and $\widetilde{f}$ in (2.21) can be chosen so that $\mathcal{S}$ is an open subset of a Euclidean space and $\widetilde{f}$ is of class $\mathcal{C}^k$.*

If Σ is a system with outputs, it is required in addition that $\mathcal{Y}$ be an open subset of some Euclidean space $\mathbb{K}^p$, and that $h(t, \cdot)$ be of class $\mathcal{C}^k$ for each t. □

Every continuous-time system is of "class $\mathcal{C}^0$" since $\widetilde{f}$ is continuous by definition. When $k = 1$ or ∞, one also says that Σ is *differentiable* or *smooth*, respectively. Note that, in particular, $f(t, \cdot, \cdot)$ is of class $\mathcal{C}^k$ when the system is.

Note that a time-invariant continuous-time system Σ is of class $\mathcal{C}^k$ precisely if $\mathcal{U}$ is an open subset of $\mathbb{K}^n$ and f is k-times continuously differentiable.

It is useful sometimes to consider also the "time reversal" of a system. Assume that Σ is a continuous-time system, and pick any $\mu \in \mathbb{R}$. Then

$$g(t,x,u) := -f(\mu - t, x, u)$$

is also a rhs, since

$$g(t,x,u) = \widetilde{g}(\beta(t), x, u)$$

under the definition

$$\widetilde{g} : \mathcal{S} \times \mathcal{X} \times \mathcal{U} \to \mathbb{K}^n, \ \widetilde{g}(s,x,u) := -\widetilde{f}(s,x,u)$$

and $\beta(t) := -\pi(\mu - t)$, where f, $\widetilde{f}$, $\mathcal{S}$, and π are as in the definition of rhs.

Definition 2.6.7 *If Σ and μ are as above, the continuous-time system with rhs*

$$-f(\mu - t, x, u)$$

is the **reversal of** Σ **at** μ, *and is denoted Σ_μ^-. If f is independent of t, then Σ_μ^- is time-invariant and independent of the particular μ taken, and is denoted just as Σ^-.* □

Lemma 2.6.8 Assume that $\sigma, \tau \in \mathbb{R}, \sigma < \tau$, let Σ be as above, and let ϕ^- denote the transition map of $\Sigma_{\sigma+\tau}^-$. Then, for any $x, z \in \mathcal{X}$, and any $\omega \in \mathcal{U}^{[\sigma,\tau)}$,

$$\phi(\tau, \sigma, x, \omega) = z \quad \text{iff} \quad \phi^-(\tau, \sigma, z, \nu) = x \,,$$

where $\nu(t) := \omega(\sigma + \tau - t)$.

Proof. Consider the trajectory (ξ,ω) with $\xi(\sigma) = x, \xi(\tau) = z$. Then, the function

$$\zeta(t) := \xi(\sigma + \tau - t)$$

satisfies the differential equation with control ν and rhs

$$-f(\sigma + \tau - t, x, u)$$

and has initial state z and final state x. Uniqueness of solutions gives the desired conclusion. ∎

Definition 2.6.9 *A* **continuous-time behavior** *is one for which*

- $\mathcal{T} = \mathbb{R}$;
- $\mathcal{U}$ *is a metric space;*
- $\mathcal{Y}$ *is a metric space;*

and for each $\sigma < \tau$ it holds that the domain of each $\lambda^{\sigma,\tau}$ is an open subset of $\mathcal{L}^{\infty}_{\mathcal{U}}(\sigma,\tau)$ and $\lambda^{\sigma,\tau}$ is continuous. □

For any continuous-time system Σ and each initial state x^0, the i/s response map

$$\omega \mapsto \phi(\tau, \sigma, x^0, \omega)$$

will be shown to be continuous in Theorem 1 (p. 57) (see also Remark 2.8.1 for the non-time-invariant case) for each fixed σ, τ. By composition, the following result then holds:

Proposition 2.6.10 If (Σ, x^0) is an initialized continuous-time system, then its behavior Λ is a continuous-time behavior. □

Remark 2.6.11 It is also true that the complete i/o map $\omega \mapsto \bar{\lambda}_\Sigma(\tau, \sigma, \omega)$ is continuous as a mapping into $\mathcal{C}^0([\sigma,\tau],\mathcal{Y})$, the space of continuous functions from $[\sigma,\tau]$ to $\mathcal{Y}$ with the uniform metric. This is because the map ψ that sends initial states and controls into the complete path $\xi(\cdot)$ is also shown to be continuous as a map into $\mathcal{C}^0([\sigma,\tau],\mathcal{X})$ in the same Theorem. □

2.7 Linear Continuous-Time Systems

In this section, $\mathbb{K} = \mathbb{R}$ or $\mathbb{C}$. Appendix C.4 on linear differential equations summarizes needed background material.

Lemma/Exercise 2.7.1 Let m, n be nonnegative integers, $\mathcal{X} := \mathbb{K}^n, \mathcal{U} := \mathbb{K}^m$. Assume that

$$f(t, x, u) = A(t)x + B(t)u, \tag{2.25}$$

where $A(t)$ is an $n \times n$ matrix and $B(t)$ is an $n \times m$ matrix, each of whose entries is a locally essentially bounded (measurable) function $\mathbb{R} \to \mathbb{K}$. Equivalently,

$$f : \mathbb{R} \times \mathcal{X} \times \mathcal{U} \to \mathcal{X}$$

is linear in (x, u) for each fixed t, and is locally bounded in t for each fixed (x, u). Conclude that f is a rhs. □

Thus, the following definition is justified:

Definition 2.7.2 *The continuous-time system* Σ *is* **linear (and finite dimensional)** *(over* $\mathbb{K} = \mathbb{R}$ *or* $\mathbb{C}$*) if it is of class* C^1 *and:*

- $\mathcal{X} = \mathbb{K}^n$ *and* $\mathcal{U} = \mathbb{K}^m$*; and*
- *Its local-in-time description satisfies that* $f(t, \cdot, \cdot)$ *is linear for each* $t \in \mathbb{R}$*.*

The system with outputs Σ *is linear if in addition:*

- $\mathcal{Y} = \mathbb{K}^p$*; and*
- $h(t, \cdot)$ *is linear for each* $t \in \mathbb{R}$ *.*

The **dimension** *of* Σ *is the dimension of* $\mathcal{X}$*.* □

Thus, there exist matrix representations as in (2.13) and (2.14). As in the discrete-time case, we just say "the system $(A(t), B(t))$" (or $(A(t), B(t), C(t))$) in order to refer to this system. By definition, we are considering only finite dimensional systems; to define the general infinite dimensional case would require notions of semigroup theory. We are allowing the degenerate cases $n = 0$ (system of dimension 0) and, more interestingly, $m = 0$; in the latter case, we interpret Σ as a system with no controls and rhs $f(t, x) = A(t)x$.

A *linear system* is by definition one that is either a discrete-time or a continuous-time linear system.

Lemma/Exercise 2.7.3 Prove that every linear continuous-time system is complete (with respect to the class of all measurable essentially bounded controls). (*Hint:* Use the result given in the Appendix which assumes a global Lipschitz condition on the rhs.) □

Lemma 2.7.4 Let Σ be a linear continuous-time system, and let $\sigma < \tau$ be in $\mathbb{R}$. Then, $\phi(\tau, \sigma, \cdot, \cdot)$ is linear as a map $\mathcal{X} \times \mathcal{L}_m^\infty \to \mathcal{X}$.

Proof. Note first that ϕ is indeed defined on all of $\mathcal{X} \times \mathcal{L}_m^\infty$, by completeness. Now assume that $x, z \in \mathcal{X}$ and $\omega, \nu \in \mathcal{U}^{[\sigma,\tau)}$, and pick any $r \in \mathbb{K}$. Let ξ, ζ satisfy

$$\begin{aligned} \dot{\xi}(t) &= A(t)\xi(t) + B(t)\omega(t) \\ \dot{\zeta}(t) &= A(t)\zeta(t) + B(t)\nu(t), \end{aligned}$$

with $\xi(\sigma) = x, \zeta(\sigma) = z$. Then, $\xi + r\zeta$ satisfies a similar equation, with control $\omega + r\nu$, and

$$\xi(\sigma) + r\zeta(\sigma) = x + rz\,.$$

It follows by uniqueness of solutions that

$$\phi(\tau, \sigma, x + rz, \omega + r\nu) \;=\; \xi(\tau) + r\zeta(\tau) \;=\; \phi(\tau, \sigma, x, \omega) + r\phi(\tau, \sigma, z, \nu)$$

and hence ϕ is linear. ∎

This result is also an immediate consequence of the variation of parameters formula, which here becomes

$$\xi(t) = \Phi(t, \sigma^0)x^0 \;+\; \int_{\sigma^0}^{t} \Phi(t,s)B(s)\omega(s)\,ds\,. \tag{2.26}$$

Exercise 2.7.5 This problem illustrates that one may be able to consider larger classes of controls than just bounded ones, provided that the system has a special structure (for instance, if it is linear).

(a) Prove that if f is as in equation (2.25) then also arbitrary integrable inputs may be applied. More precisely, show that in this case equation (2.24) has a solution, at least for small $\delta > \sigma$, for any $\sigma < \tau, x \in \mathbb{K}^n$, and any integrable $\omega \in \mathcal{U}^{[\sigma,\tau)}$.

(b) Give an example of a (necessarily *non*linear) rhs (with $\mathcal{U} = \mathcal{X} = \mathbb{R}$), an $x \in \mathbb{R}$, and an integrable $\omega \in \mathcal{U}^{[\sigma,\tau)}$, for some σ and τ, such that (2.24) has no possible solution on any nontrivial interval $[\sigma, \delta]$, no matter how small $\delta - \sigma$ is. (*Hint:* by definition of absolute continuity, $f(t, \xi(t), \omega(t))$ must be integrable as a function of t if ξ is a solution. Find an example with $\omega(t) = t^{-1/2} \in \mathcal{U}^{(0,1]}$ and $\omega(0)$ arbitrary.) □

Remark 2.7.6 The above exercise justifies the definition, for any given continuous-time linear system Σ, of a "different" system Σ_f^1 for which integrable controls are allowed, or for that matter, say, the system Σ_f^2 corresponding to using only $\mathcal{L}^2$ controls. The systems Σ_f^1 and Σ_f^2 are basically the same as Σ_f, since ϕ^1 and ϕ^2 are just extensions of ϕ. In such cases one uses the expression "Σ_f with integrable (or, square integrable, etc.) controls" to refer to such an extended system. The $\mathcal{L}^2$ case is especially important, since a Hilbert space of controls is useful in order to understand quadratic linear optimal control problems. In other contexts, for instance when defining the impulse response of a system, often it is useful also to introduce even more general controls, such as distributional inputs. □

Linear Continuous-Time Behaviors

Definition 2.7.7 *The continuous-time i/o behavior Λ is* **linear** *(over the field $\mathbb{K}$) if:*

- *It is complete (with respect to essentially bounded controls);*
- $\mathcal{U} = \mathbb{K}^m$ *and* $\mathcal{Y} = \mathbb{K}^p$*;*
- *For every* $\sigma \leq \tau$ *in* $\mathbb{R}$, $\lambda(\sigma, \sigma, \diamond) = 0$ *and the mapping* $\lambda^{\sigma,\tau}$ *is a linear (bounded) operator; and*
- *For each* $\sigma < \mu < \tau$ *in* $\mathbb{R}$ *and each measurable bounded input* $\omega \in \mathcal{U}^{[\mu,\tau)}$ *it holds that*
$$\lambda(\tau, \sigma, \mathbf{0}\omega) = \lambda(\tau, \mu, \omega),$$
where $\mathbf{0}$ *is the input identically equal to* 0 *on* $[\sigma, \mu)$. □

Linear continuous-time behaviors can be characterized in terms of integral operators with respect to finitely additive measures. We do not pursue such a characterization here, and concentrate instead on a subclass that corresponds to those operators for which the complete i/o maps $\bar{\lambda}$ are continuous with respect to the $\mathcal{L}^1$ topology on controls. This subclass will be general enough for our intended applications. The next definition should be compared to the conclusion of Lemma 2.4.5; of course, one could just as well start with an abstract notion of integral operator and then make the following a conclusion, as in the discrete-time case.

Definition 2.7.8 *The linear behavior* Λ *is an* **integral** *behavior if there exists a measurable locally essentially bounded mapping*

$$\widetilde{K} : \{(\tau, \sigma) \in \mathbb{R}^2 \mid \sigma \leq \tau\} \to \mathbb{K}^{p\times m}$$

such that

$$\lambda^{\sigma,\tau}(\omega) = \int_{\sigma}^{\tau} \widetilde{K}(\tau, s)\omega(s)\, ds$$

for any $\sigma < \tau$ *and any* $\omega \in \mathcal{U}^{[\sigma,\tau)}$. □

Lemma/Exercise 2.7.9 The kernel $\widetilde{K}$ is uniquely determined (up to a set of measure zero) by the behavior Λ. □

Lemma/Exercise 2.7.10 If Σ is linear and Λ is an integral behavior, then

$$\Lambda_{\Sigma,0} = \Lambda$$

if and only if

$$\widetilde{K}(t, \tau) = C(t)\Phi(t, \tau)B(\tau)$$

for almost all (t, τ). Thus, the integral behavior Λ is realizable by some linear system if and only if there exist time-varying matrices

$$(A(t), B(t), C(t))$$

such that the above conditions hold. □

Lemma/Exercise 2.7.11 The linear integral behavior Λ is time-invariant if and only if there exists a matrix of locally bounded maps

$$K \in \mathcal{L}^{\infty,loc}_{p\times m}(0,\infty)$$

such that

$$\widetilde{K}(t,\tau) = K(t-\tau)$$

for almost all $\tau \leq t$. □

Thus, in the time-invariant case, a bounded input $\omega \in \mathcal{L}^{\infty}_{m}(\sigma,\tau)$ produces the final output value

$$y(\tau) = \int_{\sigma}^{\tau} K(\tau - s)\omega(s)\,ds\,. \tag{2.27}$$

As with discrete-time systems, the output is a convolution of the input with the impulse response.

The map K is once again called the *impulse response*, since just as in the discrete-time case it is the output that results from an impulsive input at time 0, assuming that one interprets impulsive inputs as limits of step functions of unit area and support tending to zero:

Exercise 2.7.12 Assume that K in formula (2.27) is continuous. Pick any fixed $i \in \{1,\ldots,m\}$ and any $T > 0$. For each $\varepsilon > 0$, let ω_ε be the control defined on the interval $[0,T]$ by

$$\omega_\varepsilon(t) := \begin{cases} \frac{1}{\varepsilon}e_i & \text{if } 0 \leq t < \varepsilon, \\ 0 & \text{otherwise,} \end{cases}$$

where $e_i = (0,\ldots,0,1,0,\ldots,0)'$ is the ith canonical basis vector. Prove that

$$\lim_{\varepsilon\to 0} y_\varepsilon(T) = K_i(T)\,,$$

where $y_\varepsilon(T) = \lambda^{0,T}(\omega_\varepsilon)$ and K_i is the ith column of K. □

The condition $\Lambda_{\Sigma,0} = \Lambda$ is equivalent in the time-invariant case to

$$K(t) = Ce^{tA}B\,.$$

If $K(\cdot)$ is known to be analytic about $t = 0$, we can expand into a (matrix) power series

$$K(t) \;=\; \sum_{i=1}^{\infty} \mathcal{A}_i \frac{t^{i-1}}{(i-1)!} \tag{2.28}$$

and so we conclude:

Lemma 2.7.13 The integral behavior with analytic kernel K as in (2.28) is realized by the linear system (A,B,C) if and only if the factorization conditions (2.18) hold. □

This means that *the realization theories of (finite dimensional) linear continuous-time and discrete-time time-invariant systems reduce to the same algebraic problem.*

Linearization

The definition of linearization is entirely analogous to the one given for discrete-time systems. Again, we only consider the linearization of systems over $\mathbb{R}$.

Definition 2.7.14 *Let Σ be a $\mathcal{C}^1$ continuous-time system over $\mathbb{R}$, and assume that $\Gamma = (\bar{\xi}, \bar{\omega})$ is a trajectory for Σ on an interval $\mathcal{I}$. The* **linearization of Σ along Γ** *is the continuous-time linear system $\Sigma_*[\Gamma]$ with local-in-time description f_*, where*

$$f_*(t,\cdot,\cdot) := \begin{cases} f(t,\cdot,\cdot)_*[\bar{\xi}(t), \bar{\omega}(t)] & \text{if } t \in \mathcal{I}, \\ 0 & \text{if } t \notin \mathcal{I}. \end{cases} \tag{2.29}$$

If Σ is a system with outputs, then $\Sigma_*[\Gamma]$ is the continuous-time linear system with outputs having the above f, $\mathcal{Y} = \mathbb{R}^p$, and the readout map

$$h_*(t,\cdot) := \begin{cases} h(t,\cdot)_*[\bar{\xi}(t)] & \text{if } t \in \mathcal{I}, \\ 0 & \text{if } t \notin \mathcal{I}. \end{cases} \tag{2.30}$$

When $\mathcal{I} = \mathbb{R}$ and $\bar{\xi}$ and $\bar{\omega}$ are both constant, $\bar{\xi}(t) \equiv x$ and $\bar{\omega}(t) \equiv u$, then $\Sigma_*[\Gamma]$ is the **linearization of Σ at (x, u)**. □

As before, when Γ is clear from the context, we write simply Σ_*. Now for a pair (x, u) to give rise to a constant trajectory for Σ is equivalent to the property

$$f(t, x, u) \equiv 0 \text{ for all } t.$$

We again call such a pair $(x, u) \in \mathcal{X} \times \mathcal{U}$ an **equilibrium pair** and the corresponding x an **equilibrium state**. As in discrete-time, if (x, u) is an equilibrium pair, then Σ_* is again time-invariant, provided that Σ was time-invariant.

The local description of the linearization can be given in terms of partial derivatives of f and h with respect to x and u, just as in the discrete-time case. Linearization was discussed in detail in the introductory Chapter.

As an example, consider again the controlled pendulum (with $m = g = 1$); this is a time-invariant continuous-time system, with $\mathcal{U} = \mathbb{R}$, $\mathcal{X} = \mathbb{R}^2$, and

$$f(x, u) = \begin{pmatrix} x_2 \\ -\sin x_1 + u \end{pmatrix}.$$

If only positions could be measured, we could model that fact by using instead the system with outputs having $\mathcal{Y} = \mathbb{R}$ and $h(x) = x_1$.

Given any fixed trajectory $(\bar{\xi}, \bar{\omega})$ on an interval $\mathcal{I}$, the linearization becomes the system with matrices

$$A(t) = \begin{pmatrix} 0 & 1 \\ -\cos \bar{\xi}_1(t) & 0 \end{pmatrix} \quad B(t) = \begin{pmatrix} 0 \\ 1 \end{pmatrix}.$$

In particular, we may consider the equilibrium pair $(0,0)$; then the linearization becomes the time-invariant system (A,B), where

$$A(t) = \begin{pmatrix} 0 & 1 \\ -1 & 0 \end{pmatrix} \quad B(t) = \begin{pmatrix} 0 \\ 1 \end{pmatrix} \quad . \tag{2.31}$$

Or we may take the equilibrium pair $((\pi,0)',0)$, and one gets the state space model of the linearized pendulum for which we designed controllers in the Introduction.

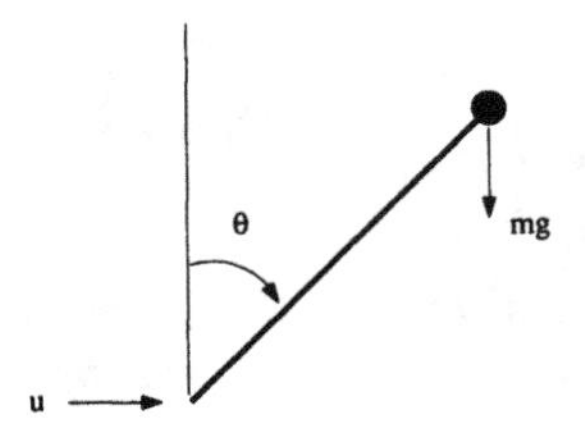

Figure 2.1: *Stick-balancing example.*

Exercise 2.7.15 Assume that we try to control a stick by applying a horizontal force at one tip (see Figure 2.1). Under simplifying assumptions, and choosing appropriate units and time scales, we have an equation

$$\ddot{\theta} = \sin\theta - u\cos\theta$$

for the orientation of the stick. (We are ignoring the position of the center of gravity, and assume that the motion occurs in a plane.) With $x_1 = \theta, x_2 = \dot{\theta}$, this is a continuous-time, time-invariant system ($\mathcal{X} = \mathbb{R}^2, \mathcal{U} = \mathbb{R}$).

(a) Give the form of the linearized system along a general trajectory.

(b) Consider now the same example but in a horizontal plane. After simplification, we use as a model

$$\ddot{\theta} = u\cos\theta\,.$$

Find the linearization of this system at the equilibrium point $x_1 = x_2 = u = 0$.

(c) Idem about $x_1 = \pi/2$, $x_2 = u = 0$.

(d) In what sense is the linear system that you obtained in (b) "nicer" than the one in (c)? Intuitively, how should this relate to the behavior of the original nonlinear system near the two different equilibrium pairs? □

Exercise 2.7.16 Consider the ("bilinear") system $\dot{x} = xu$, where $\mathcal{X} = \mathcal{U} = \mathbb{R}$.

(a) Give the linearized system at $x = 2$, $u = 0$.

(b) Idem along the trajectory $\xi(t) = e^t$, $t \geq 0$, $\omega \equiv 1$. □

2.8 Linearizations Compute Differentials

For any system Σ, and any fixed pair of times $\sigma < \tau$ in $\mathcal{T}$, we denote

$$\mathcal{D}_{\sigma,\tau} := \{(x,\omega) \mid (\tau,\sigma,x,\omega) \in \mathcal{D}_\phi\} \subseteq \mathcal{X} \times \mathcal{U}^{[\sigma,\tau)},$$

and for each $x \in \mathcal{X}$ we let

$$\mathcal{D}_{\sigma,\tau,x} := \{\omega | (\tau,\sigma,x,\omega) \in \mathcal{D}_\phi\} \subseteq \mathcal{U}^{[\sigma,\tau)}.$$

Note that for some σ, τ, x, these sets may of course be empty.

The main objective of this section is to establish conditions under which the mapping

$$\phi(\tau,\sigma,\cdot,\cdot),$$

which has $\mathcal{D}_{\sigma,\tau}$ as its domain, is continuous and/or differentiable in an appropriate sense. More importantly, we prove that its derivatives are computed by linearizations along trajectories. This last fact underlies the mathematical importance of linearizations, and is also the basis of gradient descent techniques for numerical nonlinear control (cf. Section 9.4).

To develop some intuition, consider first the case of a time-invariant discrete-time system of class $\mathcal{C}^1$ having $\mathcal{X} = \mathbb{R}^n$, $\mathcal{U} = \mathbb{R}^m$, and equations

$$x^+ = P(x,u), \tag{2.32}$$

and for which $(0,0)$ is an equilibrium point, that is, $P(0,0) = 0$. Assume for simplicity that we are only interested in trajectories of length 2, and consider the map

$$\alpha = \phi(2,0,\cdot,\cdot).$$

Writing elements of

$$\mathcal{U}^{[0,2)} = \mathcal{U}^2 = \mathbb{R}^{2m}$$

as pairs (u_1, u_2), the map

$$\alpha : \mathcal{X} \times \mathcal{U}^2 \to \mathcal{X} : (x,(u_1,u_2)) \mapsto P(P(x,u_1),u_2)$$

has as its Jacobian at $x = u_1 = u_2 = 0$ a matrix (of size n by $n + 2m$) which can be partitioned as

$$J = (\alpha_x, \alpha_{u_1}, \alpha_{u_2}).$$

(The partial Jacobians are evaluated at $x = u_1 = u_2 = 0$.) By the chain rule, this is the same as

$$(A^2, AB, B)$$

where

$$A = P_x(0,0) \text{ and } B = P_u(0,0)$$

in terms of the partial Jacobians of P with respect to x and u. Thus, the linear mapping induced by J is

$$\alpha_*[0,(0,0)] : \mathbb{R}^{n+2m} \to \mathbb{R}^n : \begin{pmatrix} x \\ u_1 \\ u_2 \end{pmatrix} \mapsto J \begin{pmatrix} x \\ u_1 \\ u_2 \end{pmatrix} = A^2x + ABu_1 + Bu_2 . \tag{2.33}$$

This mapping is the differential, or Jacobian, of α at the point $x = u_1 = u_2 = 0$.

Alternatively, assume that we start again with the system Σ, and take its linearization along the trajectory $\Gamma = (\xi, \omega)$ induced by the same initial state $x = 0$ and control ω, that is, $\xi(0) = \xi(1) = \omega(0) = \omega(1) = 0$. By definition, this is the linear system (time-invariant, because it corresponds to an equilibrium pair,) with the same matrices A, B,

$$x^+ = Ax + Bu .$$

The main observation is that the 2-step "α" map for this system is precisely the same as the map in equation (2.33). That is, we have in informal terms,

linearization map of transition = transition map of linearization

To continue the intuitive discussion, consider next time-invariant continuous-time systems $\dot{x} = f(x,u)$ over $\mathbb{K} = \mathbb{R}$. Suppose given a state $x^0 \in \mathcal{X}$, an input $\widetilde{\omega} \in \mathcal{L}^\infty_{\mathcal{U}}(\sigma,\tau)$ which is admissible for x^0, and denote the corresponding solution as $\widetilde{x}(t) = \phi(t,\sigma,x^0,\widetilde{\omega})$. We will show in Theorem 1 that the same input $\widetilde{\omega}$ is also admissible for all initial states x near enough x^0, that is, the domain of admissibility is open. Thus, we may consider the mapping $\Theta(x) := \phi(\tau,\sigma,x,\widetilde{\omega})$, which assigns the final state to each initial state, given this control $\widetilde{\omega}$, and the mapping Θ is defined on some open subset of $\mathcal{X}$ and takes values in $\mathcal{X}$. Theorem 1 will show that this map is continuously differentiable. For purposes only of simplifying this preliminary discussion, we make the additional assumption that $\phi(t,\sigma,x,\widetilde{\omega})$ is *twice* continuously differentiable jointly as a function of (t,x); this allows us to exchange orders of derivation. We now compute the Jacobian at $x = x^0$ of the map Θ. From the differential equation

$$\frac{\partial \phi}{\partial t}(t,\sigma,x,\widetilde{\omega}) = f(\phi(t,\sigma,x,\widetilde{\omega}),\widetilde{\omega}) ,$$

we have, taking derivatives with respect to x,

$$\frac{\partial}{\partial t}\left(\frac{\partial \phi}{\partial x}\right) = \frac{\partial}{\partial x}\left(\frac{\partial \phi}{\partial t}\right) = f_x(\phi,\widetilde{\omega})\left(\frac{\partial \phi}{\partial x}\right)$$

(where f_x indicates the Jacobian of $f(x,u)$ with respect to x, and we omitted the arguments $(t,\sigma,x,\widetilde{\omega})$ in order to not clutter the formula). In addition, from the identity $\phi(\sigma,\sigma,x,\widetilde{\omega}) = x$ we also have that

$$\frac{\partial \phi}{\partial x}(\sigma,\sigma,x,\widetilde{\omega}) = I$$

($n \times n$ identity matrix) for all x near x^0. Therefore $\frac{\partial \phi}{\partial x}(t,\sigma,x^0,\widetilde{\omega})$ is the fundamental solution associated to the linearized (or "variational") system along the pair $(\widetilde{x},\widetilde{\omega})$ with zero input perturbation, that is, the solution of the matrix equation

$$\dot{\Lambda}(t) = A(t)\Lambda(t)\,, \quad \Lambda(\sigma) = I\,,$$

where $A(t) = f_x(\widetilde{x}(t),\widetilde{\omega}(t))$ for all $t \in [\sigma,\tau]$. The Jacobian of $\Theta(x) = \phi(\tau,\sigma,x,\widetilde{\omega})$ evaluated at x^0 is then the final value of this solution:

$$\Theta_*(x^0) = \Lambda(\tau)\,.$$

An equivalent way to state this formula, which is in the form given in Theorem 1, is to say that the directional derivative of $\Theta(x)$ at $x = x^0$ in the direction of an arbitrary vector λ_0, that is, $\Theta_*(x^0)(\lambda_0)$, is given by the solution at final time τ, $\lambda(\tau)$, of the variational equation $\dot{\lambda}(t) = A(t)\lambda(t)$ with initial condition $\lambda(\sigma) = \lambda_0$.

Moreover, if the control-value set is an open subset of $\mathbb{R}^m$ and $f(x,u)$ is differentiable in u, then the linearized system also provides a way to compute the Jacobian *with respect to input perturbations*. To be precise, take any fixed input $\mu \in \mathcal{L}^\infty_{\mathcal{U}}(\sigma,\tau)$, and consider $\phi(\tau,\sigma,x^0,\widetilde{\omega}+h\mu)$. If h is small enough, the control $\widetilde{\omega}+h\mu$ is admissible for x^0, because the set of admissible controls is open in the topology of uniform convergence. (This is also proved in Theorem 1; observe that $\widetilde{\omega}+h\mu \in \mathcal{L}^\infty_{\mathcal{U}}(\sigma,\tau)$ for all sufficiently small h, because there is some compact subset C of $\mathcal{U}$ so that $\widetilde{\omega}(t) \in C$ for almost all t, and there is some $h_1 > 0$ so that the h_1-neighborhood of C is included in $\mathcal{U}$.) Let us compute the partial derivative of $\phi(t,\sigma,x^0,\widetilde{\omega}+h\mu)$ with respect to h, evaluated at $h = 0$, making again a simplifying assumption, in this case that this map is C^2 in (t,h). From

$$\frac{\partial \phi}{\partial t}(t,\sigma,x,\widetilde{\omega}+h\mu) \;=\; f(\phi(t,\sigma,x,\widetilde{\omega}+h\mu),\widetilde{\omega}+h\mu),$$

we get

$$\frac{\partial}{\partial t}\left(\frac{\partial \phi}{\partial h}\right) \;=\; \frac{\partial}{\partial h}\left(\frac{\partial \phi}{\partial t}\right) \;=\; f_x(\phi,\widetilde{\omega}+h\mu)\left(\frac{\partial \phi}{\partial h}\right) \;+\; f_u(\phi,\widetilde{\omega}+h\mu)\,\mu\,.$$

In particular, when evaluating at $h = 0$ we obtain the following formula:

$$\frac{\partial}{\partial t}\left(\left.\frac{\partial \phi}{\partial h}\right|_{h=0}\right) \;=\; A(t)\left(\left.\frac{\partial \phi}{\partial h}\right|_{h=0}\right) \;+\; B(t)\,\mu$$

where $B(t) = f_u(\widetilde{x}(t),\widetilde{\omega}(t))$ for all $t \in [\sigma,\tau]$. Also, $\phi(\sigma,\sigma,x^0,\widetilde{\omega}+h\mu) \equiv x^0$ implies that $\frac{\partial \phi}{\partial h}(t,\sigma,x^0,\widetilde{\omega}+h\mu) = 0$ when $t = \sigma$. In other words,

$$\left.\frac{\partial \phi}{\partial h}(\tau,\sigma,x^0,\widetilde{\omega}+h\mu)\right|_{h=0}$$

equals the solution $\lambda(\tau)$ of the linearized equation $\dot{\lambda} = A(t)\lambda + B(t)\mu$ with initial condition $\lambda(\sigma) = 0$.

A stronger statement can be made if one considers the total, not merely directional, derivative of $\phi(\tau,\sigma,x^0,\omega)$ with respect to ω, or, more precisely, since we are dealing with elements of an infinite dimensional space, the Fréchet differential (cf. Appendix B) of $\alpha = \alpha_{x^0,\sigma,\tau} : \omega \to \phi(\tau,\sigma,x^0,\omega)$, seen as a map from an open subset of $\mathcal{L}_{\mathcal{U}}^{\infty}(\sigma,\tau)$ into $\mathbb{R}^n$. This map is shown below to be continuously differentiable. The differential evaluated at $\widetilde{\omega}$, $\alpha_*[\widetilde{\omega}]$, is the linear map which, when applied to any perturbation μ, is computed as the solution of the linearized system shown above.

In summary, also for continuous-time systems one has the principle:

linearization map of transition = transition map of linearization.

The main purpose of this section is to establish this fact as a general theorem. We provide details only for continuous-time systems. A similar result holds for differentiable discrete-time systems, but this is trivial to establish and will be left as an exercise. Also, for simplicity, we deal first with time-invariant systems. For such systems, properties (2.22) and (2.23) hold, and the system is of class $\mathcal{C}^k$ if $\mathcal{U}$ is an open subset of $\mathbb{R}^m$ and f is k-times continuously differentiable. The time-varying case is dealt with as a simple consequence of the time-invariant case, in Remark 2.8.1.

From now on, Σ will always denote a fixed continuous-time time-invariant system, $\sigma < \tau$ are in $\mathbb{R}$, and $T := \tau - \sigma$.

As in the Appendix on differential equations, we write $\mathcal{L}_{\mathcal{U}}^{\infty}(\sigma,\tau)$ for the set of all essentially bounded measurable controls

$$\omega : [\sigma,\tau] \to \mathcal{U}$$

or just $\mathcal{L}_{\mathcal{U}}^{\infty}$, since the interval will be fixed. We think of $\mathcal{L}_{\mathcal{U}}^{\infty}$ as a metric space under uniform convergence, that is, with the metric

$$\mathrm{d}_{\infty}(\omega,\nu) := \text{ess.sup.}\left\{\mathrm{d}_{\mathcal{U}}(\omega(t),\nu(t)) \mid t \in [\sigma,\tau]\right\}.$$

When $\mathcal{U} = \mathbb{R}^m$, we write also $\mathcal{L}_m^{\infty}$; this is the Banach space of essentially bounded measurable functions $[\sigma,\tau] \to \mathbb{R}^m$ endowed with the sup norm

$$\|\omega\|_{\infty} := \text{ess.sup.}\left\{|\omega(t)|, t \in [\sigma,\tau]\right\}.$$

In this section, we use the notation $|x|$ (rather than $\|x\|$) for the Euclidean norm in the state space $\mathcal{X} \subseteq \mathbb{R}^n$, or when $\mathcal{U} \subseteq \mathbb{R}^m$ in the control-value space $\mathcal{U}$, to avoid confusion in arguments in which both Euclidean and function space norms appear. For matrices $A \in \mathbb{R}^{k\times l}$ (such as Jacobians), we use also $|A|$ instead of $\|A\|$, for the matrix norm corresponding to Euclidean norms in $\mathbb{R}^l$ and $\mathbb{R}^k$.

The main result on continuity and differentiability is given below. We include a weak convergence statement that is useful in optimal control theory. By *weak convergence* of a sequence $\omega_j \to \omega$, with all ω_j's as well as ω in $\mathcal{L}_m^{\infty}(\sigma,\tau)$, we mean that

$$\int_{\sigma}^{\tau} \varphi(s)'\omega_j(s)\,ds \to \int_{\sigma}^{\tau} \varphi(s)'\omega(s)\,ds$$

for every integrable function $\varphi : [\sigma, \tau] \to \mathbb{R}^m$; see Chapter 10 for more discussion of weak convergence.

By a system *affine in controls* (or *control-affine*) we mean a system over $\mathbb{R}$ for which f is affine in $u \in \mathcal{U} = \mathbb{R}^m$, so that its equations take the form

$$\dot{x} = g_0(x) + \sum_{i=1}^{m} u_i g_i(x) . \tag{2.34}$$

This includes in particular linear systems.

Theorem 1 *Let Σ be a (time-invariant, continuous-time) system, and pick any interval $[\sigma, \tau]$. Consider the map that gives the final state given an initial state and control,*

$$\alpha : \mathcal{D}_{\sigma,\tau} \to \mathcal{X} : \quad (x, \omega) \mapsto \xi(\tau) ,$$

as well as the mapping into the entire path

$$\psi : \mathcal{D}_{\sigma,\tau} \to \mathcal{C}_n^0 : \quad (x, \omega) \mapsto \xi ,$$

where $\xi(t) = \phi(t, \sigma, x, \omega|_{[\sigma,t)})$. If the system is of class $\mathcal{C}^1$ and (ξ, ω) is any trajectory, and if

$$\lambda_0 \in \mathbb{R}^n \quad and \quad \mu \in \mathcal{L}_m^\infty ,$$

consider also the solution $\lambda : [\sigma, \tau] \to \mathbb{R}^n$ of the variational equation

$$\dot{\lambda}(t) = f_x(\xi(t), \omega(t))\lambda(t) + f_u(\xi(t), \omega(t))\mu(t), \tag{2.35}$$

with initial condition $\lambda(\sigma) = \lambda_0$. The following conclusions then hold:

1. *The set $\mathcal{D}_{\sigma,\tau}$ is an open subset of $\mathcal{X} \times \mathcal{L}_{\mathcal{U}}^\infty$, and both ψ and α are continuous.*

2. *Take any x and any ω admissible for x, and denote $\xi := \psi(x, \omega)$. Let $\{\omega_j\}_{j=1}^\infty$ be an equibounded sequence of controls (that is, there is some fixed compact $K \subseteq \mathcal{U}$ such that $\omega_j(t) \in K$ for all j and almost all $t \in [\sigma, \tau]$) and*

   $$\lim_{j \to \infty} x_j = x .$$

   If either one of the following conditions hold:

   (i) $\omega_j \to \omega$ as $j \to \infty$ pointwise almost everywhere, or

   (ii) $\omega_j \to \omega$ as $j \to \infty$ weakly and Σ is affine in controls,

   then $\xi_j := \psi(x_j, \omega_j)$ is defined for all large j and

   $$\lim_{j \to \infty} \|\xi_j - \xi\|_\infty = 0 .$$

3. If Σ is of class C^1, then α is of class C^1 and

$$\alpha_*[x,\omega](\lambda_0,\mu) = \lambda(\tau)$$

when λ is as in (2.35). That is, $\alpha_*[x,\omega]$ is the same as the map α_* corresponding to the linearization $\Sigma_*[\xi,\omega]$.

In particular, for systems of class C^1,

$$\alpha(x,\cdot) : \mathcal{D}_{\sigma,\tau,x} \to \mathcal{X}$$

has full rank at ω if and only if the linear map

$$\mathcal{L}_m^\infty \to \mathcal{X} : \mu \mapsto \int_\sigma^\tau \Phi(\tau,s)B(s)\mu(s)\,ds$$

(the map $\alpha(0,\cdot)$ for the linearization $\Sigma_*[\Gamma]$ along $\Gamma = (\xi,\omega)$, seen as a time-varying linear system) is onto.

Proof. First note that conclusion 1 follows from statement 2(i); this is because uniform convergence $\omega_j \to \omega$ in $\mathcal{L}_\mathcal{U}^\infty(\sigma,\tau)$ implies in particular pointwise convergence, as well as — cf. Remark C.1.3 in the Appendix on ODE's — equiboundedness of the sequence $\{\omega_j\}$. We will show how the continuity statements 2(i) and 2(ii) are both easy consequences of Theorem 55 (p. 486) found in that Appendix. Assume given any $(x,\omega) \in \mathcal{D}_{\sigma,\tau}$ and any sequence $\omega_j \to \omega$ converging in either of the two senses, and let $\xi = \psi(x,\omega)$. Pick any $\varepsilon > 0$. We wish to show that $\|\xi_j - \xi\|_\infty < \varepsilon$ for all large j.

We start by choosing an open subset $\mathcal{X}_0$ of $\mathcal{X}$ whose closure is a compact subset of $\mathcal{X}$ and such that $\xi(t) \in \mathcal{X}_0$ for all $t \in \mathcal{I} := [\sigma,\tau]$. Let $K \subseteq \mathcal{U}$ be a compact set so that $\omega_j(t) \in K$ and $\omega(t) \in K$ for almost all $t \in \mathcal{I}$. Introduce the functions $\mathcal{I} \times \mathcal{X} \to \mathbb{R}^n$ given by

$$h_j(t,z) := f(z,\omega_j(t)) - f(z,\omega(t))$$

for each $j = 1,2,\ldots$, and

$$\widetilde{f}(t,z) := f(z,\omega(t))\,.$$

Note that all these functions satisfy the hypotheses of the existence Theorem 54 (namely (H1), (H2), and the local Lipschitz and integrability properties) found in the Appendix on ODE's. For each j let $g_j := f + h_j$, so $g_j(t,z) = f(z,\omega_j(t))$. Since the Jacobian $\|f_x\|$ is bounded on the compact set $\operatorname{clos}\mathcal{X}_0 \times K$, there is a constant $a > 0$ so that

$$\|g_j(t,x_1) - g_j(t,x_2)\| \le a\,\|x_1 - x_2\|$$

for all $t \in \mathcal{I}$, all j, and all $x_1, x_2 \in \mathcal{X}_0$. Let

$$H_j(t) := \int_\sigma^t h_j(s,\xi(s))\,ds\,, \quad t \in \mathcal{I}$$

and $\underline{H}_j := \sup_{t\in\mathcal{I}} \|H_j(t)\|$.

Assume for now that we have already proved the following fact:

$$\lim_{j\to\infty} \underline{H}_j = 0 \tag{2.36}$$

(which amounts to saying that the functions H_j converge uniformly to zero). We wish to apply Theorem 55 (p. 486), with "$\mathcal{X}$" there being $\mathcal{X}_0$, "f" being $\widetilde{f}$, $\underline{\alpha}(t) \equiv a$, and z^0 and h equal to x_j and h_j respectively, for all large enough j. Take any $0 < D \leq \varepsilon$ so that

$$\{z \mid \|z - \xi(t)\| \leq D \text{ for some } t \in [\sigma, \tau]\}$$

is included in $\mathcal{X}_0$, and choose an integer j_0 so that, for each $j \geq j_0$, $\|x - x_j\|$ and $\underline{H}_j$ are both less than

$$\frac{\varepsilon}{2} e^{-a(\tau-\sigma)} \leq \frac{D}{2} e^{-a(\tau-\sigma)}.$$

For any such j, using $z^0 = x_j$ and $h = h_j$ in Theorem 55 (p. 486) implies that

$$\|\xi - \xi_j\|_\infty \leq \left(\|x - x_j\| + \underline{H}_j\right) \leq \varepsilon$$

as desired.

We now show (2.36). Suppose first that $\omega_j(t) \to \omega(t)$ a.e. as $j \to \infty$. Then $h_j(t, \xi(t)) \to 0$ a.e. as well. Moreover, letting c be an upper bound on the values of $\|f\|$ on the compact set $\operatorname{clos} \mathcal{X}_0 \times K$,

$$\|h_j(t, \xi(t))\| \leq 2c$$

for all $t \in \mathcal{I}$. The Lebesgue Dominated Convergence Theorem ensures that

$$\int_\sigma^t \|h_j(s, \xi(s))\| \, ds \to 0,$$

from which (2.36) follows.

Suppose now that Σ is affine in controls and convergence is in the weak sense. We may write

$$h_j(s, \xi(s)) = G(\xi(s)) \left[\omega_j(s) - \omega(s)\right],$$

where $G(z)$ is the matrix whose columns are $g_1(z), \ldots, g_m(z)$. Thus the ith coordinate of $h_j(s, \xi(s))$, $i = 1, \ldots, n$, has the inner product form

$$\varphi_i(s)' \left[w_j(s) - \omega(s)\right]$$

where $\varphi_i(s)$, the transpose of the ith row of $G(\xi(s))$, is a continuous and hence integrable function of s. By the weak convergence assumption,

$$\int_\sigma^t \varphi_i(s)' \left[w_j(s) - \omega(s)\right] ds \to 0 \text{ as } j \to \infty$$

for each fixed $t \in \mathcal{I}$ and each $i = 1, \ldots, n$. Moreover, the functions H_j form an equicontinuous family, because $\|H_j(t_1) - H_j(t_2)\| \leq 2c\,|t_1 - t_2|$ (recall that the norm of h_j is upper bounded by $2c$). Pointwise convergence of an equicontinuous function implies uniform convergence, so also in this case we have proved (2.36).

We next prove differentiability. (The proof is organized in such a manner as to make a generalization in Section 2.9 very simple.) Let (x, ω) be given, and $\xi := \psi(x, \omega)$. Note that now $\mathcal{X} = \mathbb{R}^n$ and $\mathcal{U} = \mathbb{R}^m$. We first multiply f by a smooth function $\mathbb{R}^n \times \mathbb{R}^m \to \mathbb{R}$ which is equal to one in a neighborhood of the set of values $(\xi(t), \omega(t))$ and has compact support. Since differentiability is a local property, it is sufficient to prove the result for the new f obtained in this manner. Thus from now on we assume that f is bounded and globally Lipschitz:

$$|f(x,u) - f(z,v)| \leq M\,(|x - z| + |u - v|)$$

for all x, z in $\mathbb{R}^n$ and all u, v in $\mathbb{R}^m$. In particular, solutions are globally defined.

As a consequence of the Mean Value Theorem, we can write

$$f(x+a, u+b) - f(x,u) - f_x(x,u)a - f_u(x,u)b \;=\; N(x,u,a,b) \qquad (2.37)$$

for all $x, a \in \mathbb{R}^n$ and $u, b \in \mathbb{R}^m$, where $N(x,u,a,b) =$

$$\int_0^1 [(f_x(x+ta, u+tb) - f_x(x,u))a \;+\; (f_u(x+ta, u+tb) - f_u(x,u))b]\; dt \qquad (2.38)$$

(see, for instance, the proof of [264], Chapter V.4, Corollary 2). In particular, N is jointly continuous in its arguments and vanishes when $a = b = 0$. Observe that, because N has compact support, it is uniformly continuous; thus, the supremum $C(a,b)$ over all $x \in \mathbb{R}^n, u \in \mathbb{R}^m$ of

$$\int_0^1 |f_x(x+ta, u+tb) - f_x(x,u)| \;+\; |f_u(x+ta, u+tb) - f_u(x,u)|\; dt \quad (2.39)$$

is continuous and it vanishes at $a, b = 0$.

Along the given trajectory (ξ, ω), consider the linearization matrices

$$\begin{aligned} A(t) &= f_x(\xi(t), \omega(t)) \\ B(t) &= f_u(\xi(t), \omega(t)) \,. \end{aligned}$$

For any other control $\nu \in \mathcal{L}_m^\infty(\sigma, \tau)$ sufficiently near ω, and each z near x, we consider the corresponding trajectory (ζ, ν), where $\zeta = \psi(z, \nu)$. Introduce

$$\delta(t) := \zeta(t) - \xi(t)$$

and

$$\mu(t) := \nu(t) - \omega(t) \,.$$

From (2.37) and (2.39) we conclude that

$$\dot{\delta}(t) = A(t)\delta(t) + B(t)\mu(t) + \gamma(t)$$

with

$$|\gamma(t)| \leq (|\delta(t)| + |\mu(t)|)\; C(\delta(t), \mu(t))\,.$$

Thus, if λ solves (2.35) with

$$\lambda(\sigma) = \delta(\sigma) = \lambda_0 = x - z\,,$$

it follows that

$$\delta(\tau) = \lambda(\tau) + \int_\sigma^\tau \Phi(\tau, s)\gamma(s)\, ds\,,$$

where Φ is the fundamental solution associated to $A(\cdot)$. So

$$|\delta(\tau) - \lambda(\tau)| \;\leq\; c_1\, (\|\delta\|_\infty + \|\mu\|_\infty)\;\sup_{t\in[\sigma,\tau]} C(\delta(t), \mu(t))$$

for some constant c_1 (where "sup" means "essential sup").

By a Bellman-Gronwall argument as in the proof of continuity, one has an estimate

$$\|\delta\|_\infty \leq c_2\, (|\lambda_0| + \|\mu\|_\infty)\,,$$

so we can conclude finally an estimate of the type

$$|\delta(\tau) - \lambda(\tau)| \;\leq\; c\; (|\lambda_0| + \|\mu\|_\infty)\;\sup_{t\in[\sigma,\tau]} C(\delta(t), \mu(t))$$

for some constant c. Since the supremum is small for z near x and ν near ω, this establishes differentiability.

Finally, we must prove that the map α is *continuously* differentiable. Let (x, ω) be in the domain $\mathcal{D}_{\sigma,\tau}$. We must prove that, for each $\varepsilon > 0$ there is some $\delta > 0$, such that for each $(\widetilde{x}, \widetilde{\omega}) \in \mathcal{D}_{\sigma,\tau}$ for which

$$\|\omega - \widetilde{\omega}\| + |x - \widetilde{x}| < \delta\,, \tag{2.40}$$

necessarily

$$\|\alpha_*[x, \omega] - \alpha_*[\widetilde{x}, \widetilde{\omega}]\| < \varepsilon\,.$$

By definition of the operator norm, and using the characterization of the derivative just obtained, this is the same as asking that for each ε there be a δ such that, for each $\lambda_0 \in \mathbb{R}^n$ and each $\mu \in \mathbb{R}^m$,

$$\begin{aligned}\left|\Big(\Phi(\tau, \sigma) - \widetilde{\Phi}(\tau, \sigma)\Big)\,\lambda_0 \right.\;\; &+ \;\; \left.\int_\sigma^\tau \Big(\Phi(\tau, s)B(s) - \widetilde{\Phi}(\tau, s)\widetilde{B}(s)\Big)\,\mu(s)\, ds\right| \\ &\leq \;\; \varepsilon\,(|\lambda_0| + \|\mu\|)\end{aligned} \tag{2.41}$$

when (2.40) holds, where $\Phi(\tau, s)$ is the solution at time s of the matrix initial value problem

$$\dot{X}(t) = f_x(\xi(t), \omega(t))X(t), \;\; X(\tau) = I \tag{2.42}$$

and similarly with $\widetilde{\Phi}$, and where

$$B(t) := f_u(\xi(t), \omega(t))$$

and similarly for $\widetilde{B}$.

Note that the left-hand side of (2.41) is always bounded by

$$|\Phi(\tau,\sigma) - \widetilde{\Phi}(\tau,\sigma)||\lambda_0| + \int_\sigma^\tau |Q(s)||\mu(s)|\,ds\,, \tag{2.43}$$

where

$$Q(s) := \Phi(\tau,s)B(s) - \widetilde{\Phi}(\tau,s)\widetilde{B}(s)\,, \tag{2.44}$$

and also that $|Q(s)|$ is bounded by

$$|\Phi(\tau,s)||B(s) - \widetilde{B}(s)| + c\,|\Phi(\tau,s) - \widetilde{\Phi}(\tau,s)|\,, \tag{2.45}$$

where c is any constant bounding the values

$$f_u(x,u)$$

whenever x is in a neighborhood of the set

$$\{\xi(t), t \in [\sigma,\tau]\}$$

and u is in a neighborhood of the set

$$\{\omega(t), t \in [\sigma,\tau]\}\,.$$

This implies that (the essential supremum of) $|Q(s)|$ is small when (x,ω) is near $(\widetilde{x},\widetilde{\omega})$: Indeed, $|\Phi(\tau,s) - \widetilde{\Phi}(\tau,s)|$ is small (consider the variational equation seen as a control system, and apply the continuity part of the proof), and $|B(s) - \widetilde{B}(s)|$ is small, too, since it is obtained as an evaluation of a continuous matrix function along nearby trajectories. This completes the proof. ∎

The reader may wish to consult Corollary 9.1.1 for a (weaker) version of this theorem which is stated in terms of directional derivatives, in the style of the discussion that preceded the proof. Also, later, in Section 4.2 (page 147), we establish several additional properties regarding joint continuity and differentiability of $\phi(t,\sigma,x,\omega)$ on (t,x). It should be noted that, in the last part of the theorem, we actually established a more general fact on differentiability than stated, namely, that the map ψ is differentiable.

Remark 2.8.1 The results can be generalized to time-varying systems in a straightforward manner. Recalling that the definition of rhs involved the auxiliary function $\widetilde{f}$, it is only necessary to apply the Theorem to the system

$$\dot{x} = \widetilde{f}(\pi,x,u)$$

with π thought of as a new control. Fixing the value of π at that of the given function that provides the time variation, results on differentials of the original dynamics provide results for $\dot{x} = f(t,x,u)$. For instance, the mappings α and ψ are still continuous, and are differentiable in the case of systems of class $\mathcal{C}^1$ with respect to the uniform convergence norm. □

A very useful Corollary is that piecewise constant inputs approximate the effect of arbitrary controls. By a *piecewise constant* control

$$\omega \in \mathcal{U}^{[\sigma,\tau)}$$

we mean a control for which there exists an integer $l > 0$, a sequence of times

$$\sigma = t_1, \ldots, t_l = \tau\,,$$

and a sequence of values

$$u_1, \ldots, u_l$$

in $\mathcal{U}$ so that

$$\omega(t) \equiv u_i \;\text{ if }\; t \in [t_i, t_{i+1})$$

for each $i = 1, \ldots, l-1$.

Lemma 2.8.2 Assume that Σ is a time-invariant continuous-time system and that $\mathcal{U}$ is separable. (This is the case, for instance, if $\mathcal{U}$ is a subset of a Euclidean space, with the induced metric.) Suppose that

$$\phi(\tau, \sigma, x, \omega) = z$$

for some $x, z \in \mathcal{X}$ and $\omega \in \mathcal{U}^{[\sigma,\tau)}$. Then there exists a sequence of piecewise constant controls

$$\omega_1, \omega_2, \ldots, \omega_k, \ldots$$

so that $(x, \omega_j) \in \mathcal{D}_{\sigma,\tau}$ for all j and

$$z_j := \phi(\tau, \sigma, x, \omega_j) \;\rightarrow\; z \quad \text{as} \quad t \rightarrow \infty\,.$$

Further, there exists a sequence of states $\{x_j\}$ so that $(x_j, \omega_j) \in \mathcal{D}_{\sigma,\tau}$ for all j,

$$\phi(\tau, \sigma, x_j, \omega_j) = z \;\text{ for all }\; j\,,$$

and

$$x_j \rightarrow x \quad \text{as} \quad t \rightarrow \infty\,.$$

Proof. The first conclusion is immediate from Remark C.1.2 in Appendix C, plus Part 2(i) of Theorem 1. The second conclusion follows from the same argument applied to the time-reversed system

$$\dot{x} = -f(x, u)$$

(cf. Lemma 2.6.8). ∎

The following discrete-time result is easy to prove by induction on the length of inputs:

Lemma/Exercise 2.8.3 Let Σ be a time-invariant discrete-time system of class $\mathcal{C}^0$, and denote

$$\mathcal{D}_{\sigma,\tau} := \{(x,\omega) \mid (\tau,\sigma,x,\omega) \in \mathcal{D}\} \subseteq \mathcal{X} \times \mathcal{U}^{[\sigma,\tau)}.$$

Then the following conclusions hold, when $\mathcal{U}^{[\sigma,\tau)}$ is identified with an open subset of a Euclidean space $\mathbb{K}^{m(\tau-\sigma)}$.

As a subset of $\mathcal{X} \times \mathcal{U}^{[\sigma,\tau)}$, the set $\mathcal{D}_{\sigma,\tau}$ is open, and the mapping

$$\alpha : \mathcal{D}_{\sigma,\tau} \to \mathcal{X}, \quad \alpha(x,\omega) := \phi(\tau,\sigma,x,\omega)$$

is continuous. If the system is of class $\mathcal{C}^1$, then α is differentiable; moreover,

$$\alpha_*[x,\omega](\lambda_0,\mu)$$

is the solution $\lambda(\tau)$ of the variational equation

$$\lambda(t+1) = f_x(\xi(t),\omega(t))\lambda(t) + f_u(\xi(t),\omega(t))\mu(t), \quad t \in [\sigma,\tau) \tag{2.46}$$

with initial condition $\lambda(\sigma) = \lambda_0$, where f_x and f_u denote Jacobians of f with respect to the first n variables and the last m variables, respectively, and where $\lambda(t) \in \mathbb{R}^n$ and $\mu(t) \in \mathbb{R}^m$ for each t. Here ξ denotes the path $\psi(x,\omega)$.

In particular,

$$\alpha(x,\cdot) : \mathcal{D}_{\sigma,\tau,x} \to \mathcal{X}$$

has full rank at ω if and only if the linear map $\alpha(0,\cdot)$ for the linearization $\Sigma_*[\Gamma]$ along $\Gamma = (\xi,\omega)$, seen as a time-varying linear system, is onto. □

2.9 More on Differentiability*

When $\mathcal{U} \subseteq \mathbb{R}^m$, we may consider for each $p \geq 1$ and each $\sigma < \tau$ the space of p-integrable functions on $[\sigma,\tau]$. Since the interval of definition is finite, this contains $\mathcal{L}^\infty_m$. We shall be interested in the latter space, endowed with the norm of the first:

$$B^p_m$$

will denote $\mathcal{L}^\infty_m$ with the norm

$$\|\omega\|_p := \left(\int_\sigma^\tau |\omega(t)|^p \, dt\right)^{\frac{1}{p}}$$

for each $p \in [1,\infty)$. For simplicity in statements, we also let B^∞_m be the same as $\mathcal{L}^\infty_m$ (with the sup norm). The normed spaces B^p_m, $p < \infty$, are not Banach, being dense subspaces of the respective Banach space of p-integrable maps.

It is a standard real-variables fact that for each pair p,q so that

$$1 \leq p < q \leq \infty$$

* This section can be skipped with no loss of continuity.

there exists a constant c_1 so that

$$\|\omega\|_p \leq c_1 \|\omega\|_q$$

for all $\omega \in \mathcal{L}_m^\infty$. In fact, as a consequence of Hölder's inequality this holds with

$$c_1 = (\tau - \sigma)^{\frac{1}{p} - \frac{1}{q}}$$

(see, for instance, [190], Theorem 13.17). Conversely, if $1 \leq p < q < \infty$ and if k is any given constant, then there is another constant c_2 such that

$$\|\omega\|_q^q \leq c_2 \|\omega\|_p^p$$

whenever $\|\omega\|_\infty \leq k$. This is proved by writing

$$|\omega(t)|^q = |\omega(t)|^{q-p}|\omega(t)|^p \leq k^{q-p}|\omega(t)|^p$$

and integrating. In particular, all of the p-topologies ($p < \infty$) are equivalent on bounded subsets of $\mathcal{L}_m^\infty$ (but not for $p = \infty$).

The results on continuity and differentiability will depend on the precise value of p. To understand why, it is worth thinking about the most simple case of a system with equation

$$\dot{x} = g(u)$$

and $\mathcal{X} = \mathcal{U} = \mathbb{R}$. Here

$$\phi(\tau, \sigma, 0, \omega) = \int_\sigma^\tau g(\omega(t))\, dt\,.$$

If ω_n converges uniformly to ω and g is continuous, then also

$$\phi(\tau, \sigma, 0, \omega_n) \to \phi(\tau, \sigma, 0, \omega)\,. \tag{2.47}$$

But if the convergence $\omega_n \to \omega$ is weaker, then (2.47) will in general fail to follow, unless either one knows that the convergence is dominated or if some other condition on g, such as linearity on u, is satisfied.

We introduce a technical condition that will be needed later.

Definition 2.9.1 *Let the system Σ be of class C^1. We will say that Σ has* **linear growth on controls** *if $\mathcal{U}$ is a convex subset of $\mathbb{R}^m$ and it holds that, for each compact subset $K \subseteq \mathcal{X}$, there is some constant k so that the rhs of Σ satisfies*

$$\|f_u(x, u)\| \leq k \tag{2.48}$$

whenever $x \in K$ and $u \in \mathcal{U}$. If in addition Σ is of class C^2 and it also holds that the second partial derivative

$$f_{xu}(x, u) \tag{2.49}$$

is bounded in norm uniformly on $K \times \mathcal{U}$ for each compact subset $K \subseteq \mathcal{X}$, meaning that the second mixed partials of f with respect to all variables (x_i, u_j) are all bounded, then we will say that Σ has **strongly linear growth in controls.** □

There are two main cases of interest in control theory in which both of these properties are satisfied. The first is that in which the system is of class $\mathcal{C}^2$ and the control value set $\mathcal{U}$ is (convex and) bounded and f extends as a $\mathcal{C}^2$ function in a neighborhood of $\operatorname{clos} \mathcal{U} \subseteq \mathbb{R}^m$. The second is that of systems affine in controls as described in Equation (2.34).

Lemma 2.9.2 Assume that Σ is of linear growth in controls, $\mathcal{X} = \mathbb{R}^n$, and there exists some compact subset K_1 of $\mathcal{X}$ such that $f(x,u) = 0$ for all $x \notin K_1$. Let K_2 be any compact subset of $\mathcal{U}$. Then, there exists a constant M such that

$$|f(x^0,u) - f(z^0,v)| \leq M \left(|x^0 - z^0| + |u - v|\right) \tag{2.50}$$

for every $x^0, z^0 \in \mathcal{X}$, $u \in K_2$, and $v \in \mathcal{U}$. Further, if Σ is of class $\mathcal{C}^2$ and $p \in [1,2]$, then denoting

$$g(x,\beta) := f(x,\beta) - f(x^0,u) - f_x(x^0,u)(x - x^0) - f_u(x^0,u)(\beta - u)$$

there exists a constant N so that

$$|g(z^0,v)| \leq N \left(|x^0 - z^0|^2 + |u - v|^p\right) \tag{2.51}$$

for every $x^0, z^0 \in \mathcal{X}$, $u \in K_2$, and $v \in \mathcal{U}$.

Proof. Property (2.50) is obtained by separately bounding

$$|f(x^0,u) - f(z^0,u)|$$

and

$$|f(z^0,u) - f(z^0,v)|$$

using property (2.48) and the Mean Value Theorem (convexity of $\mathcal{U}$ and the compact support property are used here). We now prove property (2.51). Assume first that $|u - v| \leq 1$. Then v is in the compact set

$$K_3 := \{v \mid |v - \beta| \leq 1 \text{ for some } \beta \in K_2\}.$$

By Taylor's formula with remainder (recall that f is twice differentiable), we know that, for all x^0, z^0, and $u, v \in K_3$,

$$|g(z^0,v)| \leq a|x^0 - z^0|^2 + b|u - v|^2$$

for some constants a, b (which depend on K_1 and K_3 and hence also depend on K_2). Since $|u - v| \leq 1$ and $p \leq 2$, this means that also

$$|g(z^0,v)| \leq a|x^0 - z^0|^2 + b|u - v|^p. \tag{2.52}$$

Next note that $g_\beta(x,\beta) = f_u(x,\beta) - f_u(x^0,u)$, so that the system having g as a rhs also has linear growth in controls (in equation (2.48), for the same compact one must use $2k$ instead of k). We can then apply the first conclusion, (2.50), to this other system to conclude, for z^0, u, v as in the statement, that

$$|g(z^0, v) - g(z^0, u)| \leq M'|u - v|$$

for a constant M' that depends only on K_1 and K_2. If $|u - v| > 1$, then also $|u - v| < |u - v|^p$; thus,

$$\begin{aligned} |g(z^0, v)| &\leq |g(z^0, u)| + |g(z^0, v) - g(z^0, u)| \\ &\leq a|x^0 - z^0|^2 + M'|u - v|^p. \end{aligned}$$

Choosing $N := \max\{a, b, M'\}$, the result follows. ∎

We now provide a p-norm version of Theorem 1 (p. 57).

Proposition 2.9.3 Let the notations be as in the statement of Theorem 1. Pick any $p \in [1, \infty)$. Then:

1. If S has linear growth on controls, then the set $\mathcal{D}_{\sigma,\tau}$ is open in $\mathcal{X} \times B^p_m$ and both ψ and α are continuous (with respect to the p-topology on controls).

2. Take any x and any ω admissible for x, and denote $\xi := \psi(x, \omega)$. Assume that
$$\lim_{j\to\infty} x_j = x \quad \text{and} \quad \lim_{j\to\infty} \omega_j = \omega\,,$$
where the sequence $\{\omega_j\}$ is equibounded and the convergence is now in B^p_m. Then $\xi_j := \psi(x_j, \omega_j)$ is defined for all large j and
$$\lim_{j\to\infty} \|\xi_j - \xi\|_\infty = 0\,.$$

3. If $p > 1$ and Σ has strongly linear growth on controls, then the same differentiability conclusion as in Theorem 1, Part 3, holds with respect to the p-topology.

Proof. Consider first the continuity statements. We assume first that $\mathcal{X} = \mathbb{R}^n$ and that f is globally Lipschitz in the following sense: For each compact subset K_2 of $\mathcal{U}$, there is a constant M such that

$$|f(x, u) - f(z, v)| \leq M\left(|x - z| + |u - v|\right)$$

for all x, z in $\mathbb{R}^n$, u in K_2, and v in $\mathcal{U}$. So solutions are globally defined. Assume that (x, ω) gives rise to the trajectory (ξ, ω); we wish to prove continuity at (x, ω). Let (ζ, ν) be any other trajectory. We have that, for each $\sigma \leq t \leq \tau$,

$$\xi(t) - \zeta(t) = \xi(\sigma) - \zeta(\sigma) + \int_\sigma^t \left(f(\xi(s), \omega(s)) - f(\zeta(s), \nu(s))\right) ds\,.$$

By the Lipschitz condition, using for K_2 any compact containing all values of ω, there is an M so that

$$|\xi(t)-\zeta(t)| \leq |\xi(\sigma)-\zeta(\sigma)| + M\,\|\omega-\nu\|_1 + M\int_\sigma^\tau |\xi(s)-\zeta(s)|\,ds\,.$$

By the Bellman-Gronwall Lemma, we conclude that

$$|\xi(t)-\zeta(t)| \leq (|\xi(\sigma)-\zeta(\sigma)| + M\,\|\omega-\nu\|_1)\,e^{Mt}$$

for all $\sigma \leq t \leq \tau$. Thus, for each $1 \leq p < \infty$ there are constants a, b such that

$$|\phi(t,\sigma,x,\omega)-\phi(t,\sigma,z,\nu)| \leq a|x-z| + b\,\|\omega-\nu\|_p \tag{2.53}$$

for all x, z in $\mathbb{R}^n$, all ν in B_m^p, and all $\sigma \leq t \leq \tau$. This shows the continuity of ψ and α at (x,ω) under the above global Lipschitz assumption.

Assume now that $\mathcal{X}$, $\mathcal{U}$, and f are arbitrary. Pick any element (x,ω) in $\mathcal{D}_{\sigma,\tau}$. Choose open subsets $\mathcal{V}$ and $\mathcal{W}$ of $\mathcal{X}$ such that

$$\mathcal{V} \subseteq \text{clos}\,(\mathcal{V}) \subseteq \mathcal{W}$$

and so that

$$\xi(t) \in \mathcal{V} \text{ for all } t \in [\sigma,\tau]$$

and $\mathcal{W}$ has a compact closure. Let $\theta : \mathbb{R}^n \to \mathbb{R}$ be any smooth function that is identically 1 on clos$\,(\mathcal{V})$ and vanishes outside $\mathcal{W}$.

Consider now the system obtained with $\mathcal{X}' = \mathbb{R}^n$, same $\mathcal{U}$, and f replaced by h, where

$$h(x,u) := \theta(x)f(x,u)$$

on $\mathcal{X} \times \mathcal{U}$ and $h \equiv 0$ if $x \notin \mathcal{X}$. If f has linear growth in u then h also does, and hence since θ has compact support we are in the situation of (2.50) in Lemma 2.9.2. Thus, h is globally Lipschitz in the above sense, and we can apply the first case of the proof to this h.

We let $\widetilde{\phi}$ be the transition map ϕ for this new system. By (2.53), there is then a neighborhood $\mathcal{V}'$ of x and an $\varepsilon > 0$ such that $\widetilde{\phi}(t,\sigma,z,\nu)$ is in $\mathcal{V}$ for all $\sigma \leq t \leq \tau$ whenever $z \in \mathcal{V}'$ and $\|\omega-\nu\|_p < \varepsilon$.

The maps h and f coincide in a neighborhood of the points of the form $(\xi(t),\omega(t))$, $t \in [\sigma,\tau]$, because θ is independent of u. Therefore $\phi(t,\sigma,z,\nu)$ solves the original differential equation, i.e. it equals $\widetilde{\phi}(t,\sigma,z,\nu)$ for these t, z, ν. In particular, $\mathcal{D}_{\sigma,\tau}$ contains a neighborhood of (x,ω) and is therefore open. Continuity of ψ and α then follows from (2.53).

For the dominated convergence statement, one may apply the previous case. The hypothesis of linear growth can be assumed, because we can always modify the rhs of the system in such a manner that its values remain the same in a neighborhood of the values of the corresponding trajectories but so that the growth condition is satisfied. For instance, one may multiply f by a function $\theta(u)$ which vanishes for $|u| > C$.

We now prove differentiability, assuming that the system is of strong linear growth in controls. Since differentiability is a local property, we may again modify f (by multiplication by a scalar function) and assume that $\mathcal{X} = \mathbb{R}^n$ and f is globally Lipschitz in the previous sense.

As in the proof for $p = \infty$, let (ξ, ω) be any fixed trajectory, and consider the linearization matrices $A(t), B(t)$ along it. For any other control ν (of length T) sufficiently near ω in $\mathcal{B}^p_m$, and each z near x, we consider the corresponding trajectory (ζ, ν), where $\zeta = \psi(z, \nu)$. By continuity, we may choose a neighborhood of ω (in the p-norm) so that this trajectory stays always in a given compact convex neighborhood of x in $\mathcal{X}$, say K_1. Let K_2 be any compact set such that the (essentially bounded) control ω satisfies $\omega(t) \in K_2$ for almost all t. As before, introduce $\delta(t) := \zeta(t) - \xi(t)$ and $\mu(t) := \nu(t) - \omega(t)$. From the conclusion (2.51) in Lemma 2.9.2, it follows that, if $1 \leq p \leq 2$, then

$$\dot{\delta}(t) = A(t)\delta(t) + B(t)\mu(t) + \gamma(t)\,,$$

where

$$|\gamma(t)| \leq N\left(|\delta(t)|^2 + |\mu(t)|^p\right)$$

for a suitable constant N. So if λ solves (2.35) with

$$\lambda(\sigma) = \delta(\sigma) = \lambda_0 = x - z\,,$$

one has again the explicit form

$$\delta(\tau) = \lambda(\tau) + \int_\sigma^\tau \Phi(\tau, s)\gamma(s)\,ds\,.$$

So

$$|\delta(\tau) - \lambda(\tau)| \leq M\left(\|\mu\|_p^p + \int_\sigma^\tau |\delta(t)|^2\,dt\right)$$

for some constant M. Applying (2.53) to the second term there results that there is a constant M' such that

$$|\delta(\tau) - \lambda(\tau)| \leq M'\left(\|\mu\|_p^p + |\lambda_0|^2 + \|\mu\|_p^2\right)\,. \tag{2.54}$$

Since $p > 1$, it follows that $|\delta(\tau) - \lambda(\tau)|$ is indeed

$$o(|\lambda_0| + \|\mu\|_p)$$

as required to establish differentiability. If, instead, $2 < p$, we consider equation (2.54) for the case $p = 2$. Since

$$\|\mu\|_2 \leq c\,\|\mu\|_p$$

for some constant, it follows that $|\delta(\tau) - \lambda(\tau)|$ is majorized by an expression

$$M''\left(|\lambda_0|^2 + 2\,\|\mu\|_p^2\right)\,,$$

again as desired.

The proof of continuity of the derivative is as follows: As in the proof of Theorem 1, for each (x,ω) in the domain $\mathcal{D}_{\sigma,\tau}$ and each $\varepsilon > 0$ we need to show that there is some $\delta > 0$ such that, for each $(\widetilde{x},\widetilde{\omega}) \in \mathcal{D}_{\sigma,\tau}$ so that

$$\|\omega - \widetilde{\omega}\|_p + |x - \widetilde{x}| < \delta\,, \tag{2.55}$$

necessarily

$$\|\alpha_*[x,\omega] - \alpha_*[\widetilde{x},\widetilde{w}]\| < \varepsilon\,.$$

As in the proof for the sup norm, this means that for each ε there should be some δ such that, for each $\lambda_0 \in \mathbb{R}^n$ and each $\mu \in \mathbb{R}^m$, Equation (2.41) must hold, except that the term $\|\mu\|$ now should be $\|\mu\|_p$. We again have the bound (2.43), with Q as in (2.44), as well as the bound (2.45), with c now being valid for arbitrary u.

Bounding Q and applying Hölder's inequality, we conclude from (2.44) and (2.45) that the left-hand side of (2.43) is bounded by

$$k|\lambda_0| + k' \left\|B(\cdot) - \widetilde{B}(\cdot)\right\|_q \|\mu\|_p + (\tau-\sigma)^{1/q} ck \|\mu\|_p\,, \tag{2.56}$$

where

$$\begin{aligned} k &:= \sup_{s\in[\sigma,\tau]} |\Phi(\tau,s) - \widetilde{\Phi}(\tau,s)|\,,\\ k' &:= \sup_{s\in[\sigma,\tau]} |\Phi(\tau,s)|\,, \end{aligned}$$

and $q = \frac{p}{p-1}$.

Note that k' is finite because Φ is continuous in s; thus, the proof will be completed if we show that both $\left\|B(\cdot) - \widetilde{B}(\cdot)\right\|_q$ and k are small for $(\widetilde{x},\widetilde{\omega})$ near (x,ω).

Consider equation (2.42). This can be thought of as a system in dimension n^2 whose controls are ξ,ω. Note that f_{xu} is bounded, because we multiplied f by a function of x with compact support, and the original system has strong linear growth. In particular, then, this n^2-dimensional system has itself linear growth. It follows from Part 1 that k in equation (2.56) is indeed small provided that $\|\omega - \widetilde{\omega}\|_p$ and $\left\|\xi - \widetilde{\xi}\right\|$ be small, but the second of these tends to zero as the first does and $|x - \widetilde{x}| \to 0$, by continuity.

Finally, we need to establish that if $|x_n - x| \to 0$ and $\|\omega_n - \omega\|_p \to 0$ then, with $B_n(t) = f_u(\xi(t),\omega(t))$, also $\|B_n(\cdot) - B(\cdot)\|_q \to 0$. Since under the assumptions also $\|\xi_n - \xi\|_\infty \to 0$ and therefore $\|\xi_n - \xi\|_p \to 0$, where $\xi_n = \psi(x_n,\omega_n)$, and the values of all ξ_n may be assumed to stay in a compact subset of the state space, it will be enough to prove the following fact (for each entry b of the matrix B): If

$$b : \mathbb{R}^l \to \mathbb{R}$$

is a bounded continuous function and if

$$\|\nu_n - \nu\|_p \to 0\,,$$

then also

$$\|b \circ \nu_n - b \circ \nu\|_q \to 0\,.$$

Note that $\|\omega_n - \omega\|_p \to 0$ implies that ω_n converges to ω in measure, that is,

$$\lim_{n\to\infty} \text{meas}\,\{t \in [\sigma,\tau] \mid |\omega_n(t) - \omega(t)| \geq \delta\} \to 0$$

for each $\delta > 0$ (see, for instance, Exercise 13.33 in [190]). Since b is uniformly continuous, also $b\circ\omega_n$ converges in measure to $b\circ\omega$. Because the sequence $\{b\circ\omega_n\}$ is uniformly bounded, it follows from the Dominated Convergence Theorem and the Riesz Theorem (see, e.g., Exercise 12.57 in [190]) that the convergence is also in q-norm, as desired. ∎

Remark 2.9.4 The differentiability result is false if $p = 1$. For instance, let $\sigma = 0, \tau = 1$, and consider the system

$$\dot{x} = u^2$$

with $\mathcal{X} = \mathbb{R}$ and $\mathcal{U} = (-1,1)$. We claim that the map

$$\beta := \alpha(0,\cdot) = \phi(1,0,0,\cdot) :\; \nu \to \int_0^1 \nu(t)^2\,dt$$

is not differentiable at $\omega = 0$, from which it follows that neither is α differentiable at $(0,0)$. To see this by contradiction, write

$$L := \beta_*[0]$$

and note that, because $\beta(\nu) = \beta(-\nu)$ and $L\nu = -L(-\nu)$ for each ν,

$$2\,\|\beta(\nu)\| \leq \|\beta(\nu) - L\nu\| + \|\beta(-\nu) - L(-\nu)\|\,. \tag{2.57}$$

In particular, consider the family of controls ω_ε:

$$\omega_\varepsilon(t) = \begin{cases} 1 & \text{if } t \in [0,\varepsilon], \\ 0 & \text{otherwise} \end{cases}$$

and note that $\|\omega_\varepsilon - \omega\|_1 \to 0$ when $\varepsilon \to 0$. Thus, it should hold that each term in the right-hand side of (2.57), and hence also $\beta(\omega_\varepsilon)$, is of order

$$o(\|\omega_\varepsilon\|_1) = o(\varepsilon)\,.$$

Since $\beta(\omega_\varepsilon) = \varepsilon$, this is a contradiction.

Note that, on the other hand, for $p > 1$ one has for this example that

$$\|\omega_\varepsilon\|_p = \varepsilon^{1/p},$$

and there is no contradiction in that case. □

Remark 2.9.5 For $p < \infty$ either the linear growth condition or the dominated convergence assumption is essential. Otherwise, not even continuity holds. This is because convergence in the p-topology does not imply convergence in the q-topology, for $q > p$. Indeed, take any finite p and consider the equation $\dot{x} = u^q$, where $q > p$ is arbitrary. Pick any r with $q > (1/r) > p$. The control ω_ε defined now by

$$\omega_\varepsilon(t) := \varepsilon^{-r} \text{ for } t \leq \varepsilon$$

and zero otherwise, has $\|\omega_\varepsilon\|_p \to 0$ as $\varepsilon \to 0$, but

$$|\xi_\varepsilon(1)| = \varepsilon^{1-rq} \to \infty$$

for the corresponding solution ξ_ε. □

2.10 Sampling

In Section 1.3 we discussed the motivation behind the process of sampling and digital control. We now define this precisely.

Definition 2.10.1 *Let $\Sigma = (\mathbb{R}, \mathfrak{X}, \mathfrak{U}, \phi)$ be a continuous-time system, and pick any real number $\delta > 0$. The δ-***sampled system associated to** *Σ is the discrete-time system $\Sigma_{[\delta]}$ defined as follows. Let $\mathcal{E}_\delta$ be the subset of*

$$\mathbb{Z} \times \mathfrak{X} \times \mathfrak{U}$$

consisting of those triples (k, x, u) that satisfy that ω is admissible for x, where

$$\omega \in \mathfrak{U}^{[k\delta,(k+1)\delta)}$$

is defined by

$$\omega(t) \equiv u \text{ for all } t \in [k\delta, (k+1)\delta)\,.$$

The state space of $\Sigma_{[\delta]}$ is then the set of all $x \in \mathfrak{X}$ for which there is some k, ω with $(k, x, \omega) \in \mathcal{E}_\delta$, and the local-in-time dynamics are given on $\mathcal{E}_\delta$ by

$$\mathcal{P}(k, x, u) := \phi((k+1)\delta, k\delta, x, \omega)\,.$$

For systems with outputs, $\Sigma_{[\delta]}$ has the same $\mathcal{Y}$ and h as Σ. □

Note that, in this definition, the effect of the control u on the sampled system is *undefined* if the solution to the differential equation does not exist *for the entire interval* $[k\delta, (k+1)\delta]$. We think of the sampled system as providing a "snapshot" view of the original system every δ seconds, with constant control applied in between; if the original system had an explosion time during the interval, the state at the end of the interval would be undefined.

It is easy to show that if Σ is linear (respectively, time-invariant), then $\Sigma_{[\delta]}$ is also, for any δ. For time-invariant systems, $\mathcal{P}(x, u)$ is just the result of integrating $\dot{\xi} = f(\xi, \omega)$ on $[0, \delta]$, with initial state $\xi(0) = x$ and control $\omega \equiv u$. (And $\mathcal{P}(x, u)$ is undefined if the solution does not exist.)

Exercise 2.10.2 Find $\Sigma_{[\delta]}$ for the following Σ and $\delta = 1$ (all examples have $\mathcal{X} = \mathcal{U} = \mathbb{R}$):

(a) $\dot{x} = xu$;

(b) $\dot{x} = ax + bu$ (a, b are constants); and

(c) $\dot{x} = (x + u)^2$ (be sure to specify the state space and the domain of $\mathcal{P}$ for this example). □

2.11 Volterra Expansions*

For continuous-time systems, certain representations of input to state and response maps are particularly useful. Before describing one of these, we review the Peano-Baker formula and a close relative.

Recall that a fundamental solution of a linear differential equation $\Phi(t, \sigma)$ admits a power series expansion, the Peano-Baker formula (C.27) discussed in Appendix C.4. Substituting the power series (C.27) into the variation of parameters formula gives an expression for $\phi(\tau, \sigma, x, \omega)$ in terms of a series of integrals, for linear systems, as shown in Exercise C.4.2 in that Appendix. We now describe an alternative such series, assuming for simplicity that $x_0 = 0$. Replace

$$\begin{aligned} \dot{\xi}(t) &= A(t)\xi(t) + B(t)\omega(t) \\ \xi(\sigma) &= 0 \end{aligned} \tag{2.58}$$

by the corresponding integral equation

$$\xi(t) = \int_\sigma^t B(s)\omega(s)ds + \int_\sigma^t A(s)\xi(s)\,ds\,,$$

and solve this by iteration. There results a series representation for the solution as follows:

$$\begin{aligned} \xi(t) \;=\; & \int_\sigma^t B(s_1)\omega(s_1)\,ds_1 \;+\; \int_\sigma^t \int_\sigma^{s_1} A(s_1)B(s_2)\omega(s_2)\,ds_2 ds_1 \\ +\; & \int_\sigma^t \int_\sigma^{s_1} \int_\sigma^{s_2} A(s_1)A(s_2)B(s_3)\omega(s_3)\,ds_3 ds_2 ds_1 \;+\ldots \\ +\; & \int_\sigma^t \int_\sigma^{s_1} \cdots \int_\sigma^{s_{l-1}} A(s_1)A(s_2)\ldots A(s_{l-1})B(s_l)\omega(s_l)\,ds_l \ldots ds_2 ds_1 \\ +\; & \ldots\,. \end{aligned} \tag{2.59}$$

Note that this series is slightly different from that given in Exercise C.4.2: the bottom limits of integration are now all equal to σ.

* This section can be skipped with no loss of continuity.

Lemma/Exercise 2.11.1 Prove that both the series (2.59) and the series obtained by taking termwise derivatives converge uniformly for t on bounded intervals, and that therefore the series (2.59) provides a solution to the initial value problem (2.58). □

Using the above we now show, for a simple type of nonlinear system, how to obtain an integral representation similar to the variation of parameters formula. The resulting representation is called a **Volterra Series.**

A *bilinear continuous-time system* is one with $\mathcal{X} = \mathbb{R}^n$, $\mathcal{U} = \mathbb{R}^m$, and equations

$$\dot{\xi}(t) = \left(A(t) + \sum_{i=1}^{m} \omega_i(t)E_i(t)\right)\xi(t) + B(t)\omega(t)\,, \tag{2.60}$$

where $E_1, \ldots, E_m$, as well as A and B, are matrices of locally essentially bounded functions. Note that linear systems appear as the special case when all E_i vanish. A bilinear system with outputs is one with a linear measurement map

$$y(t) = C(t)\xi(t) \tag{2.61}$$

and $\mathcal{Y} = \mathbb{R}^p$. (One could equally well define complex-valued bilinear systems, of course.)

For notational simplicity, we restrict attention to single-input bilinear systems, those for which $\mathcal{U} = \mathbb{R}$ and

$$\dot{\xi}(t) = A(t)\xi(t) + \omega(t)E(t)\xi(t) + B(t)\omega(t) \tag{2.62}$$

(here B is a column vector), but the multi-input case is entirely analogous.

We will find a series representation for solutions of this equation that exhibits the dependence on powers of the control ω. To start, we consider the solution ξ of (2.62) having initial condition $\xi(\sigma) = 0$. Introduce the function

$$\zeta(t) := \Phi(\sigma, t)\xi(t)\,,$$

where Φ is the fundamental solution associated to $A(t)$. Then, ζ satisfies the differential equation

$$\dot{\zeta} = \omega F\zeta + G\omega\,, \tag{2.63}$$

where we introduced

$$G(t) := \Phi(\sigma, t)B(t)$$

and

$$F(t) := \Phi(\sigma, t)E(t)\Phi(t, \sigma)\,.$$

Using the expression (2.59), with ωF in place of A and G in the place of B, in order to give a series representation for the solution of (2.63), we obtain that $\zeta(t)$ equals

$$\int_\sigma^t G(s_1)\omega(s_1)\,ds_1 \;+\; \int_\sigma^t \int_\sigma^{s_1} F(s_1)G(s_2)\omega(s_1)\omega(s_2)\,ds_2ds_1$$

$$+\int_\sigma^t\int_\sigma^{s_1}\int_\sigma^{s_2} F(s_1)F(s_2)G(s_3)\omega(s_1)\omega(s_2)\omega(s_3)\,ds_3ds_2ds_1 \;+\ldots$$

$$+\int_\sigma^t\int_\sigma^{s_1}\cdots\int_\sigma^{s_{l-1}} F(s_1)F(s_2)\ldots F(s_{l-1})G(s_l)\omega(s_1)\ldots\omega(s_l)\,ds_l\ldots ds_2ds_1$$

$$+\ldots. \tag{2.64}$$

We now multiply both sides of this equation by $\Phi(t,\sigma)$ and use that $\xi(t) = \Phi(t,\sigma)\zeta(t)$ and the formulas for F and G in terms of Φ, B, E, to conclude that there is a *Volterra expansion*

$$\xi(t) = \sum_{l=1}^{\infty}\int_\sigma^t\int_\sigma^{s_1}\cdots\int_\sigma^{s_l} W_l(t,s_1,\ldots,s_l)\omega(s_1)\ldots\omega(s_l)ds_l\ldots ds_1 \tag{2.65}$$

where $W_l(t,s_1,\ldots,s_l) :=$

$$\Phi(t,s_1)E(s_1)\Phi(s_1,s_2)E(s_2)\ldots E(s_{l-1})\Phi(s_{l-1},s_l)B(s_l)\,. \tag{2.66}$$

The case when $\xi(\sigma) = x^0 \neq 0$ can be handled as follows: Introducing the variable

$$\widetilde{\xi}(t) := \xi(t) - \Phi(t,\sigma)x^0\,,$$

there results for $\widetilde{\xi}$ the equation

$$\dot{\widetilde{\xi}}(t) = A(t)\widetilde{\xi}(t) + \omega(t)E(t)\widetilde{\xi}(t) + \widetilde{B}(t)\omega(t)$$

having $\widetilde{\xi}(\sigma) = 0$, where

$$\widetilde{B}(t) := E(t)\Phi(t,\sigma)x^0 + B(t)\,,$$

and a series can be obtained for $\widetilde{\xi}$ as above.

Now it is possible to obtain a representation for the output y as

$$y(t) = C(t)\Phi(t,\sigma)x^0 + V(t)\,,$$

where $V(t)$ is given by

$$\sum_{l=1}^{\infty}\int_\sigma^t\int_\sigma^{s_1}\cdots\int_\sigma^{s_l} C(t)W_l(t,s_1,\ldots,s_l)\omega(s_1)\ldots\omega(s_l)ds_l\ldots ds_1 \tag{2.67}$$

for the series obtained for $\widetilde{\xi}$.

Exercise 2.11.2 The transmission of FM signals is based on the modulation of frequency based on an input voltage. A simple circuit generating such modulated signals is the one satisfying the differential equation

$$\ddot{y}(t) + \left[\lambda^2 + u(t)\right]y(t) = 0\,,$$

which corresponds to a harmonic oscillator with frequency controlled about the nominal value λ. We model this as a time-invariant bilinear system

$$\begin{aligned} \dot{x} &= \begin{pmatrix} 0 & 1 \\ -\lambda^2 & 0 \end{pmatrix} x + u \begin{pmatrix} 0 & 0 \\ -1 & 0 \end{pmatrix} x \\ y &= (1 \quad 0)\, x \end{aligned} \tag{2.68}$$

with $\mathcal{X} = \mathbb{R}^2$, $\mathcal{U} = \mathbb{R}$, and $\mathcal{Y} = \mathbb{R}$. (The "B" matrix is identically zero.) When $u \equiv 0$ one has a standard harmonic oscillator, and

$$x(t) = \begin{pmatrix} \sin \lambda t \\ \lambda \cos \lambda t \end{pmatrix}$$

is a solution, with

$$x(0) = \begin{pmatrix} 0 \\ \lambda \end{pmatrix}.$$

Find the first three terms of the Volterra representation of $y(t)$ for (2.68), with this initial state. □

2.12 Notes and Comments

Basic Definitions

Various definitions of "system" or "machine" have been proposed in the literature, all attempting to model basically the same notion, but differing among themselves in mathematical technicalities. Some early references are [114], [230], and [444], but of course similar transition formalisms (with no controls nor outputs) have long been used in the theory of dynamical systems. Category-theoretic definitions also have been suggested, which make the definition of "linear" system or certain classes of "topological" systems very easy; see, for instance, [22] for pointers to that literature, as well as [314], Section 8.5. The area called *general systems theory* concerns itself with abstract systems defined by transition maps, typically with no extra algebraic or topological structure; see, for instance, [300] and [429].

Recently there has been a resurgence of interest in control theory in the use of automata models that had originated in computer science, logic, and operations research. In most of these studies, time corresponds not to "real" clock time but to instants at which the system changes in some special manner, as in queuing models when a new customer arrives. Work in this area falls under the general label of "control of *discrete-event systems*"; some references are [90], [405], and [433]. An approach combining discrete-event ideas together with more classical linear design, in the context of a special-purpose computer language, has been recently proposed in [46]; this paper models some of the logical operations as discrete-time systems over a finite field. More generally, the area of *hybrid systems* has become active in the mid 1990s, see for instance the conference proceedings [15]; one possible approach to combining automata and linear systems was studied in [362] (see also the author's paper in [15]).

A conceptually very different foundation for system and control theory has been also proposed. Rosenbrock in [331] emphasized the fact that physical principles sometimes lead to mixtures of differential and algebraic equations, and state space descriptions are not necessarily the most appropriate; implicit differential equations are then used. Related to this is the literature on *singular systems*, or *descriptor systems*, defined (in the linear case) by equations of the type $E\dot{x} = Ax + Bu$ with the matrix E not necessarily invertible; see, e.g., [83], [94], [192], and [277], as well as [82] for the related differential equation theory. The recent work of Willems has carried out this reasoning to the limit and proposed a completely different formalism for systems, based on not distinguishing a priori between inputs and outputs. Instead of "input/output" data, one then has an "observed time-series" which summarizes all i/o data without distinguishing between the two; the papers [309], [424], [425], and [426] provide an introduction to this literature.

The definitions given may be generalized to include what are often called *multidimensional systems*, for which the time set $\mathcal{T}$ is $\mathbb{R}^k$ or $\mathbb{Z}^k$, with $k > 1$. (The terminology is most unfortunate, since we normally use the term "dimension" to refer to the number of dependent variables in the equations defining the dynamics, rather than to the number of independent variables.) Multidimensional systems are useful in areas such as picture processing or seismic data analysis, where different space directions may be thought of as "time" coordinates. Multidimensional systems are considerably more difficult to handle, since the difference and differential equations that appear with ordinary systems are replaced by *partial* difference and differential equations. For references to this area, see the book [57].

There is also a variation of the definition of system that allows for the possibility of the readout map h to depend directly on $\omega(t)$. In other words, one has $h : \mathcal{T} \times \mathcal{X} \times \mathcal{U} \to \mathcal{Y}$. In the context of such more general systems, one would call **state-output** systems those that we have defined here. The use of non-state output systems causes some technical difficulties which are avoided with our definition, but one could model such more general systems using pairs (Σ, α), where Σ is a system with outputs as defined in this Section and $\alpha : \mathcal{U} \times \mathcal{Y} \to \mathcal{Y}$ is a map.

In practice, it may often happen that control of a large system can only be achieved through local effects; this gives rise to the theory of *large scale systems* and corresponding *decentralized control* objectives; see [350] for a detailed treatment of many of the issues that arise there, including the use of decomposition techniques, as well as many references.

Continuous-time systems $\dot{x} = f(x, u)$ can be studied as *differential inclusions* ([26], [111], [93]). In general, given a set-valued map $F : \mathcal{X} \to 2^{\mathbb{R}^n}$, that is, a map that assigns to each point x in $\mathcal{X} \subseteq \mathbb{R}^n$ a *subset* $F(x) \subseteq \mathbb{R}^n$, a *solution* of the differential inclusion $\dot{x} \in F(x)$ means, by definition, an absolutely continuous function $x : \mathcal{I} \to \mathcal{X}$, defined on some interval, so that $\dot{x}(t) \in F(x(t))$ for almost all $t \in \mathcal{I}$. To a system $(\Sigma)\ \dot{x} = f(x,u)$ one may associate the differential inclusion with $F_\Sigma(x) := \{f(x,u) \mid u \in \mathcal{U}\}$, and paths of the system become solutions of the latter. Conversely, there is a rich literature dealing with *selection theorems* which, under appropriate regularity assumptions on F_Σ, guarantee that every solution of $\dot{x} = F_\Sigma(x)$ arises in this manner, from some control u.

I/O Maps

There are sometimes cases in which the input affects the output with no delay, for instance in the "i/o behavior" of more general systems such as those mentioned above (h depends in u) or in dealing with a *memoryless i/o map*, a map induced by an arbitrary mapping

$$h : \mathcal{U} \to \mathcal{Y}$$

between any two sets by the obvious rule

$$\lambda(\tau,\sigma,\omega)(t) := h(\omega(t)).$$

Unfortunately, there are technical difficulties in formulating this consistently in terms of response maps. For instance, when treating continuous-time systems and behaviors one identifies controls that are equal almost everywhere, and this means that the instantaneous value $\omega(t)$ is not well-defined. Also, for the empty control the output would be undefined, meaning that one cannot really talk about "instantaneous measurements" of states. Thus, a concept analogous to that of response cannot be used in this case. It is possible, however, to generalize the idea of i/o map, where strict causality is now weakened to the requirement (*causality*) that

$$\bar{\lambda}(\mu,\sigma,\omega|_{[\sigma,\mu)})(t) = \bar{\lambda}(\tau,\sigma,\omega)(t)$$

only for $t \in [\sigma,\mu)$, rather than for all $t \in [\sigma,\mu]$. Then a satisfactory theory can be developed for causal mappings. We prefer to deal here with the concept of a response, or equivalently, a (strictly causal) i/o map.

Linear Discrete-Time Systems

In control applications, linear systems are of interest mainly in the case in which the field is $\mathbb{K} = \mathbb{R}$, and sometimes, mostly for technical reasons, $\mathbb{K} = \mathbb{C}$. However, finite fields do appear in communications and signal processing theory, as illustrated by the error detection example; in particular the theory of what are called *convolutional codes* is to a great extent the study of linear systems over such fields (see, for instance, [118] and [297]). Recently, work has been done on systems over fields of formal Laurent series, which provide a good model for certain perturbation results; see, for instance, [313].

The algebraic theory of linear systems is no more difficult when dealing with arbitrary fields than with $\mathbb{R}$ or $\mathbb{C}$, and so we will give most results for general $\mathbb{K}$.

On the other hand, it is also possible to define *linear systems over rings* rather than fields, with the same definition but using "module" instead of "vector space"; these are useful in various applications such as modeling the effect of fixed precision integer arithmetic or delay systems (see remarks below). Some results valid over fields generalize, typically with much harder proofs, to systems over rings, but many others do not. For references on systems over rings see, for instance, [61], as well as the early papers [332] and [333] that started most of this research.

Linear Continuous-Time Systems and Behaviors

It is also of great interest to consider linear (or nonlinear) *infinite dimensional systems*, also called *distributed systems.* For instance, transmission delays give rise to what are often called *delay-differential* or *hereditary* models such as

$$\dot{x}(t) = x(t-1) + u(t)\,,$$

which reflects the effect of a retarded feedback; or one may study systems defined by partial differential equations, such as a heat equation with boundary control. Many such examples can be modeled by equations of the type $\dot{x} = Ax + Bu$ evolving in Banach spaces. There is a rich and extensive literature on these matters; see, for instance, [281] for optimal control, as well as the books [107], [133], or [147], or the paper [437], for more "system theoretic" questions, each using a somewhat different approach.

Sometimes it is possible to model usefully infinite dimensional systems as "systems over rings of operators." For instance, one may write the above delay system as the "one dimensional" system

$$\dot{x} = \sigma x + u\,,$$

where "σ" is a shift operator on an appropriate space of functions. Viewing this as the linear system $(\sigma, 1)$ over the polynomial ring $\mathbb{R}[\sigma]$ permits using techniques —like Pole Shifting— from the theory of systems over rings, resulting in constructive synthesis procedures. This approach is due to Kamen; see, for instance, [61], [62], [79], [81], [233], [234], [305], [349], [357], [396], and [397] for more on the topic and related results, and [74] for the study of certain systems defined by partial differential equations in this manner.

Volterra Expansions

The Volterra expansion can be generalized to bilinear systems with more controls, and locally or in asymptotic senses to more general classes of nonlinear systems; see, for instance, [336] for details on this and the above example. (One approach for general nonlinear systems relies in first approximating nonlinear systems by bilinear ones, a process analogous to taking truncations of a Taylor expansion. There is also an alternative approach to bilinear approximation, valid in compact-open topologies of controls and based on the Stone-Weierstrass Theorem; see [381].)

Volterra series are important because, among other reasons, they exhibit the various homogeneous nonlinear effects separately. In the context of identification, they allow the separation of higher harmonic responses to periodic signals. When the original system is linear, every term except that for $l = 1$ vanishes, and the Volterra series (2.65) is nothing more than the variation of parameters formula; more generally, the solution of the linearized system along $\xi \equiv 0, \omega \equiv 0$ is obtained by simply dropping all terms except the first one. It is also possible to give Volterra expansions for smooth discrete-time systems, as also discussed in [336].

Volterra series were one of the most popular engineering methods for nonlinear systems modeling during the 1950s. Their theoretical study, especially in relation to state space models, was initiated during the early 1970s; see, for instance, [69] for a survey of the early literature.

An alternative approach to Volterra series is based on what are called *generating series*, or *Fliess series*. These are formal power series that essentially correspond to expanding the "kernels" W_l, and can be interpreted in terms of mappings from jets of germs of differentiable inputs into jets of outputs. They are closely related to work on the representation of solutions of differential equations, as in [89]; see [140] and references therein, as well as [199], [373], [241], and the excellent book [327].

Chapter 3

Reachability and Controllability

3.1 Basic Reachability Notions

In all of the definitions to follow, $\Sigma = (\mathcal{T}, \mathcal{X}, \mathcal{U}, \phi)$ is an arbitrary system.

Definition 3.1.1 *An* **event** *is a pair* $(x, t) \in \mathcal{X} \times \mathcal{T}$.

- *The event* (z, τ) **can be reached from** *the event* (x, σ) *iff there is a path of* Σ *on* $[\sigma, \tau]$ *whose initial state is* x *and final state is* z*, that is, if there exists an* $\omega \in \mathcal{U}^{[\sigma,\tau)}$ *such that*
$$\phi(\tau, \sigma, x, \omega) = z\,.$$
One says also that (x, σ) **can be controlled to** (z, τ).

- *If* $x, z \in \mathcal{X}, T \geq 0 \in \mathcal{T}$*, and there exist* $\sigma, \tau \in \mathcal{T}$ *with* $\tau - \sigma = T$ *such that* (z, τ) *can be reached from* (x, σ)*, then* z **can be reached from** x **in time** T. *Equivalently,* x **can be controlled to** z **in time** T.

- z **can be reached from** x *(or* x **can be controlled to** z*) if this happens for at least one* T. □

Note that, by the identity axiom in the definition of system, every event and every state can be reached from itself. Also, note that if Σ is time-invariant, then z can be reached from x in time T iff $(z, t+T)$ can be reached from (x, t) for any $t \in \mathcal{T}$; thus, there is no need to consider explicitly the notion of event for time-invariant systems.

We use the notation $(x, \sigma) \rightsquigarrow (z, \tau)$ to indicate that the event (z, τ) can be reached from (x, σ). Note that the inequality $\sigma \leq \tau$ is implicit in this notation. For states, we write $x \underset{T}{\rightsquigarrow} z$ if z can be reached from x in time T, and just $x \rightsquigarrow z$ to indicate that z can be reached from x.

Lemma/Exercise 3.1.2 Prove the following statements:

(a) If $(x,\sigma) \rightsquigarrow (z,\tau)$ and $(z,\tau) \rightsquigarrow (y,\mu)$, then $(x,\sigma) \rightsquigarrow (y,\mu)$.

(b) If $(x,\sigma) \rightsquigarrow (y,\mu)$ and if $\sigma < \tau < \mu$, then there exists a $z \in \mathcal{X}$ such that $(x,\sigma) \rightsquigarrow (z,\tau)$ and $(z,\tau) \rightsquigarrow (y,\mu)$.

(c) If $x \underset{T}{\rightsquigarrow} y$ for some $T > 0$ and if $0 < t < T$, then there is some $z \in \mathcal{X}$ such that $x \underset{t}{\rightsquigarrow} z$ and $z \underset{T-t}{\rightsquigarrow} y$.

(d) If $x \underset{t}{\rightsquigarrow} z$, $z \underset{s}{\rightsquigarrow} y$, *and* Σ *is time-invariant*, then $x \underset{t+s}{\rightsquigarrow} y$.

(e) If $x \rightsquigarrow z$, $z \rightsquigarrow y$, *and* Σ *is time-invariant*, then $x \rightsquigarrow y$. □

Exercise 3.1.3 Give examples to show that:

- Properties (d) and (e) in Lemma 3.1.2 may be false if Σ is not time-invariant, and
- Even for time-invariant systems, it is not necessarily true that $x \rightsquigarrow z$ implies that $z \rightsquigarrow x$ (so, "$\rightsquigarrow$" is not an equivalence relation). □

Remark 3.1.4 Let Σ be a continuous-time system as in Definition 2.6.7 and let $\sigma < \tau$. With the present terminology, Lemma 2.6.8 says that $(x,\sigma) \rightsquigarrow (z,\tau)$ for the system Σ iff $(z,\sigma) \rightsquigarrow (x,\tau)$ for the system $\Sigma^-_{\sigma+\tau}$. This remark is sometimes useful in reducing many questions of control *to* a given state to analogous questions (for the time-reversed system) of control *from* that same state, and vice versa. □

Recall that a **linear** system is one that is either as in Definition 2.4.1 or as in Definition 2.7.2.

Lemma 3.1.5 If Σ is linear, then:

(a) If $(x_1,\sigma) \rightsquigarrow (z_1,\tau)$ and $(x_2,\sigma) \rightsquigarrow (z_2,\tau)$, then, for each $r \in \mathbb{K}$, $(x_1 + rx_2,\sigma) \rightsquigarrow (z_1 + rz_2,\tau)$.

(b) $(x,\sigma) \rightsquigarrow (z,\tau)$ iff $(0,\sigma) \rightsquigarrow (z-\phi(\tau,\sigma,x,\mathbf{0}),\tau)$. (Here $\mathbf{0}$ denotes the control $\omega \equiv 0$.)

(c) If Σ is time-invariant, $x \underset{T}{\rightsquigarrow} z$ iff $0 \underset{T}{\rightsquigarrow} (z - \phi(T,0,x,\mathbf{0}))$.

(d) If Σ is time-invariant, $x_1 \underset{T}{\rightsquigarrow} z_1$ and $x_2 \underset{T}{\rightsquigarrow} z_2$ imply that $(x_1 + rx_2) \underset{T}{\rightsquigarrow} (z_1 + rz_2)$ for all $r \in \mathbb{K}$.

Proof. Part (a) follows by linearity of $\phi(\tau,\sigma,\cdot,\cdot)$ (see Lemmas 2.4.2 and 2.7.4), and part (d) is a consequence of this. The equivalence

$$\phi(\tau,\sigma,x,\omega) = z \;\text{ iff }\; z - \phi(\tau,\sigma,x,\mathbf{0}) = \phi(\tau,\sigma,0,\omega)$$

implies both parts (b) and (c). ∎

Definition 3.1.6 *The system Σ is* **(completely) controllable on the interval** $[\sigma, \tau]$ *if for each $x, z \in \mathcal{X}$ it holds that $(x, \sigma) \rightsquigarrow (z, \tau)$. It is* **(completely) controllable in time** T *if for each $x, z \in \mathcal{X}$ it holds that $x \underset{T}{\rightsquigarrow} z$. It is just* **(completely) controllable** *if $x \rightsquigarrow z$ for all x, z.* □

Lemma 3.1.7 Let Σ be a linear system, and pick any $\sigma, \tau, T \in \mathcal{T}$.

(a) Σ is controllable on $[\sigma, \tau]$ iff $(0, \sigma) \rightsquigarrow (y, \tau)$ for all $y \in \mathcal{X}$.

(b) If Σ is time-invariant, then Σ is controllable in time T iff $0 \underset{T}{\rightsquigarrow} y$ for all $y \in \mathcal{X}$.

(c) If Σ is continuous-time, then it is controllable on $[\sigma, \tau]$ iff $(x, \sigma) \rightsquigarrow (0, \tau)$ for all $x \in \mathcal{X}$.

(d) If Σ is time-invariant and continuous-time, then Σ is controllable in time T iff $x \underset{T}{\rightsquigarrow} 0$ for all $x \in \mathcal{X}$.

(e) The conclusions in (c) and (d) hold also if "continuous-time" is replaced by "discrete-time and $A(k)$ is invertible for all $k \in [\sigma, \tau]$" (in (d), "A is invertible").

Proof. To prove part (a), pick any events (x, σ) and (z, τ). Let

$$y := z - \phi(\tau, \sigma, x, \mathbf{0}).$$

By assumption, $(0, \sigma) \rightsquigarrow (y, \tau)$. Thus, by Lemma 3.1.5(b), $(x, \sigma) \rightsquigarrow (z, \tau)$, as desired. Assume now that Σ is time-invariant and that

$$0 \underset{T}{\rightsquigarrow} y$$

for all $y \in \mathcal{X}$. Then, $(0, 0) \rightsquigarrow (y, T)$ for all $y \in \mathcal{X}$, so by part (a) Σ is controllable (on $[0, T]$), and this establishes part (b).

Assume now that Σ is continuous-time, or that it is discrete-time and the invertibility condition in (e) holds. Pick any $y \in \mathcal{X}$. By part (a), we must prove for (c) that $(0, \sigma) \rightsquigarrow (y, \tau)$. We claim that there is some $x \in \mathcal{X}$ such that

$$\phi(\tau, \sigma, x, \mathbf{0}) = -y.$$

In continuous-time we may take $x := -\Phi(\sigma, \tau)y$, and in discrete-time we take $x := -P^{-1}y$, where P is the linear transformation

$$\Phi(\tau, s) = A(\tau - 1)A(\tau - 2) \ldots A(\sigma + 1)A(\sigma),$$

which is invertible under the assumption that each $A(k)$ is. By hypothesis, $(x, \sigma) \rightsquigarrow (0, \tau)$. So by Lemma 3.1.5(b),

$$(0, \sigma) \rightsquigarrow (-\phi(\tau, \sigma, x, \mathbf{0}), \tau) = (y, \tau),$$

as desired. Finally, part (d) follows from (b) because the hypothesis gives now that $(x,0) \rightsquigarrow (0,T)$ for all $x \in \mathcal{X}$. ∎

The property in (a) and (b), that any state be reachable from zero, is sometimes called **reachability**. We proved that, for linear systems, *reachability and controllability are equivalent.* Note also that the property in (c), (d), and (e), that one be able to control every state to the origin, is called "controllability" by some authors (a more precise term, also used sometimes, is *null-controllability*). We prefer to reserve the term controllability for the property defined in 3.1.6.

Remark 3.1.8 Part (e) of the above result illustrates a principle that is rather general in control theory: Finite dimensional smooth discrete-time systems with "reversible" or "invertible" dynamics tend to have a theory very analogous to that of continuous-time systems. (Essentially, because group actions can be associated to systems for which the maps $\phi(\tau, \sigma, \cdot, \omega)$ are invertible.) Except for the algebraic structure theory of *time-invariant linear* systems, general (non-invertible) discrete-time systems behave in a much more complicated manner than continuous-time systems. One way in which distributed (i.e., infinite dimensional) continuous-time systems behave very differently from continuous-time (finite dimensional) systems is precisely in the sense that they tend not to have invertible dynamics; for instance, a system that evolves according to a diffusion equation will smooth out initial conditions. □

3.2 Time-Invariant Systems

Throughout this section, unless otherwise stated, Σ is an arbitrary *time-invariant* system.

Definition 3.2.1 *Let $T \in \mathcal{T}$ and $x \in \mathcal{X}$. The* **reachable set from x in time T** *is*

$$\mathcal{R}^T(x) := \{z \in \mathcal{X} \mid x \underset{T}{\rightsquigarrow} z\}.$$

The **reachable set from x** *is*

$$\mathcal{R}(x) := \bigcup_{T \in \mathcal{T}_+} \mathcal{R}^T(x) = \{z \in \mathcal{X} \mid x \rightsquigarrow z\}.$$

If $\mathcal{S}$ is a subset of $\mathcal{X}$, we also write

$$\mathcal{R}^T(\mathcal{S}) := \bigcup_{x \in \mathcal{S}} \mathcal{R}^T(x),$$

$$\mathcal{R}(\mathcal{S}) := \bigcup_{x \in \mathcal{S}} \mathcal{R}(x),$$

for the sets reachable from the subset $\mathcal{S}$. □

Note that $\mathcal{R}^0(\mathcal{S}) = \mathcal{S}$ for all subsets $\mathcal{S}$, and that Σ is controllable (respectively, controllable in time T) iff $\mathcal{R}(x)$ (respectively, $\mathcal{R}^T(x)$) $= \mathcal{X}$ for all $x \in \mathcal{X}$.

From Lemmas 2.4.2 and 2.7.4, we have the following:

Lemma 3.2.2 If Σ is linear, then $\mathcal{R}^T(0)$ is a subspace, for all $T \in \mathcal{T}$. □

If Σ is linear and x is now an arbitrary state, the equality

$$\phi(T,0,x,\omega) = \phi(T,0,x,\mathbf{0}) + \phi(T,0,0,\omega)$$

shows that

$$\mathcal{R}^T(x) = \phi(T,0,x,\mathbf{0}) + \mathcal{R}^T(0)$$

and hence that $\mathcal{R}^T(x)$ is a linear submanifold (translation of a subspace) in $\mathcal{X}$.

Lemma 3.2.3 For each $t, s \in \mathcal{T}$,

$$\mathcal{R}^{s+t}(x) = \mathcal{R}^s(\mathcal{R}^t(x))$$

for each $x \in \mathcal{X}$.

Proof. Lemma 3.1.2, Part (b) implies that $\mathcal{R}^{s+t}(x) \subseteq \mathcal{R}^s(\mathcal{R}^t(x))$, while the reverse inclusion follows by Part (d). ∎

It is of interest to know whether reachability can be tested in finitely many steps. In automata theory, one frequently uses the following observation:

Lemma 3.2.4 Assume that Σ is a discrete-time system and $\operatorname{card}(\mathcal{X}) = n < \infty$. Then,

$$\bigcup_{\{k=0,\ldots,n-1\}} \mathcal{R}^k(x) = \mathcal{R}(x)$$

for all $x \in \mathcal{X}$.

Proof. Pick x and a $z \in \mathcal{R}(x)$. Assume that k is the smallest integer such that $x \underset{k}{\leadsto} z$, and assume by way of contradiction that $k \geq n$. By part (c) of Lemma 3.1.2, there exist elements

$$x = z_0, z_1, \ldots, z_k = z$$

with

$$z_i \underset{1}{\leadsto} z_{i+1}, i = 0, \ldots, k-1\,.$$

Since $k \geq n$, there are $0 \leq i < j \leq k$ such that $z_i = z_j$. Since

$$x \underset{i}{\leadsto} z_i \text{ and } z_j \underset{k-j}{\leadsto} z\,,$$

it follows by part (d) of Lemma 3.1.2 that $x \underset{l}{\leadsto} z, l = i + k - j < k$, which contradicts minimality of k. ∎

Recall the notion of equilibrium state $x \in \mathcal{X}$; for a time-invariant system, this is equivalent to the requirement that there exists a $u \in \mathcal{U}$ such that for all $T \in \mathcal{T}_+$, $\phi(T, 0, x, \omega) = x$, where $\omega \equiv u$. For discrete-time systems this is equivalent to

$$\mathcal{P}(x, u) = x$$

and for continuous-time systems to

$$f(x, u) = 0 \, .$$

If x is an equilibrium state, then $x \in \mathcal{R}^T(x)$ for all T. Thus, for all $S \in \mathcal{T}_+$,

$$\mathcal{R}^S(x) \subseteq \mathcal{R}^S(\mathcal{R}^T(x)) \, ,$$

and so by Lemma 3.2.3:

Corollary 3.2.5 If x is an equilibrium state, then

$$\mathcal{R}^S(x) \subseteq \mathcal{R}^{S+T}(x)$$

for each $S, T \in \mathcal{T}_+$. □

Lemma 3.2.6 Let x be an equilibrium state. If there are any $S, T \in \mathcal{T}_+, T > 0$, such that

$$\mathcal{R}^S(x) = \mathcal{R}^{S+T}(x) \, ,$$

then necessarily $\mathcal{R}^S(x) = \mathcal{R}(x)$.

Proof. First note that

$$\mathcal{R}^{S+kT}(x) = \mathcal{R}^S(x)$$

for all positive integers k. This can be proved inductively: The case $k = 1$ is given, and

$$\begin{aligned} \mathcal{R}^{S+(k+1)T}(x) &= \mathcal{R}^T(\mathcal{R}^{S+kT}(x)) \\ &= \mathcal{R}^T(R^S(x)) \\ &= \mathcal{R}^{S+T}(x) \\ &= \mathcal{R}^S(x) \, , \end{aligned}$$

where the second and last equalities hold by inductive hypothesis. Now assume given any $z \in \mathcal{R}(x)$. So there is some $t \in \mathcal{T}$ so that $z \in \mathcal{R}^t(x)$. Find an integer k such that $S + kT > t$. By Corollary 3.2.5, $z \in \mathcal{R}^{S+kT}(x) = \mathcal{R}^S(x)$. ∎

Corollary 3.2.7 Assume that x is an equilibrium state, that $\mathcal{X}$ is a vector space over a field $\mathbb{K}$ with $\dim \mathcal{X} = n < \infty$, and that $\mathcal{R}^T(x)$ is a subspace for each $T \in \mathcal{T}_+$. (By Lemma 3.2.2, this always happens if Σ is a finite dimensional linear system and $x = 0$.) Then:

(a) If $\mathcal{T} = \mathbb{Z}$, $\mathcal{R}^n(x) = \mathcal{R}(x)$.

(b) If $\mathcal{T} = \mathbb{R}$, then $\mathcal{R}^\varepsilon(x) = \mathcal{R}(x)$ for all $\varepsilon > 0$.

Proof. Pick any strictly increasing sequence

$$0 = \tau_0 < \tau_1 < \tau_2 < \ldots < \tau_{n+1}$$

of elements of $\mathcal{T}$, and let

$$\mathcal{X}_i := \mathcal{R}^{\tau_i}(x)$$

for each i. If it were true that $\mathcal{X}_i$ were properly included in X_{i+1} for all these i, then it would follow that $\dim \mathcal{X}_{i+1} > \dim \mathcal{X}_i$ for all i, and hence that $\dim \mathcal{X}_{n+1} \geq n+1 > \dim \mathcal{X}$, a contradiction. Thus, in each such chain of reachability sets there is some i such that $\mathcal{X}_i = \mathcal{X}_{i+1}$. By Lemma 3.2.6, we conclude that

$$\mathcal{R}^{\tau_i}(x) = \mathcal{R}(x)\,.$$

Now in case (a) we apply the above argument with $\tau_i = i$, and in case (b) we use instead $\tau_i = i\varepsilon/n$. ∎

The conclusion that, for linear continuous-time systems, if one can reach z from 0, then one can do so in arbitrarily small time, is non-intuitive. Of course, this can never happen in a physical system. Underlying the above proof is the fact that the control-values space is all of $\mathbb{K}^m$. In order to obtain smaller ε, one will in general need larger values for the controls (see, for instance, Exercise 3.5.7). Later, in the Chapter on optimal control, we will state an optimization problem that imposes a cost associated to the magnitudes of controls.

Lemma/Exercise 3.2.8 Let

$$\mathcal{C}^T(x) = \{z \in \mathcal{X} \mid z \underset{T}{\leadsto} x\}, \quad \mathcal{C}(x) = \bigcup_{T \in \mathcal{T}_+} \mathcal{C}^T(x)\,.$$

Prove Lemmas 3.2.3, 3.2.4, and 3.2.6, and Corollaries 3.2.5 and 3.2.7 for the statements that result when "$\mathcal{C}$" is substituted for "$\mathcal{R}$" throughout. □

Exercise 3.2.9 Consider a *discrete-time bilinear* system $x(t+1) = u(t)Ex(t)$, with $\mathcal{X} = \mathbb{R}^2, \mathcal{U} = \mathbb{R}$, and

$$E = \begin{pmatrix} 1 & 0 \\ 1 & 1 \end{pmatrix}.$$

Show that

$$\mathcal{R}^T\begin{pmatrix} 1 \\ 0 \end{pmatrix} = \left\{ \begin{pmatrix} x_1 \\ x_2 \end{pmatrix} \middle|\, Tx_1 = x_2 \right\}$$

for each positive integer T. This is a subspace for each such T, but $R^T\binom{1}{0} \neq R\binom{1}{0}$ for all T. Does this contradict Corollary 3.2.7? □

Corollary 3.2.10 (a) An n-dimensional discrete-time linear system is controllable if and only if

$$\mathcal{R}(0) = \mathcal{R}^n(0) = \mathcal{X}.$$

(b) A continuous-time linear system is controllable if and only if

$$\mathcal{R}(0) = \mathcal{R}^\varepsilon(0) = \mathcal{X} \text{ for all } \varepsilon > 0$$

if and only if $\mathcal{C}(0) = \mathcal{C}^\varepsilon(0) = \mathcal{X}$ for all $\varepsilon > 0$.

Proof. If Σ is controllable, then $\mathcal{R}(x) = \mathcal{X}$ for all $x \in \mathcal{X}$, so by Corollary 3.2.7 it holds in particular that $\mathcal{R}^n(0) = \mathcal{X}$ in the discrete-time case, and that $\mathcal{R}^\varepsilon(0) = \mathcal{X}$ for all $\varepsilon > 0$ in the continuous-time case. Conversely, if $\mathcal{R}^T(0) = \mathcal{X}$ for some $T > 0$, then Σ is controllable, by Lemma 3.1.7(b). In continuous-time, $\mathcal{C}^T(0) = \mathcal{X}$ implies the same conclusion, by part (d) of the same Lemma. ∎

Exercise 3.2.11 Give an example of a discrete-time (time-invariant) linear system of dimension 1 for which $\mathcal{C}(0) = \mathcal{X}$ but $\mathcal{R}(0) = \{0\}$. □

Let Σ be an n-dimensional discrete-time time-invariant linear system (A, B). From the formula

$$\phi(n, 0, 0, \omega) = \sum_{i=1}^{n} A^{i-1} B\omega(n-i),$$

it follows that z is in $\mathcal{R}^n(0)$ iff it is in the image of the linear map

$$\boxed{\mathbf{R} = \mathbf{R}(A, B) := [B, AB, A^2B, \ldots, A^{n-1}B]} \tag{3.1}$$

which maps

$$\mathcal{U}^n \to \mathcal{X}$$

sending

$$(u_1, \ldots, u_n) \mapsto \sum_{i=1}^{n} A^{i-1} B u_i .$$

When $\mathcal{U} = \mathbb{K}^m$ and $\mathcal{X} = \mathbb{K}^n$, we identify $\mathbf{R}(A, B)$ with an n by nm matrix whose columns are the columns of $B, AB, \ldots, A^{n-1}B$ in that order. By Corollary 3.2.10, Σ is controllable iff the image of the linear map $\mathbf{R}$ is all of $\mathcal{X}$, and this happens iff its rank (i.e., the dimension of its image) is n. Thus:

Theorem 2 *The n-dimensional discrete-time linear system Σ is controllable if and only if* $\operatorname{rank} \mathbf{R}(A, B) = n$. □

Note that from the Cayley-Hamilton Theorem one knows that

$$A^n = \alpha_1 I + \alpha_2 A + \ldots + \alpha_n A^{n-1} \tag{3.2}$$

where

$$\chi_A(s) = \det(sI - A) = s^n - \alpha_n s^{n-1} - \ldots - \alpha_2 s - \alpha_1 \quad (3.3)$$

is the characteristic polynomial of A. Multiplying both sides of (3.2) by A, and using again (3.2) to substitute for the A^n that results in the right-hand side, it follows that A^{n+1} is a linear combination of the matrices $A^i, i = 0, \ldots, n-1$. Recursively, the same holds for all powers of A. Thus, for any vector v, the span of $\{A^i v, i \geq 0\}$ coincides with the span of $\{A^i v, i < n\}$. Applying this to each of the columns of B, we conclude that the condition $\operatorname{rank} \mathbf{R}(A, B) = n$ fails if and only if there is some n-dimensional row vector $\rho \neq 0$ such that

$$\rho A^i B = 0 \text{ for all } i \geq 0, \quad (3.4)$$

which means that there exists a linear function (namely, $x \mapsto \rho x$) on the state space that vanishes at all states reachable from the origin.

If now Σ is an n-dimensional *continuous-time* system, we still may consider the matrix $\mathbf{R}(A, B)$ as in (3.1). A fact that is at first sight remarkable — and which is ultimately due to the common properties shared by difference and differential time-invariant linear equations— is that the *same* algebraic condition as for discrete-time is necessary and sufficient for continuous-time. We now prove this fact. (Later, cf. Remark 3.5.22, we derive it from a general theorem about time-varying systems.) The critical observation is that, for any row vector ρ,

$$\rho e^{tA} B = \sum_{i=0}^{\infty} \rho A^i B \frac{t^i}{i!} \equiv 0$$

if and only if (3.4) holds, because of analyticity of the exponential.

Theorem 3 *The n-dimensional continuous-time linear system Σ is controllable if and only if* $\operatorname{rank} \mathbf{R}(A, B) = n$.

Proof. If the rank condition fails, there exists some row vector $\rho \neq 0$ so that (3.4) holds. Since every element of $\mathcal{R}(0) = \mathcal{R}^1(0)$ has the form

$$x = \int_0^1 e^{(1-t)A} B\omega(t)\, dt = \int_0^1 e^{tA} B\omega(1-t)\, dt, \quad (3.5)$$

also $\rho x = 0$ for all such x. Thus, $\mathcal{R}(0) \neq \mathcal{X}$, and the system is not controllable.

Conversely, if controllability fails, then there is some $\rho \neq 0$ so that $\rho x = 0$ for every x in the subspace $\mathcal{R}(0)$. In particular, consider the control

$$\omega(t) := B^* e^{(1-t)A^*} \rho^*$$

on the interval $[0, 1]$, where "*" indicates conjugate transpose. From (3.5),

$$0 = \rho x = \int_0^1 \rho e^{tA} B B^* e^{tA^*} \rho^*\, dt = \int_0^1 \left\| B^* e^{tA^*} \rho^* \right\|^2 dt,$$

and hence, $\rho e^{tA}B \equiv 0$. ∎

Take as an illustration the linearized pendulum (harmonic oscillator). With the matrices in equation (2.31),

$$\mathbf{R} = \begin{pmatrix} 0 & 1 \\ 1 & 0 \end{pmatrix},$$

and the system is controllable. This means that every configuration of position and velocity can be changed into any other such configuration through the application of a suitable control. We later compute an explicit example of such a control.

For the original nonlinear pendulum, the use of feedback transformations as discussed in Chapter 1 allows establishing that

$$\dot{x}_1 = x_2,\ \dot{x}_2 = -\sin x_1 + u$$

is also controllable; it is only necessary to modify whatever control is used to transfer x to z in the linear system $\dot{x}_1 = x_2, \dot{x}_2 = u$ by addition of the term $\sin \xi_1(t)$ evaluated along the ensuing trajectory. This is just a property of this simple example; in general, for nonlinear systems controllability is far more difficult to characterize. The linearization principle for controllability studied later is useful in reducing questions of "local" controllability to the linear case, however.

Exercise 3.2.12 Assume that the body of an airplane is slanted ϕ radians (its "pitch angle") with respect to the horizontal. It is flying at a constant (nonzero) ground speed of c meters per second, and its flight path forms an angle of α radians with the horizontal (for $\alpha > 0$ the plane is gaining altitude, and for $\alpha < 0$ it is descending). Denoting the plane's altitude in meters by h and assuming the angles are small, the above quantities are related by the following linearized differential equations:

$$\begin{aligned}
\dot{\alpha} &= a(\phi - \alpha) \\
\ddot{\phi} &= -\omega^2(\phi - \alpha - bu) \\
\dot{h} &= c\alpha
\end{aligned}$$

where $\omega > 0$ is a constant representing a natural oscillation frequency and a, b are positive constants. (See Figure 3.1(a).) The control u is proportional to the position of the elevators. (Elevators are movable surfaces located in the tail. A more complete space and orientation model would include other such surfaces: For instance, the rudder, also in the tail, provides directional control or "yaw," and the ailerons in the wing create opposing torques to affect lateral attitude or "roll." In addition, engine thrust affects speed.)

Model the above as a linear system Σ with $n = 4, m = 1$, using $x_1 = \alpha$, $x_2 = \phi$, $x_3 = \dot{\phi}$, and $x_4 = h$. Prove that Σ is controllable. □

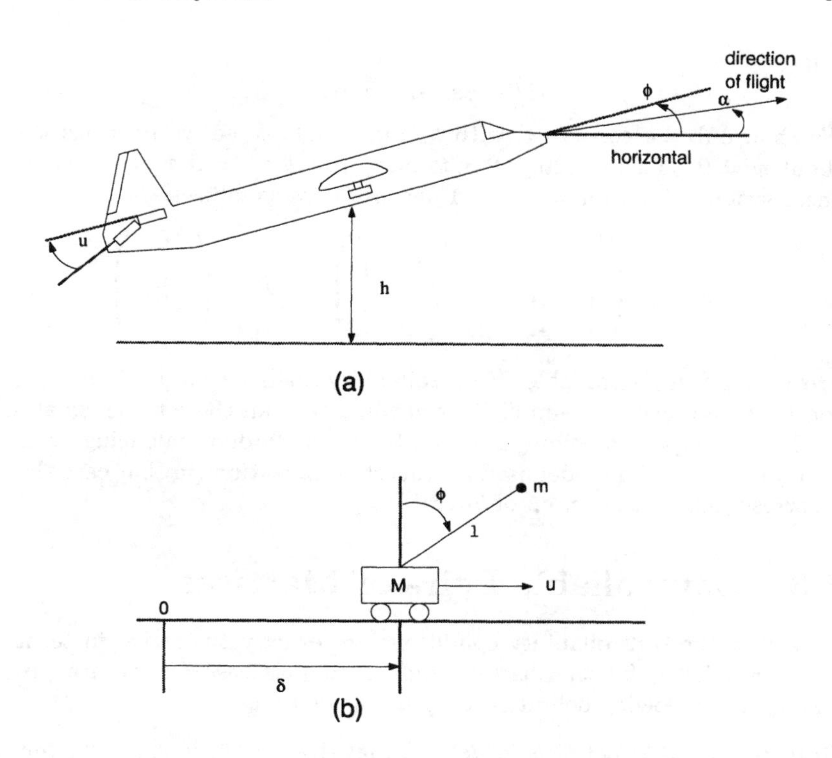

Figure 3.1: *(a) Airplane example. (b) Inverted pendulum.*

Exercise 3.2.13 Consider a system consisting of a cart to the top of which an inverted pendulum has been attached through a frictionless pivot. The cart is driven by a motor which at time t exerts a force $u(t)$, taken as the control. (See Figure 3.1(b).) We assume that all motion occurs in a plane, that is, the cart moves along a straight line. We use ϕ to denote the angle that the pendulum forms with the vertical, δ for the displacement of the center of gravity of the cart with respect to some fixed point, $F \geq 0$ for the coefficient of friction associated with the motion of the cart, g for the acceleration of gravity, $l > 0$ for the length of the pendulum, $M > 0$ for the mass of the cart, and $m \geq 0$ for the mass of the pendulum, which we'll assume is concentrated at the tip. (If the mass is not so concentrated, elementary physics calculations show that one may replace the model by another one in which this does happen, using a possibly different length l. We allow the case $m = 0$ to model the situation where this mass is negligible.)

Newton's second law of motion applied to linear and angular displacements gives the two second order nonlinear equations

$$(M+m)\ddot{\delta} + ml\ddot{\phi}\cos\phi - ml\dot{\phi}^2\sin\phi + F\dot{\delta} = u$$

and

$$l\ddot{\phi} - g\sin\phi + \ddot{\delta}\cos\phi = 0\,.$$

We shall only be concerned with a small angle ϕ, so we linearize the model about $\phi = 0$. This results, after taking $x_1 = \delta, x_2 = \dot{\delta}, x_3 = \phi, x_4 = \dot{\phi}$, in a linear system Σ with $n = 4, m = 1$ and matrices as follows:

$$A = \begin{pmatrix} 0 & 1 & 0 & 0 \\ 0 & -\frac{F}{M} & -\frac{mg}{M} & 0 \\ 0 & 0 & 0 & 1 \\ 0 & \frac{F}{lM} & \frac{g(m+M)}{lM} & 0 \end{pmatrix}, \quad B = \begin{pmatrix} 0 \\ \frac{1}{M} \\ 0 \\ -\frac{1}{lM} \end{pmatrix}.$$

Prove that Σ is controllable. (Controllability holds for all possible values of the constants; however, for simplicity you could take all these to be equal to one.)

This example, commonly referred to as the "broom balancing" example, is a simplification of a model used for rocket stabilization (in that case the control u corresponds to the action of lateral jets). □

3.3 Controllable Pairs of Matrices

Note that the controllability conditions are exactly the same, in terms of the matrices (A, B), for the discrete and continuous case; so we are justified in making the following definition for *pairs of matrices*:

Definition 3.3.1 *Let $\mathbb{K}$ be a field and let $A \in \mathbb{K}^{n\times n}, B \in \mathbb{K}^{n\times m}$, for positive integers n, m. The pair (A, B) is* **controllable** *or* **reachable** *if* $\operatorname{rank} \mathbf{R}(A, B) = n$. □

We call $\mathbf{R} = \mathbf{R}(A, B)$ the *reachability* (or *controllability*) *matrix of* (A, B); the image of the corresponding map, i.e., the column space of $\mathbf{R}$, is the *reachable* (or *controllable*) *space* of (A, B), and we denote it by

$$\mathcal{R}(A, B)$$

or just $\mathcal{R}$. For discrete-time or continuous-time linear systems, $\mathcal{R}$ is then the same as the set of states reachable from the origin, as is clear for discrete-time and follows from the proof of Theorem 3 (p. 89) for continuous-time. Whenever we refer below to a "pair" (A, B), we mean an (A, B) as above. The Cayley-Hamilton Theorem implies the following observation:

Lemma 3.3.2 Let (A, B) be a pair as in Definition 3.3.1, and let

$$b_j := j\text{th column of B}, j = 1, \ldots, m.$$

Then the controllable space $\mathcal{R}(A, B)$ is the span of

$$\{A^i b_j, i \geq 0, j = 1, \ldots, m\}\,.$$

Thus, $\mathcal{R}$ is the smallest A-invariant subspace of $\mathcal{X} = \mathbb{K}^n$ that contains the columns of B. In particular, (A, B) is controllable iff this span is all of $\mathcal{X}$. □

When $m = 1$ this says that controllability of (A, B) is equivalent to the requirement that B, seen as a (column) vector in $\mathbb{K}^n$, be *cyclic* for the matrix A. In this sense controllability generalizes the classical notion of cyclicity in linear algebra.

We let $GL(n)$ denote the group of all invertible $n \times n$ matrices over $\mathbb{K}$. The following result, often called the *Kalman controllability decomposition*, is extremely useful in giving simple proofs of facts about controllability.

Lemma 3.3.3 Assume that (A, B) is not controllable. Let $\dim \mathcal{R}(A, B) = r < n$. Then there exists a $T \in GL(n)$ such that the matrices $\widetilde{A} := T^{-1}AT$ and $\widetilde{B} := T^{-1}B$ have the block structure

$$\boxed{\widetilde{A} = \begin{pmatrix} A_1 & A_2 \\ 0 & A_3 \end{pmatrix} \qquad \widetilde{B} = \begin{pmatrix} B_1 \\ 0 \end{pmatrix}} \tag{3.6}$$

where A_1 is $r \times r$ and B_1 is $r \times m$. (If $r = 0$, the decomposition is trivial, and A_1, A_2, B_1 are not there.)

Proof. Pick any subspace $\mathcal{S}$ such that

$$\mathcal{R} \oplus \mathcal{S} = \mathbb{K}^n$$

and let $\{v_1, \ldots, v_r\}$ be a basis of $\mathcal{R}$ and $\{w_1, \ldots, w_{n-r}\}$ be a basis of $\mathcal{S}$. Then, with

$$T := (v_1, \ldots, v_r, w_1, \ldots, w_{n-r}) \,,$$

the desired forms for $\widetilde{A}$ and $\widetilde{B}$ follow from the facts that $\mathcal{R}$ is A-invariant and that it contains the image of B. ∎

Lemma/Exercise 3.3.4 Prove that (A_1, B_1) is itself a controllable pair. □

For a continuous-time system (A, B), we can interpret the above decomposition as follows: With the change of variables

$$x(t) := Tz(t)$$

the equations $\dot{x} = Ax + Bu$ are transformed into

$$\begin{aligned} \dot{z}_1 &= A_1 z_1 + A_2 z_2 + B_1 u \\ \dot{z}_2 &= A_3 z_2 \end{aligned}$$

where z_1, z_2 are r- and $(n - r)$-dimensional respectively. It is then clear that the z_2 component of the state cannot be controlled in any way. An analogous observation applies to discrete-time systems. Note that the "only if" part of Theorem 3 (p. 89) is an immediate consequence of this decomposition result.

The characteristic polynomial of A splits as

$$\chi_A = \chi_c \chi_u$$

where

$$\chi_c = \chi_{A_1} \text{ and } \chi_u = \chi_{A_3}.$$

Note that A_1 is a matrix representation for the restriction of A to $\mathcal{R}$, so χ_c is independent of the choice of basis on $\mathcal{R}$. The same is true therefore for $\chi_u = \chi_A/\chi_c$. (Independence of choice of basis for the latter also follows from the fact that A_3 is a matrix for the linear map induced by A on the quotient $\mathbb{K}^n/\mathcal{R}$ and hence is uniquely defined up to similarity.) If $r = 0$, we make the convention that $\chi_c = 1$.

Definition 3.3.5 *The polynomials χ_c and χ_u are, respectively, the* **controllable** *and the* **uncontrollable parts** *of the characteristic polynomial χ_A (with respect to the pair (A, B)).* □

Sometimes one refers to the eigenvalues (or the eigenvectors) of these as the "controllable modes" and "uncontrollable modes," respectively, of the system (or matrix pair) (A, B).

Exercise 3.3.6 Prove that if $m = 1$ and A is a diagonal matrix

$$A = \begin{pmatrix} \lambda_1 & 0 & \dots & 0 & 0 \\ 0 & \lambda_2 & \dots & 0 & 0 \\ \vdots & \vdots & \ddots & \vdots & \vdots \\ 0 & 0 & \dots & \lambda_{n-1} & 0 \\ 0 & 0 & \dots & 0 & \lambda_n \end{pmatrix} \qquad B = \begin{pmatrix} b^1 \\ b^2 \\ \vdots \\ b^{n-1} \\ b^n \end{pmatrix}$$

then (A, B) is controllable iff $\lambda_i \neq \lambda_j$ for each $i \neq j$ and all $b^i \neq 0$. □

We will refer to the following result as the *Hautus Lemma*:

Lemma 3.3.7 Let $\widetilde{\mathbb{K}}$ be the algebraic closure of the field $\mathbb{K}$. (In most applications, $\mathbb{K} = \mathbb{R}$ or $\mathbb{C}$, so $\widetilde{\mathbb{K}} = \mathbb{C}$.) The following properties are equivalent for the pair (A, B):

(a) (A, B) is controllable.

(b) $\operatorname{rank}[\lambda I - A, B] = n$ for all $\lambda \in \widetilde{\mathbb{K}}$.

(c) $\operatorname{rank}[\lambda I - A, B] = n$ for each eigenvalue λ of A.

Proof. Note first that (b) and (c) are equivalent, since the first $n \times n$ block of the $n \times (n + m)$ matrix $[\lambda I - A, B]$ already has full rank whenever λ is not an eigenvalue of A. Also, since controllability is characterized by a rank condition, and hence by the nonvanishing of certain determinants, the pair (A, B) is controllable iff it is controllable as a pair over the field $\widetilde{\mathbb{K}}$. It follows that we may assume that $\widetilde{\mathbb{K}} = \mathbb{K}$.

We prove next that (a) implies (b). Assume that the rank is less than n for some $\lambda \in \mathbb{K}$. Thus, the row space of $[\lambda I - A, B]$ has dimension less than n, so there exists some nonzero vector $p \in \mathbb{K}^n$ and some λ such that

$$p'[\lambda I - A, B] = 0\,. \tag{3.7}$$

Thus, $p'A = \lambda p'$ (λ is a *left eigenvalue* of A), and $p'B = 0$. Therefore also $p'A^kB = \lambda^k p'B = 0$ for all k, and hence

$$p'\mathbf{R}(A, B) = 0\,,$$

contradicting controllability.

Finally, we establish that (b) implies (a). Assume that (a) does not hold, so there is a decomposition as in (3.6), with $r < n$. Let λ, v be an eigenvalue/eigenvector pair of the transposed matrix A_3', so that

$$v'(\lambda I - A_3) = 0\,.$$

It follows that the nonzero n-vector

$$w := \begin{pmatrix} 0 \\ v \end{pmatrix}$$

is a left eigenvector of $\widetilde{A}$, with $w'\widetilde{B} = 0$. Hence $p := (T')^{-1}w \neq 0$ satisfies

$$p'[(\lambda I - A)T, B] = 0\,.$$

Since the column space of the matrix $[(\lambda I - A)T, B]$ coincides with that of $[\lambda I - A, B]$, also (3.7) holds, and hence (b) cannot be true. ∎

Corollary 3.3.8 If (A, B) is controllable, then $\operatorname{rank}[A, B] = n$. □

Recall that the *geometric multiplicity* of an eigenvalue λ of a matrix A is the dimension of the nullspace $\ker(\lambda I - A)$.

Lemma/Exercise 3.3.9 Assume that (A, B) is controllable and that $\operatorname{rank} B = q$. Then, the geometric multiplicity of each eigenvalue of A is at most q. □

The uncontrollable modes can be characterized elegantly using the Hautus condition, as follows.

Lemma/Exercise 3.3.10 The zeros of χ_u are precisely the complex numbers λ for which

$$\operatorname{rank}[\lambda I - A, B] < n\,.$$ □

It would seem to be natural also to study a slightly more general class of "linear" (or more precisely, "affine") systems, namely, the class of non-homogeneous linear systems

$$\dot{x} = Ax + Bu + c\,, \tag{3.8}$$

$c \neq 0$ (or their analogue in discrete-time). At least for controllable systems however, the following simple consequence of Lemma 3.3.7 shows that, except for a translation of coordinates in state and input-value spaces, $x \to x - x_0$, $u \to u - u_0$, homogeneous linear systems result again. By $\operatorname{col} W$ we denote the *column space* of the matrix W, i.e., the span of the columns of W.

Exercise 3.3.11 Assume that (A, B) is a pair as before and $c \in \mathbb{K}^n$ is an arbitrary vector. Prove that:

(a) If there is any $\lambda \in \mathbb{K}$ such that $c \notin \operatorname{col}[\lambda I - A, B]$, then there is a $T \in GL(n)$ such that, with the change of variables $z := Tx$, equation (3.8) has the partitioned form

$$\begin{aligned} \dot{z}_1 &= A_1 z_1 + B_1 u, \\ \dot{z}_2 &= \lambda z_2 + 1 \end{aligned}$$

(z_1 of size $n-1$ and z_2 scalar) and similarly in discrete-time. It follows that the system (3.8) and its discrete-time analogue are not controllable.

(b) If $c \in \operatorname{col}[A, B]$, then there are x_0, u_0 such that

$$Ax + Bu + c = A(x - x_0) + B(u - u_0)$$

and if $c \in \operatorname{col}[I - A, B]$ there are x_0, u_0 such that

$$Ax + Bu + c = A(x - x_0) + B(u - u_0) + x_0\,.$$

Interpret in terms of a differential (respectively, difference) equation for $x - x_0$ with control $u - u_0$.

(c) Conclude that a nonhomogeneous linear system as above is controllable if and only if the pair (A, B) is controllable and that in that case the system is, up to a change of variables, a linear system. □

Let $\mathbb{K} = \mathbb{R}$ or $\mathbb{C}$. One important characteristic of the notion of controllability is that it is a *generic* property, in the sense that the set of controllable pairs is an open and dense subset of the set of all pairs of a given size. We formulate this statement as follows. For each fixed pair of positive integers n, m, let

$$\mathcal{S}_{n,m} := \{(A, B) \mid A \in \mathbb{K}^{n\times n}, B \in \mathbb{K}^{n\times m}\}\,,$$

identified with $\mathbb{K}^{n^2+nm}$ by listing all coefficients of A, B in some fixed order. We let $\mathcal{S}^c_{n,m}$ be the subset of $\mathcal{S}_{n,m}$ consisting of all controllable pairs.

Proposition 3.3.12 If $\mathbb{K} = \mathbb{R}$ or $\mathbb{C}$, the set $\mathcal{S}^c_{n,m}$ is open and dense in $\mathcal{S}_{n,m}$.

Proof. The complement $\mathcal{S}^u_{n,m}$ of $\mathcal{S}^c_{n,m}$ is an *algebraic subset* of $\mathcal{S}_{n,m}$, that is, the set of common zeros of a set of polynomials $\{P_1, \ldots, P_r\}$ in $n^2 + nm$ variables. Indeed, we may take as the P_i the set of all possible determinants of $n \times n$ submatrices of $\mathbf{R}(X, Y)$, where X, Y are matrices of indeterminates. Then

$$\operatorname{rank} \mathbf{R}(A, B) < n\,,$$

that is,

$$(A, B) \in \mathcal{S}^u_{n,m}$$

iff all these minors vanish. Note that $\mathcal{S}^u_{n,m}$ is closed, since it is the intersection of the zero sets of the P_i, each of which is closed by continuity of the P_i. Moreover, $\mathcal{S}^u_{n,m}$ is proper, since for any fixed n, m there exists at least one controllable system.

It is only left to prove that any proper algebraic subset of a Euclidean space $\mathbb{K}^q$ must have a dense complement, or equivalently that this set cannot have any interior. Since an algebraic set is an intersection of sets of zeros of polynomials, it is only necessary to see that a set of the type $\{y | P(y) = 0\}$ (P polynomial) cannot have a nonempty interior unless $P \equiv 0$. This is clear by the principle of analytic continuation, or explicitly for polynomials simply by noting that all derivatives of P must be zero at an interior point, and hence that all coefficients of P are zero. (Alternatively, one could argue purely algebraically, using inductively the fact that a single-variable polynomial cannot have infinitely many zeros.) ∎

One may also prove that $\mathcal{S}^u_{n,m}$ has Lebesgue measure zero. Genericity of controllability means that, if all the coefficients of A and B are experimentally measured quantities, then the resulting system in all likelihood will be controllable. However, if a system is near $\mathcal{S}^u_{n,m}$, it will be "hard" to control, and in fact the "true" system we are trying to model may be uncontrollable, with measurement errors accounting for the discrepancy. It is important in practice to estimate how close a system is to being uncontrollable. Various recent papers deal with this issue, and numerically robust methods based on this research have been designed. In particular, one has the following characterization.

For any controllable pair (A, B) over $\mathbb{C}$, we let

$$\delta(A, B) := \operatorname{dist}\left((A, B), \mathcal{S}^u_{n,m}\right)\,,$$

where the distance is measured with respect to the operator norm, i.e.,

$$\operatorname{dist}(F, G) \;=\; \|F - G\|$$

for any two matrices F, G of size $n \times (n + m)$. Since for each fixed n, m the set of uncontrollable systems is closed —in fact it is an algebraic set, as remarked in the above proof— it follows that the distance is always achieved, i.e., there is some $(\widetilde{A}, \widetilde{B}) \in \mathcal{S}^u_{n,m}$ so that

$$\delta(A, B) = \operatorname{dist}\left((A, B), (\widetilde{A}, \widetilde{B})\right) . \tag{3.9}$$

This distance is not easy to compute, as it involves a minimization over all uncontrollable pairs. On the other hand, we may consider the quantity

$$\delta'(A, B) := \min_{\lambda \in \mathbb{C}} \sigma_{\min}(\lambda I - A, B) ,$$

where $\sigma_{\min}$ denotes smallest singular value, which involves just a scalar minimization. (Singular values are discussed in Appendix A.2.) We are justified in writing "minimum" as opposed to just "infimum" because the function

$$f(\lambda) := \sigma_{\min}(\lambda I - A, B)$$

is nonnegative and continuous (continuity of singular values on matrix entries, Corollary A.4.5 in Appendix A.4, using the fact that the rank of $(\lambda I - A, B)$ is always n, by controllability), and $\lim_{|\lambda| \to \infty} f(\lambda) = \infty$ because

$$\sigma_{\min}(\lambda I - A, B) = |\lambda| \, \sigma_{\min}\left(I - \frac{1}{\lambda} A, \frac{1}{\lambda} B\right)$$

and $\sigma_{\min}(I - \frac{1}{\lambda} A, \frac{1}{\lambda} B) \to \sigma_{\min}(I, 0) = 1$.

Proposition 3.3.13 For each controllable pair (A, B), $\delta(A, B) = \delta'(A, B)$.

Proof. Let $(\widetilde{A}, \widetilde{B}) \in \mathcal{S}^u_{n,m}$ be as in (3.9), and write

$$\Delta_A := \widetilde{A} - A, \, \Delta_B := \widetilde{B} - B .$$

By the Hautus condition, there is some λ so that

$$[\lambda I - \widetilde{A}, \widetilde{B}] = [\lambda I - A, B] + [-\Delta_A, \Delta_B]$$

has rank less than n. By Corollary A.2.5 applied to $(\lambda I - A, B)$ and $\Delta = (-\Delta_A, \Delta_B)$,

$$\delta'(A, B) \leq \sigma_{\min}(\lambda I - A, B) \leq \|(-\Delta_A, \Delta_B)\| = \|(\Delta_A, \Delta_B)\| = \delta(A, B) .$$

To prove the other inequality, let λ be such that $\delta'(A, B) = \sigma_{\min}(\lambda I - A, B)$. Again by Corollary A.2.5, there must exist matrices (Δ_A, Δ_B) of norm equal to $\sigma_{\min}(\lambda I - A, B)$ so that, defining $\widetilde{A}, \widetilde{B}$ as above, this is a pair at distance $\sigma_{\min}(\lambda I - A, B)$ from (A, B), which is not controllable. So

$$\delta(A, B) \leq \sigma_{\min}(\lambda I - A, B) = \delta'(A, B)$$

holds too. ∎

Exercise 3.3.14 Let (A, B) correspond to a time-invariant *discrete-time* linear system Σ. Recall that null-controllability means that every state can be controlled to zero. Prove that the following conditions are all equivalent:

1. Σ is null-controllable.
2. The image of A^n is contained in the image of $\mathbf{R}(A, B)$.
3. In the decomposition in Lemma 3.3.3, A_3 is nilpotent.
4. $\operatorname{rank}[\lambda I - A, B] = n$ for all nonzero $\lambda \in \widetilde{\mathbb{K}}$.
5. $\operatorname{rank}[I - \lambda A, B] = n$ for all $\lambda \in \widetilde{\mathbb{K}}$.

□

3.4 Controllability Under Sampling

When using digital control, inputs are restricted in various ways. As discussed at length in Chapter 1, one way of modeling the restriction to piecewise constant controls is via the notion of sampling. It is therefore of interest to know when a continuous-time system that is known to be controllable remains so if we can only apply controls that are constant on sampling intervals $[k\delta, (k+1)\delta]$. Mathematically, the question is whether the sampled system $\Sigma_{[\delta]}$ is controllable, and the answer will depend on the sampling time δ as well as the system Σ. The material in Appendix A.3 is used here.

If Σ is a continuous-time linear (still time-invariant) system, $u \in \mathcal{U}$ is a control value, and $\delta > 0$ is a real number, the variation of parameters formula gives that, for $\omega \equiv u$,

$$\begin{aligned}\phi(\delta, 0, x, \omega) &= e^{\delta A}x + \int_0^\delta e^{(\delta-s)A}Bu\, ds \\ &= Fx + Gu\,,\end{aligned}$$

where

$$F := e^{\delta A}\,, \quad G := A^{(\delta)}B\,, \quad A^{(\delta)} := \int_0^\delta e^{(\delta-s)A}\, ds. \tag{3.10}$$

So $\Sigma_{[\delta]}$ is the discrete-time time-invariant linear system (F, G), and it is controllable iff

$$\begin{aligned}\mathbf{R}(F, G) &= [G, FG, \ldots, F^{n-1}G] \\ &= A^{(\delta)}[B, e^{\delta A}B, \ldots, e^{(n-1)\delta A}B] \\ &= A^{(\delta)}\mathbf{R}(e^{\delta A}, B)\end{aligned} \tag{3.11}$$

has rank n. The second equality follows from the fact that the matrices $A^{(\delta)}$ and $e^{\delta A}$ commute, being both functions of A. Note that $A^{(\delta)} = f(A)$, where f is the entire function

$$f(s) \;=\; \int_0^\delta e^{(\delta-t)s}dt \;=\; \sum_{n=0}^{\infty} \frac{\delta^{n+1}}{(n+1)!}s^n \;=\; \frac{e^{\delta s}-1}{s}.$$

Since $sf(s) = e^{\delta s} - 1$, one also has the useful identity

$$AA^{(\delta)} = e^{\delta A} - I\,.$$

We shall say that the continuous-time system Σ is **δ-sampled controllable** if the discrete-time system $\Sigma_{[\delta]}$ is controllable. Of course, δ-sampled controllability for *any* $\delta > 0$ implies controllability, since for any states x, z, $x \leadsto z$ in $\Sigma_{[\delta]}$ also implies that $x \leadsto z$ in the original system Σ (using piecewise constant controls). The interest is in determining conditions for the converse to hold. From equation (3.11) it follows that Σ is δ-sampled controllable iff both $A^{(\delta)}$ and $\mathbf{R}(e^{\delta A}, B)$ have rank n.

Lemma 3.4.1 Σ is δ-sampled controllable if and only if the pair $(e^{\delta A}, B)$ is controllable and A has no eigenvalues of the form $2k\pi i/\delta$ for any nonzero integer k.

Proof. By the above remarks, it suffices to prove that $A^{(\delta)}$ is invertible if and only if the eigenvalue condition holds. This follows from the Spectral Mapping Theorem, since its eigenvalues are the possible values $f(\lambda)$, where f is the above function and λ is an eigenvalue of A. If $\lambda = 0$, $f(\lambda) = \delta \neq 0$; otherwise,

$$f(\lambda) = \frac{e^{\delta\lambda} - 1}{\lambda}\,,$$

the numerator being nonzero for all $\lambda \neq 0$ if and only if the eigenvalue condition holds. ∎

Example 3.4.2 Consider, as an example, the system Σ corresponding to the linearized pendulum (2.31), which was proved earlier to be controllable. In Appendix C.4 we compute

$$e^{tA} = \begin{pmatrix} \cos t & \sin t \\ -\sin t & \cos t \end{pmatrix}.$$

Thus, for any $\delta > 0$,

$$\mathbf{R}(e^{\delta A}, B) = \begin{pmatrix} 0 & \sin\delta \\ 1 & \cos\delta \end{pmatrix},$$

which has determinant $(-\sin\delta)$. By Lemma 3.4.1, Σ is δ-sampled controllable iff

$$\sin\delta \neq 0 \;\text{ and }\; 2k\pi i \neq \pm i\delta\,,$$

i.e., if and only if δ is not a multiple of π.

Take, for instance, the sampling time $\delta = 2\pi$. From the explicit form of e^{tA}, we know that $e^{\delta A} = I$. Thus,

$$A^{(\delta)} = A^{-1}(e^{\delta A} - I) = 0\,,$$

so $G = 0$. This means that the discrete-time system $\Sigma_{[\delta]}$ has the evolution equation

$$x(t+1) = x(t)\,.$$

No matter what (constant) control is applied during the sampling interval $[0, \delta]$, the state (position and velocity) is the same at the end of the interval as it was at the start of the period. (Intuitively, say for the linearized pendulum, we are acting against the natural motion for half the interval duration, and with the natural motion during the other half.) Consider now the case when $\delta = \pi$, which according to the above Lemma should also result in noncontrollability of $\Sigma_{[\delta]}$. Here

$$F = e^{\delta A} = -I$$

and

$$A^{(\delta)} = A^{-1}(e^{\delta A} - I) = -2A^{-1} = 2A\,,$$

so

$$G = 2AB = \begin{pmatrix} 2 \\ 0 \end{pmatrix}.$$

Thus, the discrete-time system $\Sigma_{[\delta]}$ has the evolution equations:

$$\begin{aligned} x_1(t+1) &= -x_1(t) + 2u(t) \\ x_2(t+1) &= -x_2(t)\,. \end{aligned}$$

This means that we now can partially control the system, since the first coordinate (position) can be modified arbitrarily by applying suitable controls u. On the other hand, the value of the second coordinate (velocity) cannot be modified in any way, and in fact at times $\delta, 2\delta, \ldots$ it will oscillate between the values $\pm x_2(0)$, independently of the (constant) control applied during the interval. □

Exercise 3.4.3 Consider the system (with $\mathcal{U} = \mathbb{R}, \mathcal{X} = \mathbb{R}^2$)

$$\begin{aligned} \dot{x}_1 &= x_2 \\ \dot{x}_2 &= -x_1 - x_2 + u\,, \end{aligned}$$

which models a linearized pendulum with damping. Find explicitly the systems $\Sigma_{[\delta]}$, for each δ. Characterize (without using the next Theorem) the δ's for which the system is δ-sampled controllable. □

The example that we discussed above suggests that controllability will be preserved provided that we sample at a frequency $1/\delta$ that is larger than twice the natural frequency (there, $1/2\pi$) of the system. The next result, sometimes known as the "Kalman-Ho-Narendra" criterion, and the Lemma following it, make this precise.

Theorem 4 *Let $\mathbb{K} = \mathbb{R}$ and let $\Sigma = (A, B)$ be a controllable continuous-time (time-invariant) linear system. If $\delta > 0$ is such that*

$$\boxed{\delta(\lambda - \mu) \neq 2k\pi i, \quad k = \pm 1, \pm 2, \ldots,}$$

for every two eigenvalues λ, μ of A, then Σ is also δ-sampled controllable.

Proof. We will prove that $(e^{\delta A}, B)$ is a controllable pair. The conclusion then will be a consequence of Lemma 3.4.1, because if A were to have any eigenvalue λ of the form $2k\pi i/\delta, k =$ nonzero integer, then, with this λ,

$$\delta(\lambda - \bar{\lambda}) = 4k\pi i\,,$$

contradicting the hypothesis.

We now establish that the Hautus criterion must hold for $(e^{\delta A}, B)$. By the Spectral Mapping Theorem, all eigenvalues of $e^{\delta A}$ are of the form $e^{\delta \lambda}$, where λ is an eigenvalue of A. So we must check that

$$\operatorname{rank}\,[e^{\delta\lambda}I - e^{\delta A}, B] = n$$

for each such λ. Equivalently, we must prove that for each nonzero vector p which is an eigenvector of $(e^{\delta A})'$, necessarily

$$p'B \neq 0\,. \tag{3.12}$$

By controllability of (A, B) the conclusion (3.12) holds if p is an eigenvector of A'. So the only step left is to show that A' and $(e^{\delta A})' = e^{\delta A'}$ have the same eigenvectors. Lemma A.3.3 in the Appendix on linear algebra implies this, provided that $f(\lambda) := e^{\delta\lambda}$ is one-to-one on the spectrum of A. But this is precisely what the hypothesis of the Theorem asserts. ∎

The condition in the above Theorem is not necessary except when the number of controls $m = 1$ (see below). A simpler, though stronger, sufficient condition is as follows. For any given system $\Sigma = (A, B)$, a *frequency of* Σ is any number of the form

$$\frac{|\operatorname{Im}\lambda|}{2\pi}\,,$$

where λ is an eigenvalue of A. For instance, for the system in Example 3.4.2 there is only one frequency, $1/2\pi$. This is the frequency of oscillation of the solutions of the uncontrolled system $\dot{x} = Ax$, which are all combinations of $\sin t$ and $\cos t$ (hence of period 2π). Note that B does not affect the frequencies.

Lemma/Exercise 3.4.4 A controllable continuous-time time-invariant system $\Sigma = (A, B)$ remains controllable provided that the sampling frequency $1/\delta$ is larger than 2ω for every frequency ω of Σ. □

Remark 3.4.5 When $m = 1$, the eigenvalue condition stated in Theorem 4 is also necessary. This is proved as follows: Assume that (A, B) is δ-sampled controllable but that $\lambda \neq \mu$ are eigenvalues of A such that

$$e^{\delta\lambda} = e^{\delta\mu} = \alpha .$$

Choose two (necessarily linearly independent) eigenvectors:

$$Av = \lambda v, \quad Aw = \mu w .$$

Then also

$$e^{\delta A}v = \alpha v \quad \text{and} \quad e^{\delta A}w = \alpha w ,$$

from which it follows that α is an eigenvalue of $e^{\delta A}$ of geometric multiplicity greater than one, contradicting Lemma 3.3.9. □

Exercise 3.4.6 Give an example to show that the eigenvalue condition in Theorem 4 is not necessary (if $m > 1$). □

Remark 3.4.7 There is a somewhat different way of proving Theorem 4, which is more conceptual. We give it now. The main point is that under the hypotheses of the Theorem it can be proved that A is an analytic function of $e^{\delta A}$, so that A is a linear combination of the matrices

$$e^{k\delta A}, k = 0, \ldots, n-1 .$$

Once this is known, it follows that each column of $\mathbf{R}(A, B)$ is a linear combination of the columns of $\mathbf{R}(e^{\delta A}, B)$, and hence the second matrix cannot have rank less than n, since the first one has rank n by the controllability hypothesis. It then follows that the pair $(e^{\delta A}, B)$ is controllable, as desired.

We now show that A is an analytic function of $e^{\delta A}$. (The proof requires the concept of analytic but nonentire functions of matrices.) Consider the exponential function $f(s) = e^{\delta s}$. The hypothesis means precisely that f is one-to-one on the set

$$\Lambda = \{\lambda_1, \ldots, \lambda_l\}$$

of eigenvalues of A. Pick disjoint neighborhoods

$$\mathcal{O}_1, \mathcal{O}_2, \ldots, \mathcal{O}_l$$

of

$$f(\lambda_1), f(\lambda_2), \ldots, f(\lambda_l)$$

respectively. Since f is everywhere nonsingular, we may apply the Inverse Function Theorem to conclude that, taking smaller $\mathcal{O}_k$'s if necessary, there exist disjoint neighborhoods

$$\mathcal{O}_1^0, \mathcal{O}_2^0, \ldots, \mathcal{O}_l^0$$

of

$$\lambda_1, \lambda_2, \ldots, \lambda_l$$

and diffeomorphisms

$$g_k : \mathcal{O}_k \to \mathcal{O}_k^0$$

such that

$$g_k \circ f_k = \text{ identity on each } \mathcal{O}_k^0 .$$

Finally, let $\mathcal{O}$ (resp. $\mathcal{O}^0$) be the union of the sets $\mathcal{O}_k$ (resp. $\mathcal{O}_k^0$), and let g be the extension of the g_k's. Then $g \circ f$ equals the identity in the open set $\mathcal{O}^0$ which contains the eigenvalues of A, and hence

$$g(f(A)) = A ,$$

which proves that indeed A is a function of $e^{\delta A} = f(A)$. □

3.5 More on Linear Controllability

This section develops some of the basic facts about controllability of linear systems, possibly time-varying. The focus is on the continuous-time case; the discrete-time situation is in part analogous and far simpler (and left as an exercise), although some aspects cannot be generalized.

We start with a review of some basic properties of generalized inverses of operators. Appendix A.2 reviews pseudoinverses of matrices; since here the interest is on the operator from controls to states, and controls belong to an infinite dimensional space, a somewhat more general result is needed. On the other hand, we restrict for simplicity of exposition solely to the case of onto operators.

Pseudoinverses

Let $\mathbb{K} = \mathbb{R}$ or $\mathbb{C}$. Recall that an *inner product space* H over $\mathbb{K}$ is a vector space together with a binary operation $\langle x, y\rangle$ such that $\langle x, y\rangle \in \mathbb{K}$ for all $x, y \in H$ and the following properties hold for all $x, y, z \in H$ and all $\alpha, \beta \in \mathbb{K}$:

(a) $\langle x, y\rangle = \overline{\langle y, x\rangle}$ (overbars denote complex conjugation; thus, in the case $\mathbb{K} = \mathbb{R}$, we are simply saying that inner products are symmetric);

(b) $\langle x, \alpha y + \beta z\rangle = \alpha\langle x, y\rangle + \beta\langle x, z\rangle$; and

(c) $\langle x, x\rangle > 0$ for $x \neq 0$ and $\langle 0, 0\rangle = 0$.

Given such an inner product, one defines an associated norm $\|x\| := \sqrt{\langle x, x\rangle}$, and H becomes a metric space with the distance function $d(x, y) := \|x - y\|$. If H is complete as a metric space with this distance, then H is called a *Hilbert space*.

There are two particular types of Hilbert spaces that we use. For any fixed n, we view $\mathbb{K}^n$ as the set of all column n-vectors with the inner product

$$\langle x, y\rangle := x^* y = \sum_{i=1}^{n} \overline{x}_i y_i .$$

(Here "*" denotes conjugate transpose.) The other type of Hilbert space that we work with is infinite dimensional, and is defined as follows. For each positive integer m, we let $\mathcal{L}^2_m(\sigma,\tau)$ be the set of all square integrable functions $\omega : [\sigma,\tau) \to \mathbb{K}^m$. Such an ω can be thought of as a column m-vector of square integrable functions. The set $\mathcal{L}^2_m(\sigma,\tau)$ is a Hilbert space with the inner product

$$\langle \omega, \nu \rangle := \int_\sigma^\tau \omega(t)^* \nu(t)\, dt\,. \tag{3.13}$$

Let Ω and $\mathcal{X}$ be two Hilbert spaces. Given a continuous linear mapping (in functional-analytic terms, a *bounded operator*)

$$L : \Omega \to \mathcal{X}\,,$$

there exists always an *adjoint* operator L^*, defined by the property that

$$\langle L\omega, x\rangle_{\mathcal{X}} = \langle \omega, L^* x\rangle_\Omega \quad \text{for all } x \in \mathcal{X}, \omega \in \Omega\,.$$

For instance, if $\Omega = \mathbb{K}^m$ and $\mathcal{X} = \mathbb{K}^n$, and L is thought of as an $n \times m$ matrix (with respect to the canonical bases of $\mathbb{K}^m, \mathbb{K}^n$) then L^* is just the conjugate transpose of L.

Example 3.5.1 As another example, consider the case $\Omega = \mathcal{L}^2_m(\sigma,\tau), \mathcal{X} = \mathbb{K}^n$, and

$$L\omega := \int_\sigma^\tau k(t)^* \omega(t)\, dt\,, \tag{3.14}$$

where k is a fixed $m \times n$ matrix of elements of $\mathcal{L}^2(\sigma,\tau)$. Thus, if we denote by k_i the ith column of $k, i = 1,\ldots,n$, then each k_i is in $\mathcal{L}^2_m(\sigma,\tau)$ and

$$L\omega = \begin{pmatrix} \langle k_1, \omega\rangle \\ \vdots \\ \langle k_n, \omega\rangle \end{pmatrix}.$$

We now compute the adjoint of L. Pick any $\omega \in \Omega$ and $x \in \mathcal{X}$. Since $\langle L\omega, x\rangle$ equals

$$\left(\int_\sigma^\tau k(t)^*\omega(t)\, dt\right)^* x = \left(\int_\sigma^\tau \omega(t)^* k(t)\, dt\right) x = \int_\sigma^\tau \omega(t)^*(k(t)x)\, dt\,,$$

we have that L^*x is the element of $\mathcal{L}^2_m$ given by the function $(L^*x)(t) = k(t)x$. □

From now on we assume that $\mathcal{X}$ is finite dimensional. (Most of what we do generalizes easily to the case of arbitrary $\mathcal{X}$ but L of closed range. The latter property is automatic in the finite dimensional case.)

The following Lemma is basic. Here "im" denotes image, "ker" kernel or nullspace, and "$\perp$" indicates orthogonal complement:

$$S^\perp = \{z \mid \langle x, z\rangle = 0 \text{ for all } x \in S\}\,.$$

Since $\mathcal{X}$ was assumed finite dimensional, it holds that $(S^\perp)^\perp = S$ for all subspaces S of $\mathcal{X}$.

Lemma 3.5.2 For any L as above, $\operatorname{im} LL^* = (\ker L^*)^\perp = \operatorname{im} L$.

Proof. It is clear that $\operatorname{im} LL^* \subseteq \operatorname{im} L$. Since "$\perp$" reverses inclusions, it will be enough to prove that

$$(\operatorname{im} LL^*)^\perp \subseteq \ker L^* \subseteq (\operatorname{im} L)^\perp .$$

Pick any $z \in (\operatorname{im} LL^*)^\perp$. Then, $\langle LL^*x, z\rangle = 0$ for all $x \in \mathcal{X}$, so in particular for $x = z$:

$$0 = \langle LL^*z, z\rangle = \langle L^*z, L^*z\rangle = \|L^*z\|^2 ,$$

which implies that $z \in \ker L^*$. Hence also $z \in (\operatorname{im} L)^\perp$, as desired, because for $\omega \in \Omega$, $\langle L\omega, z\rangle = \langle \omega, L^*z\rangle = 0$. ∎

We let

$$W : \mathcal{X} \to \mathcal{X},\ W := LL^* . \tag{3.15}$$

When $\mathcal{X} = \mathbb{K}^n$, we identify W with an $n \times n$ matrix. Note that

$$W^* = (LL^*)^* = L^{**}L^* = LL^* = W , \tag{3.16}$$

i.e., W is *self-adjoint.* (When $\mathcal{X} = \mathbb{K}^n$, W is a Hermitian matrix.) Moreover, W is positive semidefinite, because

$$\langle x, Wx\rangle = \|L^*x\|^2$$

for all x. Since $\mathcal{X}$ is finite dimensional, the following are equivalent: W is onto, one-to-one, invertible, positive definite. We can then conclude as follows.

Corollary 3.5.3 The following statements are equivalent for L, W as above:

(a) L is onto.

(b) L^* is one-to-one.

(c) W is onto.

(d) $\det W \neq 0$.

(e) W is positive definite. □

Consider again the situation in Example 3.5.1. Here L is onto iff the matrix

$$W = \int_\sigma^\tau k(t)^* k(t)\, dt > 0 . \tag{3.17}$$

Equivalently, L is onto iff L^* is one-to-one, i.e.,

$$\text{there is no } p \neq 0 \text{ in } \mathcal{X} \text{ with } k(t)p = 0 \text{ for almost all } t \in [\sigma, \tau) , \tag{3.18}$$

or, with a slight rewrite and $k_i :=$ ith column of k^*:

$$\langle p, k_i\rangle = 0 \text{ for all } i \text{ and almost all } t \Rightarrow p = 0 . \tag{3.19}$$

Proposition 3.5.4 Assume that L is onto, and let

$$L^{\#} := L^*(LL^*)^{-1} : \mathcal{X} \to \Omega . \tag{3.20}$$

Then, $L^{\#}x$ is the unique solution of $L\omega = x$ of smallest possible norm, i.e.,

- $L(L^{\#}x) = x$ for all $x \in \mathcal{X}$, and
- $\left\|L^{\#}x\right\| < \|\omega\|$ for every ω for which $L\omega = x$ and $\omega \neq L^{\#}x$.

Proof. Since

$$LL^{\#} = (LL^*)(LL^*)^{-1} = \text{identity} ,$$

the first property is clear. Assume now that $L\omega = x$. We will prove that $\langle \omega - L^{\#}x, L^{\#}x\rangle = 0$, so

$$\|\omega\|^2 = \left\|\omega - L^{\#}x\right\|^2 + \left\|L^{\#}x\right\|^2$$

will give the desired conclusion. But

$$\langle \omega - L^{\#}x, L^{\#}x\rangle = \langle \omega, L^{\#}x\rangle - \left\|L^{\#}x\right\|^2 ,$$

so we prove that the two terms on the right are equal:

$$\begin{aligned}
\langle \omega, L^{\#}x\rangle &= \langle \omega, L^*(LL^*)^{-1}x\rangle = \langle L\omega, (LL^*)^{-1}x\rangle = \langle x, (LL^*)^{-1}x\rangle \\
&= \langle LL^*(LL^*)^{-1}x, (LL^*)^{-1}x\rangle = \langle L^*(LL^*)^{-1}x, L^*(LL^*)^{-1}x\rangle \\
&= \left\|L^{\#}x\right\|^2 .
\end{aligned}$$

This completes the proof. ∎

The operator $L^{\#}$ is the *generalized inverse* or *(Moore-Penrose) pseudoinverse* of L. When L is not onto, one may define an operator with the minimizing property in the above result restated in a least-squares sense; we omit the details here since only the case when L is onto will be needed (see Appendix A.2 for the finite dimensional case).

In Example 3.5.1,

$$(L^{\#}x)(t) = k(t)\left[\int_\sigma^\tau k(s)^*k(s)\,ds\right]^{-1} x .$$

Note that if the entries of k happen to be essentially bounded (not just square integrable), then $L^{\#}x$ is also essentially bounded.

It is easy to compute the operator norm of $L^{\#}$ in terms of the matrix W. We are defining the operator norm as

$$\left\|L^{\#}\right\| := \sup_{\|x\|=1} \left\|L^{\#}x\right\|$$

with respect to the $\mathcal{L}^2_m$ norm. From the definitions (3.15) and (3.20), it follows that, for each vector x,

$$\begin{aligned}\left\|L^{\#}x\right\|^2 &= \left\|L^*W^{-1}x\right\|^2 \\ &= \langle L^*W^{-1}x, L^*W^{-1}x\rangle \\ &= \langle LL^*W^{-1}x, W^{-1}x\rangle \\ &= \langle x, W^{-1}x\rangle .\end{aligned}$$

Therefore,

$$\left\|L^{\#}\right\| = \left\|W^{-1}\right\|^{1/2} ; \tag{3.21}$$

that is, the norm of the pseudoinverse is the square root of $1/\sigma_{\min}$, where $\sigma_{\min}$ is the smallest eigenvalue of the positive definite matrix W.

Application to Controllability

Recall from Lemma 3.1.7(a) that the continuous-time linear system Σ is controllable on $[\sigma, \tau]$ iff it holds that $(0, \sigma) \rightsquigarrow (y, \tau)$ for all $y \in \mathcal{X}$. Thus, controllability means precisely that the following operator is onto:

$$N : \mathcal{L}^{\infty}_m(\sigma,\tau) \to \mathbb{K}^n : \omega \mapsto \phi(\tau,\sigma,0,\omega) = \int_{\sigma}^{\tau} \Phi(\tau,s)B(s)\omega(s)\, ds .$$

Thus, the extension of N to $\mathcal{L}^2_m$, with $k(s) := B(s)^*\Phi(\tau,s)^*$:

$$L : \mathcal{L}^2_m(\sigma,\tau) \to \mathbb{K}^n : \omega \mapsto \int_{\sigma}^{\tau} k(s)^*\omega(s)\, ds , \tag{3.22}$$

is also onto. Note that entries of B are essentially bounded, by the definition of a continuous-time linear system, and that the entries of $\Phi(\tau, s)$ are (absolutely) continuous in s. Thus, each entry of k is in $\mathcal{L}^{\infty}_m(\sigma,\tau)$ and hence also in $\mathcal{L}^2_m(\sigma,\tau)$, so L is an operator as in Example 3.5.1. (When considering the extension L of N, we really are dealing with the system Σ "with $\mathcal{L}^2$ controls" in the sense of Exercise 2.7.5, but we do not need to use this fact explicitly.)

Given any Σ, we consider the operator L defined by formula (3.22). If Σ is controllable, L is onto. Conversely, assume that L is onto. Then for each $x \in \mathcal{X}$, $\omega := L^{\#}x$ is such that $L\omega = x$. Since the entries of k are essentially bounded, it follows that ω is also essentially bounded. So ω is in the domain of N, and hence x is in the image of N. We conclude that L is onto iff N is, that is, iff Σ is controllable. (Alternatively, we could argue that L is onto iff N is, using the fact that $\mathcal{L}^{\infty}_m$ is dense in $\mathcal{L}^2_m$ and that both N and L are continuous.) We apply the above results to this L, to obtain the following conclusion. For the last part, recall that

$$\phi(\tau,\sigma,x,\omega) = z \;\text{ iff }\; z - \phi(\tau,\sigma,x,\mathbf{0}) = N\omega .$$

Theorem 5 *Assume that Σ is a continuous-time linear system, $\sigma < \tau \in \mathbb{R}$, and denote by b_i the ith column of B. The following statements are equivalent:*

1. *Σ is controllable in $[\sigma, \tau]$;*
2. *$W(\sigma, \tau) = \int_\sigma^\tau \Phi(\tau, s)B(s)B(s)^*\Phi(\tau, s)^*\, ds\ > 0$;*
3. *There is no nonzero vector $p \in \mathcal{X}$ such that $\langle p, \Phi(\tau, s)b_i(s)\rangle = 0$ for almost all $s \in [\sigma, \tau]$ and all i.*

If the above are satisfied, then, given any two $x, z \in \mathcal{X}$, the control given by the formula

$$\omega(t) = B(t)^*\Phi(\tau, t)^*W(\sigma, \tau)^{-1}(z - \Phi(\tau, \sigma)x) \tag{3.23}$$

is the unique one that minimizes $\|\omega\|$ among all those controls satisfying

$$\phi(\tau, \sigma, x, \omega) \;=\; z\,. \qquad \square$$

Exercise 3.5.5 Let Σ be a continuous-time linear system, and pick $\sigma < \tau \in \mathbb{R}$. Consider the *controllability Gramian*

$$W_c(\sigma, \tau) \;:=\; \int_\sigma^\tau \Phi(\sigma, s)B(s)B(s)^*\Phi(\sigma, s)^*\, ds\,.$$

Show: Σ is controllable in $[\sigma, \tau]$ if and only if $W_c(\sigma, \tau)$ has rank n, and, in that case, the unique control of minimum square norm that steers x to 0 is given by the formula $\omega(t) = -B(t)^*\Phi(\sigma, t)^*W_c(\sigma, \tau)^{-1}x$. $\square$

Example 3.5.6 Consider the harmonic oscillator (linearized pendulum)

$$\begin{aligned} \dot{x}_1 &= x_2 \\ \dot{x}_2 &= -x_1 + u\,. \end{aligned}$$

Assume that we wish to find a control transferring

$$x = \begin{pmatrix} 1 \\ 0 \end{pmatrix}$$

to $z = 0$ in time 2π while minimizing the "energy"

$$\int_0^{2\pi} u(t)^2 dt\,.$$

We apply the Theorem. First note that, using e^{tA} as computed in Example 3.4.2, there results

$$\begin{aligned} W &= \int_0^{2\pi} e^{(2\pi - s)A}\begin{pmatrix} 0 \\ 1 \end{pmatrix}(0\ 1)e^{(2\pi - s)A^*}\, ds \\ &= \int_0^{2\pi} \begin{pmatrix} \sin^2 s & -\sin s\cos s \\ -\sin s\cos s & \cos^2 s \end{pmatrix} ds \\ &= \pi I\,, \end{aligned}$$

so

$$W^{-1}(z - \Phi(2\pi, 0)x) = (-1/\pi)\begin{pmatrix}1\\0\end{pmatrix}.$$

Therefore,

$$\omega(t) = -\frac{1}{\pi}(0\ 1)\begin{pmatrix}\cos(2\pi - t) & -\sin(2\pi - t)\\ \sin(2\pi - t) & \cos(2\pi - t)\end{pmatrix}\begin{pmatrix}1\\0\end{pmatrix} = \frac{1}{\pi}\sin t$$

gives the desired control. □

The above Theorem provides to some extent a computable criterion for controllability in $[\sigma, \tau]$. One may obtain numerically the fundamental solution and then perform the required integration to get the constant matrix $W(\sigma, \tau)$, whose determinant is then checked to be nonzero. Below we develop a criterion, applicable when the entries of A and B are smooth as functions of time, that does not require integration of the differential equation. In practice, of course, round-off errors will make the determinant of $W(\sigma, \tau)$ always nonzero. Thus, controllability cannot be decided precisely. Furthermore, if the determinant of W turns out to be small, even if the system is controllable, it is not so in a practical sense, since the necessary controls will tend to be very large, as suggested by the term W^{-1} that appears in the expression of the minimum norm control. This is related to the fact that even though most systems are controllable (at least in the time-invariant case), many may be "close" to being uncontrollable; recall the discussion after the proof of Proposition 3.3.12. It is also related to the fact that trying to control in too short an interval will require large controls, no matter how controllable the system is (since the integral defining W is taken over a small interval). The next exercise deals with a particular case of an estimate of how large such controls must be.

Exercise 3.5.7 For the system

$$\begin{aligned}\dot{x}_1 &= x_2\\ \dot{x}_2 &= u\end{aligned}$$

show that the operator $L^{\#}$ corresponding to $\sigma = 0$, $\tau = \varepsilon$ satisfies

$$\|L^{\#}\| = O(\varepsilon^{-\frac{3}{2}})$$

as $\varepsilon \to 0$. (*Hint:* Use (3.21) by showing that the smallest eigenvalue of $W(0, \varepsilon)$ is of the form

$$\frac{\varepsilon^3}{12} + o(\varepsilon^3).$$

The power series expansion for $\sqrt{1 + \alpha}$ may be useful here.) □

We may restate condition (3) in a slightly different form, which will be useful for our later study of observability. Transposing the conclusion (d) of Lemma C.4.1, one has the following:

Lemma 3.5.8 For any $A(\cdot)$, the fundamental solution $\Psi(t,s)$ associated to the adjoint equation

$$\dot{p}(t) = -A(t)^* p(t) \tag{3.24}$$

is $\Psi(t,s) := \Phi(s,t)^*$. □

Corollary 3.5.9 The continuous-time linear system Σ is controllable on $[\sigma,\tau]$ if and only if there does not exist a nonzero solution $p(t)$ of the adjoint equation on $[\sigma,\tau]$ such that

$$\langle p(t), b_i(t)\rangle = 0 \quad \text{for almost all } t \in [\sigma,\tau]$$

for all $i = 1,\ldots,m$ (recall that b_i is the ith column of B).

Proof. We show that the existence of such a solution would be equivalent to the property in (3) of Theorem 5 (p. 109). For this, note simply that if p is a solution of the adjoint equation on $[\sigma,\tau]$ then $p(t) = \Psi(t,\tau)p(\tau) = \Phi(\tau,t)^* p(\tau)$, so

$$\langle p(t), b\rangle = \langle p(\tau), \Phi(\tau,t)b\rangle$$

for any vector b, and that $p(t)$ is nonzero for all t if and only if $p(\tau) \neq 0$. ∎

The last part of Theorem 5 provides a solution to a particular type of optimal control problem, that of minimizing the norm of the control that affects a given state transfer. In applications, this norm is typically proportional to some measure of the energy used in controlling the system. There is a variation of this problem that can be solved with precisely the same techniques. Assume that the coordinates of the control u are assigned different costs. For instance, we may want to consider a situation in which there are two independent controls and it is 10 times more expensive to apply a unit of the first control as it is of the second. In that case, it would make sense to try to minimize the cost $(10u_1)^2 + u_2^2 = 100u_1^2 + u_2^2$. More generally, we may have a cost of the type $\sum_i \sum_j a_{ij} u_i^2 u_j^2$:

Exercise 3.5.10 Let Σ be a controllable continuous-time linear system, and let Q be a real symmetric positive definite $m \times m$ matrix. Pick any $x, z \in \mathcal{X}$, and $\sigma, \tau \in \mathbb{R}$. Find a formula for a control $\omega \in \mathcal{L}_m^\infty(\sigma,\tau)$ which gives $\phi(\tau,\sigma,x,\omega) = z$ while minimizing

$$\int_\sigma^\tau \omega(s)^* Q\omega(s)\, ds\,.$$

Prove that for each pair x, z there is a unique such control. Do this in two alternative ways: (1) Applying again the material about pseudoinverses, but using a different inner product in the set $\mathcal{L}_m^2$. (2) Factoring $Q = Q_1^* Q_1$ and observing that the same result is obtained after a change of variables. □

Remark 3.5.11 Optimal control problems for linear systems with quadratic cost result typically in solutions that are linear in a suitable sense. For instance, the optimal control in Theorem 5 (p. 109) is linear in both x and z. This is not surprising, since in an abstract sense minimizing a quadratic function (here, the norm of the control) subject to linear constraints (the linear differential equation) can be thought of, after elimination, as an unconstrained minimization of a quadratic function, which in turn can be solved by setting derivatives to zero; since derivatives of quadratic functions are linear, the solution to the original problem becomes linear in the original data. (This kind of reasoning can be made precise through the study of abstract quadratic optimization problems in infinite dimensional spaces.) The linearity property means that in general, linear systems problems with quadratic costs are far easier to solve than, say, linear problems with nonquadratic costs or nonlinear control problems with quadratic costs. Chapter 8 studies other linear-quadratic problems, in which the final state is not constrained and costs are imposed along trajectories. □

Remark 3.5.12 Theorem 5 (p. 109) shows that among all controls effecting the transfer $(x,\sigma) \rightsquigarrow (z,\tau)$ there is a unique one of minimum norm. Of course, if one does not impose the minimality requirement, there are many such controls, since N maps an infinite dimensional space into a finite dimensional one. Many dense subsets of $\mathcal{L}^{\infty}_{m}$ will map onto all of the state space. For instance, there are always in the controllable case piecewise constant controls so that $(x,\sigma) \rightsquigarrow (z,\tau)$. (See more about this in Section 3.9.)

Observe also that the explicit formula for the minimum norm control shows that this control is of class $\mathcal{C}^k$ provided that $A(\cdot)$ and $B(\cdot)$ are of class $\mathcal{C}^k$. □

Exercise 3.5.13 Let Σ be a discrete-time linear system with $\mathcal{U} = \mathbb{K}^m, \mathcal{X} = \mathbb{K}^n, \mathbb{K} = \mathbb{R}$ or $\mathbb{C}$. State and prove the analogue of Theorem 5 (p. 109) for Σ. (*Hint:* Instead of $\Phi(\sigma,\tau)$ use now the matrix product

$$A(\tau-1)A(\tau-2)\ldots A(\sigma)\,.$$

Let $\Omega := \mathbb{K}^{mT}, T = \tau - \sigma$, and apply the pseudoinverse results again.) □

Exercise 3.5.14 For this problem, we call a time-invariant system *output controllable* if it holds that for each $x \in \mathcal{X}$ and each $y \in \mathcal{Y}$ there exists a $T \geq 0$ and a control u such that $h(\phi(T,x,u)) = y$. Prove that a time-invariant continuous-time linear system (A, B, C) is output controllable iff

$$\operatorname{rank} C\mathbf{R}(A,B) = \operatorname{rank}\,[CB,\ldots,CA^{n-1}B] = p.$$ □

Exercise 3.5.15 (Optional, computer-based.) In Exercise 3.2.12, calculate numerically, using the pseudoinverse formula, controls of minimal energy transferring one given state to another in time $T = 1$. (Unrealistically, but for mathematical simplicity, take all constants equal to one.) □

A Rank Condition

The characterization of controllability given in Theorem 5 involves integration of the differential equation of the system. We now look for a simpler condition, analogous to the Kalman rank condition for time-invariant systems.

If Σ is a continuous-time linear system and $\mathcal{I}$ is an interval in $\mathbb{R}$, we say that Σ is **smoothly-varying on** $\mathcal{I}$ iff $A(t)$ and $B(t)$ are smooth (infinitely differentiable) as functions of t for $t \in \mathcal{I}$, and that Σ is **analytically varying on** $\mathcal{I}$ iff $A(t)$ and $B(t)$ also are analytic on $\mathcal{I}$. Analyticity means that for each $t \in \mathcal{I}$ each entry of A and B can be expanded in a power series about t, convergent in some nontrivial interval $(t-\varepsilon, t+\varepsilon)$ (see Appendix C).

The matrix function $\Phi(s,t)$ is smooth (respectively, analytic,) as a function of $t \in \mathcal{I}$ and $s \in \mathcal{I}$ when Σ is smoothly (respectively, analytically) varying on $\mathcal{I}$. Fix Σ and $\mathcal{I}$ so that Σ is smoothly varying on $\mathcal{I}$, and fix a nontrivial subinterval $[\sigma, \tau] \subseteq \mathcal{I}$. Introduce the (smooth) $n \times m$ matrix function

$$M_0(t) := \Phi(\tau, t)B(t),$$

and let

$$M_k(t) := (d^k M_0/dt^k)(t), \quad k \geq 1.$$

Consider also the matrix of functions obtained by listing all the columns of the $M_i, i = 0, \ldots, k$:

$$M^{(k)}(t) := (M_0(t), M_1(t), \ldots, M_k(t)).$$

Proposition 3.5.16 Let Σ be a continuous-time linear system.

1. Assume that Σ is smoothly varying on $\mathcal{I}$, and pick any $\sigma < \tau$ so that $[\sigma, \tau] \subseteq \mathcal{I}$. If there exists a $t_0 \in [\sigma, \tau]$ and a nonnegative integer k such that
$$\operatorname{rank} M^{(k)}(t_0) = n,$$
then Σ is controllable.

2. Assume now that Σ is analytically varying on $\mathcal{I}$, and let t_0 be any fixed element of $\mathcal{I}$. Then, Σ is controllable on every nontrivial subinterval of $\mathcal{I}$ if and only if
$$\operatorname{rank} M^{(k)}(t_0) = n$$
for some integer k.

Proof. If Σ were not controllable on $[\sigma, \tau]$, there would exist by condition (3) in Theorem 5 (p. 109) a nonzero vector p such that $p^* M_0(t) = 0$ for almost all $t \in [\sigma, \tau]$, and hence by continuity of M_0, for all such t. It follows that all derivatives are also identically zero, that is,

$$p^* M_k(t) \equiv 0$$

for all $t \in [\sigma, \tau]$ and all $k \geq 0$. Thus, the rows of each matrix $M^{(k)}(t)$ are linearly dependent, so $\operatorname{rank} M^{(k)}(t) < n$ for all $t \in [\sigma, \tau]$. This proves the first statement.

We now prove sufficiency in the analytic case. If Σ were not controllable on $[\sigma, \tau]$, then, as in the smooth case, $p^* M_k(t) \equiv 0$ on $[\sigma, \tau]$, for all k. But then, by the principle of analytic continuation, also

$$p^* M_k(t) \equiv 0 \text{ for all } t \in \mathcal{I}.$$

Thus, if the rank is n for some $t \in \mathcal{I}$ and some k, we obtain a contradiction.

Conversely, suppose now that Σ is controllable and analytically varying on $\mathcal{I}$. Assume that there is some $t_0 \in \mathcal{I}$ such that

$$\operatorname{rank} M^{(k)}(t_0) < n$$

for all k. For this t_0, let

$$P_k := \{p \in \mathbb{K}^n | p^* M^{(k)}(t_0) = 0\}.$$

By assumption, the subspaces P_k are all nonzero. Since

$$P_0 \supseteq P_1 \supseteq \ldots,$$

the dimensions of the P_k are nonincreasing. Let P_K have minimal dimension; then

$$P_k = P_K \text{ for } k \geq K.$$

Pick $p \in P_K$. It follows that $p^* M_k(t_0) = 0$ for all k. By analyticity, $p^* M_0(t) \equiv 0$ for t in an interval about t_0 (expand in a power series about t_0), and so, by analytic continuation, this holds on all of $\mathcal{I}$ and in particular on all of $[\sigma, \tau]$. This would contradict the assumed controllability of Σ. ∎

The above result can be restated somewhat more elegantly in terms of the infinite matrix

$$M(t) := (M_0(t), M_1(t), \ldots, M_k(t), \ldots).$$

The system Σ is controllable if $M(t_0)$ has rank n for some t_0, and in the analytic case this condition is also necessary for every t_0.

The condition would seem still to involve integration, since the transition matrix Φ is used in defining the M_k's. But we can obtain an equivalent condition as follows: Let $B_0 := B$ and, for each $i \geq 1$,

$$B_{i+1}(t) := A(t)B_i(t) - \frac{d}{dt}B_i(t).$$

Note that the B_i can be obtained directly from the data (A, B).

Lemma 3.5.17 For all i and all $t \in \mathcal{I}$, $M_i(t) = (-1)^i \Phi(\tau, t) B_i(t)$.

Proof. By induction. The case $i = 0$ holds by definition. Assume the Lemma proved for i, and consider M_{i+1}. Then,

$$\begin{aligned} M_{i+1}(t) &= \frac{d}{dt}M_i(t) = (-1)^i \frac{d}{dt}[\Phi(\tau,t)B_i(t)] \\ &= (-1)^i \left\{ \left[\frac{\partial}{\partial t}\Phi(\tau,t)\right] B_i(t) + \Phi(\tau,t)\frac{d}{dt}B_i(t) \right\} \\ &= (-1)^i \left\{ -\Phi(\tau,t)A(t)B_i(t) + \Phi(\tau,t)\frac{d}{dt}B_i(t) \right\}, \end{aligned}$$

where the last equality holds by Lemma C.4.1(d). But the last term is equal to $(-1)^{i+1}\Phi(\tau,t)B_{i+1}(t)$, as desired. ∎

Since $\Phi(\tau,t)$ is invertible for all τ, t, $p^*M_i(t_0) = 0$ if and only if $q^*B_i(t) = 0$, with $q = \Phi(\tau,t)^*p$. We conclude:

Corollary 3.5.18 The conclusions of Proposition 3.5.16 hold when

$$\operatorname{rank}\,[B_0(t_0), B_1(t_0), \ldots, B_k(t_0)] = n \tag{3.25}$$

holds instead of the rank condition on $M^{(k)}(t_0)$. □

Remark 3.5.19 In Proposition 3.5.16 and Corollary 3.5.18, infinite differentiability is not needed for the necessity statements, in the following sense: If $A(\cdot)$ and $B(\cdot)$ are k-times differentiable and if $M^{(k)}(t_0)$ or (3.25) have rank n at some t_0, then controllability already follows by the same argument. □

Exercise 3.5.20 Show by counterexample that without analyticity the rank condition is not necessary. More precisely, give two examples as follows:

(i) A smooth system with $n = m = 1$ that is controllable in some nontrivial interval but for which there is some t_0 so that $M^{(k)}(t_0) \equiv 0$ for all k; and

(ii) A smooth system with $n = m = 2$ that is controllable in some nontrivial interval but for which the rank is one for *every* t_0 and k. □

Example 3.5.21 Consider the system with $\mathbb{K} = \mathbb{R}$, $n = 3$, $m = 1$, and matrices

$$A(t) = \begin{pmatrix} t & 1 & 0 \\ 0 & t^3 & 0 \\ 0 & 0 & t^2 \end{pmatrix} \quad B(t) = \begin{pmatrix} 0 \\ 1 \\ 1 \end{pmatrix}.$$

This system is smoothly (in fact, analytically) varying on $(-\infty, \infty)$. Since

$$[B_0(0), B_1(0), B_2(0), B_3(0)] = \begin{pmatrix} 0 & 1 & 0 & -1 \\ 1 & 0 & 0 & 0 \\ 1 & 0 & 0 & 2 \end{pmatrix},$$

and this matrix has rank 3, the system is controllable on every nontrivial interval $[\sigma, \tau]$. □

Remark 3.5.22 In the time-invariant case, $B_i(t) \equiv A^{i-1}B$, so the above condition says that Σ is controllable on any nontrivial interval $[\sigma,\tau]$ if and only if the columns of $(B, AB, A^2B, \ldots)$ span an n-dimensional space, which as remarked in Lemma 3.3.2 is equivalent to the condition $\operatorname{rank}\mathbf{R}(A,B) = n$. This provides yet another proof of Theorem 3 (p. 89). □

In general, for any fixed t_0, it may be necessary to check the rank of the matrix in Corollary 3.5.18 for arbitrarily large k. As an illustration, consider the analytic system

$$\dot{x} = t^k u\,,$$

where k is any integer. This is controllable on any nontrivial interval (just check rank condition at 1, for instance), but with $t_0 = 0$ we would have to check up to the kth matrix. For *generic* t_0, $k = n$ is enough, however, as shown in the following problem. (So, even though there is no "Cayley-Hamilton Theorem" to apply, as was the case for time-invariant systems, the situation is still good enough for almost every t_0.)

Exercise 3.5.23 Let Σ be a continuous-time linear system, analytically varying on $\mathcal{I}$. Prove that if Σ is controllable on any nontrivial subinterval $[\sigma,\tau]$ then

$$\operatorname{rank}\left[B_0(t), B_1(t), \ldots, B_{n-1}(t)\right] = n$$

for almost all $t \in \mathcal{I}$. (*Hint:* First prove that if $\operatorname{rank} M^{(k)}(t) = \operatorname{rank} M^{(k+1)}(t)$ for t in an open interval $J \subseteq \mathcal{I}$, then there must exist another subinterval $J' \subseteq J$ and analytic matrix functions

$$V_0(t), \ldots, V_k(t)$$

on J' such that

$$M_{k+1}(t) = \sum_{i=0}^{k} M_i(t)V_i(t)$$

on J'. Conclude that then $\operatorname{rank} M^{(k)}(t) = \operatorname{rank} M^{(l)}(t)$ for all $l > k$ on J'. Argue now in terms of the sequence $n_k := \max\{\operatorname{rank} M^{(k)}(t), t \in \mathcal{I}\}$.) □

Exercise 3.5.24 Consider the continuous-time linear system over $\mathbb{K} = \mathbb{R}$ with $n = 2, m = 1$, and matrices

$$A(t) = \begin{pmatrix} 0 & 1 \\ -1 & 0 \end{pmatrix} \quad B(t) = \begin{pmatrix} \cos t \\ -\sin t \end{pmatrix}.$$

Prove that the system is *not* controllable over any fixed interval, but that for every fixed σ in $\mathbb{R}$, the ("frozen") time-invariant linear system $(A(\sigma), B(\sigma))$ is controllable. Show noncontrollability of the time-varying system in two different ways:

(i) Via the rank test; and

(ii) Alternatively by explicitly calculating solutions and noticing that the reachable set from the origin at any fixed instant is a line. □

Exercise 3.5.25 Refer to Exercise 3.2.12. Assume here that the ground speed is not constant but is instead a function of time, $c = c(t)$ (take c to be smooth, though much less is needed). Apply the rank criteria for time-varying linear systems to conclude that the system is controllable in any nontrivial interval in which c is not identically zero. □

3.6 Bounded Controls*

In this section, we deal with continuous-time time-invariant systems of the form $\dot{x} = Ax + Bu$ for which $\mathcal{U}$ is a subset of $\mathbb{R}^m$. Linear systems have, by definition, $\mathcal{U} = \mathbb{R}^m$, so we call such systems "linear systems with constrained controls". For instance, take the system $\dot{x} = -x + u$ ($n = m = 1$), with $\mathcal{U} = (-1, 1)$. The pair $(A, B) = (1, 1)$ is controllable, but the system with restricted controls is not, since it is impossible to transfer the state $x = 0$ to $z = 2$ (since $\dot{x}(t) < 0$ whenever $x(t) \in (1, 2)$).

In order to avoid confusion with reachability for the associated linear systems with unconstrained controls, in this section we will say that $z \in \mathbb{R}^n$ can be *$\mathcal{U}$-reached* from $x \in \mathbb{R}^n$ (or that *x can be $\mathcal{U}$-controlled to z*) *in time T* if there is some input $\omega : [0, T] \to \mathbb{R}^m$ so that $\phi(T, 0, x, \omega) = z$ and $\omega(t) \in \mathcal{U}$ for (almost) all $t \in [0, T]$. We define the reachable set in time $T \geq 0$

$$\mathcal{R}^T_{\mathcal{U}}(x) := \{z \in \mathbb{R}^n \ \mathcal{U}\text{-reachable from } x \text{ in time } T\}$$

and $\mathcal{R}_{\mathcal{U}}(x) := \bigcup_{T \geq 0} \mathcal{R}^T_{\mathcal{U}}(x)$. (Thus $\mathcal{R}_{\mathbb{R}^m}(x)$ is the same as what we earlier called $\mathcal{R}(x)$.)

In this section we will establish the following result.

Theorem 6 *Let $\mathcal{U}$ be a bounded neighborhood of zero. Then, $\mathcal{R}_{\mathcal{U}}(0) = \mathbb{R}^n$ if and only if*

(a) the pair (A, B) is controllable, and

(b) the matrix A has no eigenvalues with negative real part.

Observe that the necessity of controllability for the pair (A, B) is obvious, since $\mathcal{R}_{\mathcal{U}}(0) \subseteq \mathcal{R}(0)$.

We prove Theorem 6 after a series of preliminary results.

Lemma 3.6.1 Let $\mathcal{U} \subseteq \mathbb{R}^m$ and pick any two $S, T \geq 0$. Then

$$\mathcal{R}^T_{\mathcal{U}}(0) + e^{TA}\mathcal{R}^S_{\mathcal{U}}(0) = \mathcal{R}^{S+T}_{\mathcal{U}}(0)\,.$$

* This section can be skipped with no loss of continuity.

Proof. Pick

$$x_1 = \int_0^T e^{(T-\tau)A} B\omega_1(\tau)d\tau = \int_S^{S+T} e^{(S+T-\tau)A} B\omega_1(\tau - S)d\tau$$

and

$$x_2 = \int_0^S e^{(S-\tau)A} B\omega_2(\tau)d\tau$$

with the inputs ω_i $\mathcal{U}$-valued. Note that

$$e^{TA}x_2 = \int_0^S e^{(S+T-\tau)A} B\omega_2(\tau)d\tau \,.$$

Thus

$$x_1 + e^{TA}x_2 = \int_0^{S+T} e^{(S+T-\tau)A} B\omega(\tau)d\tau$$

where

$$\omega(s) = \begin{cases} \omega_2(\tau) & 0 \le \tau \le S \\ \omega_1(\tau - S) & S \le \tau \le S+T. \end{cases}$$

Note that $\omega(t) \in \mathcal{U}$ for all $t \in [0, S+T]$. Thus $\mathcal{R}_{\mathcal{U}}^T(0) + e^{TA}\mathcal{R}_{\mathcal{U}}^S(0) \subseteq \mathcal{R}_{\mathcal{U}}^{S+T}(0)$. The converse inclusion follows by reversing these steps. ∎

By induction on q we then conclude:

Corollary 3.6.2 Let $\mathcal{U} \subseteq \mathbb{R}^m$ and pick any $T \ge 0$ and any integer $q \ge 1$. Then

$$\mathcal{R}_{\mathcal{U}}^T(0) + e^{TA}\mathcal{R}_{\mathcal{U}}^T(0) + \ldots + e^{(q-1)TA}\mathcal{R}_{\mathcal{U}}^T(0) = \mathcal{R}_{\mathcal{U}}^{qT}(0) \,.$$

Proposition 3.6.3 (1) If $\mathcal{U} \subseteq \mathbb{R}^m$ is convex, then $\mathcal{R}_{\mathcal{U}}(0)$ is a convex subset of $\mathbb{R}^n$. (2) If (A, B) is controllable and $\mathcal{U} \subseteq \mathbb{R}^m$ is a neighborhood of $0 \in \mathbb{R}^m$, then $\mathcal{R}_{\mathcal{U}}(0)$ is an open subset of $\mathbb{R}^n$.

Proof. Convexity of each $\mathcal{R}_{\mathcal{U}}^T(0)$ follows from linearity of $\phi(T, 0, 0, u)$ on u and convexity of $\mathcal{U}$. This proves that the (increasing) union $\mathcal{R}_{\mathcal{U}}(0)$ is convex, when $\mathcal{U}$ is convex.

Assume now that (A, B) is controllable and $\mathcal{U} \subseteq \mathbb{R}^m$ is a neighborhood of 0. We first prove that, for each $T > 0$, $\mathcal{R}_{\mathcal{U}}^T(0)$ is a neighborhood of $0 \in \mathbb{R}^n$. Fix such a T. Pick a subset $\mathcal{U}_0 \subseteq \mathcal{U}$ which is a convex neighborhood of 0. The desired conclusion will follow if we show that $0 \in \mathbb{R}^n$ is in the interior of $\mathcal{R}_{\mathcal{U}_0}^T(0)$, since $\mathcal{R}_{\mathcal{U}_0}^T(0) \subseteq \mathcal{R}_{\mathcal{U}}^T(0)$. So without loss of generality, for the rest of this paragraph we replace $\mathcal{U}$ by $\mathcal{U}_0$ and hence assume that $\mathcal{U}$ is also convex. Pick any basis $e_1, \ldots, e_n$ of $\mathbb{R}^n$. Let $e_0 := -\sum_{i=1}^n e_i$. For each $i = 0, \ldots, n$ there is an input ω_i, not necessarily $\mathcal{U}$-valued, so that

$$e_i = \phi(T, 0, 0, \omega_i) \,.$$

Let $\mu > 0$ be so that, with $\omega_i' := \frac{1}{\mu}\omega_i$, $i = 0, \ldots, n$, it holds for all i that $\omega_i'(t) \in \mathcal{U}$ for (almost) all $t \in [0, T]$. (There exists some such μ because $\mathcal{U}$ is a neighborhood of 0 and the ω_i are essentially bounded.) Thus $e_i' := \frac{1}{\mu}e_i = \phi(T, 0, 0, \omega_i') \in \mathcal{R}_{\mathcal{U}}^T(0)$ for each i.

Pick any $\varepsilon_1, \ldots, \varepsilon_n$ such that $|\varepsilon_i| \leq \frac{1}{2(n+1)}$ for all i. Then

$$\frac{\varepsilon_1}{\mu}e_1 + \ldots + \frac{\varepsilon_n}{\mu}e_n = \sum_{i=1}^{n} \left(\frac{1-\varepsilon}{n+1} + \varepsilon_i\right) e_i' + \frac{1-\varepsilon}{n+1}e_0'$$

with $\varepsilon = \sum_i \varepsilon_i$. This is a convex combination. Since all $e_i' \in \mathcal{R}_{\mathcal{U}}^T(0)$, and this set is convex, it follows that $\frac{\varepsilon_1}{\mu}e_1 + \ldots + \frac{\varepsilon_n}{\mu}e_n \in \mathcal{R}_{\mathcal{U}}^T(0)$ for all small enough $\varepsilon_1, \ldots, \varepsilon_n$, which shows that $\mathcal{R}_{\mathcal{U}}^T(0)$ is a neighborhood of 0.

Finally, we show that $\mathcal{R}_{\mathcal{U}}(0)$ is open. Pick any $S > 0$. By the previous discussion, there is some open subset $V \subseteq \mathcal{R}_{\mathcal{U}}^S(0)$ containing zero. Pick any $x \in \mathcal{R}_{\mathcal{U}}(0)$; we wish to show that some neighborhood of x is included in $\mathcal{R}_{\mathcal{U}}(0)$. Let $\omega : [0, T] \to \mathcal{U}$ be so that $x = \phi(T, 0, 0, \omega)$. The set $W := e^{TA}V$ is open, because e^{TA} is nonsingular. For each $y = e^{TA}v \in W$,

$$y + x = e^{TA}v + \phi(T, 0, 0, \omega) \in \mathcal{R}_{\mathcal{U}}^T(V).$$

Thus $x + W$ is an open subset of $\mathcal{R}_{\mathcal{U}}^T(V) \subseteq \mathcal{R}_{\mathcal{U}}^{S+T}(0) \subseteq \mathcal{R}_{\mathcal{U}}(0)$ which contains x. ∎

For each eigenvalue λ of A and each positive integer k we let

$$J_{k,\lambda} := \ker(\lambda I - A)^k$$

(a subspace of $\mathbb{C}^n$) and the set of real parts

$$J_{k,\lambda}^{\mathbb{R}} := \operatorname{Re}(J_{k,\lambda}) = \{\operatorname{Re} v \mid v \in J_{k,\lambda}\}$$

(a subspace of $\mathbb{R}^n$). Observe that if $v \in J_{k,\lambda}$, $v = v_1 + iv_2$ with $v_j \in \mathbb{R}^n$, $j = 1, 2$, then $v_1 \in J_{k,\lambda}^{\mathbb{R}}$, by definition, but also the imaginary part $v_2 \in J_{k,\lambda}^{\mathbb{R}}$, because $(-iv)$ belongs to the subspace $J_{k,\lambda}$. We also let $J_{0,\lambda} = J_{0,\lambda}^{\mathbb{R}} = \{0\}$.

Let L be the sum of the various spaces $J_{k,\lambda}^{\mathbb{R}}$, with $\operatorname{Re}\lambda \geq 0$, and let M be the sum of the various spaces $J_{k,\lambda}^{\mathbb{R}}$, with $\operatorname{Re}\lambda < 0$. Each of these spaces is A-invariant, because if v is an eigenvector of A, and $v = v_1 + iv_2$ is its decomposition into real and imaginary parts, then the subspace of $\mathbb{R}^n$ spanned by v_1 and v_2 is A-invariant. From the Jordan form decomposition, we know that every element in $\mathbb{C}^n$ can be written as a sum of elements in the various "generalized eigenspaces" $J_{k,\lambda}$, so taking real parts we know that $\mathbb{R}^n$ splits into the direct sum of L and M. (In fact, L is the largest invariant subspace on which all eigenvalues of A have nonnegative real parts, and analogously for M.)

We will need this general observation:

Lemma 3.6.4 If C is an open convex subset of $\mathbb{R}^n$ and L is a subspace of $\mathbb{R}^n$ contained in C, then $C + L = C$.

Proof. Clearly $C = C + 0 \subseteq C + L$, so we only need prove the other inclusion. Pick any $x \in C$ and $y \in L$. Then, for all $\varepsilon \neq 0$:

$$x + y = \left(\frac{1}{1+\varepsilon}\right)[(1+\varepsilon)x] + \left(\frac{\varepsilon}{1+\varepsilon}\right)\left[\left(\frac{1+\varepsilon}{\varepsilon}\right)y\right].$$

Since C is open, $(1+\varepsilon)x \in C$ for some sufficiently small $\varepsilon > 0$. Since L is a subspace, $\left(\frac{1+\varepsilon}{\varepsilon}\right) y \in L \subseteq C$. Thus $x + y \in C$, by convexity. ∎

The main technical fact needed is as follows. Fix any eigenvalue $\lambda = \alpha + i\beta$ of A with real part $\alpha \geq 0$, and denote for simplicity $J_k^{\mathbb{R}} := J_{k,\lambda}^{\mathbb{R}}$.

Lemma 3.6.5 Assume that (A, B) is controllable and $\mathcal{U} \subseteq \mathbb{R}^m$ is a neighborhood of 0. Then $J_k^{\mathbb{R}} \subseteq \mathcal{R}_{\mathcal{U}}(0)$ for all k.

Proof. First replacing if necessary $\mathcal{U}$ by a convex subset, we may assume without loss of generality that $\mathcal{U}$ is a convex neighborhood of 0. We prove the statement by induction on k, the case $k = 0$ being trivial. So assume that $J_{k-1}^{\mathbb{R}} \subseteq \mathcal{R}_{\mathcal{U}}(0)$, and take any $\widetilde{v} \in J_{k,\lambda}$, $\widetilde{v} = \widetilde{v}_1 + i\widetilde{v}_2$. We must show that $\widetilde{v}_1 \in \mathcal{R}_{\mathcal{U}}(0)$.

First pick any $T > 0$ so that $e^{\lambda Tj} = e^{\alpha Tj}$ for all $j = 0, 1, \ldots$. (If $\beta = 0$ one may take any $T > 0$; otherwise, we may use for instance $T = \frac{2\pi}{|\beta|}$.) Next choose any $\delta > 0$ with the property that $v_1 := \delta\widetilde{v}_1 \in \mathcal{R}_{\mathcal{U}}^T(0)$. (There is such a δ because $\mathcal{R}_{\mathcal{U}}(0)$ contains 0 in its interior, by Proposition 3.6.3.) Since $v \in \ker(\lambda I - A)^k$, where $v = \delta\widetilde{v}$,

$$e^{(A-\lambda I)t}v = \left(I + t(A - \lambda I) + \frac{t^2}{2}(A - \lambda I)^2 + \ldots\right)v = v + w \quad \forall t,$$

where $w \in J_{k-1}$. Thus

$$e^{\alpha t}v = e^{\lambda t}v = e^{tA}v - e^{\lambda t}w = e^{tA}v - e^{\alpha t}w \quad \forall t = jT, j = 0, 1, \ldots. \tag{3.26}$$

Decomposing into real and imaginary parts $w = w_1 + iw_2$ and taking real parts in Equation (3.26),

$$e^{\alpha t}v_1 = e^{tA}v_1 - e^{\alpha t}w_1 \quad \forall t = jT, j = 0, 1, \ldots.$$

Now pick any integer $q \geq 1/\delta$. Then

$$\left(\sum_{j=0}^{q-1} e^{\alpha jT}\right) v_1 = \sum_{j=0}^{q-1} e^{jTA}v_1 + w'$$

where $w' = -\sum e^{\alpha jT}w_1$ belongs to the subspace $J_{k-1}^{\mathbb{R}}$. Applying first Corollary 3.6.2 and then Lemma 3.6.4, we conclude that

$$pv_1 \in \mathcal{R}_{\mathcal{U}}^{qT}(0) + J_{k-1}^{\mathbb{R}} \subseteq \mathcal{R}_{\mathcal{U}}(0)$$

where

$$p = \sum_{j=0}^{q-1} e^{\alpha jT} \geq \sum_{j=0}^{q-1} 1 = q \geq \frac{1}{\delta}\,.$$

(Here is precisely where we used that $\alpha \geq 0$.) Therefore $\delta p \widetilde{v}_1 = p v_1 \in R_{\mathcal{U}}(0)$. On the other hand, $\delta p \geq 1$ means that

$$\widetilde{v}_1 = \frac{1}{\delta p}\delta p \widetilde{v}_1 + \left(1 - \frac{1}{\delta p}\right) 0$$

is a convex combination. Since $\delta p \widetilde{v}_1$ and 0 both belong to $\mathcal{R}_{\mathcal{U}}(0)$, we conclude by convexity of the latter that indeed $\widetilde{v}_1 \in \mathcal{R}_{\mathcal{U}}(0)$. ∎

Corollary 3.6.6 Assume that (A,B) is controllable and $\mathcal{U} \subseteq \mathbb{R}^m$ is a neighborhood of 0. Then $L \subseteq \mathcal{R}_{\mathcal{U}}(0)$.

Proof. As before, we may assume without loss of generality that $\mathcal{U}$ is convex. We have that L is the sum of the spaces $J^{\mathbb{R}}_{k,\lambda}$, over all eigenvalues λ with real part nonnegative, and each of these spaces is included in $\mathcal{R}_{\mathcal{U}}(0)$. In general, if L_1 and L_2 are two subspaces of a convex set C, $L_1 + L_2 \subseteq C$ (since $x + y = \frac{1}{2}(2x) + \frac{1}{2}(2y)$), so the sum of the L's is indeed included in $\mathcal{R}_{\mathcal{U}}(0)$. ∎

The next result says that the reachable set from zero is a "thickened linear subspace":

Corollary 3.6.7 Assume that (A,B) is controllable and $\mathcal{U} \subseteq \mathbb{R}^m$ is a convex and bounded neighborhood of 0. Then there exists a set $\mathcal{B}$ such that $\mathcal{R}_{\mathcal{U}}(0) = \mathcal{B} + L$ and $\mathcal{B}$ is bounded, convex, and open relative to M.

Proof. We claim that $\mathcal{R}_{\mathcal{U}}(0) = (\mathcal{R}_{\mathcal{U}}(0) \bigcap M) + L$. One inclusion is clear from

$$\left(\mathcal{R}_{\mathcal{U}}(0) \bigcap M\right) + L \;\subseteq\; \mathcal{R}_{\mathcal{U}}(0) + L \;=\; \mathcal{R}_{\mathcal{U}}(0)$$

(applying Lemma 3.6.4). Conversely, any $v \in \mathcal{R}_{\mathcal{U}}(0)$ can be decomposed as $v = x + y \in M + L$; we need to show that $x \in R_{\mathcal{U}}(0)$. But $x = v - y \in R_{\mathcal{U}}(0) + L = R_{\mathcal{U}}(0)$ (applying the same Lemma yet again). This establishes the claim.

We let $\mathcal{B} := \mathcal{R}_{\mathcal{U}}(0) \bigcap M$. This set is convex and open in M because $\mathcal{R}_{\mathcal{U}}(0)$ is open and convex. We only need to prove that it is bounded.

Let $P : \mathbb{R}^n \to \mathbb{R}^n$ be the projection on M along L, that is, $P(x+y) = x$ if $x \in M$, $y \in L$. Observe that $PA = AP$ because each of L and M are A-invariant (so $v = x + y$, $Ax \in M$, $Ay \in L$, imply $PAv = Ax = APv$). Pick any $x \in \mathcal{R}_{\mathcal{U}}(0) \bigcap M$. Since $x \in \mathcal{R}_{\mathcal{U}}(0)$, there are some T and some ω so that

$$x = \int_0^T e^{(T-\tau)A} B\omega(\tau)\, d\tau\,.$$

On the other hand, since $x \in M$, $x = Px$. Thus:

$$x = Px = \int_0^T Pe^{(T-\tau)A}B\omega(\tau)\,d\tau = \int_0^T e^{(T-\tau)A}x(\tau)\,d\tau\,,$$

where $x(\tau) = PB\omega(\tau) \in M \bigcap PB(\mathcal{U})$ for all τ.

Since the restriction of A to M has all its eigenvalues with negative real part, there are positive constants $c, \mu > 0$ such that $\left\|e^{tA}x\right\| \leq ce^{-\mu t}\left\|x\right\|$ for all $t \geq 0$ and all $x \in M$. Since $PB(\mathcal{U})$ is bounded, there is then some constant c' such that, if x is also in $PB(\mathcal{U})$, $\left\|e^{tA}x\right\| \leq c'e^{-\mu t}$ for all $t \geq 0$. So, for x as above we conclude

$$\|x\| \leq c' \int_0^T e^{-\mu(T-\tau)}\,d\tau \leq \frac{c'}{\mu}(1 - e^{-\mu T}) \leq \frac{c'}{\mu}\,,$$

and we proved that $\mathcal{B}$ is bounded. ∎

Proof of Theorem 6

Assume first that $\mathcal{R}_\mathcal{U}(0) = \mathbb{R}^n$. We already remarked that the pair (A, B) must be reachable. If the eigenvalue condition (b) does not hold, then L is a proper subspace of $\mathbb{R}^n$ and $M \neq 0$. Enlarging $\mathcal{U}$ if necessary, we may assume that $\mathcal{U}$ is convex and bounded. Lemma 3.6.7 claims that $\mathbb{R}^n = \mathcal{R}_\mathcal{U}(0)$ is then a subset of $L + \mathcal{B}$, with bounded $\mathcal{B}$, a contradiction. Conversely, assume that (a) and (b) hold. By Corollary 3.6.6, $\mathbb{R}^n = L \subseteq \mathcal{R}_\mathcal{U}(0)$. ∎

We also say that the system Σ is *$\mathcal{U}$-controllable* if $\mathcal{R}_\mathcal{U}(x) = \mathbb{R}^n$ for all $x \in \mathbb{R}^n$.

Exercise 3.6.8 Let $\mathcal{U}$ be a bounded neighborhood of zero. Show:

1. Every state can be $\mathcal{U}$-controlled to zero if and only if (a) the pair (A, B) is controllable and (b) the matrix A has no eigenvalues with positive real part.

2. Σ is completely $\mathcal{U}$-controllable (that is, $\mathcal{R}_\mathcal{U}(x) = \mathbb{R}^n$ for all $x \in \mathbb{R}^n$) if and only if (a) the pair (A, B) is controllable, and (b) all eigenvalues of the matrix A are purely imaginary. ∎

3.7 First-Order Local Controllability

For nonlinear systems, the best one often can do regarding controllability notions is to characterize local controllability —and even that problem is not totally understood. In order to talk about local notions, a topology must be introduced in the state space. Even though the only nontrivial results will be proved for discrete-time and continuous-time systems of class C^1, it is more natural to provide the basic definitions in somewhat more generality. A class called here *topological systems* is introduced, which captures the basic continuity properties needed. In Chapter 5, a few basic facts about stability are proved in the general context of such systems. From now on in this Chapter, and unless otherwise stated, *system* means *topological system* in the sense of Definition 3.7.1 below.

Topological Systems

If $\mathcal{X}$ is a metric space, we use $d(a,b)$ to denote the distance between elements of $\mathcal{X}$, and as earlier, we let d_∞ be the uniform distance between two time functions into $\mathcal{X}$; that is, if $\gamma_1, \gamma_2 : \mathcal{I} \to \mathcal{X}$, for some interval $\mathcal{I} \subseteq \mathcal{T}$, then

$$d_\infty(\gamma_1, \gamma_2) := \sup\{d(\gamma_1(t), \gamma_2(t)), t \in \mathcal{I}\}$$

(essential supremum when dealing with measurable functions and $\mathcal{T} = \mathbb{R}$). We also use the notation

$$\mathcal{B}_\rho(x) := \{y \in \mathcal{X} \mid d(x,y) < \rho\}$$

for the open ball of radius ρ centered at x, and

$$\overline{\mathcal{B}}_\rho(x) := \{y \in \mathcal{X} \mid d(x,y) \leq \rho\}$$

for its closure.

Definition 3.7.1 *A* **topological system** Σ *is an object*

$$(\mathcal{T}, \mathcal{X}, \mathcal{U}, \phi)$$

such that $\mathcal{X}$ is a metric space; Σ is a system when $\mathcal{X}$ is thought of just as a set; and for each $\sigma < \tau$ in $\mathcal{T}$ and each $\omega \in \mathcal{U}^{[\sigma,\tau)}$,

$$\psi(\tau, \sigma, \cdot, \omega)$$

has an open domain and is continuous there as a map into $\mathcal{X}^{[\sigma,\tau]}$ (with metric d_∞). □

In other words, the following property must hold for each pair $\sigma < \tau$ in $\mathcal{T}$: If ω is admissible for the state x, and if

$$x_n \longrightarrow x$$

in $\mathcal{X}$, then there is some integer N such that ω is admissible for x_n for all $n > N$, and

$$\lim_{n\to\infty} d(\phi(t,\sigma,x_n,\omega), \phi(t,\sigma,x,\omega)) = 0$$

uniformly on $t \in [\sigma, \tau]$.

From Theorem 1 (p. 57) and Lemma 2.8.3, the following is then immediate:

Proposition 3.7.2 If Σ is either a continuous-time system, or a discrete-time system of class $\mathcal{C}^0$, then Σ is topological when $\mathcal{X}$ is given the induced topology of $\mathbb{R}^n$. □

Remark 3.7.3 If Σ is a topological system, then the map $\phi(t, \sigma, x, \omega)$ is jointly continuous on (t, x), provided only that $\phi(\cdot, \sigma, x, \omega)$ be continuous for each fixed σ, x, ω (as happens with continuous-time systems). Indeed, given any σ, τ, ω, any ξ for which ω is admissible, any $t \in [\sigma, \tau]$, and any number $\varepsilon > 0$, there is always, because of uniform continuity of ϕ in x, a $\delta > 0$ so that

$$d(\phi(t, \sigma, x, \omega), \phi(t, \sigma, y, \omega)) < \varepsilon/2$$

whenever $d(x, y) < \delta$. Furthermore, continuity of $\phi(\cdot, \sigma, x, \omega)$ ensures that there is a $\delta' > 0$ so that

$$d(\phi(s, \sigma, x, \omega), \phi(t, \sigma, x, \omega)) < \varepsilon/2$$

whenever $s \in [\sigma, \tau]$ is such that $|t - s| < \delta'$. Therefore, also

$$d(\phi(s, \sigma, x, \omega), \phi(t, \sigma, y, \omega)) < \varepsilon$$

if (s, x) is close to (t, y), as required. □

An equilibrium state x^0 is one for which there exists a control value $u^0 \in \mathcal{U}$ such that

$$\phi(\tau, \sigma, x^0, \omega^0) = x^0$$

for each $\sigma \leq \tau, \sigma, \tau \in \mathcal{T}$, where ω^0 is the control with $\omega^0(t) \equiv u^0$. When the system Σ is linear, we always take $x^0 = 0$ and $u^0 = 0$. Unless otherwise stated, x^0 is a fixed but arbitrary equilibrium state.

Definition 3.7.4 *Let Σ be a topological system. Let ξ be any path on an interval $\mathcal{I} = [\sigma, \tau]$, and denote*

$$x_0 := \xi(\sigma), x_1 := \xi(\tau).$$

The system Σ is **locally controllable along** *ξ if for each $\varepsilon > 0$ there is some $\delta > 0$ such that the following property holds: For each $z, y \in \mathcal{X}$ with $d(z, x_0) < \delta$ and $d(y, x_1) < \delta$ there is some path ζ on $\mathcal{I}$ such that*

$$\zeta(\sigma) = z\,,\ \zeta(\tau) = y$$

and

$$d_\infty(\zeta, \xi) < \varepsilon\,.$$

When x is an equilibrium state and $\xi \equiv x$ on the interval $[\sigma, \tau]$, $T = \tau - \sigma$, one simply says that Σ is **locally controllable (in time T) at** *x.* □

Thus, local controllability along a trajectory corresponds to the possibility of controlling every initial state near the original initial state to every final state near the original final state, and being able to do so without deviating far from the original trajectory. For equilibrium states, the initial and final states are the same, and the definition can be put in slightly different terms, using the following terminology:

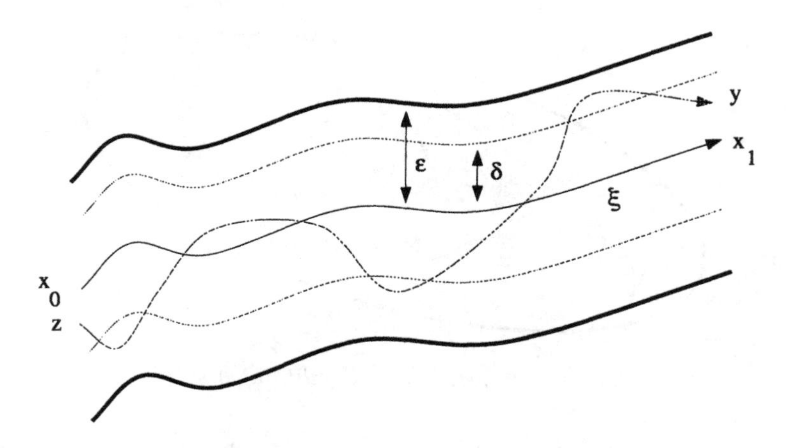

Figure 3.2: *Local controllability along trajectory.*

Definition 3.7.5 *Let Σ be a topological system, and let y, z be in $\mathcal{X}$. If $\mathcal{V}$ is a subset of $\mathcal{X}$ containing y and z, we say that z* **can be controlled to y without leaving $\mathcal{V}$ [in time T]** *(or, that y can be reached from z inside $\mathcal{V}$) if there exists some path ζ on some interval $[\sigma, \tau]$ [respectively, with $T = \tau - \sigma$] such that*

$$\zeta(\sigma) = z \quad \textit{and} \quad \zeta(\tau) = y$$

and also

$$\zeta(t) \in \mathcal{V}$$

for all $t \in [\sigma, \tau]$. □

Remark 3.7.6 Using the above definition, one can restate: The system Σ is locally controllable in time T at the equilibrium state x if and only if for every neighborhood $\mathcal{V}$ of x there is another neighborhood $\mathcal{W}$ of x such that, for any pair of elements z, y in $\mathcal{W}$, z can be controlled to y inside $\mathcal{V}$ in time T (see Figure 3.3). □

For linear systems, local controllability is not an interesting notion:

Lemma/Exercise 3.7.7 For any discrete-time or continuous-time linear system Σ and each pair $\sigma < \tau$ in $\mathcal{T}$, the following properties are equivalent:

(a) Σ is locally controllable along some trajectory Γ on $[\sigma, \tau]$.

(b) Σ is locally controllable along every trajectory Γ on $[\sigma, \tau]$.

(c) Σ is controllable on $[\sigma, \tau]$. □

The following is an example of the "linearization principle" mentioned in Chapter 1. In particular, for continuous-time systems $\dot{x} = f(x, u)$, with $(0, 0) \in$

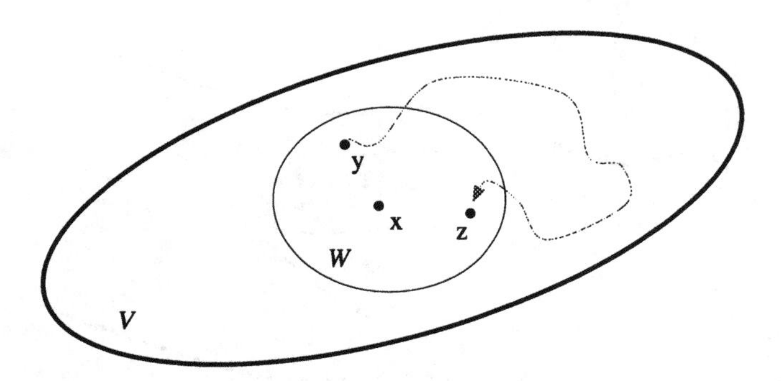

Figure 3.3: *Local controllability at a state.*

$\mathcal{X} \times \mathcal{U}$ and $f(0,0) = 0$, the result asserts that controllability of the pair (A, B) obtained by expanding to first order $\dot{x} = Ax + Bu + o(x, u)$ is sufficient to insure local controllability about $x = 0$.

Theorem 7 *Assume that Σ is a continuous-time system of class C^1. Then,*

1. *Let $\Gamma = (\xi, \omega)$ be a trajectory for Σ on an interval $\mathcal{I} = [\sigma, \tau]$. Then, a sufficient condition for Σ to be locally controllable along ξ is that $\Sigma_*[\Gamma]$ be controllable on $[\sigma, \tau]$.*

2. *Let Σ be time-invariant, and assume that x is an equilibrium state. Pick any $\varepsilon > 0$ and any $u \in \mathcal{U}$ such that $f(x, u) = 0$. Then, a sufficient condition for Σ to be locally controllable at x in time ε is that the linearization of Σ at (x, u) be controllable.*

Proof. The second part is a special case of the first, so we prove part (1). Let $x_0 := \xi(\sigma)$ and $x_1 := \xi(\tau)$. Consider the mapping

$$\alpha : \mathcal{D}_{\sigma,\tau} \to \mathcal{X}, \quad \alpha(x, \nu) := \phi(\tau, \sigma, x, \nu) .$$

Note that $\alpha(x_0, \omega) = x_1$. By Theorem 1 (p. 57), Part 3, this mapping is continuously differentiable, and its partial differential with respect to ν evaluated at (x_0, ω) is the mapping $\phi_\Gamma(\tau, \sigma, 0, \cdot)$, which is onto by the controllability assumption. It follows using the Implicit Mapping Theorem for maps from a normed space into a finite dimensional space, as in Theorem 53 (p. 465) in Appendix B, that there exists an $\varepsilon_1 > 0$ and a continuously differentiable mapping

$$j : \mathcal{B}_{\varepsilon_1}(x_0) \times \mathcal{B}_{\varepsilon_1}(x_1) \to \mathcal{L}_m^\infty$$

such that

$$\alpha(z, j(z, y)) = y$$

for all (z, y) in the domain of j and such that $j(x_0, x_1) = \omega$. Given now any $\varepsilon > 0$, pick $\delta > 0$ small enough so that

$$d_\infty(\xi, \psi(z, j(z, y))) < \varepsilon$$

whenever $z \in \mathcal{B}_\delta(x_0)$ and $y \in \mathcal{B}_\delta(x_1)$; such a δ exists by continuity of j and of ψ, and it is as desired for local controllability. ∎

Lemma/Exercise 3.7.8 For discrete-time systems of class $\mathcal{C}^1$, the same conclusions as in Theorem 7 hold. □

Remark 3.7.9 The definition of local controllability can be weakened in many ways while still preserving the idea that every state close to the original state can be controlled to every state close to the final one. One such weakening is to allow for large excursions. For instance, consider the following bilinear system in $\mathbb{R}^2$, with $\mathcal{U} = \mathbb{R}$:

$$\dot{x} = \left[\begin{pmatrix} 0 & -1 \\ 1 & 0 \end{pmatrix} + u \begin{pmatrix} 1 & 0 \\ 0 & 1 \end{pmatrix}\right] x$$

or in polar coordinates away from the origin,

$$\begin{aligned} \dot{\theta} &= 1 \\ \dot{\rho} &= u\rho \,. \end{aligned}$$

Pick any nonzero state, say, $x := (1,0)'$. From the polar coordinate description, it is clear that the system restricted to $\mathbb{R}^2 - \{0\}$ is controllable and, in particular, that every state near x can be controlled to every other state near x. However, the only way to control x to $(1+\varepsilon, 0)'$, for ε small, is through a motion of time at least 2π that circles about the origin, and in particular a motion that leaves the ball of radius 1 about x. Therefore, this system is not locally controllable in our sense. □

The following are some examples of locally controllable systems and the application of the linearization principle.

Example 3.7.10 With $X = \mathbb{R}^2$ and $\mathcal{U} = \mathbb{R}$, consider the system

$$\begin{aligned} \dot{x}_1 &= x_1 + \sin x_2 + x_1 e^{x_2} \\ \dot{x}_2 &= x_2^2 + u. \end{aligned}$$

This is locally controllable at $x = 0$, because the linearization at $(0,0)$,

$$A = \begin{pmatrix} 2 & 1 \\ 0 & 0 \end{pmatrix} \quad B = \begin{pmatrix} 0 \\ 1 \end{pmatrix}$$

satisfies the controllability rank condition. When u^2 is substituted for u in the second equation, the linear test fails, since now $B = 0$. In fact, the system is not locally controllable there, since starting at 0 it is impossible to reach any states of the form $\binom{0}{x_2}$ with $x_2 < 0$. On the other hand, the linear test is not definitive: For instance, if we have u^3 instead of u in the second equation, then the linear test fails (again $B = 0$) but the system is still locally controllable. This last fact follows from the observation that, given any trajectory (ξ, ω) of the original system, then (ξ, ν) is a trajectory of the system that has u^3 in the second equation, if we define $\nu(t) := \omega(t)^{1/3}$. □

Example 3.7.11 Consider the system with $\mathcal{X} = \mathcal{U} = \mathbb{R}$ and equations

$$\dot{x} = 1 + u\sin^2 x\,.$$

We claim that this system is controllable along every trajectory $\Gamma = (\xi, \omega)$ with ω smooth defined on any $\mathcal{I} = [\sigma, \tau]$ (with $\sigma < \tau$). Indeed, the linearization matrices (here 1×1) are

$$A(t) = (\omega(t)\sin 2\xi(t)) \quad B(t) = \left(\sin^2 \xi(t)\right)\,.$$

So $\operatorname{rank} B(t_0) = 1$ for at least one $t_0 \in [\sigma, \tau]$ if and only if $\sin\xi(t)$ is not identically equal to zero. But if the latter happens, then

$$\dot{\xi} \equiv 1\,,$$

which implies that ξ cannot be constant.

The assumption that ω is smooth is not necessary, as it is not hard to prove controllability of the linearization directly, but this assumption serves to illustrate the use of the controllability criteria for linear time-varying systems. In this example, it is not hard to establish global controllability of the original system. □

Exercise 3.7.12 Prove that, for the above example,

$$\text{rank}\ (B_0(t), B_1(t), B_2(t)) = 1$$

for *all* t. □

3.8 Controllability of Recurrent Nets*

We turn to studying controllability of what are often called "recurrent neural networks". These constitute a class of nonlinear systems which, although formally analogous to linear systems, exhibit interesting nonlinear characteristics, and arise often in applications, and for which strong conclusions can be derived.

In this section, we use θ to denote the hyperbolic tangent function

$$\theta(x) = \tanh x \,:\, \mathbb{R} \to \mathbb{R} \,:\, x \mapsto \frac{e^x - e^{-x}}{e^x + e^{-x}}\,.$$

This is sometimes called the "sigmoid" or "logistic" map; see Figure 3.4.

For each positive integer n, we let $\vec{\theta}_n$ denote the diagonal mapping

$$\vec{\theta}_n \,:\, \mathbb{R}^n \to \mathbb{R}^n \,:\, \begin{pmatrix} x_1 \\ \vdots \\ x_n \end{pmatrix} \mapsto \begin{pmatrix} \theta(x_1) \\ \vdots \\ \theta(x_n) \end{pmatrix}. \tag{3.27}$$

* This section can be skipped with no loss of continuity.

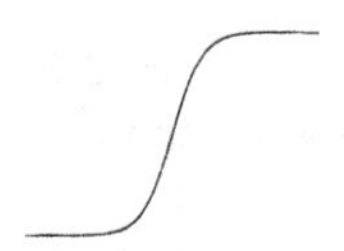

Figure 3.4: tanh

Definition 3.8.1 *An n-dimensional, m-input* **recurrent net** *is a continuous-time time-invariant system with $\mathcal{X} = \mathbb{R}^n$, $\mathcal{U} = \mathbb{R}^m$, and equations of the form*

$$\dot{x}(t) \;=\; \vec{\theta}_n \left(Ax(t) + Bu(t)\right) , \tag{3.28}$$

where $A \in \mathbb{R}^{n \times n}$ and $B \in \mathbb{R}^{n \times m}$. □

Since $\vec{\theta}_n$ is a globally Lipschitz map, such a system is complete (every input is admissible for every state). Observe that, if in place of $\theta = \tanh$ we had $\theta =$ the identity function, we would be studying continuous-time time-invariant linear systems. Recurrent nets arise when the rates of change of state variables are bounded ($|\dot{x}_i| < 1$), and one models this limitation by means of $\vec{\theta}_n$, which provides a smooth distortion, or saturation, of the rates of change. In system-theoretic terms, we may represent a net by means of a block diagram as in Figure 3.5.

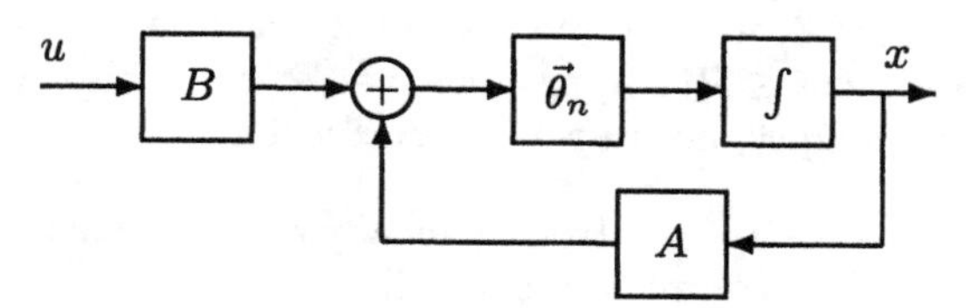

Figure 3.5: *Block diagram of recurrent net.*

We need the following concept, which we define for more general systems.

Definition 3.8.2 *A system Σ, with state-space $\mathcal{X} \subseteq \mathbb{R}^n$, is* **strongly locally controllable** *around $x^0 \in \mathcal{X}$ if for each neighborhood $\mathcal{V}$ of x^0 there is some neighborhood W of x^0, included in $\mathcal{V}$, so that, for every pair of states y and z in W, y can be controlled to z without leaving $\mathcal{V}$.* □

Related notions are studied in Section 3.7; intuitively, we are asking that any two states that are sufficiently close to x^0 can be steered to one another without large excursions.

For each pair of positive integers n and m, we let

$$\mathbf{B}_{n,m} := \left\{B \in \mathbb{R}^{n \times m} \mid \forall i,\, \text{row}_i(B) \neq 0 \text{ and } \forall i \neq j,\, \text{row}_i(B) \neq \pm\text{row}_j(B)\right\} ,$$

where $\text{row}_i(\cdot)$ denotes the ith row of the given matrix. In the special case $m = 1$, a vector $b \in \mathbf{B}_{n,1}$ if and only if all its entries are nonzero and have different absolute values. Asking that $B \in \mathbf{B}_{n,m}$ is a fairly weak requirement: the complement of $\mathbf{B}_{n,m}$ is an algebraic subset of $\mathbb{R}^{n\times m}$, so $\mathbf{B}_{n,m}$ is generic in the same sense as in Proposition 3.3.12.

In this section, we will prove the following result:

Theorem 8 *The following two properties are equivalent:*

1. $B \in \mathbf{B}_{n,m}$.
2. *The system (3.28) is strongly locally controllable around every state.*

If a system Σ is strongly locally controllable around x^0, in particular there exists some neighborhood W of x^0 so that every $x \in W$ is both controllable to x^0 and reachable from x^0, that is,

$$x^0 \in \text{int}\left(\mathcal{R}(x^0)\bigcap \mathcal{C}(x^0)\right) . \tag{3.29}$$

If Σ is strongly locally controllable around *every state* x^0, then the reachable sets $\mathcal{R}(x^0)$ are both open and closed, for each x^0. (Pick any $\bar{x} \in \mathcal{R}(x^0)$; by (3.29) some neighborhood of $\bar{x}$ is included in $\mathcal{R}(\bar{x})$, and hence also in $\mathcal{R}(x^0)$, so $\mathcal{R}(x^0)$ is indeed open. Next take any $\bar{x}$ in the closure of $\mathcal{R}(x^0)$; now we know that there is some neighborhood W of $\bar{x}$ such that every $x \in W$ can be steered to $\bar{x}$. Since $W \cap \mathcal{R}(x^0) \neq \emptyset$, we can find a point $\hat{x}$ which is reachable from x^0 and can be steered to $\bar{x}$, so $\bar{x} \in \mathcal{R}(x^0)$, proving that $\mathcal{R}(x^0)$ is also closed.) Note that $\mathcal{R}(x^0)$ is always nonempty, because $x^0 \in \mathcal{R}(x^0)$. Thus:

Lemma 3.8.3 If a system Σ is strongly locally controllable around every state and $\mathcal{X}$ is connected, Σ is completely controllable. □

Putting together Theorem 8 and this Lemma, we have, then:

Corollary 3.8.4 If $B \in \mathbf{B}_{n,m}$, the system (3.28) is completely controllable. □

Theorem 8 will be proved after some generalities are established.

A Local Controllability Lemma

We start with an observation that is of independent interest. By $\text{co}\, V$ we mean the convex hull of a set V.

Lemma 3.8.5 Consider a continuous-time time-invariant system $\dot{x} = f(x,u)$ and suppose that $x^0 \in \mathcal{X}$ is a state such that

$$0 \in \text{int}\,\text{co}\left\{f(x^0,u) \mid u \in \mathcal{U}\right\} . \tag{3.30}$$

Then

$$x^0 \in \text{int}\left(\mathcal{R}(x^0)\bigcap \mathcal{C}(x^0)\right) .$$

Proof. Pick a finite subset $\mathcal{U}_0 \subseteq \mathcal{U}$ so that $0 \in \operatorname{int}\operatorname{co}\left\{f(x^0,u) \mid u \in \mathcal{U}_0\right\}$. We assume, without loss of generality, that $x^0 = 0$. Let $\mathbb{S}^1$ be the unit sphere $\{\lambda \in \mathbb{R}^n \mid \|\lambda\| = 1\}$. Define a function $\varphi : \mathcal{X} \mapsto \mathbb{R}$ by letting

$$\varphi(x) = \max\{\mathsf{h}(\lambda, x) \mid \lambda \in \mathbb{S}^1\},$$

where

$$\mathsf{h}(\lambda, x) = \min\{\langle\lambda, f(x,u)\rangle \mid u \in \mathcal{U}_0\}.$$

The function h is a minimum of a finite collection of continuous functions, so h is continuous; thus, φ is well-defined and continuous as well.

Since 0 is an interior point of the convex hull of the set $\{f(0,u) \mid u \in \mathcal{U}_0\}$, there is some $\delta > 0$ such that, for every $\lambda \in \mathbb{S}^1$, the vector $-4\delta\lambda$ is a convex combination of the $f(0,u)$, $u \in \mathcal{U}_0$. Given any $\lambda \in \mathbb{S}^1$, the number -4δ is equal to $\langle\lambda, -4\delta\lambda\rangle$, which is a convex combination of the numbers $\langle\lambda, f(0,u)\rangle$, $u \in \mathcal{U}_0$. So at least one of these numbers is $\leq -4\delta$. So $\mathsf{h}(\lambda, 0) \leq -4\delta$. Since this is true for all $\lambda \in \mathbb{S}^1$, we conclude that $\varphi(0) \leq -4\delta$. Since φ is continuous, there exists $\alpha > 0$ such that, if $\overline{\mathcal{B}} = \overline{\mathcal{B}}_\alpha(x) = \{x \in \mathbb{R}^n \mid \|x\| \leq \alpha\}$, then $\overline{\mathcal{B}} \subseteq \mathcal{X}$ and $\varphi(x) \leq -2\delta$ whenever $x \in \overline{\mathcal{B}}$.

Now fix a point $\bar{x} \in \overline{\mathcal{B}}$, and let $\mathcal{S}$ be the set consisting of all those pairs $(\mathcal{I}, \omega)$ such that $\mathcal{I} \subseteq [0, \infty)$ is an interval containing 0, and $\omega : \mathcal{I} \to \mathcal{U}_0$ is a measurable function (i.e., a control with values in $\mathcal{U}_0$) for which there is a solution $\xi : \mathcal{I} \to \mathcal{X}$ (defined on the complete interval) of $\dot{\xi}(t) = f(\xi(t), \omega(t))$ with initial state $\xi(0) = \bar{x}$, and this solution satisfies:

$$\frac{d\,\|\xi(t)\|}{dt} \leq -\delta \tag{3.31}$$

for almost every $t \in \mathcal{I}$. (Note that $\|\xi(\cdot)\|$ is absolutely continuous, because ξ is, as is clear from the definition of absolute continuity.)

We order $\mathcal{S}$, by letting

$$(\mathcal{I}_1, \omega_1) \preceq (\mathcal{I}_2, \omega_2)$$

iff $\mathcal{I}_1 \subseteq \mathcal{I}_2$ and ω_1 is the restriction to $\mathcal{I}_1$ of ω_2. Every totally ordered nonempty subset of $\mathcal{S}$ has an upper bound in $\mathcal{S}$ (take the union of the intervals, and a common extension of the ω's). By Zorn's Lemma, $\mathcal{S}$ has a maximal element $(\mathcal{I}, \omega)$.

In general, for any pair $(\mathcal{J}, \nu) \in \mathcal{S}$ and its corresponding path ζ, Equation (3.31) implies that $\|\zeta(t)\| \leq \|\zeta(0)\| - \delta t$ for all $t \in \mathcal{J}$, and, in particular, that $\|\zeta(t)\| \leq \|\zeta(0)\| \leq \alpha$ for all $t \in \mathcal{J}$. So ζ is entirely contained in $\overline{\mathcal{B}}$. As $t \leq \frac{\alpha}{\delta}$ for all $t \in \mathcal{J}$, $\mathcal{J}$ must be bounded. Furthermore, $(\operatorname{clos}\mathcal{J}, \nu)$ is also in $\mathcal{S}$, when we view ν also as a control defined on the closure of $\mathcal{J}$, because the maximal solution for ν seen as such a control cannot be ζ, but must extend to the closure (since $\zeta(\cdot)$ takes values in a compact, cf. Proposition C.3.6), and Property (3.31) still holds a.e. on the closure. Applied to $\mathcal{J} = \mathcal{I}$, and since $\mathcal{I}$

is maximal, this argument shows that $\mathcal{I} = \operatorname{clos}\mathcal{I}$. Thus, $\mathcal{I} = [0, T]$ for some T. Let ξ be the path corresponding to $(\mathcal{I}, \omega)$. We claim that

$$\xi(T) = 0. \tag{3.32}$$

Indeed, suppose this claim would not be true, so $\hat{x} := \xi(T) \in \overline{\mathcal{B}}$ is nonzero. Since $\varphi(\hat{x}) \leq -2\delta$, we have $\mathsf{h}(\hat{\lambda}, \hat{x}) \leq -2\delta$, where $\hat{\lambda} := -\hat{x}/\|\hat{x}\|$. So there exists some $u_0 \in \mathcal{U}_0$ such that $\langle \hat{\lambda}, f(\hat{x}, u_0) \rangle \leq -2\delta$. Let $\zeta : [T, T+\beta] \to \mathcal{X}$ be the path, with $\zeta(T) = \hat{x}$, which corresponds to the control that is constantly equal to u_0, for some $\beta > 0$. We may assume that $\zeta(t) \neq 0$ for all t (otherwise, we could extend by ζ and already have a control that drives $\bar{x}$ to zero). Using that the gradient of $x \mapsto \|x\|$ is $\frac{x'}{\|x\|}$, we have

$$\rho(t) := \frac{d}{dt}\|\zeta(t)\| = \langle \lambda(t), f(\zeta(t), u_0) \rangle,$$

where

$$\lambda(t) = \frac{\zeta(t)}{\|\zeta(t)\|}$$

is a continuous function with $\lambda(T) = \hat{\lambda}$. Since $\rho(T) = \langle \hat{\lambda}, f(\hat{x}, u_0) \rangle \leq -2\delta$, we can assume, by taking β small enough, that $\rho(t) \leq -\delta$ for $t \in [T, T+\beta]$. Letting $\tilde{\mathcal{I}} = [0, T+\beta]$, we extend ω to a control $\tilde{\omega} : \tilde{\mathcal{I}} \to \mathcal{U}_0$, defining $\tilde{\omega}(t) := u_0$ for $t \in (T, T+\beta]$. The path with initial state $\bar{x}$ is $\tilde{\xi} : \tilde{\mathcal{I}} \to \mathcal{X}$, whose restrictions to $[0, T]$ and $[T, T+\beta]$ are ξ and ζ respectively. As $(\tilde{\mathcal{I}}, \tilde{\omega}) \in \mathcal{S}$ and $(\mathcal{I}, \omega) \prec (\tilde{\mathcal{I}}, \tilde{\omega})$ but $(\mathcal{I}, \omega) \neq (\tilde{\mathcal{I}}, \tilde{\omega})$, we have contradicted the maximality of $(\mathcal{I}, \omega)$. This contradiction proves claim (3.32).

In conclusion, we established that every $\bar{x} \in \overline{\mathcal{B}}$ can be controlled to 0. We may also apply this argument to the reversed system $\dot{x} = -f(x, u)$, whose paths are those of the original system run backward in time (cf. Lemma 2.6.8). The hypothesis applies to the reversed system, because if the convex hull of the set $\{f(0, u) \mid u \in \mathcal{U}_0\}$ contains zero in its interior, then the same is true for the convex hull of $\{-f(0, u) \mid u \in \mathcal{U}_0\}$. Thus, we find some other ball $\overline{\mathcal{B}}_-$ with the property that every $x \in \overline{\mathcal{B}}_-$ can be reached from 0. Then $W = \overline{\mathcal{B}} \cap \overline{\mathcal{B}}_-$ is so that every state in W is both controllable to 0 and reachable from 0. ∎

Corollary 3.8.6 Under the same assumptions as Lemma 3.8.5, the system is strongly locally controllable at x^0.

Proof. Given x^0, and given any neighborhood $\mathcal{V}$ of x^0, we may consider the new continuous-time system with state space $\mathcal{V}$, same input value set, and same equations, but restricted to $\mathcal{V}$. The restricted system has the property that there is an open set W which contains x^0 and is so that every state in W is both controllable to and reachable from x^0. Thus any two states in W can be steered to each other, and by definition the paths stay in the restricted state space, namely, the desired neighborhood $\mathcal{V}$. ∎

Remark 3.8.7 The argument used in the proof of Lemma 3.8.5 is analogous to the methods of proof used in Section 5.7, relying upon a "control-Lyapunov function" $V(x)$ (the function $V(x) = \|x\|^2$) so that $\min_u \nabla V(x) \cdot f(x,u) < -c$ for all x near zero, for some $c > 0$. If this property is weakened to merely asking that, for all nonzero x near zero, $\min_u \nabla V(x) \cdot f(x,u) < 0$, then only asymptotic controllability (cf. Chapter 5) will be guaranteed to hold, instead of controllability. □

A Result on Convex Hulls

The relevant properties of θ are summarized follows.

Lemma 3.8.8 The function $\theta = \tanh$ satisfies:

1. θ is an odd function, i.e. $\theta(-s) = -\theta(s)$ for all $s \in \mathbb{R}$;
2. $\theta_\infty = \lim_{s\to+\infty} \theta(s)$ exists and is > 0;
3. $\theta(s) < \theta_\infty$ for all $s \in \mathbb{R}$;
4. for each $a, b \in \mathbb{R}$, $b > 1$,

$$\lim_{s\to+\infty} \frac{\theta_\infty - \theta(a+bs)}{\theta_\infty - \theta(s)} = 0\,. \tag{3.33}$$

Proof. The first three properties are clear, with $\theta_\infty = 1$, so we need to prove the limit property (3.33). Note that $\sigma(x) = (1+e^{-x})^{-1}$ satisfies

$$\frac{\sigma(r)}{\sigma(t)} = \sigma(r) + e^{r-t}\sigma(-r) \tag{3.34}$$

for all r, t, and that $1 - \tanh x = 2\sigma(-2x)$ for all $x \in \mathbb{R}$. Thus

$$\frac{1-\tanh(a+bs)}{1-\tanh s} = \underbrace{\sigma(-2a-2bs)}_{\to 0} + e^{-2a}\underbrace{e^{2(1-b)s}}_{\to 0}\underbrace{\sigma(2a+2bs)}_{\to 1} \to 0$$

as desired. ∎

For each vector $\mathbf{a} \in \mathbb{R}^n$ and matrix $B \in \mathbb{R}^{n\times m}$, we write

$$S_{\mathbf{a},B} := \left\{\vec{\theta}_n(\mathbf{a}+Bu)\,,\; u \in \mathbb{R}^m\right\}.$$

Lemma 3.8.9 If $B \in \mathbf{B}_{n,m}$ and $\mathbf{a} \in \mathbb{R}^n$, then $0 \in \operatorname{int}\operatorname{co} S_{\mathbf{a},B}$.

Proof. Since $B \in \mathbf{B}_{n,m}$, there is some $u_0 \in \mathbb{R}^m$ such that the numbers $b_i := \mathrm{row}_i(B)u_0$ are all nonzero and have distinct absolute values. (Because the set of u's that satisfy at least one of the equations

$$\mathrm{row}_i(B)u = 0\,,\;\; \mathrm{row}_i(B)u + \mathrm{row}_j(B)u = 0\,,\;\; \mathrm{row}_i(B)u - \mathrm{row}_j(B)u = 0$$

for some $i \neq j$ is a finite union of hyperplanes in $\mathbb{R}^m$.) Fix one such u_0. As $S_{\mathbf{a},Bu_0} \subseteq S_{\mathbf{a},B}$, it is enough to show the result for Bu_0 instead of B. So we assume from now on that $B = (b_1, \ldots, b_n)'$ has just one column, and all its entries b_i are nonzero and have distinct absolute values.

Assume by way of contradiction that $0 \notin \operatorname{int} \operatorname{co} S_{\mathbf{a},B}$. Using a separating hyperplane†, we know that there is a nonzero row vector $c = (c_1, \ldots, c_n)$ such that $c\vec{\theta}_n(\mathbf{a} + Bu) \geq 0$ for all $u \in \mathbb{R}$. Writing $\mathbf{a} = (a_1, \ldots, a_n)'$, this means that

$$\sum_{i=1}^{n} c_i\, \theta(a_i + b_i u) \;\geq\; 0 \quad \forall u \in \mathbb{R}. \tag{3.35}$$

We now prove that such an inequality cannot hold, if the b_i are nonzero and have distinct absolute values, unless all the c_i are equal to 0. Since θ is odd, we may assume that each $b_i > 0$, since each term $c_i\theta(a_i + b_i u)$ for which $b_i < 0$ can be rewritten as $(-c_i)\theta(-a_i + (-b_i)u)$. Thus, reordering if needed, we assume that $0 < b_1 < \ldots < b_n$. Finally, dropping all those terms in the sum for which $c_i = 0$, we may assume that $c_1 \neq 0$ (using smaller n, if necessary). Taking the limit in (3.35) as $u \to -\infty$ we obtain $\sum_{i=1}^{n} c_i(-\theta_\infty) \geq 0$. So from (3.35) we obtain

$$\sum_{i=1}^{n} c_i\,(\theta_\infty - \theta(a_i + b_i u)) \;\leq\; 0 \quad \forall u \in \mathbb{R}. \tag{3.36}$$

Therefore

$$c_1 + \sum_{i=2}^{n} c_i \frac{\theta_\infty - \theta(a_i + b_i u)}{\theta_\infty - \theta(a_1 + b_1 u)} \;\leq\; 0 \quad \forall u \in \mathbb{R}. \tag{3.37}$$

If we prove that each term in the sum converges to zero as $u \to +\infty$ then it will follow that $c_1 \leq 0$. But this fact follows from property (3.33) (applied with $a = a_i - b_i a_1/b_1$, $b = b_i/b_1$, and noting that $s = a_1 + b_1 u \to \infty$ as $u \to \infty$).

If we take instead the limit in (3.35) as $u \to +\infty$, we find that $\sum_{i=1}^{n} c_i\theta_\infty \geq 0$. We therefore also obtain from (3.35) that:

$$\sum_{i=1}^{n} c_i\,(\theta_\infty + \theta(a_i + b_i u)) \;\geq\; 0 \quad \forall u \in \mathbb{R}. \tag{3.38}$$

Letting $v = -u$ and $\widetilde{a}_i = -a_i$, and using that θ is odd,

$$c_1 + \sum_{i=2}^{n} c_i \frac{\theta_\infty - \theta(\widetilde{a}_i + b_i v)}{\theta_\infty - \theta(\widetilde{a}_1 + b_1 v)} \;\geq\; 0 \quad \forall v \in \mathbb{R}. \tag{3.39}$$

Taking the limit as $v \to +\infty$ and appealing again to property (3.33), we conclude that also $c_1 \geq 0$. Thus $c_1 = 0$, contradicting the assumption made earlier. ∎

---

†Several basic facts about separation of convex sets are reviewed in the chapter on linear time-optimal control, see page 431.

A Necessity Statement

Lemma 3.8.10 If $B \notin \mathbf{B}_{n,m}$ then there exists a nonzero $\lambda \in \mathbb{R}^n$ so that

$$\operatorname{sign} \lambda' \vec{\theta}_n(Ax + Bu) \quad = \quad \operatorname{sign} \lambda' Ax$$

for all $(x, u) \in \mathbb{R}^n \times \mathbb{R}^m$ (with the convention $\operatorname{sign} 0 = 0$).

Proof. We denote, for each i, $b_i := \operatorname{row}_i(B)$ and $a_i := \operatorname{row}_i(A)$. There are three cases to consider: (1) some $b_i = 0$, (2) $b_i - b_j = 0$ for some $i \neq j$, and (3) $b_i + b_j = 0$ for some $i \neq j$.

In the first case, we let $\lambda := e_i$ (ith canonical basis vector). The conclusion follows from the equality

$$e_i' \vec{\theta}_n(Ax + Bu) = \theta(a_i x)$$

and the fact that $\operatorname{sign} \theta(v) = \operatorname{sign} v$. In the second case, we let $\lambda := e_i - e_j$. The expression

$$(e_i - e_j)' \vec{\theta}_n(Ax+Bu) \; = \; \theta(a_i x + b_i u) - \theta(a_j x + b_j u) \; = \; \theta(a_i x + b_i u) - \theta(a_j x + b_i u)$$

is nonnegative if and only if $a_i x \geq a_j x$ (using that θ is monotonic), that is, when $\lambda' A x = a_i x - a_j x \geq 0$, so also in this case we have the conclusion. Finally, if $b_i + b_j = 0$, we pick $\lambda := e_i + e_j$. Then

$$(e_i + e_j)' \vec{\theta}_n(Ax+Bu) = \theta(a_i x + b_i u) + \theta(a_j x + b_j u) = \theta(a_i x + b_i u) - \theta(-a_j x + b_i u)$$

(since θ is odd), which is nonnegative precisely when $(a_i + a_j)x \geq 0$. ∎

Proof of Theorem 8

From Corollary 3.8.6 and Lemma 3.8.9, since $f(x^0, u) = \vec{\theta}_n(Ax^0 + Bu)$, we have that $B \in \mathbf{B}_{n,m}$ implies strong local controllability at each x^0. To show the converse, suppose that $B \notin \mathbf{B}_{n,m}$. Let λ be as in Lemma 3.8.10 and let $\mathcal{V}$ be $\mathbb{R}^n$ if $\lambda' A = 0$ and $\{x \mid \lambda' A x > 0\}$ otherwise. Pick any state $x^0 \in \mathcal{V}$. Let $\mathcal{R}_{\mathcal{V}}(x^0)$ be the set of states reachable from x^0 without leaving $\mathcal{V}$. Then $\mathcal{R}_{\mathcal{V}}(x^0)$ is included in the half space

$$\{x \mid \lambda' x \geq \lambda' x^0\} \, .$$

This is because, if $\dot{\xi} = \vec{\theta}_n(A\xi + B\omega)$ and $\xi(t) \in \mathcal{V}$ for all t, then

$$\frac{d}{dt} \lambda' \xi(t) \; = \; \lambda' \vec{\theta}_n(A\xi(t) + B\omega(t)) \; \geq \; 0 \, ,$$

so the function $\varphi(t) := \lambda' \xi(t)$ is nondecreasing. Thus there cannot be a neighborhood of x^0 which is reachable from x^0 without leaving $\mathcal{V}$. This completes the proof of the Theorem. ∎

Exercise 3.8.11 One could also define a class of systems as in (3.28) with other choices of θ. Theorem 8 may not be true for such other choices. For instance, the theorem fails for $\theta =$ identity (why?). It also fails for $\theta =$ arctan: Show that the 4-dimensional, single-input system

$$\begin{aligned}\dot{x}_1 &= \arctan(x_1 + x_2 + x_3 + x_4 + 2u)\\ \dot{x}_2 &= \arctan(x_1 + x_2 + x_3 + x_4 + 12u)\\ \dot{x}_3 &= \arctan(-3u)\\ \dot{x}_4 &= \arctan(-4u)\end{aligned}$$

satisfies that $B \in \mathbf{B}_{n,m}$ but is not controllable. Explain exactly where the argument given for $\theta = \tanh$ breaks down. □

3.9 Piecewise Constant Controls

Often it is of interest to know how much more restrictive it is to consider only controls that are of a particular type, such as polynomial in time or piecewise constant. In general, for any m, σ, τ, let $\mathcal{A}$ be any subspace of $\mathcal{L}_m^\infty(\sigma, \tau)$ that satisfies:

1. For each $\omega \in \mathcal{L}_m^\infty$ there is an equibounded sequence in $\mathcal{A}$ converging to ω; and

2. it contains all constant controls.

For instance, one may take polynomial or piecewise constant controls (see Remark C.1.2 in Appendix C). For linear systems, controls in $\mathcal{A}$ are rich enough:

Proposition 3.9.1 Assume that Σ is a linear continuous-time system, $\sigma < \tau$, and $\mathcal{A}$ is as above. Then the following conditions are equivalent:

(a) Σ is controllable on $[\sigma, \tau]$.

(b) For each $x, z \in \mathcal{X}$, there exists some control in $\mathcal{A}$ such that $\phi(\tau, \sigma, x, \omega) = z$.

In particular, for time-invariant systems, controllability is equivalent to controllability using controls in $\mathcal{A}$.

Proof. Fix any $x \in \mathcal{X}$. By Theorem 1 (p. 57), Part 2(i), the set

$$\{\phi(\tau, \sigma, x, \omega) \mid \omega \in \mathcal{A}\}$$

is dense in $\mathcal{X}$. But this set is an affine subspace, since this is the set of all expressions of the form

$$\phi(\tau, \sigma, x, 0) + \phi(\tau, \sigma, 0, \omega),$$

and the map $\phi(\tau, \sigma, 0, \cdot)$ is linear. Thus, it must be all of $\mathcal{X}$. ∎

Proposition 3.9.2 Let Σ be a time-invariant continuous-time system of class $\mathcal{C}^1$ with $\mathcal{U} = \mathbb{R}^m$, and assume that there is some equilibrium pair (x^0, u^0) such that the linearization of Σ at (x^0, u^0) is controllable. Pick any $\sigma < \tau$ and any $\mathcal{A}$ as above. Then there exists a neighborhood V of x^0 such that for each $x, z \in V$ there exists some control $\omega \in \mathcal{A}$ such that $\phi(\tau, \sigma, x, \omega) = z$.

Proof. This is proved in exactly the same form as the local controllability Theorem 7 (p. 126), by application of the Implicit Mapping Theorem to the mapping α, with controls given the sup norm. The only modification is that one must restrict controls to $\mathcal{A}$, seen as a (not necessarily dense) subset of $\mathcal{L}_m^\infty(\sigma, \tau)$. But full rank is already achieved on these, because of Proposition 3.9.1 and the fact that the differential is nothing other than the corresponding reachability map for the linearization. ∎

Note that the above argument used the fact that α is $\mathcal{C}^1$ with respect to the sup norm and that the Implicit Function Theorem was applied about the control constantly equal to u^0 (which is in $\mathcal{A}$ by hypothesis).

Corollary 3.9.3 If Σ is as in Proposition 3.9.2, $\mathcal{U} = \mathbb{R}^m$, and Σ is controllable, then for each $x, z \in \mathcal{X}$ there exists some $\sigma \leq \tau$ and some piecewise constant control $\omega \in \mathcal{U}^{[\sigma,\tau)}$ such that $\phi(\tau, \sigma, x, \omega) = z$.

Proof. Pick first any $\sigma < \tau$ and apply the Proposition with $\mathcal{A}$ = family of all piecewise constant controls, to obtain the open set V. Let x, z be as in the statement. Because of the controllability assumption, there is some control mapping x into x^0. By Lemma 2.8.2 there exists some piecewise constant control ω_1 mapping x to some element x' in V, and similarly there exists a piecewise constant control ω_2 mapping some z' from V into z. Concatenating ω_1 with a piecewise constant control sending x' to z' and this in turn with ω_2, the conclusion follows. ∎

3.10 Notes and Comments

Controllability of Time-Invariant Systems

Since the early work on state-space approaches to control systems analysis, it was recognized that certain nondegeneracy assumptions were useful, in particular in the context of optimality results. However, it was not until R.E. Kalman's work (see, e.g., [215], [216], [218], and [231]) that the property of controllability was isolated as of interest in and of itself, as it characterizes the degrees of freedom available when attempting to control a system.

The study of controllability for linear systems has spanned a great number of research directions, and topics such as testing degrees of controllability, and their numerical analysis aspects, are still the subject of intensive research.

The idea in the proof of Corollary 3.2.7 was to use an ascending chain condition on the spaces $\mathcal{X}_i$ to conclude a finite-time reachability result. Similar arguments are used often in control theory, under other hypotheses than finite dimensionality (or finiteness, as in Lemma 3.2.4). For instance, for discrete-time linear systems over Noetherian rings, one may use the same idea in order to conclude that $\mathcal{R}(x) = \mathcal{R}^T(x)$ for large enough T.

Controllability of continuous-time piecewise linear systems is studied in [267] and [411].

There is also an extensive literature on the controllability properties of infinite dimensional continuous-time or discrete-time linear systems; see, for instance, [107], [147], and [291]. For such systems it is natural to characterize "almost" reachability, where the set of states reachable from a given state is required only to be dense in the state space.

Algebraic Facts About Controllability

What we called the "Hautus condition" actually appeared first in [321], Theorem 1 on page 320, and in a number of other references, including [40]. However, [175] was the first to stress its wide applicability in proving results for linear systems, as well as in extending the criterion to stabilizability (asymptotic controllability) in [176]. Sometimes the condition is also referred to as the "PBH condition" because of the Belevitch and Popov contributions.

It should be emphasized that genericity of controllability was established only with respect to the set of *all* possible pairs. Often one deals with restricted classes of systems, and among these, controllable systems may or may not form an open dense subset. This gives rise to the study of *structural controllability*; see, for example, [298], and more generally the book [350] and references therein, for related questions.

In structural controllability studies, classes of systems are defined typically by constraints on the matrices (A, B). For example, any system obtained from an equation of the type

$$\ddot{x} + \alpha\dot{x} + \beta x = \gamma u$$

(as arises with a damped spring-mass system) via the introduction of state variables $x_1 := x, x_2 := \dot{x}$ will have $A_{11} = B_{11} = 0$ and $A_{12} = 1$. These coefficients $0, 1$ are independent of experimental data. Graph-theoretic methods are typically used in structural controllability.

Controllability Under Sampling

Necessary and sufficient conditions for the preservation of controllability under sampling for multivariable ($m > 1$) systems are known, but are more complicated to state than the sufficient condition in Theorem 4 (p. 102). See, for instance, [154].

The dual version of the Theorem, for observability (see Chapter 6), is very closely related to Shannon's Sampling Theorem in digital signal processing.

Some generalizations to certain types of nonlinear systems are also known (see, for instance, [208] and [371]).

More on Controllability of Linear Systems

The result in Exercise 3.5.7 is a particular case of the more general fact, for time-invariant continuous-time single-input ($m = 1$) systems, that the operator that gives the optimal transfer from 0 to a state x in time ε has an operator norm

$$\left\|L^{\#}\right\| = O(\varepsilon^{-n+\frac{1}{2}})$$

for small ε. This is proved in [345], where the multiple-input case is also characterized.

Bounded Controls

Theorem 6 (p. 117) can be found in [343]. A particular case had earlier been covered in [266], and the general case, as well as the obvious discrete-time analogue, were proved in an abstract algebraic manner in [364]. We based our presentation on [170], which in turn credits the result, in the stronger form given in Corollary 3.6.7, to the earlier thesis [196].

It is also possible to give characterizations of controllability under other constraints on control values. For instance, positive controls are treated in [60].

First-Order Local Controllability

The problem of characterizing local controllability when the first-order test given in Theorem 7 fails is extremely hard, and constitutes one of the most challenging areas of nonlinear control research. It is often of interest to strengthen the definition of local controllability: For instance, about an equilibrium state x one might require that any state close to x be reachable from x in small time (*small-time local controllability*). Such stronger notions give rise to interesting variations of the nonlinear theory; see, e.g., [388] and references there for this and related issues.

Recurrent Nets

The systems studied in Section 3.8 are often called "continuous-time recurrent neural networks". The motivation for the term comes from an interpretation of the vector equations for x in (3.28) as representing the evolution of an ensemble of n "neurons," where each coordinate x_i of x is a real-valued variable which represents the internal state of the ith neuron, and each coordinate $u_i, i = 1, \ldots, m$ of u is an external input signal. The coefficients A_{ij}, B_{ij} denote the weights, intensities, or "synaptic strengths," of the various connections. The choice $\theta = \tanh$, besides providing a real-analytic and globally Lipschitz right-hand side in (3.28), has major advantages in numerical computations, due to the fact that its derivative can be evaluated from the function itself ($\theta' = 1 - \theta^2$). (Sometimes, however, other "activation functions" are used. One common choice is the function $\sigma(x) = (1 + e^{-x})^{-1}$, which amounts to a rescaling of θ to the range $(0, 1)$.) Among the variants of the basic model (3.28) that have been studied in the literature are systems of the general form $\dot{x} = -x + \vec{\theta}_n(Ax + Bu)$, where the term $-x$ provides stability. Recurrent nets (or these variants) arise in digital signal processing, control, design of associative memories ("Hopfield nets"), language inference, and sequence extrapolation for time series prediction, and can be shown to approximate a large class of nonlinear systems; see e.g. [237].

The implication $2 \Rightarrow 1$ in Theorem 8 is from [375], where the reader may also find a solution to Exercise 3.8.11. The necessity of the condition $B \in \mathbf{B}_{n,m}$ was shown to the author by Y. Qiao. A characterization of a weak controllability property for the discrete-time analogue of these systems had been given earlier, in [12], which had also obtained partial results for the continuous-time problem.

Piecewise Constant Controls

Results similar to Proposition 3.9.1 are in fact true for much larger classes of systems. It is possible to establish, for instance, that piecewise constant controls are enough for any analytic continuous-time system. In [367], it is shown that under one weak extra assumption, polynomial controls can be used for controllable analytic systems. See [163] for a survey of such results. This section gave only very simple versions of these more general facts.

Chapter 4

Nonlinear Controllability

In this chapter* we study controllability questions for time-invariant continuous-time systems $\dot{x} = f(x, u)$.

The property that is easiest to characterize completely for nonlinear continuous-time systems is the *accessibility* property, sometimes also referred to as *weak controllability*. This is the property that one should be able to reach from any given state a set of full dimension. However, for certain restricted classes of systems, results on complete controllability are also available.

To begin, we establish some basic facts regarding Lie algebras of vector fields. (Working on open subsets of Euclidean spaces, we do not make any explicit use of notions of differential geometry.)

4.1 Lie Brackets

If $f = (f_1, \ldots, f_n)' : \mathcal{O} \to \mathbb{R}^n$ is a continuously differentiable map defined on some open subset $\mathcal{O} \subseteq \mathbb{R}^p$, f_* denotes the Jacobian of f, thought of as a matrix function on $\mathcal{O}$. That is, for each $x^0 \in \mathcal{O}$, $f_*(x^0) \in \mathbb{R}^{n \times p}$ is the Jacobian of f evaluated at x^0, the matrix whose (i,j)-th entry is $\frac{\partial f_i}{\partial x_j}|_{x=x^0}$.

In the special case when $p = n$, one calls a continuously differentiable map $f : \mathcal{O} \to \mathbb{R}^n$ a *vector field* defined on the open subset $\mathcal{O}$ of $\mathbb{R}^n$. A *smooth* vector field is a smooth (infinitely differentiable) $f : \mathcal{O} \to \mathbb{R}^n$; from now on, and unless otherwise stated, when we say vector field we will mean "smooth vector field". The set of all (smooth) vector fields on a given $\mathcal{O} \subseteq \mathbb{R}^n$ is denoted by $\mathbb{V}(\mathcal{O})$. It is a real vector space under pointwise operations, that is, $(rf + g)(x) := rf(x) + g(x)$ for all $r \in \mathbb{R}$, $f, g \in \mathbb{V}(\mathcal{O})$, and $x \in \mathcal{O}$.

When $\varphi : \mathcal{O} \to \mathbb{R}$ is differentiable, $\varphi_*(x) = \nabla\varphi(x)$ is a row vector, the gradient of φ evaluated at x. The set of smooth functions $\mathcal{O} \to \mathbb{R}$ is denoted by $\mathbb{F}(\mathcal{O})$. This is also a real vector space under pointwise operations.

*The rest of the book, except for Section 5.3, is independent of the material in this chapter.

For each $f \in \mathbb{V}(\mathcal{O})$ and each $\varphi \in \mathbb{F}(\mathcal{O})$, $L_f\varphi \in \mathbb{F}(\mathcal{O})$ is the directional or Lie derivative of φ along f:

$$(L_f\varphi)(x) := \varphi_*(x) f(x).$$

One may view L_f as a linear operator $\mathbb{F}(\mathcal{O}) \to \mathbb{F}(\mathcal{O})$. Two vector fields $f, g \in \mathbb{V}(\mathcal{O})$ are equal if and only if $L_f = L_g$. (Because, if $L_f\varphi = L_g\varphi$ holds for all φ, then it holds in particular for each of the n coordinate functions $\pi_i(x) = x_i$, and hence the ith coordinates of f and g coincide: $f_i(x) = (L_f\pi_i)(x) = (L_g\pi_i)(x) = g_i(x)$.) Note that L_f is a first-order differential operator, while the composition $L_f \circ L_g$, which we write simply as L_fL_g, is a second-order operator: one can verify easily that

$$L_fL_g\,\varphi = g'H_\varphi f + \varphi_* g_* f\,, \tag{4.1}$$

where H_φ is the Hessian matrix $\left(\frac{\partial^2\varphi}{\partial x_i \partial x_j}\right)$. Since H_φ is symmetric,

$$L_fL_g\,\varphi - L_gL_f\,\varphi = L_{g_*f - f_*g}\,\varphi\,. \tag{4.2}$$

Definition 4.1.1 *The* **Lie bracket** *of $f, g \in \mathbb{V}(\mathcal{O})$ is $[f, g] := g_*f - f_*g \in \mathbb{V}(\mathcal{O})$.* □

Equation (4.2) says that

$$L_{[f,g]} = L_fL_g - L_gL_f \tag{4.3}$$

for all vector fields f and g. The binary operation $f, g \mapsto [f, g]$ is skew-symmetric: $[f, g] = -[g, f]$ and bilinear: $[rf_1 + f_2, g] = r[f_1, g] + [f_2, g]$ and $[f, rg_1 + g_2] = r[f, g_1] + [f, g_2]$. It is convenient to write

$$\mathrm{ad}_f g := [f, g] \tag{4.4}$$

and to think of ad_f, for each fixed $f \in \mathbb{V}(\mathcal{O})$, as a linear operator $\mathbb{V}(\mathcal{O}) \to \mathbb{V}(\mathcal{O})$. This operator is a *differentiation* operator with respect to the Lie bracket:

$$\mathrm{ad}_f[g, h] = [\mathrm{ad}_f g, h] + [g, \mathrm{ad}_f h]$$

for all f, g, h, a formula also known as the *Jacobi identity*, especially when written in the equivalent form

$$[f, [g, h]] + [h, [f, g]] + [g, [h, f]] = 0\,.$$

This is immediate from the equality

$$L_{[f,[g,h]]} = L_fL_gL_h - L_fL_hL_g - L_gL_hL_f + L_hL_gL_f$$

obtained by using Equation (4.3) twice and using similar expansions for the other two terms.

Some properties of, and relations between, the operators just introduced are as follows.

Lemma/Exercise 4.1.2 For any $f, g \in \mathbb{V}(\mathcal{O})$ and any $\varphi, \psi \in \mathbb{F}(\mathcal{O})$,

- $L_f(\varphi\psi) = (L_f\varphi)\psi + \varphi(L_f\psi)$,
- $L_{\varphi f}\psi = \varphi L_f\psi$,
- $[\varphi f, \psi g] = \varphi\psi[f,g] + (L_f\psi)\varphi g - (L_g\varphi)\psi f$.

Definition 4.1.3 *A* **Lie algebra** *(of vector fields on $\mathcal{O}$) is a linear subspace $S \subseteq \mathbb{V}(\mathcal{O})$ that is closed under the Lie bracket operation, that is, $[f,g] \in S$ whenever f and g are in S.* □

For any subset $\mathcal{A} \subseteq \mathbb{V}(\mathcal{O})$, we define $\mathcal{A}_{\text{LA}}$, the *Lie algebra generated by $\mathcal{A}$*, as the intersection of all the Lie algebras of vector fields which contain $\mathcal{A}$. (The set of all such algebras is nonempty, since it includes $\mathbb{V}(\mathcal{O})$.) An intersection of any family of Lie algebras is also a Lie algebra; thus,

$$\mathcal{A}_{\text{LA}} = \text{smallest Lie algebra of vector fields which contains } \mathcal{A}.$$

Lemma 4.1.4 Let $\mathcal{A}$ be a subset of $\mathbb{V}(\mathcal{O})$. Denote $\mathcal{A}_0 := \mathcal{A}$, and, recursively,

$$\mathcal{A}_{k+1} := \{[f,g] \mid f \in \mathcal{A}_k, g \in \mathcal{A}\}, \quad k = 0, 1, 2, \ldots,$$

as well as $\mathcal{A}_\infty := \bigcup_{k \geq 0} \mathcal{A}_k$. Then, $\mathcal{A}_{\text{LA}}$ is equal to the linear span of $\mathcal{A}_\infty$.

Proof. Any Lie algebra which contains $\mathcal{A}$ must contain $\mathcal{A}_\infty$ (because, inductively, it contains each $\mathcal{A}_k$), and hence also contains its linear span, which we will denote by $\widetilde{\mathcal{A}}$. So we must only show that $\widetilde{\mathcal{A}}$ is a Lie algebra, i.e., that it is closed under the Lie bracket operation. Since $[X, \cdot]$ is linear, it suffices to show that

$$X \in \widetilde{\mathcal{A}} \text{ and } Y \in \mathcal{A}_k \Rightarrow [X,Y] \in \widetilde{\mathcal{A}} \qquad (P_k)$$

holds for every k. Observe that, since each $X \in \widetilde{\mathcal{A}}$ can be written as a linear combination of elements of $\mathcal{A}_\infty$, and since $[\cdot, Y]$ is linear, property (P_k) is equivalent to the statement that $[X,Y] \in \widetilde{\mathcal{A}}$ whenever $X \in \mathcal{A}_\infty$ and $Y \in \mathcal{A}_k$. We show (P_k) by induction on k.

The case $k = 0$ is clear since, for any $f \in \mathcal{A}_0 = \mathcal{A}$, by definition $[X, f] \in \mathcal{A}_{\ell+1}$ if $X \in \mathcal{A}_\ell$. Assume now that the result has been proved for all indices less than or equal to k, and pick any $X \in \widetilde{\mathcal{A}}$ and any $Y \in \mathcal{A}_{k+1}$. Thus we may write $Y = [Y_0, f]$, for some $f \in \mathcal{A}$ and $Y_0 \in \mathcal{A}_k$. The Jacobi identity gives:

$$[X,Y] = [X, [Y_0, f]] = [[X, Y_0], f] - [[X, f], Y_0].$$

By inductive assumption, $[X, Y_0] \in \widetilde{\mathcal{A}}$, so (induction again) $[[X, Y_0], f] \in \widetilde{\mathcal{A}}$. Similarly, by induction we know that $[[X, f], Y_0] \in \widetilde{\mathcal{A}}$. As $\widetilde{\mathcal{A}}$ is a subspace, $[X,Y] \in \widetilde{\mathcal{A}}$, as desired. ∎

The Lemma says that every element in the Lie algebra generated by the set $\mathcal{A}$ can be expressed as a linear combination of iterated brackets of the form

$$[[\cdots[f_1, f_2], f_3], \ldots, f_\ell],$$

for some $f_i \in \mathcal{A}$. (We make the convention that an "iterated bracket of length $\ell = 1$" is just an element f_1 of the generating set $\mathcal{A}$.) Equivalently, by skew symmetry, one can write any such element as a combination of brackets $[f_\ell, \ldots, [f_3, [f_2, f_1]]]$, that is, as $\mathrm{ad}_{f_\ell} \ldots \mathrm{ad}_{f_2} f_1$. So $\mathcal{A}_{\mathrm{LA}}$ is the smallest linear subspace of $\mathbb{V}(\mathcal{O})$ which includes $\mathcal{A}$ and is invariant under the linear operators ad_f, $f \in \mathcal{A}$.

A Digression on Linear Algebra

Before proceeding, we make a simple but useful observation in linear algebra. We fix any $k = 0, 1, \ldots, \infty$ or $k = \omega$ (as usual, we let $\mathcal{C}^\omega$ denote the set of analytic functions).

Lemma 4.1.5 Suppose that $Q : W \to \mathbb{R}^{n \times p}$ is a matrix function of class $\mathcal{C}^k$, defined on an open subset $W \subseteq \mathbb{R}^m$, which satisfies that, for some q, $\operatorname{rank} Q(w) = q$ for all $w \in W$. Then, for each $w_0 \in W$, there is some neighborhood W_0 of w_0, and there are two matrix functions $C : W_0 \to \mathbb{R}^{n \times n}$ and $D : W_0 \to \mathbb{R}^{p \times p}$, both of class $\mathcal{C}^k$ and nonsingular for each $w \in W_0$, such that

$$C(w)\,Q(w)\,D(w) = \begin{pmatrix} I & 0 \\ 0 & 0 \end{pmatrix} \quad \text{for all } w \in W_0\,, \tag{4.5}$$

where I is the $q \times q$ identity matrix.

Proof. Pick any $w_0 \in W$. Since $Q(w_0)$ has rank q, there exist two permutation matrices $P_1 \in \mathbb{R}^{n \times n}$ and $P_2 \in \mathbb{R}^{p \times p}$ so that the submatrix obtained from the first q rows and columns of $P_1 Q(w_0) P_2$ is nonsingular.

Now let $\Delta(w)$ be the submatrix obtained by selecting the first q rows and columns of $P_1 Q(w) P_2$, for each $w \in W$, that is,

$$P_1 Q(w) P_2 = \begin{pmatrix} \Delta(w) & * \\ * & * \end{pmatrix}$$

(where asterisks indicate arbitrary functions of w of class $\mathcal{C}^k$). Thus $\Delta(w)$ is a matrix function of class $\mathcal{C}^k$, and is nonsingular at $w = w_0$. Since the determinant of $\Delta(w)$ is a continuous function of w, there is some open subset W_0 of W such that $\det \Delta(w) \neq 0$ for all $w \in W_0$. On W_0, we consider the matrix function

$$Q_1(w) := P_1 Q(w) P_2 \begin{pmatrix} \Delta(w)^{-1} & 0 \\ 0 & I \end{pmatrix} = \begin{pmatrix} I & X(w) \\ Y(w) & * \end{pmatrix}$$

where I is an identity matrix of size $q \times q$ and $X(w)$ and $Y(w)$ are of class C^k. Note that $\operatorname{rank} Q_1(w) = \operatorname{rank} Q(w) = q$ for all w. So this matrix has rank $\equiv q$:

$$\begin{pmatrix} I & 0 \\ -Y(w) & I \end{pmatrix} Q_1(w) \begin{pmatrix} I & -X(w) \\ 0 & I \end{pmatrix} = \begin{pmatrix} I & 0 \\ 0 & * \end{pmatrix}$$

and therefore the "*" block is identically zero. In conclusion,

$$\left[\begin{pmatrix} I & 0 \\ -Y(w) & I \end{pmatrix} P_1\right] Q(w) \left[P_2 \begin{pmatrix} \Delta(w)^{-1} & 0 \\ 0 & I \end{pmatrix} \begin{pmatrix} I & -X(w) \\ 0 & I \end{pmatrix}\right] = \begin{pmatrix} I & 0 \\ 0 & 0 \end{pmatrix},$$

and the proof is complete. ∎

Corollary 4.1.6 Suppose that $Q : W \to \mathbb{R}^{n \times p}$ and $f : W \to \mathbb{R}^n$ are a matrix and a vector function, both of class C^k, defined on the same open subset of some space $\mathbb{R}^m$, and that $Q(w)$ has constant rank on W. Then, the following two properties are equivalent:

- For each $w \in W$, $f(w)$ belongs to the subspace of $\mathbb{R}^n$ spanned by the columns of $Q(w)$.
- For each $w_0 \in W$, there exist a neighborhood W_0 of w_0 and a class C^k function $\alpha : W_0 \to \mathbb{R}^p$, so that

$$f(w) = Q(w)\alpha(w) \quad \text{for all } w \in W_0 . \tag{4.6}$$

Proof. The second property obviously implies the first, so we need only to prove the other implication. Thus, assume that $\operatorname{rank}(Q(w), f(w)) = \operatorname{rank} Q(w) = q$ for all $w \in W$, and pick any $w_0 \in W$. By Lemma 4.1.5, there exist a neighborhood W_0 of w_0, and $C(w)$ and $D(w)$ of class C^k on W_0, so that Equation (4.5) holds. Note that the matrix $(C(w)Q(w)D(w), C(w)f(w))$ has the same rank as the matrix $(Q(w), f(w))$ (since $C(w)$ and $D(w)$ are nonsingular), so

$$\operatorname{rank} \left(\begin{array}{cc|c} I & 0 & \\ 0 & 0 & C(w)f(w) \end{array}\right) = q$$

for all w. We conclude that the last $n-q$ coordinates of $C(w)f(w)$ must vanish, that is, there is a class C^k function $\alpha_0 : W_0 \to \mathbb{R}^q$ so that $C(w)f(w) = \begin{pmatrix} \alpha_0(w) \\ 0_{n-q} \end{pmatrix}$ on W_0, where 0_{n-q} is the zero vector in $\mathbb{R}^{n-q}$. We conclude that

$$C(w)Q(w)D(w) \begin{pmatrix} \alpha_0(w) \\ 0_{p-q} \end{pmatrix} = C(w)f(w),$$

and therefore, since $C(w)$ is nonsingular, that $\alpha(w) := D(w) \begin{pmatrix} \alpha_0(w) \\ 0_{p-q} \end{pmatrix}$ is as desired. ∎

Tangent Vectors

Let $W \subseteq \mathbb{R}^p$ and $\mathcal{O} \subseteq \mathbb{R}^n$ be open subsets of Euclidean spaces. We will call a smooth map

$$M : W \to \mathcal{O}$$

a *slice* if its Jacobian $M_*(w)$ has rank p at each $w \in W$. The vector field $f \in \mathbb{V}(\mathcal{O})$ is said to be *tangent to* M if, for all $w \in W$, $f(M(w))$ is in the subspace of $\mathbb{R}^n$ spanned by the columns of $M_*(w)$.

Remark 4.1.7 Geometrically, we think of the image of M as a "p-dimensional smooth slice" of $\mathcal{O}$. In differential-geometric terms, when M is one-to-one we are providing a chart for the submanifold represented by this image. Tangent vector fields in the sense just defined are precisely those which, when restricted to the submanifold, are tangent to it in the usual sense. □

By Corollary 4.1.6, applied to the composition $f \circ M : W \to \mathbb{R}^n$, f is tangent to M if and only if for each $w_0 \in W$ there is some neighborhood W_0 of w_0 and a smooth function $\alpha : W_0 \to \mathbb{R}^p$ so that

$$f(M(w)) = M_*(w)\alpha(w) \quad \text{for all } w \in W_0 . \tag{4.7}$$

It is clear that the set of all vector fields tangent to M forms a linear subspace of $\mathbb{V}(\mathcal{O})$. One of the most useful facts about Lie brackets is expressed in the next result.

Lemma 4.1.8 Let $M : W \to \mathcal{O}$ be a slice. Then the set of all vector fields tangent to M is a Lie algebra.

Proof. Pick any two tangent vector fields f, g and any $w_0 \in W$. We must show that $[f,g](M(w_0))$ belongs to the subspace of $\mathbb{R}^n$ spanned by the columns of $M_*(w_0)$. Pick a (common) neighborhood W_0 of w_0 and smooth functions $\alpha : W_0 \to \mathbb{R}^p$ and $\beta : W_0 \to \mathbb{R}^p$ so that $f(M(w)) = M_*(w)\alpha(w)$ and $g(M(w)) = M_*(w)\beta(w)$ for all $w \in W_0$. We will prove the following formula, from which the conclusion will follow:

$$[f,g](M(w)) \;=\; M_*(w)\,[\alpha,\beta](w) \quad \text{for all } w \in W_0 . \tag{4.8}$$

(Note that $[f,g]$ is a Lie bracket of vector fields in $\mathbb{R}^n$, and $[\alpha,\beta]$ is a Lie bracket of vector fields in $\mathbb{R}^p$. The equation amounts to the statement that vector fields transform covariantly under smooth maps.) From now on, when we write w we mean "for each $w \in W_0$". Consider:

$$\widetilde{f}(w) := f(M(w)) = M_*(w)\alpha(w)\,, \quad \widetilde{g}(w) := g(M(w)) = M_*(w)\beta(w)\,. \tag{4.9}$$

The chain rule gives

$$(\widetilde{f})_*(w) = f_*(M(w))M_*(w)\,, \quad (\widetilde{g})_*(w) = g_*(M(w))M_*(w)\,. \tag{4.10}$$

We pick any $i \in \{1, \ldots, n\}$, let e_i be the ith canonical basis vector, and consider the smooth function

$$\varphi_i := e_i' M : W_0 \to \mathbb{R}.$$

From (4.9) we have that

$$e_i'\widetilde{f} = L_\alpha \varphi_i \quad \text{and} \quad e_i'\widetilde{g} = L_\beta \varphi_i. \tag{4.11}$$

So

$$\begin{aligned} e_i'(g_* f)(M(w)) &= e_i' g_*(M(w)) M_*(w)\, \alpha(w) \\ &= e_i'(\widetilde{g})_*(w)\, \alpha(w) = (e_i'\widetilde{g})_*(w)\, \alpha(w) && \text{using (4.10)} \\ &= (L_\beta \varphi_i)_*(w)\, \alpha(w) = (L_\alpha L_\beta \varphi_i)(w) && \text{using (4.11).} \end{aligned}$$

An analogous argument gives that

$$e_i'(f_* g)(M(w)) = (L_\beta L_\alpha \varphi_i)(w)$$

so we conclude, from the fact that $L_{[\alpha,\beta]} = L_\alpha L_\beta - L_\beta L_\alpha$ (Equation (4.3), applied to α and β) that

$$e_i'[f,g](M(w)) = e_i' M_*(w)\, [\alpha, \beta](w).$$

As i was arbitrary, (4.8) holds. ∎

4.2 Lie Algebras and Flows

We turn next to establishing a connection between Lie algebras of vector fields and sets of points reachable by following flows of vector fields.

Some Additional Facts Concerning Differential Equations

For an open subset $\mathcal{O} \subseteq \mathbb{R}^n$ and a (continuously differentiable but not necessarily smooth) vector field $f : \mathcal{O} \to \mathbb{R}^n$, we write ϕ, or ϕ_f if we wish to emphasize f, to denote the *flow* of f. This is the map that associates to $(t, x^0) \in \mathbb{R} \times \mathcal{O}$ the solution $\phi(t, x^0) = x(t)$ at time $t \in \mathbb{R}$, if defined, of the initial value problem

$$\dot{x} = f(x), \quad x(0) = x^0.$$

(If the solution of this initial value problem does not exist on an interval containing 0 and t, $\phi(t, x^0)$ is undefined.) We denote by $\mathcal{D}_f$, or just $\mathcal{D}$ if f is clear from the context, the domain of ϕ. This is a subset of $\mathbb{R} \times \mathcal{O}$ which contains $\{0\} \times \mathcal{O}$.

Exercise 4.2.1 Let $\mathcal{O} = \mathbb{R}$ and $f(x) = x(1-x)$. Show that $\mathcal{D}$ is the union of $\mathbb{R} \times [0,1]$,

$$\left\{(t,x) \in \mathbb{R}^2 \;\middle|\; x < 0 \text{ and } t < \ln \frac{x-1}{x}\right\},$$

and

$$\left\{(t,x) \in \mathbb{R}^2 \;\middle|\; x > 1 \text{ and } t > \ln \frac{x-1}{x}\right\}.$$

Graph $\mathcal{D}$ as a subset of $\mathbb{R}^2$, and find an explicit formula for the flow $\phi(t,x)$. □

The set $\mathcal{D}$ is always open and the map $\phi : \mathcal{D} \to \mathcal{O}$ is continuous. This is a corollary of Theorem 1 (p. 57), arguing as follows. (See also Remark 3.7.3.) Let $(t, x^0) \in \mathcal{D}$, and pick any numbers $a < 0 < b$ so that $[a,b] \times \{x^0\} \subseteq \mathcal{D}$ and $t \in (a,b)$. The map

$$\mathcal{O}_0 \to C^0([a,b],\mathcal{O}) \;:\; z \mapsto \text{ restriction of } \phi(\cdot,z) \text{ to } [a,b] \tag{4.12}$$

is defined for all z in some neighborhood $\mathcal{O}_0$ of x^0, and it is continuous when $C^0([a,b],\mathcal{O})$ is endowed with the uniform convergence norm. This is shown in Theorem 1. (The proof there applies only to the interval $[0,b]$, but the result can also be applied to the "reversed-time" system $\dot{x} = -f(x)$, and this gives the conclusion for the interval $[a,0]$ as well; thus, the result is also true on $[a,b]$.) In particular, it follows that $(a,b) \times \mathcal{O}_0 \subseteq \mathcal{D}$, so the domain $\mathcal{D}$ is open. To prove continuity, assume that $(t_k, z_k) \to (t, x^0)$ as $k \to \infty$ and take any $\varepsilon > 0$. Since (4.12) is continuous, there is some K such that $k > K$ implies $|\phi(s,x^0) - \phi(s,z_k)| < \varepsilon/2$ for all $s \in [a,b]$, and since $\phi(\cdot,x^0)$ is continuous, we may also assume that $|\phi(t_k,x^0) - \phi(t,x^0)| < \varepsilon/2$ for all such k. Furthermore, we may assume that $t_k \in [a,b]$. Thus

$$|\phi(t_k,z_k) - \phi(t,x^0)| \le |\phi(t_k,z_k) - \phi(t_k,x^0)| + |\phi(t_k,x^0) - \phi(t,x^0)| < \varepsilon$$

for all $k > K$.

Since f is of class C^1, the map $\phi : \mathcal{D} \to \mathcal{O}$ is in fact continuously differentiable. Like continuity, this is also a corollary of Theorem 1 (p. 57), where we now argue as follows. It is shown in that Theorem (as before, the result was only proved for positive times, but the generalization to $\tau < 0$ is immediate) that, for each τ, the map $z \mapsto \phi(\tau,z)$ is continuously differentiable, and its partial derivative with respect to the ith coordinate, evaluated at $z = z^0$, is

$$\frac{\partial \phi}{\partial x_i}(\tau, z^0) \;=\; \lambda_{z^0}(\tau)\,, \tag{4.13}$$

where $\lambda_{z^0}(\cdot)$ is the solution of the variational equation $\dot{\lambda}(t) = f_*(\phi(t,z^0))\lambda(t)$ on $[0,\tau]$ (or $[\tau,0]$, if τ is negative) with initial condition $\lambda(0) = e_i$, the ith canonical basis vector in $\mathbb{R}^n$.

Moreover, $\frac{\partial \phi}{\partial x_i}(\tau,z^0) = \lambda_{z^0}(\tau)$ depends continuously on (τ,z^0). Indeed, suppose that $(\tau_k, z_k) \to (\tau, z^0)$. As a first step, we remark that $\lambda_{z_k}(s)$ converges uniformly to $\lambda_{z^0}(s)$, for $s \in [a,b]$. This can be proved by noticing that the same proof that was used to show that $\lambda_{z_k}(t)$ converges to $\lambda_{z^0}(t)$ for each t, that is to say the continuity of $z \mapsto \frac{\partial \phi}{\partial x_i}(t,z)$, gives automatically that $z \mapsto \frac{\partial \phi}{\partial x_i}(\cdot,z)$ is also continuous as a map into $C^0([a,b],\mathbb{R}^n)$; see the arguments after Equation (2.40). (Another proof is by viewing $\dot{\lambda} = f_*(\phi(t,x))\lambda$ as a system $\dot{\lambda} = f_*(u)\lambda$ where $u = \phi(t,x)$ is a control, and using continuity with respect to controls.) As a second step, we point out that $\lambda_{z^0}(\tau_k) \to \lambda_{z^0}(\tau)$, because λ_{z^0} is continuous, being the solution of a differential equation. Writing $\lambda_{z_k}(\tau_k) - \lambda_{z^0}(\tau) = \lambda_{z_k}(\tau_k) - \lambda_{z^0}(\tau_k) + \lambda_{z^0}(\tau_k) - \lambda_{z^0}(\tau)$, we conclude continuity as earlier.

To show that ϕ is C^1, we need to prove also that $\frac{\partial \phi}{\partial t}$ is continuous. But this is clear from the facts that $\frac{\partial \phi}{\partial t}(t,x) = f(\phi(t,x))$ and $\phi(t,x)$ is continuous on (t,x).

Observe also that, being the solution of a differential equation, $\lambda_{z^0}(\cdot)$ is differentiable, from which it follows that $\frac{\partial^2 \phi}{\partial t \partial x_i}$ exists. Similarly, $\frac{\partial^2 \phi}{\partial t^2}$ exists, because $\frac{\partial}{\partial t} f(\phi(t,x)) = f_*(\phi)\frac{\partial \phi}{\partial t}$.

Lemma 4.2.2 Pick any $k = 1, 2, \ldots, \infty$. If f is of class C^k, then ϕ is of class C^k.

Proof. The case $k = 1$ was established in the previous discussion. By induction, assume that the result has been proved for all $1 \leq \ell < k < \infty$; we now prove it for k as well. Consider the following differential equation in $\mathcal{O} \times \mathbb{R}^n$:

$$\frac{d}{dt}\begin{pmatrix} x \\ \lambda \end{pmatrix} = \begin{pmatrix} f(x) \\ f_*(x)\lambda \end{pmatrix} = F\begin{pmatrix} x \\ \lambda \end{pmatrix}. \tag{4.14}$$

Observe that F is of class C^{k-1} and $k - 1 \geq 1$. For any fixed $i = 1, \ldots, n$, letting e_i be the ith canonical basis vector,

$$\phi_F(t,(x,e_i)) = \begin{pmatrix} \phi(t,x) \\ \frac{\partial \phi}{\partial x_i}(t,x) \end{pmatrix}.$$

The inductive hypothesis applies to Equation (4.14), implying that the flow $\phi_F(\cdot,(\cdot,\cdot))$ is of class C^{k-1}, so also $\phi_F(\cdot,(\cdot,e_i))$ is C^{k-1}, from which we conclude that $\frac{\partial \phi}{\partial x_i}$ is $k-1$-times continuously differentiable. Thus every partial derivative of order k which starts with $\frac{\partial}{\partial x_i}$ exists and is continuous.

If we show that $\frac{\partial \phi}{\partial t}$ is also C^{k-1}, it will have been proved that ϕ is C^k. To see this, consider the new extended system, also in $\mathcal{O} \times \mathbb{R}^n$,

$$\frac{d}{dt}\begin{pmatrix} x \\ z \end{pmatrix} = \begin{pmatrix} f(x) \\ f_*(x)f(x) \end{pmatrix} = G\begin{pmatrix} x \\ z \end{pmatrix}. \tag{4.15}$$

Since $\dot{x} = f(x)$ implies $\ddot{x} = f_*(x)f(x)$, we have that

$$\phi_G(t,(x,f(x))) = \begin{pmatrix} \phi(t,x) \\ \frac{\partial \phi}{\partial t}(t,x) \end{pmatrix}$$

and, by induction, ϕ_G is of class C^{k-1}, so $\frac{\partial \phi}{\partial t}$ is indeed C^{k-1}. ∎

Remark 4.2.3 It is tempting to try to show the continuity of $\frac{\partial \phi}{\partial x_i}$ (case $k = 1$) by the same argument, using Equation (4.14). However, a problem with that argument is that F would only be assured of being continuous, so an additional assumption is required (for instance, that f is C^2). The argument given for $k = 1$, before the Lemma, which views x as an input to the second equation, avoids this assumption. Yet another argument would be based on the general fact that solutions of ode's depend continuously on initial data provided only that the right hand side be continuous – not necessarily Lipschitz – and that the equation be known to have unique solutions (as Equation (4.14) does, even if f is only C^1, because of its special triangular structure). □

Exercise 4.2.4 Show that, if f is of class C^k, all the partial derivatives of ϕ of order k are differentiable with respect to t. (*Hint:* Show, by induction, that every such derivative is also the solution of a differential equation.) □

Exercise 4.2.5 Find explicitly the extended flow ϕ_F associated to the system in Exercise 4.2.1. □

Lie Brackets and Accessibility: Motivational Discussion

We prefer to use, in this discussion, the alternative notation

$$e^{tf}x^0 \;:=\; \phi(t, x^0)\,, \tag{4.16}$$

which is more convenient for dealing with compositions of flows corresponding to different vector fields. For instance, we can write $e^{t_2 f}e^{t_1 g}x$ instead of the more cumbersome $\phi_f(t_2, \phi_g(t_1, x))$. (Note that, when the differential equation happens to be linear, i.e. $\dot{x} = Ax$ and A is a matrix, $e^{tf}x$ is precisely the same as $e^{tA}x$, where e^{tA} is the exponential of the matrix A; in general, however, $e^{tf}x$ is merely a useful notational convention.)

The idea behind the use of Lie algebraic techniques is as follows. Suppose that our goal is to describe the set of points z^0 that can be reached, from a given starting point x^0, by following the flow of two vector fields f and g, which may represent the motions corresponding to two different constant controls u_1 and u_2, for different intervals of time. For instance, one may follow f forward for 7 units of time, then g forward for 2 units, and finally f backward for 3 units. (That is, we go from x^0 to $e^{-3f}e^{2g}e^{7f}x^0$.) The question of reachability is nontrivial, because the motions along f and g need not commute. (The resulting state is not the same, in general, as $e^{2g}e^{4f}x^0$.)

An intuitive and geometric example of this phenomenon is as follows: let f induce a counterclockwise rotation (at 1 rad/sec) about the z-axis in $\mathbb{R}^3$ and g a counterclockwise rotation about the y axis (also at 1 rad/sec), that is,

$$f(x) = \begin{pmatrix} -x_2 \\ x_1 \\ 0 \end{pmatrix} \quad \text{and} \quad g(x) = \begin{pmatrix} -x_3 \\ 0 \\ x_1 \end{pmatrix}.$$

The flows are

$$e^{tf}x = \begin{pmatrix} \cos t & -\sin t & 0 \\ \sin t & \cos t & 0 \\ 0 & 0 & 1 \end{pmatrix} \begin{pmatrix} x_1 \\ x_2 \\ x_3 \end{pmatrix}$$

and

$$e^{tg}x = \begin{pmatrix} \cos t & 0 & -\sin t \\ 0 & 1 & 0 \\ \sin t & 0 & \cos t \end{pmatrix} \begin{pmatrix} x_1 \\ x_2 \\ x_3 \end{pmatrix}.$$

Then, the sequence of rotations: f for $\pi/2$ sec, g for $\pi/2$ sec, and finally f for $-\pi/2$ sec (i.e., rotate clockwise) has a different net effect than that of simply rotating about the y axis for $\pi/2$ rad. It produces instead a clockwise $\pi/2$-rad rotation around the x-axis, as is easy to see geometrically, or algebraically, since $e^{(-\pi/2)f}e^{(\pi/2)g}e^{(\pi/2)f}x$ equals

$$\begin{pmatrix} 0 & 1 & 0 \\ -1 & 0 & 0 \\ 0 & 0 & 1 \end{pmatrix} \begin{pmatrix} 0 & 0 & -1 \\ 0 & 1 & 0 \\ 1 & 0 & 0 \end{pmatrix} \begin{pmatrix} 0 & -1 & 0 \\ 1 & 0 & 0 \\ 0 & 0 & 1 \end{pmatrix} \begin{pmatrix} x_1 \\ x_2 \\ x_3 \end{pmatrix} = \begin{pmatrix} 1 & 0 & 0 \\ 0 & 0 & 1 \\ 0 & -1 & 0 \end{pmatrix} \begin{pmatrix} x_1 \\ x_2 \\ x_3 \end{pmatrix}.$$

This example shows that a totally new motion, different from a pure rotation around the y or the z axes, results by combining the original flows.

Lie theory provides a way to approach the question, by asking the following "infinitesimal" version of the problem: In what effective directions can one move, from the given initial point x^0? Clearly, one can move in the direction of $f(x^0)$, by following the flow of f (more precisely, one can move along a curve $\gamma(t) = \phi(t, x^0)$ whose tangent $\dot{\gamma}(0)$ at the initial time is $f(x^0)$), and similarly one can move in the direction of $g(x^0)$. On further thought, it is easy to see that one may also move in the direction of $f(x^0)+g(x^0)$, in the following sense: there is a curve γ with the property that $\gamma(t)$ is in the reachable set from x^0, for all t small enough, and such that $\gamma(0) = x^0$ and $\dot{\gamma}(0) = f(x^0) + g(x^0)$. Indeed, it is enough to take the curve $\gamma(t) := e^{tg}e^{tf}x^0$, calculate

$$\begin{aligned} \dot{\gamma}(t) &= \frac{\partial \phi_g}{\partial t}\left(t, \phi_f(t, x^0)\right) + \frac{\partial \phi_g}{\partial x}\left(t, \phi_f(t, x^0)\right)\frac{\partial \phi_f}{\partial t}(t, x^0) \\ &= g\left(e^{tg}e^{tf}x^0\right) + \frac{\partial \phi_g}{\partial x}\left(t, \phi_f(t, x^0)\right) f(e^{tf}x^0), \end{aligned}$$

and use that $\frac{\partial \phi_g}{\partial x}(0, x^0) = I$. More generally, one may find curves which allow movement in the direction of any linear combination of $f(x^0)$ and $g(x^0)$. The interesting observation is that other directions, in addition to those lying in the linear span of $\{f(x^0), g(x^0)\}$, may appear, namely, directions obtained by taking Lie brackets. For example, in the case of the rotations discussed above,

$$[f, g](x) = \left[\begin{pmatrix} -x_2 \\ x_1 \\ 0 \end{pmatrix}, \begin{pmatrix} -x_3 \\ 0 \\ x_1 \end{pmatrix}\right] = \begin{pmatrix} 0 \\ x_3 \\ -x_2 \end{pmatrix},$$

which is the vector field whose flow is the one that corresponds to the clockwise rotation about the x axis which we had discovered before. The result in the following exercise shows the role of Lie brackets in general; we do not need the result in this form, though it will be implicit in several arguments.

Exercise 4.2.6 Show that, for any two vector fields f and g, and any state x^0, the curve

$$\gamma(t) := e^{-\sqrt{t}g}e^{-\sqrt{t}f}e^{\sqrt{t}g}e^{\sqrt{t}f}\,x^0 \tag{4.17}$$

(defined for all t in some neighborhood of $t = 0$) has the property that $\dot{\gamma}(0) = [f, g](x^0)$. (*Hint:* You may want to use the expansion, for $x(t) = e^{tf}x^0$,

$$x(t) = x(0) + t\dot{x}(0) + \frac{t^2}{2}\ddot{x}(0) + o(t^2) = x^0 + tf(x^0) + \frac{t^2}{2}f_*(x^0)f(x^0) + o(t^2)$$

(and similarly for g), as well as Taylor expansions to first order for each of $f(x)$ and $g(x)$, to show that $e^{-\sqrt{t}g}e^{-\sqrt{t}f}e^{\sqrt{t}g}e^{\sqrt{t}f}\,x^0 = e^{t[f,g]}x^0 + o(t)$ as $t \to 0$.) □

Nonsingular Reachability

Assume given an open set $\mathcal{O} \subseteq \mathbb{R}^n$ and a set of vector fields $\mathcal{A} \subseteq \mathbb{V}(\mathcal{O})$.

Definition 4.2.7 *A k-tuple $(f_1, \ldots, f_k)$ of elements of $\mathcal{A}$ is* **nonsingular** *at $x^0 \in \mathcal{O}$ if there exists some vector $\vec{t_0} = (t_1^0, \ldots, t_k^0) \in \mathbb{R}^k_{\geq 0}$ such that the map*

$$F^{x^0}_{f_1,\ldots,f_k} : D^{x^0}_{f_1,\ldots,f_k} \to \mathcal{O} : \vec{t} = (t_1, \ldots, t_k) \mapsto e^{t_k f_k} \ldots e^{t_2 f_2} e^{t_1 f_1} x^0 ,$$

which is defined on some open subset $D^{x^0}_{f_1,\ldots,f_k}$ of $\mathbb{R}^k$ containing $\vec{t} = (0, \ldots, 0)$, has Jacobian of rank k at $(t_1^0, \ldots, t_k^0)$. □

Notice that since, by Lemma 4.2.2, $e^{tf}x^0$ is C^∞ on (t, x^0), one knows, inductively, that each map $F^{x^0}_{f_1,\ldots,f_k}$ is C^∞.

Recall that $\mathcal{A}_{\text{LA}}$ is the Lie algebra generated by the set $\mathcal{A}$. For each $x^0 \in \mathcal{O}$, we consider the subspace

$$\boxed{\mathcal{A}_{\text{LA}}(x^0) := \{X(x^0), X \in \mathcal{A}_{\text{LA}}\}}$$

of $\mathbb{R}^n$. The key result is the next one.

Lemma 4.2.8 If $\mathcal{A}_{\text{LA}}(x^0) = \mathbb{R}^n$ then there is some nonsingular n-tuple at x^0. Moreover, for each $\varepsilon > 0$, there is some $\vec{t_0} \in R^n_{\geq 0}$ such that $t_i^0 < \varepsilon$ for all i, and there are elements $f_1, \ldots, f_n$ in $\mathcal{A}$, such that $\left(F^{x^0}_{f_1,\ldots,f_n}\right)_*(\vec{t_0})$ has rank n.

Proof. Suppose that $X_1, \ldots, X_n$ are in $\mathcal{A}_{\text{LA}}$ and the vectors $X_1(x^0), \ldots, X_n(x^0)$ are linearly independent. Then, $X_1(x), \ldots, X_n(x)$ are linearly independent for each x belonging to some open neighborhood $\mathcal{O}_0$ of x^0 in $\mathcal{O}$. Thus, replacing if necessary $\mathcal{O}$ by the subset $\mathcal{O}_0$, and restricting all elements of $\mathcal{A}$ to $\mathcal{O}_0$, we assume without loss of generality that $\mathcal{A}_{\text{LA}}(x) = \mathbb{R}^n$ for all $x \in \mathcal{O}$.

Fix an $\varepsilon > 0$. Let k be the largest possible integer for which there exist some k-tuple $(f_1, \ldots, f_k)$, and some $\vec{t_0} \in R^k_{\geq 0}$ with $t_i^0 < \varepsilon$ for all i, so that $F^{x^0}_{f_1,\ldots,f_k}$ is defined on some neighborhood of

$$[0, t_1^0] \times \ldots \times [0, t_k^0]$$

and with the property that $(F^{x^0}_{f_1,\ldots,f_k})_*(\vec{t_0})$ has rank k. Note that $k \geq 1$, because there is some $f \in \mathcal{A}$ with $f(x^0) \neq 0$ (otherwise, all brackets are zero, which would imply that $\mathcal{A}_{\text{LA}}(x^0) = \{0\}$), and, for this one-tuple, $(F^{x^0}_f)_*(0) = f(x^0)$ has rank one. To prove the lemma, we need to show that $k = n$.

Let $(f_1, \ldots, f_k)$, and $\vec{t_0} \in R^k_{\geq 0}$, all $t_i^0 < \varepsilon$, be so that rank $(F^{x^0}_{f_1,\ldots,f_n})_*(\vec{t_0}) = k$. By continuity of $(F^{x^0}_{f_1,\ldots,f_n})_*$, we may assume without loss of generality that all the entries of $\vec{t_0}$ are positive. We pick a neighborhood $W \subseteq \mathbb{R}^n_{>0}$ of $\vec{t_0}$, included in the domain $D^{x^0}_{f_1,\ldots,f_k}$, so that rank $(F^{x^0}_{f_1,\ldots,f_k})_*(\vec{t}) = k$ for all $\vec{t}$ in W, and consider the slice $M = F^{x^0}_{f_1,\ldots,f_k} : W \to \mathcal{O}$.

Claim: Every element of $\mathcal{A}$ is tangent to M. Indeed, assume that this would not be the case. Then there would be some $f \in \mathcal{A}$ so that $f(M(\vec{s_0}))$ is not in the span of the columns of $M_*(\vec{s_0})$, for some $\vec{s_0} \in W$. Consider in that case the $(k+1)$-tuple

$$(f_1, \ldots, f_k, f),$$

let G be the map $F^{x^0}_{f_1,\ldots,f_k,f}$, and take $\vec{s_0}' := (\vec{s_0}, 0) = (s_1^0, \ldots, s_k^0, 0)$. Note that G is defined on a neighborhood of

$$[0, s_1^0] \times \ldots \times [0, s_k^0] \times \{0\}.$$

We compute the Jacobian $G_*(\vec{s_0}')$. Observe that

$$G(\vec{t}, t) = e^{tf} M(\vec{t}),$$

so the Jacobian with respect to the variables $\vec{t}$ equals $Q(t)M_*(\vec{t})$, where $Q(t)$ is the differential of the map $x \mapsto e^{tf}x$ evaluated at $M(\vec{t})$. In particular, at $(\vec{t}, t) = \vec{s_0}'$, one has $M_*(\vec{s_0})$. With respect to t, the derivative is $f(e^{tf}M(\vec{t}))$, which evaluated at $\vec{s_0}'$ equals $f(M(\vec{s_0}))$. We conclude that

$$(F^{x^0}_{f_1,\ldots,f_k,f})_*(\vec{s_0}') = [M_*(\vec{s_0}), f(M(\vec{s_0}))],$$

which has rank $k+1$ (because the first k columns give a matrix of rank k, since M is a slice, and the last column is not in the span of the rest). Since $\vec{s_0}'$ has all its entries nonnegative and less than ε, this contradicts the fact that k is maximal, and the claim is established.

We proved in Lemma 4.1.8 that the set of all vector fields tangent to M is a Lie algebra. This set, as we just showed, contains $\mathcal{A}$. So $\mathcal{A}_{\text{LA}}$, being the smallest Lie algebra containing $\mathcal{A}$, must be a subset of the set of vector fields tangent to M. Pick any $\vec{t} \in W$. Then, $X(M(\vec{t}))$ must be in the column space of $M_*(\vec{t})$, for each $X \in \mathcal{A}_{\text{LA}}$. It follows that $\operatorname{rank} M_*(\vec{t}) = n$, so necessarily $k = n$, as desired. ∎

4.3 Accessibility Rank Condition

We consider time-invariant continuous-time systems

$$\dot{x} = f(x, u),$$

where states evolve in an open subset $\mathcal{X} \subseteq \mathbb{R}^n$, for some n. We assume that $f(\cdot, u)$ is C^∞ in x, for each $u \in \mathcal{U}$.

Remark 4.3.1 As done in the rest of the text, in order to be able to define and have reasonable properties for measurable controls, we suppose that the control-value set $\mathcal{U}$ is a metric space and that the Jacobian of $f(x, u)$ with respect to x is continuous on (x, u). However, for the basic accessibility result to be given, no regularity in u would be required if one would define "reachable sets" using only piecewise constant controls. □

Assume given a system $\dot{x} = f(x, u)$. We associate to it the following set of vector fields:

$$\mathcal{A} := \{f_u = f(\cdot, u), u \in \mathcal{U}\}.$$

Definition 4.3.2 *The Lie algebra of vector fields $\mathcal{A}_{\text{LA}}$ is called the* **accessibility Lie algebra** *associated to the system. The* **accessibility rank condition (ARC) at x^0** *holds if $\mathcal{A}_{\text{LA}}(x^0) = \mathbb{R}^n$.* □

An especially interesting class of systems, very common in applications, is that of *control-affine* systems, for which $\mathcal{U} \subseteq \mathbb{R}^m$, $f(x, u)$ is affine in u:

$$\dot{x} = g_0(x) + u_1 g_1(x) + \ldots + u_m g_m(x) = g_0(x) + G(x)u \tag{4.18}$$

where g_i, $i = 0, \ldots, m$, are $m + 1$ vector fields and $G = (g_1, \ldots, g_m)$, and, we assume here,

$$0 \in \mathcal{U} \quad \text{and} \quad \text{linear span of } \mathcal{U} = \mathbb{R}^m. \tag{4.19}$$

For systems of this general form, it is not necessary to use all the vector fields f_u when generating $\mathcal{A}_{\text{LA}}$:

Lemma 4.3.3 For a control-affine system, $\mathcal{A}_{\text{LA}} = \{g_0, \ldots, g_m\}_{\text{LA}}$.

Proof. It will suffice to show that the linear spans L_G of $\{g_0, \ldots, g_m\}$ and $L_{\mathcal{A}}$ of $\mathcal{A} = \{f_u = f(\cdot, u), u \in \mathcal{U}\}$ are the same. Each element $f_u = g_0 + \sum u_i g_i$ of $\mathcal{A}$ is by definition a linear combination of the g_i's, so it is clear that $L_{\mathcal{A}} \subseteq L_G$. Conversely, using $u = 0$ shows that $g_0 \in \mathcal{A}$, and thus also that $G(x)u = u_1 g_1 + \ldots + u_m g_m = f(x, u) - f(x, 0) \in L_{\mathcal{A}}$ for every $u \in \mathcal{U}$. To see that each $g_i \in L_{\mathcal{A}}$,

$i \in \{1, \ldots, m\}$, fix any such i, and write the ith canonical basis vector of $\mathbb{R}^m$ as $e_i = \sum_{j \in J} \rho_j u_j$, for some u_j's in $\mathcal{U}$ and reals ρ_j (recall that we are assuming that (4.19) holds). So, $g_i = G(x)e_i = \sum_{j \in J} \rho_j G(x) u_j \in L_{\mathcal{A}}$. ∎

In view of the characterization in Lemma 4.1.4, the accessibility rank condition amounts to asking that there exist n vector fields $X_1, \ldots, X_n$, each of which can be expressed as an iterated Lie bracket $[[\ldots[f_{u_1}, f_{u_2}], f_{u_3}], \ldots, f_{u_\ell}]$ (with possibly a different length ℓ for each of the X_i's), so that $X_1(x^0), \ldots, X_n(x^0)$ are linearly independent. In the case of control-affine systems, the condition means that there are n iterated brackets formed out of $\{g_0, \ldots, g_m\}$ which are linearly independent when evaluated at x^0.

Exercise 4.3.4 The length of the brackets needed in order to generate enough linearly independent vector fields (the "degree of nonholonomy") may be much larger than n. In other words, the ARC may hold at x^0, yet its verification requires one to compute brackets of arbitrary length. Specifically, for each positive integer ℓ give an example of a system of dimension two, and a state x^0, so that span $(\mathcal{A}_0 \bigcup \ldots \bigcup \mathcal{A}_{\ell-1})(x^0)$ (recall Lemma 4.1.4) has dimension one but span $(\mathcal{A}_0 \bigcup \ldots \bigcup \mathcal{A}_\ell)(x^0)$ has dimension two. □

Remark 4.3.5 Any time-invariant linear system $\dot{x} = Ax + Bu$ is control-affine, because it has the form (4.18) with $g_0(x)$ = the linear vector field Ax and each $g_i(x)$ = the constant vector field b_i (the ith column of B). To compute the Lie algebra $\mathcal{A}_{\text{LA}}$ for a linear system, we note that, in general, the Lie bracket of two constant vector fields is zero, and the Lie bracket of a constant and a linear vector field is a constant one: $[h, Fx] = Fh$. Also, $[Ax, Ax] = 0$. Therefore, an iterated bracket with $\ell > 1$, $[[\ldots[g_{i_1}, g_{i_2}], g_{i_3}], \ldots, g_{i_\ell}]$, is necessarily zero unless it has the form $[[\ldots[b_i, Ax], Ax], \ldots, Ax] = A^{\ell-1} b_i$ (or the form $[[\ldots[Ax, b_i], Ax], \ldots, Ax]$, which differs from it at most in its sign). Thus, the accessibility rank condition holds at a point x^0 if and only if

$$\operatorname{rank}\left(Ax^0, b_1, \ldots, b_m, Ab_1, \ldots, Ab_m, \ldots, A^{n-1}b_1, \ldots, A^{n-1}b_m\right) = n$$

(as usual, the Cayley-Hamilton Theorem implies that we do not need to include any $A^\ell b_j$ with $\ell \geq n$). This is somewhat weaker than controllability, because of the additional vector Ax^0, which does not appear in the Kalman controllability rank condition. For example, take the one-dimensional system $\dot{x} = x$ (with $B = 0$); we have that the accessibility rank condition holds at each $x^0 \neq 0$, but the set of states reachable from such a state does not even contain the initial state in its interior (the reachable set from $x^0 = 1$, for instance, is the half-line $[1, +\infty)$), much less is every other state reachable from x^0. □

For each subset $\mathcal{V} \subseteq \mathcal{X}$, initial state $x^0 \in \mathcal{X}$, and time $t > 0$, we consider the set of states that are *reachable from* x^0*, in time exactly* t*, without leaving* $\mathcal{V}$ (using measurable essentially bounded controls):

$$\mathcal{R}^t_{\mathcal{V}}(x^0) := \{z^0 \mid \exists \omega \in \mathcal{L}^\infty_{\mathcal{U}}(0,t) \text{ s.t. } \phi(s,0,x^0,\omega) \in \mathcal{V} \; \forall s \in [0,t] \text{ and } \phi(t,0,x^0,\omega) = z^0\} .$$

For each $T \geq 0$, we also consider the set of states reachable in time *at most* T, $\mathcal{R}_{\mathcal{V}}^{\leq T}(x^0) := \bigcup_{t\in[0,T]} \mathcal{R}_{\mathcal{V}}^t(x^0)$, When $\mathcal{V} = \mathcal{X}$, we drop the subscript. The whole reachable set from x^0 is $\mathcal{R}(x^0) = \bigcup_{t\geq 0} \mathcal{R}^t(x^0)$. For each $\mathcal{V}$, z^0, and $t > 0$, we also consider the set of states that are *controllable to* z^0 in time exactly t without leaving $\mathcal{V}$, namely

$$C_{\mathcal{V}}^t(z^0) \;:=\; \{x^0 \mid \exists \omega \in \mathcal{L}_{\mathcal{U}}^{\infty}(0,t) \text{ s.t. } \phi(s,0,x^0,\omega) \in \mathcal{V}\, \forall s \in [0,t] \\ \text{and } \phi(t,0,x^0,\omega) = z^0\} \,,$$

$\mathcal{C}_{\mathcal{V}}^{\leq T}(z^0) := \bigcup_{t\in[0,T]} C_{\mathcal{V}}^t(x^0)$, dropping subscripts if $\mathcal{V} = \mathcal{X}$, and $\mathcal{C}(z^0)$ is the set of states that can be controlled to z^0, i.e., $\bigcup_{t\geq 0} \mathcal{C}^t(x^0)$. The main result can now be given.

Theorem 9 *Assume that the accessibility rank condition holds at x^0. Then, for each neighborhood $\mathcal{V}$ of x^0, and each $T > 0$,*

$$\operatorname{int} \mathcal{R}_{\mathcal{V}}^{\leq T}(x^0) \neq \emptyset \tag{4.20}$$

and

$$\operatorname{int} \mathcal{C}_{\mathcal{V}}^{\leq T}(x^0) \neq \emptyset \,. \tag{4.21}$$

In particular, $\mathcal{R}(x^0)$ and $\mathcal{C}(x^0)$ have nonempty interiors.

Proof. Pick a neighborhood $\mathcal{V}$ of x^0 and a $T > 0$. Since $\mathcal{A}_{\text{LA}}(x^0) = \mathbb{R}^n$, there is a finite subset of vector fields $\mathcal{A}_0 \subseteq \mathcal{A}$ so that

$$\mathcal{A}_{0\text{LA}}(x^0) \;=\; \mathbb{R}^n \,.$$

By continuity of each $e^{tf}x$ on t and x, there exists some number $\varepsilon > 0$ with the property that, for all sequences of n elements $f_1, \ldots, f_n \in \mathcal{A}_0$, and all sequences of positive numbers $t_1, \ldots, t_n$ with $0 \leq t_i < \varepsilon$,

$$e^{t_n f_n} \ldots e^{t_2 f_2} e^{t_1 f_1} x^0 \quad \text{is defined and belongs to } \mathcal{V} \,.$$

(To see this, one may first find a neighborhood $\mathcal{V}_1$ of x^0 and an $\varepsilon > 0$ so that $e^{tf}x \in \mathcal{V}$ for all $0 \leq t < \varepsilon$, all $x \in \mathcal{V}_1$, and all the elements of $\mathcal{A}_0$; then one finds a $\mathcal{V}_2$ so that $e^{tf}x \in \mathcal{V}_1$ if t is small enough, $f \in \mathcal{A}_0$, and $x \in \mathcal{V}_2$, and so on inductively.) We pick such an ε, and assume without loss of generality that $\varepsilon < T/n$. Lemma 4.2.8 then insures that there are vector fields $f_1 = f_{u_1}, \ldots, f_n = f_{u_n}$ belonging to $\mathcal{A}_0$, and some $\vec{t_0} \in R^n_{\geq 0}$, such that $0 \leq t_i^0 < \varepsilon$ for all i and $(F^{x^0}_{f_1,\ldots,f_n})_*(\vec{t_0})$ has rank n. By continuity, we may assume that all $t_i^0 > 0$. By the Implicit Mapping Theorem, the image of $F^{x^0}_{f_1,\ldots,f_n}$, restricted to some subset of $\{(t_1, \ldots, t_n) \mid 0 < t_i < \varepsilon, i = 1, \ldots, n\}$, contains an open set. Since

$$F^{x^0}_{f_1,\ldots,f_n}(\vec{t}) \;=\; e^{t_n f_{u_n}} \ldots e^{t_2 f_{u_2}} e^{t_1 f_{u_1}} x^0 \;=\; \phi(t_1 + \ldots + t_n, 0, x^0, \omega) \,,$$

where ω is the piecewise constant control having value u_1 on $[0, t_1)$, value u_2 on $[t_1, t_1 + t_2)$, ..., and value u_n on $[t_1 + \ldots + t_{n-1}, t_1 + \ldots + t_n]$, this image is included in $\mathcal{R}_{\mathcal{V}}^{\leq T}(x^0)$.

To prove the statement for the sets $\mathcal{C}_{\mathcal{V}}^{\leq T}(x^0)$, we argue as follows. Consider the "time-reversed" system $\dot{x} = -f(x,u)$. Since the accessibility Lie algebra of the time-reversed system is obtained from the vector fields $\{-f_u, u \in \mathcal{U}\}$, and is, in particular, a subspace, it coincides with the accessibility Lie algebra of the original system. Thus, the accessibility rank condition holds for the time-reversed system. Pick any T and $\mathcal{V}$. In general, if $\phi(T, 0, x, \omega) = z$, then, for the reversed dynamics ϕ^- we clearly have $\phi^-(T, 0, z, \nu) = x$ using the control $\nu(t) := \omega(T - t)$, (cf. Lemma 2.6.7), and, moreover, if the intermediate states $\phi(t, 0, x, \omega)$ are in $\mathcal{V}$, then the same is true for the states $\phi^-(t, 0, z, \nu)$. Thus, the set $\mathcal{C}_{\mathcal{V}}^{\leq T}(z)$ for the original system coincides with $\mathcal{R}_{\mathcal{V}}^{\leq T}(z)$ for the time-reversal, and the latter has a nonempty interior (by the first part of the proof). ∎

The property that $\mathcal{R}(x^0)$ have a nonempty interior is often called *accessibility* (from x^0). The accessibility rank condition is, in general, only a sufficient condition:

Exercise 4.3.6 (a) Show that for systems of dimension one: $\dot{x} = f(x,u)$, $\mathcal{X} = \mathbb{R}$, the accessibility rank condition holds at a state x^0 if and only if $\mathcal{R}(x^0)$ has a nonempty interior.
(b) Consider the following system with $\mathcal{X} = \mathbb{R}^2$ and $\mathcal{U} = \mathbb{R}^2$:

$$\begin{aligned} \dot{x}_1 &= u_1 \\ \dot{x}_2 &= \varphi(x_1) u_2 \end{aligned}$$

where $\varphi : \mathbb{R} \to \mathbb{R}$ is a C^∞ function with the property that $\varphi(z) \neq 0$ for each $z \neq 0$ but $\frac{d^j \varphi}{dz^j}(0) = 0$ for all $j = 0, 1, 2, \ldots$ (for instance, $\varphi(z) := e^{-1/z^2}$ for $z \neq 0$, $\varphi(0) := 0$). Show that, for all x^0, $x^0 \in \operatorname{int} \mathcal{R}_{\mathcal{V}}^{\leq T}(x^0)$, but, on the other hand, the accessibility rank condition does not hold at the points of the form $(0, x_2)$. □

Exercise 4.3.7 Let $\dot{x} = A(t)x + B(t)u$ be a time-varying continuous time linear system, and assume that all the entries of A and B are smooth functions of t. Introduce the following system with state-space $\mathbb{R}^{n+1}$:

$$\begin{aligned} \dot{x}_0 &= 1 \\ \dot{x} &= A(x_0)x + B(x_0)u \,. \end{aligned}$$

Explain the relationship between the accessibility rank condition, applied to this system, and the Kalman-like condition for controllability of $\dot{x} = A(t)x + B(t)u$ studied in Corollary 3.5.18. □

The following easy fact is worth stating, for future reference. It says that the map $x \mapsto \phi(T, 0, x, \omega)$ is a local homeomorphism, for each fixed control ω.

Lemma 4.3.8 Suppose that $\omega \in \mathcal{L}_{\mathcal{U}}^{\infty}(0,T)$ is admissible for x^0, and let $z^0 = \phi(T,0,x^0,\omega)$. Then the map $\alpha = \alpha_\omega : x \mapsto \phi(T,0,x,\omega)$, which is defined on an open subset $\mathcal{D}$ of $\mathcal{X}$, is one-to-one, continuous, and has a continuous inverse. In particular, $\alpha(\mathcal{W})$ is a neighborhood of z^0, for each neighborhood $\mathcal{W}$ of x^0 included in $\mathcal{D}$.

Proof. We have that the control $\nu(t) := \omega(T-t)$ is admissible for z^0, with respect to the reversed system $\dot{x} = -f(x,u)$. Moreover, if we write $\mathcal{D}'$ for the set where $\beta(z) := \phi^-(T,0,z,\nu)$ is defined (the dynamics map for the reversed system), we have that $\alpha(\mathcal{D}) = \mathcal{D}'$ and $\alpha \circ \beta$ and $\beta \circ \alpha$ are both the identity. From Theorem 1 we know that the domain $\mathcal{D}$ is open and α is continuous, and, applied to the reversed system, also that $\mathcal{D}'$ is open and β is continuous. ∎

Reversible Systems

In general, accessibility is weaker than controllability (as $\dot{x} = 1$ illustrates). Sometimes, however, controllability can be characterized by the accessibility rank condition.

Definition 4.3.9 *A system is* **weakly reversible** *if* $\rightsquigarrow$ *is an equivalence relation, and* **strongly reversible** *if for each x^0 and each $\omega \in \mathcal{L}_{\mathcal{U}}^{\infty}(0,T)$ admissible for x^0, there is some $\nu \in \mathcal{L}_{\mathcal{U}}^{\infty}(0,T)$ which is admissible for $z^0 := \phi(T,0,x^0,\omega)$ and is such that $\phi(t,0,x^0,\omega) = \phi(T-t,0,z^0,\nu)$ for all $t \in [0,T]$.* □

That is, weakly reversible means that $z^0 \in \mathcal{R}(x^0)$ if and only if $x^0 \in \mathcal{R}(z^0)$, and strongly reversible means that the same path that takes us from x^0 to z^0 can be traveled backward. Strong reversibility implies weak reversibility, but not conversely: the system

$$\dot{x} = \begin{pmatrix} 0 & -1 \\ 1 & 0 \end{pmatrix} x,$$

whose paths are circles transversed counterclockwise, is weakly reversible but not strongly so.

A very interesting class of strongly reversible systems is obtained from those control-affine systems for which $g_0 \equiv 0$, that is, systems of the form

$$\dot{x} = u_1 g_1(x) + \ldots + u_m g_m(x) = G(x)u. \tag{4.22}$$

We assume that (4.19) holds and also that $\mathcal{U}$ is symmetric: $u \in \mathcal{U} \Rightarrow -u \in \mathcal{U}$. These are called *systems without drift* (because, when $u \equiv 0$, the state does not "drift" but, instead, remains constant) and they arise as kinematic models of mechanical systems. For such systems, we may let $\nu(t) := -\omega(T-t)$ in the definition of strongly reversible system, and verify the property using the fact that $\dot{x} = G(x)\omega$ implies $(d/dt)x(T-t) = G(x(T-t))\nu(t)$.

Remark 4.3.10 It is worth noticing that the first-order test for local controllability provides no information in the case of systems without drift. The linearization of (4.22) at any state x^0 and input value 0 is $\dot{x} = G_*(x^0)u$. This system cannot be controllable, unless $m \geq n$. □

Proposition 4.3.11 Assume that the accessibility rank condition holds at x^0. If the system $\dot{x} = f(x,u)$ is weakly reversible, then

$$x^0 \in \text{int}\left(\mathcal{R}(x^0)\bigcap\mathcal{C}(x^0)\right).$$

Moreover, if it is strongly reversible, then

$$x^0 \in \text{int}\left(\mathcal{R}_{\mathcal{V}}^{\leq T}(x^0)\bigcap\mathcal{C}_{\mathcal{V}}^{\leq T}(x^0)\right).$$

for every neighborhood $\mathcal{V}$ of x^0 and every $T > 0$.

Proof. We know from Theorem 9 that $\mathcal{R}(x^0)$ has a nonempty interior. Pick any state z^0 in the interior of $\mathcal{R}(x^0)$. Because of weak reversibility, there is some control $\omega \in \mathcal{L}_{\mathfrak{U}}^{\infty}(0,T)$, for some $T \geq 0$, such that $\phi(T,0,z^0,\omega) = x^0$. By Lemma 4.3.8, there is an open subset $\mathcal{W}$ of $\mathcal{R}(x^0)$ which contains z^0 and is such that $\mathcal{W}' := \phi(T,0,\mathcal{W},\omega)$ is an open subset containing x^0. But $\mathcal{W}' \subseteq \mathcal{R}(x^0)$, since $x \in \mathcal{W}'$ implies $x = \phi(T,0,z,\omega)$ for some $z \in \mathcal{W} \subseteq \mathcal{R}(x^0)$. The proof for $\mathcal{C}(x^0)$ is analogous.

Assume now that strong reversibility holds. Pick T_0 and $\mathcal{V}$, and let $T := T_0/2$. Again by the Theorem, we know that $\mathcal{R}_{\mathcal{V}}^{\leq T}(x^0)$ has a nonempty interior. Pick any z^0 in the interior of $\mathcal{R}_{\mathcal{V}}^{\leq T}(x^0)$, and let $0 \leq T_1 \leq T$ and $\omega \in \mathcal{L}_{\mathfrak{U}}^{\infty}(0,T_1)$ be such that $\phi(T_1,0,x^0,\omega) = z^0$ and $\phi(t,0,x^0,\omega) \in \mathcal{V}$ for all $t \in [0,T_1]$. Let $\nu \in \mathcal{L}_{\mathfrak{U}}^{\infty}(0,T_1)$ be as in the definition of strong reversibility, for this ω. By the continuity of $z \mapsto \phi(\cdot,0,z,\nu)$ (as a map into $C^0([0,T_1],\mathcal{X})$), there is an open set $\mathcal{W}$ containing z^0 and such that $\phi(t,0,z,\omega) \in \mathcal{V}$ for all $z \in \mathcal{W}$ and all $t \in [0,T_1]$, and we may assume without loss of generality that $\mathcal{W} \subseteq \mathcal{R}_{\mathcal{V}}^{\leq T}(x^0)$. Moreover, by Lemma 4.3.8, the image $\mathcal{W}' := \phi(T_1,0,\mathcal{W},\nu)$ is an open subset which contains x^0. Note that $\mathcal{W}' \subseteq \text{int}\,\mathcal{R}_{\mathcal{V}}^{\leq T_0}(x^0)$. Indeed, for each $x \in \mathcal{W}'$ there is, by definition, an element z of $\mathcal{R}_{\mathcal{V}}^{\leq T}(x^0)$ which is steered to x in time $\leq T$ and without leaving $\mathcal{V}$ (in fact, using the control ν); and, in turn, z can be reached from x^0 in time T_1 without leaving $\mathcal{V}$. The proof for $\mathcal{C}_{\mathcal{V}}^{\leq T}(x^0)$ is entirely analogous. ■

Corollary 4.3.12 For a weakly reversible system, if the accessibility rank condition holds at every state $x \in \mathcal{X}$ and $\mathcal{X}$ is connected, then the system is completely controllable.

Proof. Pick any state x^0. Pick any $z^0 \in \mathcal{R}(x^0)$. From Proposition 4.3.11, we know that $z^0 \in \text{int}\,\mathcal{R}(z^0)$; by transitivity of reachability, $\mathcal{R}(z^0) \subseteq \mathcal{R}(x^0)$, so this proves that $\mathcal{R}(x^0)$ is open. Now pick $z^0 \notin \mathcal{R}(x^0)$. Again from the Proposition, $z^0 \in \operatorname{int} \mathcal{C}(z^0)$. But $\mathcal{C}(z^0) \bigcap \mathcal{R}(x^0) = \emptyset$ (again, by transitivity of reachability). Thus both the reachable set $\mathcal{R}(x^0)$ and its complement are open; since $\mathcal{X}$ is connected and $\mathcal{R}(x^0) \neq \emptyset$, $\mathcal{R}(x^0) = \mathcal{X}$, as wanted. ∎

Example 4.3.13 (Nelson's car.) The following is a simplified model of a front-wheel drive automobile. The state space is $\mathbb{R}^4$, and the coordinates of the state

$$x = \begin{pmatrix} x_1 \\ x_2 \\ x_3 \\ x_4 \end{pmatrix} = \begin{pmatrix} x_1 \\ x_2 \\ \varphi \\ \theta \end{pmatrix}$$

denote, respectively, the position of the center of the front axle (coordinates x_1, x_2), the orientation of the car (angle φ, measured counterclockwise from the positive x-axis), and the angle of the front wheels relative to the orientation of the car (θ, also counterclockwise); see Figure 4.1.

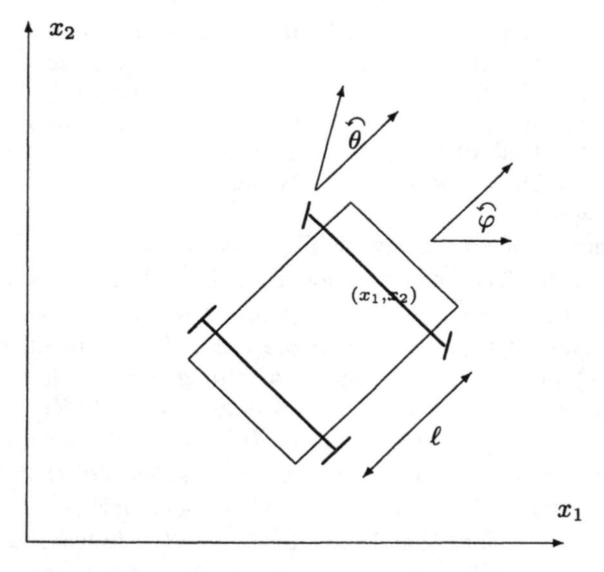

Figure 4.1: *4-dimensional car model.*

We now sketch how to derive the equations of motion. There are two constraints to take into account, corresponding respectively to the requirement that the front and back wheels do not slip (no sideways motion of tires; the only motions allowed are rolling and rotation in place). The front wheels are parallel to the vector $(\cos(\theta + \varphi), \sin(\theta + \varphi))$, so the instantaneous direction of movement of the center of the front axle is parallel to this vector:

$$\frac{d}{dt}\begin{pmatrix} x_1 \\ x_2 \end{pmatrix} = u_2(t)\begin{pmatrix} \cos(\theta + \varphi) \\ \sin(\theta + \varphi) \end{pmatrix} \tag{4.23}$$

for some time-dependent scalar $u_2(t)$. Observe that $\dot{x}_1(t)^2 + \dot{x}_2(t)^2 = u_2(t)^2$, so u_2 is the velocity of the car, which we think of as a control (engine speed). The center of the rear axle has coordinates $(x_1 - \ell\cos\varphi, x_2 - \ell\sin\varphi)'$, where ℓ is the distance between the two axles. The velocity of this point must be parallel to the orientation $(\cos\varphi, \sin\varphi)'$ of the back wheels:

$$\sin\varphi \frac{d}{dt}(x_1 - \ell\cos\varphi) - \cos\varphi\frac{d}{dt}(x_2 - \ell\sin\varphi) \equiv 0\,.$$

Substituting (4.23) into this orthogonality relation, and after performing some trigonometric simplifications, we obtain:

$$\ell\dot{\varphi} = u_2 \sin\theta\,.$$

To avoid extra notation, we assume from now on that $\ell = 1$. The angle θ changes depending on the steering wheel position; we take the velocity at which the steering wheel is being turned as another control, u_1. In summary, with the control-value space $\mathcal{U} = \mathbb{R}^2$, and writing controls as $u = (u_1, u_2)' \in \mathbb{R}^2$, we obtain the following system:

$$\dot{x} = u_1 \begin{pmatrix} 0 \\ 0 \\ 0 \\ 1 \end{pmatrix} + u_2 \begin{pmatrix} \cos(\varphi+\theta) \\ \sin(\varphi+\theta) \\ \sin\theta \\ 0 \end{pmatrix}\,. \tag{4.24}$$

Note that a control such that $u_2(t) \equiv 0$ corresponds to a pure steering move, while one with $u_1(t) \equiv 0$ models a pure driving move in which the steering wheel is fixed in one position. In general, a control is a function $u(t)$ which indicates, at each time t, the current steering velocity and engine speed.

In practice, of course, the angle θ would be restricted to some maximal interval $(-\theta_0, \theta_0)$; thus we could take the state space as $\mathbb{R}^3 \times (-\theta_0, \theta_0)$, which would not change anything to follow. More importantly, the orientation angle φ only makes sense modulo 2π, that is, angles differing by 2π correspond to the same physical orientation. Nonlinear control theory is usually developed in far more generality than here, allowing the state space $\mathcal{X}$ to be a differentiable manifold; thus, a more natural state space than $\mathbb{R}^4$ would be, for this example, $\mathbb{R}^2 \times \mathbb{S}^1 \times (-\theta_0, \theta_0)$, that is, the angle φ is thought of as an element of the unit circle.

The model is control-affine; in fact, it is a system without drift $\dot{x} = u_1 g_1 + u_2 g_2$, where the defining vector fields are $g_1 =$ *steer* and $g_2 =$ *drive*:

$$g_1 = \begin{pmatrix} 0 \\ 0 \\ 0 \\ 1 \end{pmatrix}, \quad g_2 = \begin{pmatrix} \cos(\varphi+\theta) \\ \sin(\varphi+\theta) \\ \sin\theta \\ 0 \end{pmatrix}.$$

By Proposition 4.3.11 and Corollary 4.3.12, we will have complete controllability, as well as the property that all states near a given state can be attained with small excursions and in small time, once we prove that the accessibility rank condition holds at every point. (Of course, it is quite obvious from physical reasoning, for this example, that complete controllability holds.) We compute some brackets:

$$wriggle \;:=\; [steer, drive] \;=\; \begin{pmatrix} -\sin(\varphi+\theta) \\ \cos(\varphi+\theta) \\ \cos\theta \\ 0 \end{pmatrix}$$

and

$$slide \;:=\; [wriggle, drive] \;=\; \begin{pmatrix} -\sin\varphi \\ \cos\varphi \\ 0 \\ 0 \end{pmatrix}.$$

(The bracket $[wriggle, steer]$ equals *drive*, so it is redundant, in so far as checking the accessibility rank condition is concerned.) It turns out that these four brackets are enough to satisfy the accessibility test. Indeed, one computes

$$\det(steer, drive, wriggle, slide) \;\equiv\; 1\,,$$

so the condition indeed holds.

It is instructive to consider, in particular, the problem of accessibility starting from the special state $x^0 = 0$. We may interpret this state as corresponding to a "parallel parked" car; accessibility allows us to move in various directions without leaving a small neighborhood, that is, without hitting the cars in front and behind. For $\varphi = \theta = 0$, *wriggle* is the vector $(0, 1, 1, 0)$, a mix of sliding in the x_2 direction and a rotation, and *slide* is the vector $(0, 1, 0, 0)$ corresponding to sliding in the x_2 direction. This means that one can in principle implement infinitesimally both of the above motions. The "wriggling" motion is, based on the characterization in Exercise 4.2.6, the one that arises, in a limiting sense, from fast repetitions of the following sequence of four basic actions:

$$steer \;\text{-}\; drive \;\text{-}\; reverse\ steer \;\text{-}\; reverse\ drive\,. \qquad (*)$$

This is, essentially, what one does in order to get out of a tight parking space. Interestingly enough, one could also approximate the pure sliding motion in the x_2 direction: "*wriggle, drive, reverse wriggle, reverse drive, repeat*" corresponds to the last vector field, "slide", which at $x = 0$ coincides with this motion. Note that the square roots in Equation (4.17) explain why many iterations of basic motions (*) are required in order to obtain a displacement in the wriggling direction: the order of magnitude t of a displacement is much smaller than the total time needed to execute the maneuver, $4\sqrt{t}$. □

Exercise 4.3.14 Consider the following three systems, all with state space $\mathbb{R}^3$ and control-value space $\mathbb{R}^2$:

$$\dot{x}_1 = u_1, \quad \dot{x}_2 = u_2, \quad \dot{x}_3 = x_2 u_1 - x_1 u_2 \qquad (\Sigma_1)$$
$$\dot{x}_1 = u_1, \quad \dot{x}_2 = u_2, \quad \dot{x}_3 = x_2 u_1 \qquad (\Sigma_2)$$
$$\dot{x}_1 = u_1, \quad \dot{x}_2 = u_2, \quad \dot{x}_3 = x_2 u_1 + x_1 u_2\,. \qquad (\Sigma_3)$$

Answer (justify!): which of these systems are controllable? (*Hint:* For (Σ_3), you may want to analyze the function $\ell(x_1, x_2, x_3) := x_1x_2 - x_3$.) For related material, cf. Exercises 4.3.16 and 9.4.2, and Examples 5.9.16 and 9.2.14. □

Exercise 4.3.15 Consider a rigid body which is being controlled by means of one or more applied torques (for example, a satellite in space, under the action of one or more pairs of opposing thruster jets). We only study here the effect of controls on angular momenta; more complete models incorporating orientations and even linear displacements are of course also possible. With the components of $x = (x_1, x_2, x_3)'$ denoting the angular velocity coordinates with respect to the principal axes, and the positive numbers I_1, I_2, I_3 denoting the respective principal moments of inertia, this is a system with $\mathcal{X} = \mathbb{R}^3$, $\mathcal{U} = \mathbb{R}^m$, where m is the number of torques; the evolution is represented by (the Euler equations for rotational movement):

$$I\dot{x} = S(x)Ix + Tu\,, \tag{4.25}$$

where I is the diagonal matrix with entries I_1, I_2, I_3 and where T is a matrix whose columns describe the axes along which the torques act. The matrix $S(x)$ is the rotation matrix

$$S(x) = \begin{pmatrix} 0 & x_3 & -x_2 \\ -x_3 & 0 & x_1 \\ x_2 & -x_1 & 0 \end{pmatrix}.$$

(Equivalently, the equations can be written as $I\dot{x} = I \times x + Tu$, where "$\times$" is the vector product in $\mathbb{R}^3$.)
(a) Consider the case in which there are two torques ($m = 2$), which act about the first two principal axes, that is, $T \in \mathbb{R}^{3\times 2}$ has columns $(1, 0, 0)'$ and $(0, 1, 0)'$. The equations can be written as follows:

$$\begin{aligned} \dot{x}_1 &= a_1x_2x_3 + b_1u_1 \\ \dot{x}_2 &= a_2x_1x_3 + b_2u_2 \\ \dot{x}_3 &= a_3x_1x_2 \end{aligned}$$

where $a_1 = (I_2 - I_3)/I_1$, $a_2 = (I_3 - I_1)/I_2$, and $a_3 = (I_1 - I_2)/I_3$, and b_1, b_2 are both nonzero. You may assume that $b_1 = b_2 = 1$. Show that the accessibility rank condition holds at every point, if and only if $I_1 \neq I_2$.
(b) Now consider the case in which there is only one torque, acting about a mixed axis. Taking for simplicity the case in which there is rotational symmetry, $I_1 = I_2$, the equations can be written as follows:

$$\begin{aligned} \dot{x}_1 &= ax_2x_3 + b_1u \\ \dot{x}_2 &= -ax_1x_3 + b_2u \\ \dot{x}_3 &= b_3u \end{aligned}$$

where we assume $a \neq 0$, and the b_i's are real numbers. Show that the accessibility rank condition holds at every point of the state space if and only if $b_3 \neq 0$ and $b_1^2 + b_2^2 \neq 0$. □

Exercise 4.3.16 Consider a model for the "shopping cart" shown in Figure 4.2 ("knife-edge" or "unicycle" are other names for this example). The state is given by the orientation θ, together with the coordinates x_1, x_2 of the midpoint between the back wheels.

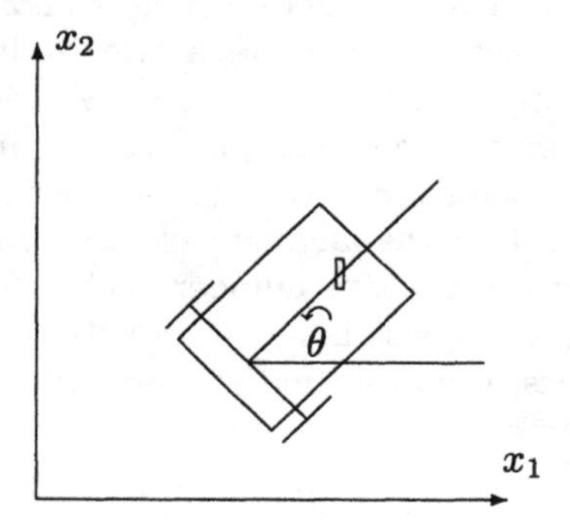

Figure 4.2: *Shopping cart.*

The front wheel is a castor, free to rotate. There is a non-slipping constraint on movement: the velocity $(\dot{x}_1, \dot{x}_2)'$ must be parallel to the vector $(\cos\theta, \sin\theta)'$. This leads to the following equations:

$$\begin{aligned} \dot{x}_1 &= u_1 \cos\theta \\ \dot{x}_2 &= u_1 \sin\theta \\ \dot{\theta} &= u_2 \end{aligned}$$

where we may view u_1 as a "drive" command and u_2 as a steering control (in practice, we implement these controls by means of differential forces on the two back corners of the cart). We view the system as having state space $\mathbb{R}^3$ (a more accurate state space would be the manifold $\mathbb{R}^2 \times \mathbb{S}^1$).
(a) Show that the system is completely controllable.
(b) Consider these new variables: $z_1 := \theta$, $z_2 := x_1 \cos\theta + x_2 \sin\theta$, $z_3 := x_1 \sin\theta - x_2 \cos\theta$, $v_1 := u_2$, and $v_2 := u_1 - u_2 z_3$. (Such a change of variables is called a "feedback transformation".) Write the system in these variables, as $\dot{z} = \widetilde{f}(z, v)$. Note that this is one of the systems Σ_i in Exercise 4.3.14. Explain why controllability can then be deduced from what you already concluded in that previous exercise. □

4.4 Ad, Distributions, and Frobenius' Theorem

The converse of Theorem 9 (p. 156) is not true: it may well be that the accessibility rank condition does not hold at a state x^0, yet $\mathcal{R}_{\mathcal{V}}^{\leq T}(x^0)$ has a nonempty interior for each neighborhood $\mathcal{V}$ of x^0 and each $T > 0$; cf. Exercise 4.3.6. In order to provide a partial converse of the Theorem, we first need to develop a few additional facts regarding flows and Lie brackets.

The "Ad" Operator

Let $X \in \mathbb{V}(\mathcal{O})$ and pick any (t, x^0) in $\mathcal{D}_X$ (the domain of definition of the flow of X). If Y is a vector field defined on an open subset $\mathcal{O}_Y$ of $\mathcal{O}$ which contains $e^{tX}x^0$, we define

$$\boxed{\mathrm{Ad}_{tX}Y\left(x^0\right) := \left(e^{-tX}\right)_* \left(e^{tX}x^0\right) \cdot Y\left(e^{tX}x^0\right)}$$

(this is a matrix, the Jacobian of e^{-tX} evaluated at $e^{tX}x^0$, multiplied by the vector $Y(e^{tX}x^0)$; we will sometimes use dots, as here, to make formulas more readable). Geometrically, the vector $\mathrm{Ad}_{tX}Y(x^0)$ is the "pull-back" to x^0, along the flow of X, of the vector $Y(e^{tX}x^0)$. For example, let, in $\mathbb{R}^2$,

$$A := \begin{pmatrix} 0 & -1 \\ 1 & 0 \end{pmatrix}, \quad x^0 := \begin{pmatrix} -1 \\ 0 \end{pmatrix}, \quad Y(x) := \begin{pmatrix} x_1 \\ -x_1 + x_2^2 \end{pmatrix}, \quad t := \frac{\pi}{2},$$

and $X(x) = Ax$. Then,

$$\mathrm{Ad}_{\frac{\pi}{2}X}Y(x^0) = e^{(-\pi/2)A}\begin{pmatrix} 0 \\ 1 \end{pmatrix} = \begin{pmatrix} 1 \\ 0 \end{pmatrix}$$

(a clockwise rotation by $\pi/2$ of $Y(e^{tX}x^0)$), see Figure 4.3.

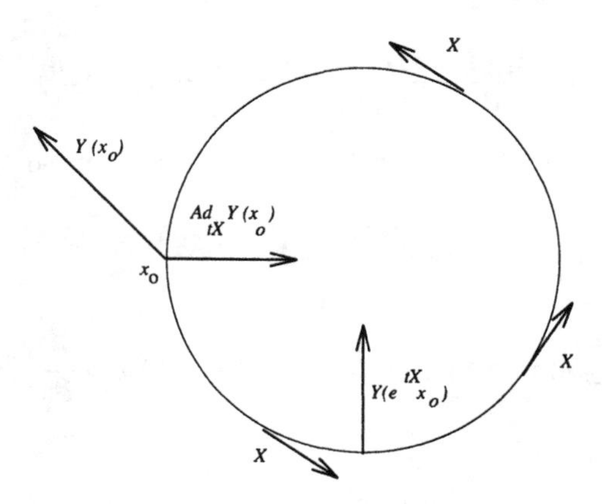

Figure 4.3: *Example of Ad operator.*

The state $e^{-tX}e^{sY}e^{tX}x^0 = \phi_X(-t, \phi_Y(s, \phi_X(t, x^0)))$ is well-defined provided that s is near zero (because $e^{-tX}z^0$ is defined for $z^0 = e^{tX}x^0$ and, thus, also for all z^0 sufficiently near $e^{tX}x^0$, and $e^{sY}e^{tX}x^0$ is defined, and is close to $e^{tX}x^0$, when s is small), and

$$\mathrm{Ad}_{tX}Y\left(x^0\right) = \left.\frac{\partial}{\partial s}\right|_{s=0} e^{-tX}e^{sY}e^{tX}x^0 .$$

We may view $\mathrm{Ad}_{tX}Y$ as a (smooth) vector field on the open subset $\widetilde{\mathcal{O}} \subseteq \mathcal{O}$ consisting of all x^0 such that $(t, x^0) \in \mathcal{D}_X$ and $e^{tX}x^0 \in \mathcal{O}_Y$. Given a third vector field $Z \in \mathbb{V}(\mathcal{O})$, and $(s, x^0) \in \mathcal{D}_Z$, we may consider $\mathrm{Ad}_{sZ}(\mathrm{Ad}_{tX}Y)(x^0)$, provided $e^{sZ}x^0 \in \widetilde{\mathcal{O}}$ (that is, if $(t, e^{sZ}x^0) \in \mathcal{D}_X$ and $e^{tX}e^{sZ}x^0 \in \mathcal{O}_Y$); we write this simply as $\mathrm{Ad}_{sZ}\mathrm{Ad}_{tX}Y(x^0)$. Similarly, one may consider iterations of applications of the Ad operator. The following is an easy consequence of the chain rule.

Lemma/Exercise 4.4.1 Let $X_1, \ldots, X_k, Y \in \mathbb{V}(\mathcal{O})$. Then,

$$\begin{aligned} &\mathrm{Ad}_{t_k X_k} \ldots \mathrm{Ad}_{t_1 X_1} Y(x^0) = \\ &\qquad \left(e^{-t_k X_k} \ldots e^{-t_1 X_1}\right)_* \left(e^{t_1 X_1} \ldots e^{t_k X_k} x^0\right) \cdot Y\left(e^{t_1 X_1} \ldots e^{t_k X_k} x^0\right) \end{aligned} \tag{4.26}$$

if $(t_k, x^0) \in \mathcal{D}_{X_k}$, $(t_{k-1}, e^{t_k X_k} x^0) \in \mathcal{D}_{X_{k-1}}, \ldots, (t_1, e^{t_2 X_2} \ldots e^{t_k X_k} x^0) \in \mathcal{D}_{X_1}$. In particular,

$$\mathrm{Ad}_{t_k X} \ldots \mathrm{Ad}_{t_1 X} Y(x^0) = \mathrm{Ad}_{(t_1 + \ldots + t_k)X} Y(x^0)$$

when all the X_i equal the same X. □

We may view Ad_{tX} as an operator on vector fields (defined on open subsets of $\mathcal{O}$). The following fact relates "Ad" to the "ad" operator introduced earlier, in Equation (4.4):

Lemma 4.4.2 Let $X \in \mathbb{V}(\mathcal{O})$ and $(t^0, x^0) \in \mathcal{D}_X$. Pick any $Y \in \mathbb{V}(\mathcal{O}_Y)$ such that $e^{t^0 X}x^0 \in \mathcal{O}_Y$. Then,

$$\left.\frac{\partial \mathrm{Ad}_{tX} Y(x^0)}{\partial t}\right|_{t=t^0} = \mathrm{Ad}_{t^0 X}[X, Y](x^0).$$

Proof. We have, for all $(t, x) \in \mathcal{D}_X$, $e^{-tX}e^{tX}x = x$, so taking $\partial/\partial x$ there results

$$\left(e^{-tX}\right)_* \left(e^{tX}x\right) \cdot \left(e^{tX}\right)_*(x) = I$$

and, taking $\partial/\partial t$,

$$\frac{\partial}{\partial t}\Big(\left(e^{-tX}\right)_* \left(e^{tX}x\right)\Big) \cdot \left(e^{tX}\right)_*(x) + \left(e^{-tX}\right)_* \left(e^{tX}x\right) \cdot \frac{\partial}{\partial t}\Big(\left(e^{tX}\right)_*(x)\Big) = 0. \tag{4.27}$$

On the other hand,

$$\begin{aligned} \frac{\partial}{\partial t}\Big(\left(e^{tX}\right)_*(x)\Big) &= \frac{\partial}{\partial t}\frac{\partial}{\partial x}\left(e^{tX}x\right) = \frac{\partial}{\partial x}\frac{\partial}{\partial t}\left(e^{tX}x\right) \\ &= \frac{\partial}{\partial x} X\left(e^{tX}x\right) = X_*\left(e^{tX}x\right) \cdot \left(e^{tX}\right)_*(x), \end{aligned}$$

so substituting in (4.27) and postmultiplying by $\left(\left(e^{tX}\right)_*(x)\right)^{-1}$ there results the identity:

$$\frac{\partial}{\partial t}\Big(\left(e^{-tX}\right)_* \left(e^{tX}x\right)\Big) = -\left(e^{-tX}\right)_* \left(e^{tX}x\right) \cdot X_*\left(e^{tX}x\right). \tag{4.28}$$

Thus,

$$
\begin{aligned}
&\frac{\partial}{\partial t}\Big(\left(e^{-tX}\right)_* \left(e^{tX}x\right) \cdot Y\left(e^{tX}x\right)\Big) \\
&\quad = \; -\left(e^{-tX}\right)_* \left(e^{tX}x\right) \cdot X_* \left(e^{tX}x\right) \cdot Y\left(e^{tX}x\right) \\
&\qquad\qquad + \left(e^{-tX}\right)_* \left(e^{tX}x\right) \cdot Y_* \left(e^{tX}x\right) \cdot X\left(e^{tX}x\right) \\
&\quad = \; \left(e^{-tX}\right)_* \left(e^{tX}x\right) \cdot [X,Y]\left(e^{tX}x\right),
\end{aligned}
$$

as required. ∎

A generalization to higher-order derivatives is as follows; it can be interpreted, in a formal sense, as saying that $\mathrm{Ad}_{tX} = e^{t\,\mathrm{ad}_X}$. For each $k = 0, 1, \ldots$, and each vector fields X, Y, we denote $\mathrm{ad}_X^0 Y = Y$ and $\mathrm{ad}_X^{k+1} Y := \mathrm{ad}_X(\mathrm{ad}_X^k Y)$.

Lemma 4.4.3 Let $X \in \mathbb{V}(\mathcal{O})$, $(t^0, x^0) \in \mathcal{D}_X$, $\mathcal{O}_Y \subseteq \mathcal{O}$, and $Y \in \mathbb{V}(\mathcal{O}_Y)$ be so that $e^{t^0 X}x^0 \in \mathcal{O}_Y$. Then, for each $k = 0, 1, 2, \ldots$,

$$\left.\frac{\partial^k \mathrm{Ad}_{tX} Y(x^0)}{\partial t^k}\right|_{t=t^0} = \mathrm{Ad}_{t^0 X}\mathrm{ad}_X^k Y(x^0).$$

In particular, for any $X, Y \in \mathbb{V}(\mathcal{O})$ and each $x^0 \in \mathcal{O}$,

$$\left.\frac{\partial^k \mathrm{Ad}_{tX} Y(x^0)}{\partial t^k}\right|_{t=0} = \mathrm{ad}_X^k Y(x^0)$$

for each k.

Proof. Fix the state x^0 and consider the function $\gamma(t) := \mathrm{Ad}_{tX} Y(x^0)$; we want to prove that $\gamma^{(k)}(t) = \mathrm{Ad}_{tX}\mathrm{ad}_X^k Y(x^0)$ for each k and for all t such that $(t, x^0) \in \mathcal{D}_X$ and $e^{tX}x^0 \in \mathcal{O}_Y$. We proceed by induction on k. The case $k = 0$ is trivial, since, by definition, $\mathrm{ad}_X^0 Y = Y$. Now, assume that the conclusion has been shown for k. Then,

$$\gamma^{(k+1)}(t) = \frac{d}{dt}\gamma^{(k)}(t) = \frac{d}{dt}\mathrm{Ad}_{tX} Z(x^0),$$

where Z is $\mathrm{ad}_X^k Y$, seen as a vector field defined on $\mathcal{O}_Y$. By Lemma 4.4.2 (the case $k = 1$), $\gamma^{(k+1)}(t) = \mathrm{Ad}_{tX}\mathrm{ad}_X Z(x^0) = \mathrm{Ad}_{tX}\mathrm{ad}_X^{k+1} Y(x^0)$, as desired.

In particular, consider any $X, Y \in \mathbb{V}(\mathcal{O})$ and $x^0 \in \mathcal{O}$. As $(0, x^0) \in \mathcal{D}_X$, the conclusions hold for $t = 0$. ∎

The next few technical results will be needed later, when we specialize our study to analytic systems.

Lemma 4.4.4 Suppose X and Y are analytic vector fields defined on $\mathcal{O}$. For any $x^0 \in \mathcal{O}$, let $\mathcal{I} = \mathcal{I}_{X,x^0} := \{t \in \mathbb{R} \mid (t, x^0) \in \mathcal{D}_X\}$. Then, the function $\gamma : \mathcal{I} \to \mathbb{R}^n : t \mapsto \mathrm{Ad}_{tX} Y(x^0)$ is analytic.

Proof. Let $\alpha(t) := \left(e^{-tX}\right)_* \left(e^{tX}x^0\right)$, seen as a function $\mathcal{I} \to \mathbb{R}^{n\times n}$. Note that $\alpha(0) = I$, and that, by Equation (4.28), the vector $(e^{tX}x^0, \alpha(t))$ is the solution of the differential equation

$$\begin{aligned} \dot{x}(t) &= X(x(t)) & x(0) &= x^0 \\ \dot{\alpha}(t) &= -\alpha(t) \cdot X_*(x(t)) & \alpha(0) &= I\,. \end{aligned}$$

This is a differential equation with analytic right-hand side, so $x(\cdot)$ and $\alpha(\cdot)$ are both analytic (see, for instance, Proposition C.3.12). Then, $\gamma(t) = \alpha(t)Y(x(t))$ is also analytic. ∎

Proposition 4.4.5 If the vector fields $Y_1, \ldots, Y_\ell$ and X are analytic, then

$$\operatorname{span}\left\{\operatorname{Ad}_{tX}Y_j(x^0),\ j = 1, \ldots \ell,\ t \in \mathcal{I}_{X,x^0}\right\}$$

equals

$$\operatorname{span}\left\{\operatorname{ad}_X^k Y_j(x^0),\ j = 1, \ldots \ell,\ k \geq 0\right\}$$

for each $x^0 \in \mathcal{O}$.

Proof. Fix $x^0 \in \mathcal{O}$. Let S_0 and S_1 be, respectively, the sets of vectors $\nu \in \mathbb{R}^n$ and $\mu \in \mathbb{R}^n$ such that

$$\nu'\operatorname{ad}_X^k Y_j(x^0) = 0\,, \quad j = 1, \ldots \ell\,,\ k \geq 0 \tag{4.29}$$

and

$$\mu'\operatorname{Ad}_{tX}Y_j(x^0) = 0\,, \quad j = 1, \ldots \ell\,,\ t \in \mathcal{I}_{X,x^0}\,. \tag{4.30}$$

Take any $\nu \in S_0$. For each $j = 1, \ldots \ell$, by Lemma 4.4.3,

$$\nu' \left.\frac{\partial^k \operatorname{Ad}_{tX}Y_j(x^0)}{\partial t^k}\right|_{t=0} = 0\,,\ \forall\, k \geq 0\,.$$

Since, by Lemma 4.4.4, $\operatorname{Ad}_{tX}Y_j(x^0)$ is analytic as a function of t, this means that $\nu'\operatorname{Ad}_{tX}Y_j(x^0) \equiv 0$, so $\nu \in S_1$. Conversely, if $\nu \in S_1$, then $\nu'\operatorname{Ad}_{tX}Y_j(x^0) \equiv 0$ implies that all derivatives at zero vanish, so $\nu \in S_0$ (analyticity is not needed here). Thus $S_0 = S_1$, and the result is proved. ∎

Corollary 4.4.6 Let the vector fields $Y_1, \ldots, Y_\ell$, and X be analytic, and pick any $(t, x^0) \in \mathcal{D}_X$. Let

$$d := \dim\operatorname{span}\left\{Y_j(e^{tX}x^0),\ j = 1, \ldots \ell\right\}\,.$$

Then,

$$d \leq \dim\operatorname{span}\left\{\operatorname{ad}_X^k Y_j(x^0),\ j = 1, \ldots \ell,\ k \geq 0\right\}\,.$$

Proof. Since $\operatorname{Ad}_{tX}Y_j(x^0) = Q \cdot Y_j\left(e^{tX}x^0\right)$, where $Q := \left(e^{-tX}\right)_* \left(e^{tX}x^0\right)$ is a nonsingular matrix, $d = \dim\operatorname{span}\{\operatorname{Ad}_{tX}Y_j(x^0), j = 1, \ldots \ell\}$. The result follows then from Proposition 4.4.5. ∎

Exercise 4.4.7 Show that Corollary 4.4.6 need not hold for smooth but non-analytic vector fields. □

Distributions

A distribution (in the sense of differential geometry) is the assignment of a subspace of tangent vectors to each point in the state space.

Definition 4.4.8 *A* **distribution** *on the open subset $\mathcal{O} \subseteq \mathbb{R}^n$ is a map Δ which assigns, to each $x \in \mathcal{O}$, a subspace $\Delta(x)$ of $\mathbb{R}^n$. A vector field $f \in \mathbb{V}(\mathcal{O})$ is* **pointwise in** Δ, *denoted $f \in_p \Delta$, if $f(x) \in \Delta(x)$ for all $x \in \mathcal{O}$. A distribution is* **invariant** *under a vector field $f \in \mathbb{V}(\mathcal{O})$ if*

$$g \in_p \Delta \quad \Rightarrow \quad [f, g] \in_p \Delta\,,$$

and it is **involutive** *if it is invariant under all $f \in_p \Delta$, that is, it is pointwise closed under Lie brackets:*

$$f \in_p \Delta \text{ and } g \in_p \Delta \quad \Rightarrow \quad [f, g] \in_p \Delta\,.$$

The distribution **generated by** *a set of vector fields $f_1, \ldots, f_r \in \mathbb{V}(\mathcal{O})$ is defined by*

$$\Delta_{f_1,\ldots,f_r}(x) \;:=\; \text{span}\,\{f_1(x), \ldots, f_r(x)\}$$

for each $x \in \mathcal{O}$. A distribution has **constant rank** *r if $\dim \Delta(x) = r$ for all $x \in \mathcal{O}$.* □

Lemma 4.4.9 Suppose that $\Delta = \Delta_{f_1,\ldots,f_r}$ is a distribution of constant rank r. Then,

1. The following two properties are equivalent, for any $f \in \mathbb{V}(\mathcal{O})$:

   (a) $f \in_p \Delta$

   (b) For each $x^0 \in \mathcal{O}$, there are a neighborhood $\mathcal{O}_0$ of x^0 and r smooth functions $\alpha_i : \mathcal{O}_0 \to \mathbb{R}$, $i = 1, \ldots, r$, so that

$$f(x) \;=\; \sum_{i=1}^{r} \alpha_i(x) f_i(x) \quad \text{for all } x \in \mathcal{O}_0\,. \tag{4.31}$$

2. The following two properties are equivalent, for any $f \in \mathbb{V}(\mathcal{O})$:

   (a) Δ is invariant under f.

   (b) $[f, f_j] \in_p \Delta$ for each $j \in \{1, \ldots, r\}$.

3. Finally, the following two properties are equivalent:

   (a) Δ is involutive.

   (b) $[f_i, f_j] \in_p \Delta$ for all $i, j \in \{1, \ldots, r\}$.

Proof. The equivalence of 1a and 1b was already established in Corollary 4.1.6 (applied with $W = \mathcal{O}$, $p = r$, and Q the matrix function whose columns are the f_i's).

The implication 2a $\Rightarrow$ 2b is clear, since each $f_j \in_p \Delta$. To prove the converse, pick any $g \in_p \Delta$, and consider any point $x^0 \in \mathcal{O}$. By the first equivalence, we can write $g(x) = \sum_{j=1}^r \alpha_j(x) f_j(x)$, on some neighborhood $\mathcal{O}_0$ of x^0, for some smooth functions α_j's. Lemma 4.1.2 gives, then, as functions on $\mathcal{O}_0$:

$$[f,g] = \sum_{j=1}^{r} \alpha_j [f, f_j] + (L_f \alpha_j) f_j \,.$$

Since $f_j(x^0) \in \Delta(x^0)$ and $[f, f_j](x^0) \in \Delta(x^0)$ for all j, we conclude that $[f,g](x^0) \in \Delta(x^0)$.

Similarly, the implication 3a $\Rightarrow$ 3b is clear, since each $f_i \in_p \Delta$. To prove the converse, we need to show that Δ is invariant under any $f \in_p \Delta$. The previous equivalence tells us that it is enough to check that 2b holds, for each such f. Since each $\Delta(x)$ is a subspace, however, this is the same as asking that $[f_i, f] \in_p \Delta$ for all i and all $f \in_p \Delta$, that is, the property that Δ is invariant under f_i; but again by the previous equivalence, this holds if $[f_i, f_j] \in_p \Delta$ for all j, which was the assumption. ∎

Exercise 4.4.10 Provide examples of distributions $\Delta_{f_1,\ldots,f_r}$ (necessarily not having constant rank) for which each of the above equivalences fail. □

Exercise 4.4.11 A distribution is said to be *smooth* if it is locally generated by sets (possibly infinite) of vector fields, that is, for each $x^0 \in \mathcal{O}$ there is a subset $\mathcal{F} \subseteq \mathbb{V}(\mathcal{O})$, and there is an open subset $\mathcal{O}_0 \subseteq \mathcal{O}$ which contains x^0, such that, for each $x \in \mathcal{O}_0$, $\Delta(x)$ is the span of the vectors $\{f(x), f \in \mathcal{F}\}$. Show that, if Δ is a smooth distribution of constant rank r, then for each $x^0 \in \mathcal{O}$ there is some open subset $\mathcal{O}_0 \subseteq \mathcal{O}$ which contains x^0, and a set of r vector fields $f_1, \ldots, f_r$, such that $\Delta = \Delta_{f_1,\ldots,f_r}$ on $\mathcal{O}_0$, that is, $\Delta(x) = \text{span}\,\{f_1(x), \ldots, f_r(x)\}$ for each $x \in \mathcal{O}_0$. □

Invariance of a distribution under f is equivalent to invariance under the linear operators Ad_{tf}, in the sense that Δ is invariant under f if and only if $\text{Ad}_{tf}\, g(x) \in \Delta(x)$ for each $(t,x) \in \mathcal{D}_f$ and each $g \in_p \Delta$. This fact, proved next, is perhaps the most important property of the Lie bracket operation. (We need a slightly stronger version, in which g is not necessarily defined globally.)

Lemma 4.4.12 Suppose that $\Delta = \Delta_{f_1,\ldots,f_r}$ has constant rank r, and let $X \in \mathbb{V}(\mathcal{O})$. Then, the following two properties are equivalent:

1. Δ is invariant under X.

2. Let $\mathcal{O}_1$ be an open subset of $\mathcal{O}$ and let $t \in \mathbb{R}$ be so that $(t,x) \in \mathcal{D}_X$ for all $x \in \mathcal{O}_1$. Define $\mathcal{O}_0 := e^{tX}\mathcal{O}_1$. Assume that $Y \in \mathbb{V}(\mathcal{O}_0)$ is such that $Y(z) \in \Delta(z)$ for each $z \in \mathcal{O}_0$. Then, $\text{Ad}_{tX} Y(x) \in \Delta(x)$ for each $x \in \mathcal{O}_1$.

Proof. We first prove that 2 $\Rightarrow$ 1. Pick any $Y \in_p \Delta$, and any $x^0 \in \mathcal{O}$. Take any open subset $\mathcal{O}_1 \subseteq \mathcal{O}$ containing x^0 and any $\varepsilon > 0$ so that $(t, x) \in \mathcal{D}_X$ for all $|t| < \varepsilon$ and all $x \in \mathcal{O}_1$. It follows from 2 that $\mathrm{Ad}_{tX} Y(x) \in \Delta(x)$ for all t near zero and all $x \in \mathcal{O}_1$, so, in particular, $\mathrm{Ad}_{tX} Y(x^0) \in \Delta(x^0)$ for small t. Then, since $\Delta(x^0)$ is a subspace, also

$$\frac{1}{t}\left(\mathrm{Ad}_{tX} Y(x^0) - Y(x^0)\right) \in \Delta(x^0)$$

and, therefore, since $\Delta(x^0)$ is closed, the limit $[X, Y](x^0)$ of this expression (cf. Lemma 4.4.2) is in $\Delta(x^0)$. So Δ is invariant under X.

We now prove the converse. We first show that, for each given $x \in \mathcal{O}$, and letting $\mathcal{I} := \{t \in \mathbb{R} \mid (t, x) \in \mathcal{D}_X\}$,

$$\alpha_i(t) := \mathrm{Ad}_{tX} f_i(x) \in \Delta(x) \quad \text{for each } i \in \{1, \ldots, r\} \text{ and each } t \in \mathcal{I}. \tag{4.32}$$

For this, it is enough to prove that, for each vector ν such that $\nu' f_j(x) = 0$ for all $j = 1, \ldots, r$, necessarily $\nu' \alpha_i \equiv 0$ for all i. So pick such a ν, and consider the row vector function $\rho(t) := \nu' M(t)$, where $M(t) := (\alpha_1(t), \ldots, \alpha_r(t))$. Note that $\rho(0) = \nu'(f_1(x), \ldots, f_r(x)) = 0$.

Assume that it is not the case that $\rho \equiv 0$. Suppose there is some $t > 0$ so that $\rho(t) \neq 0$ (if this only happens for some $t < 0$, the argument is analogous). Let

$$t_0 := \inf\{t > 0, t \in \mathcal{I}, \rho(t) \neq 0\}.$$

Note that $\rho(t_0) = 0$. Consider the point $z := e^{t_0 X} x$. For each $i = 1, \ldots, r$, since $[X, f_i] \in_p \Delta$ (invariance under X), from Lemma 4.4.9 we know that there is a neighborhood $\mathcal{O}_0$ of z and there exists a set of smooth functions $\{\gamma_{ij}, j = 1, \ldots, r\} \subseteq \mathbb{F}(\mathcal{O}_0)$ such that

$$[X, f_i](z) = \sum_{j=1}^{r} \gamma_{ij}(z) f_j(z), \quad \text{for all } z \in \mathcal{O}_0. \tag{4.33}$$

(We may assume that the neighborhood $\mathcal{O}_0$ is common to all i.) By Lemma 4.4.2 and Equation (4.33),

$$\begin{aligned}\dot{\alpha}_i(t) &= \mathrm{Ad}_{tX}[X, f_i](x) = \left(e^{-tX}\right)_* \left(e^{tX}x\right) \cdot [X, f_i]\left(e^{tX}x\right) \\ &= \sum_{j=1}^{r} \gamma_{ij}\left(e^{tX}x\right) \alpha_j(t)\end{aligned}$$

as long as $|t - t_0|$ is small enough so that $e^{tX}x \in \mathcal{O}_0$. So we have that

$$\dot{M}(t) = M(t)\Gamma(t),$$

where we are denoting $\Gamma(t) := (\gamma_{ji}(e^{tX}x))$. Then, $\dot{\rho}(t) = \rho(t)\Gamma(t)$, and $\rho(t_0) = 0$; by uniqueness of solutions of $\dot{\rho} = \rho\Gamma$ we have that $\rho \equiv 0$ for all t sufficiently near t_0. This contradicts the definition of t_0. So it must be that $\rho \equiv 0$. This concludes the proof of (4.32).

Finally, pick any $\mathcal{O}_0$, $\mathcal{O}_1$, t, and Y as in 2, and any $x^0 \in \mathcal{O}_1$. Taking if necessary a smaller neighborhood $\mathcal{O}_1$ of x^0, we know that there are smooth functions $\{\gamma_i, i = 1, \ldots, r\} \subseteq \mathbb{F}(\mathcal{O}_0)$ such that $Y = \sum_{i=1}^r \gamma_i f_i$ on $\mathcal{O}_0$. Therefore, letting $\alpha_i(t) = \mathrm{Ad}_{tX} f_i(x^0)$ for each i,

$$\mathrm{Ad}_{tX} Y(x^0) = \left(e^{-tX}\right)_* \left(e^{tX}x^0\right) \cdot Y\left(e^{tX}x^0\right) = \sum_{i=1}^{r} \gamma_i\left(e^{tX}x^0\right) \alpha_i(t),$$

which is in $\Delta(x^0)$ because each $\alpha_i(t) \in \Delta(x^0)$ by (4.32). ■

By induction, we also have then the following consequence of Lemma 4.4.12 (see Figure 4.4).

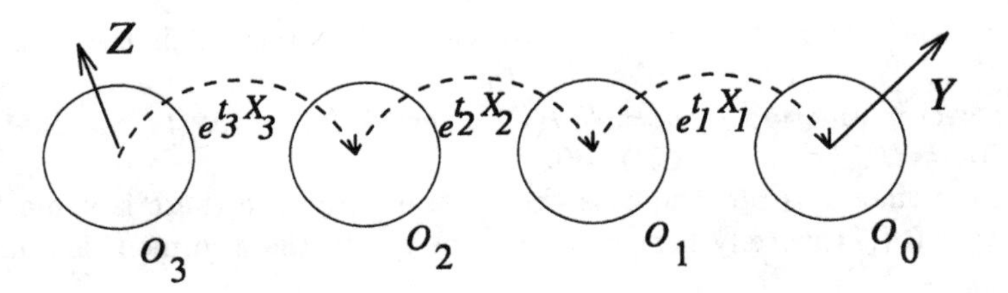

Figure 4.4: $Z = \mathrm{Ad}_{t_3X_3}\mathrm{Ad}_{t_2X_2}\mathrm{Ad}_{t_1X_1}Y$.

Corollary 4.4.13 Suppose that $\Delta = \Delta_{f_1,\ldots,f_r}$ has constant rank r and is invariant under each of $X_1, \ldots, X_k \in \mathbb{V}(\mathcal{O})$. Let $\mathcal{O}_k$ be an open subset of $\mathcal{O}$ and let $s_1, \ldots, s_k$ be real numbers with the following property:

$$(s_i, x) \in \mathcal{D}_{X_i}, \quad \forall x \in \mathcal{O}_i, \; i = 1, \ldots, k,$$

where we define

$$\mathcal{O}_{k-1} := e^{s_kX_k}\mathcal{O}_k, \ldots, \; \mathcal{O}_1 := e^{s_2X_2}\mathcal{O}_2, \mathcal{O}_0 := e^{s_1X_1}\mathcal{O}_1.$$

Assume that $Y \in \mathbb{V}(\mathcal{O}_0)$ is such that $Y(z) \in \Delta(z)$ for each $z \in \mathcal{O}_0$. Then, $\mathrm{Ad}_{s_kX_k} \cdots \mathrm{Ad}_{s_1X_1} Y(x) \in \Delta(x)$ for all $x \in \mathcal{O}_k$. □

Definition 4.4.14 *A distribution Δ of constant rank r is* **completely integrable** *if, for each $x^0 \in \mathcal{O}$ there exists some neighborhood $\mathcal{O}_0$ of x^0 and a smooth function*

$$\Psi : \mathcal{O}_0 \to \mathbb{R}^{n-r}$$

such that

$$\ker \Psi_*(x) = \Delta(x) \tag{4.34}$$

for all $x \in \mathcal{O}_0$. □

Equation (4.34) says that $\Psi_*(x)v = 0$ if and only if $v \in \Delta(x)$. Since Δ is assumed to have constant rank r, this implies that the Jacobian $\Psi_*(x)$ has constant rank $n-r$. If $\Delta = \Delta_{f_1,\ldots,f_r}$, then, since $\{f_1(x),\ldots,f_r(x)\}$ is a basis of $\Delta(x)$ for each x, we have that Equation (4.34) is equivalent to:

$$\Psi_*(x)f_j(x) = 0 \quad \text{for } j = 1,\ldots,r \tag{4.35}$$

together with:

$$\operatorname{rank}\Psi_*(x) = n-r\,. \tag{4.36}$$

In other words, if we consider the rows ψ_k, $k = 1,\ldots,n-r$, of Ψ, we have that

$$L_{f_j}\psi_k(x) = \nabla\psi_k(x)f_j(x) = 0 \quad \text{for } k = 1,\ldots,n-r,\ j = 1,\ldots,r\,, \tag{4.37}$$

and:

$$\{\nabla\psi_1(x),\ldots,\nabla\psi_{n-r}(x)\} \text{ are linearly independent.} \tag{4.38}$$

Therefore, complete integrability of Δ means that, locally about each point, the set of r first order linear partial differential equations

$$\nabla\psi(x)\,f_j(x) = 0 \;\; j = 1,\ldots,r$$

has $n-r$ independent (in the sense of Jacobians being independent) solutions.

Remark 4.4.15 The terminology arises as follows. A (smooth) function $\psi : \mathcal{O} \to \mathbb{R}$ is an integral of motion for the flow induced by a vector field f if $\psi(x(t))$ is constant along solutions. This means that $\frac{d\psi(\phi_f(t,x^0))}{dt} = 0$ for all x^0 and t, which is equivalent to the statement that $\nabla\psi(x)f(x) \equiv 0$. Thus, complete integrability means that, locally, we can find $n-r$ independent integrals of motion. In geometric terms, since the level sets of $\Psi(x) = c$ are differentiable submanifolds of dimension r (because Ψ_* has constant rank $n-r$) complete integrability means that the space can be locally partitioned ("foliated") into submanifolds of dimension r, in such a manner that the vector fields f_i are tangent to these submanifolds. □

Frobenius' Theorem

The following construction provides the main step in the proof of Frobenius' Theorem. We state it separately, for future reference.

Lemma 4.4.16 Suppose that $\Delta = \Delta_{f_1,\ldots,f_r}$ is involutive of constant rank r. Then, for each $x^0 \in \mathcal{O}$ there exists some open subset $\mathcal{O}_0$ containing x^0, an $\varepsilon > 0$, and a diffeomorphism $\Pi : \mathcal{O}_0 \to (-\varepsilon,\varepsilon)^n$, so that the following property holds. If we partition

$$\Pi = \begin{pmatrix}\Pi_1\\ \Pi_2\end{pmatrix}, \quad \Pi_2 : \mathcal{O}_0 \to (-\varepsilon,\varepsilon)^{n-r}\,,$$

then $\Delta(x) = \ker(\Pi_2)_*(x)$ for all $x \in \mathcal{O}_0$.

Proof. Pick $x^0 \in \mathcal{O}$. Let $v_{r+1}, \ldots, v_n$ be vectors so that

$$\{f_1(x^0), \ldots, f_r(x^0), v_{r+1}, \ldots, v_n\}$$

is a basis of $\mathbb{R}^n$. Introduce the vector fields $f_i \equiv v_i$, for $i = r+1, \ldots, n$. Then $\{f_1(x), \ldots, f_n(x)\}$ is a basis of $\mathbb{R}^n$ for all x in some open subset $\widetilde{\mathcal{O}} \subseteq \mathcal{O}$. Replacing $\mathcal{O}$ by $\widetilde{\mathcal{O}}$, we assume from now on that $\{f_1(x), \ldots, f_n(x)\}$ are linearly independent for all $x \in \mathcal{O}$. We consider the following map:

$$M : (-\varepsilon, \varepsilon)^n \to \mathcal{O} : \mathbf{t} = (t_1, \ldots, t_n) \mapsto e^{t_1 f_1} \ldots e^{t_n f_n} x^0 ,$$

where ε is chosen small enough so that, for all $j > 1$, $(t_{j-1}, e^{t_j f_j} \ldots e^{t_n f_n} x^0) \in \mathcal{D}_{f_{j-1}}$ whenever $\mathbf{t} \in (-\varepsilon, \varepsilon)^n$. We have, for each $j = 1, \ldots, n$:

$$\begin{aligned} \frac{\partial M}{\partial t_j}(\mathbf{t}) &= \left(e^{t_1 f_1} \ldots e^{t_{j-1} f_{j-1}}\right)_* \left(e^{t_j f_j} \ldots e^{t_n f_n} x^0\right) \cdot f_j \left(e^{t_j f_j} \ldots e^{t_n f_n} x^0\right) \\ &= \mathrm{Ad}_{(-t_1) f_1} \cdots \mathrm{Ad}_{(-t_{j-1}) f_{j-1}} f_j \left(M(\mathbf{t})\right) \end{aligned}$$

(the last equality because $e^{t_j f_j} \ldots e^{t_n f_n} x^0 = e^{-t_{j-1} f_{j-1}} \ldots e^{-t_1 f_1} M(\mathbf{t})$, and applying (4.26)). In particular,

$$M_*(0, \ldots, 0) = \left(f_1(x^0), \ldots, f_n(x^0)\right) . \tag{4.39}$$

We claim that

$$\frac{\partial M}{\partial t_j}(\mathbf{t}) \in \Delta(M(\mathbf{t})) , \quad \text{for each } j = 1, \ldots, r \text{ and each } \mathbf{t} \in (-\varepsilon, \varepsilon)^n . \tag{4.40}$$

To show this for each j, we apply Corollary 4.4.13, with $k = j-1$, $Y = f_j$, $X_1 = f_{j-1}$, ..., $X_k = f_1$, $s_1 = -t_{j-1}$, ..., $s_k = -t_1$, and $\mathcal{O}_k$ is chosen as any small enough neighborhood of $M(\mathbf{t})$ with the property that $e^{-t_{j-1} f_{j-1}} \ldots e^{-t_1 f_1} x$ is defined for every $x \in \mathcal{O}_k$. Involutivity of Δ insures that Δ is invariant under all the X_i's. As $Y \in_p \Delta$, we conclude from that Corollary that

$$\mathrm{Ad}_{(-t_1) f_1} \cdots \mathrm{Ad}_{(-t_{j-1}) f_{j-1}} f_j(z) \in \Delta(z)$$

for each $z \in \mathcal{O}_k$, and, so, in particular, for $M(\mathbf{t})$. This proves the claim.

Because of (4.39), by the Inverse Function Theorem, taking a smaller ε if necessary, and letting $\widetilde{\mathcal{O}}$ be the image of M, we may assume that M is a diffeomorphism, that is, there is a smooth function $\Pi : \widetilde{\mathcal{O}} \to (-\varepsilon, \varepsilon)^n$ such that $\Pi \circ M(\mathbf{t}) = \mathbf{t}$, for all $\mathbf{t} \in (-\varepsilon, \varepsilon)^n$. In particular,

$$\Pi_*(M(\mathbf{t}))\, M_*(\mathbf{t}) = I \quad \forall \mathbf{t} \in (-\varepsilon, \varepsilon)^n . \tag{4.41}$$

We partition Π as $\begin{pmatrix} \Pi_1 \\ \Pi_2 \end{pmatrix}$, where $\Pi_2 : \widetilde{\mathcal{O}} \to (-\varepsilon, \varepsilon)^{n-r}$. Then (4.41) implies:

$$\Pi_{2*}(M(\mathbf{t}))\, \frac{\partial M}{\partial t_j}(\mathbf{t}) = 0 \quad \forall \mathbf{t} \in (-\varepsilon, \varepsilon)^n , \quad j = 1, \ldots, r . \tag{4.42}$$

Since the vectors $\frac{\partial M}{\partial t_j}(\mathbf{t})$ are linearly independent (because of (4.41)) and they are in $\Delta(M(\mathbf{t}))$ (because of (4.40)), they form a basis of $\Delta(M(\mathbf{t}))$; thus

$$\Pi_{2*}(M(\mathbf{t}))\,\Delta(M(\mathbf{t})) \;=\; 0\,,$$

and, since $\operatorname{rank}\Pi_{2*}(M(\mathbf{t})) = n-r$, $\Delta(M(\mathbf{t})) = \ker\Pi_{2*}(M(\mathbf{t}))$. Each point of $\widetilde{\mathcal{O}}$ is of the form $M(\mathbf{t})$, for some $\mathbf{t} \in (-\varepsilon,\varepsilon)^n$, so this means that $\Delta(x) = \ker\Pi_{2*}(x)$ for all $x \in \widetilde{\mathcal{O}}$, as wanted. ∎

Theorem 10 (Frobenius) *Suppose that* $\Delta = \Delta_{f_1,\ldots,f_r}$ *has constant rank* r. *Then,* Δ *is completely integrable if and only if it is involutive.*

Proof. Sufficiency follows from Lemma 4.4.16, taking $\Psi := \Pi_2$. The necessity part is easy: Pick any $x^0 \in \mathcal{O}$, let $\mathcal{O}_0$ and Ψ be as in the definition of integrability, and consider the rows ψ_k, $k = 1,\ldots,n-r$, of Ψ, so that (4.37) holds. Thus, for each $i,j \in \{1,\ldots,r\}$,

$$L_{[f_i,f_j]}\psi_k(x) \;=\; L_{f_i}L_{f_j}\psi_k(x) - L_{f_j}L_{f_i}\psi_k(x) \;=\; 0$$

for all $k = 1,\ldots,n-r$ and all $x \in \mathcal{O}_0$. In particular, this means that

$$\Psi_*(x^0)\cdot[f_i,f_j](x^0) \;=\; 0\,,$$

that is, $[f_i,f_j](x^0) \in \ker\Psi_*(x^0) = \Delta(x^0)$, proving involutivity. ∎

If $\mathcal{O}_0$ is an open subset of $\mathcal{O}$, Δ is a distribution on $\mathcal{O}$, and $\Pi : \mathcal{O}_0 \to \mathcal{V}$ is a diffeomorphism onto some open subset $\mathcal{V} \subseteq \mathbb{R}^n$, we denote by $\Pi_*\Delta$ the distribution on $\mathcal{V}$ defined by attaching to each element of $\mathcal{V}$ the vector space associated to its preimage:

$$(\Pi_*\Delta)(z) \;:=\; \Pi_*\left(\Pi^{-1}(z)\right)\cdot\Delta\left(\Pi^{-1}(z)\right). \tag{4.43}$$

If $f \in \mathbb{V}(\mathcal{O})$, we also consider the vector field on $\mathcal{V}$:

$$(\Pi_*f)(z) \;:=\; \Pi_*\left(\Pi^{-1}(z)\right)\cdot f\left(\Pi^{-1}(z)\right). \tag{4.44}$$

One may interpret Π_*f as the vector field obtained when changing coordinates under $z = \Pi(x)$, in the sense that, if $x(\cdot)$ is a curve on $\mathcal{O}_0$ so that $\dot{x} = f(x)$, then $z(t) := \Pi(x(t))$ satisfies, clearly, the differential equation $\dot{z} = (\Pi_*f)(z)$. In other words, the flow of f maps into the flow of Π_*f. Frobenius' Theorem can be reinterpreted in terms of such coordinate changes:

Exercise 4.4.17 Suppose that $\Delta = \Delta_{f_1,\ldots,f_r}$ is a distribution of constant rank r on an open set $\mathcal{O} \subseteq \mathbb{R}^n$. Show that the following two properties are equivalent:

- Δ is involutive.

- For each $x^0 \in \mathcal{O}$ there is an open subset $\mathcal{O}_0 \subseteq \mathcal{O}$ which contains x^0 and a diffeomorphism $\Pi : \mathcal{O}_0 \to \mathcal{V}$ into some open subset $\mathcal{V} \subseteq \mathbb{R}^n$ such that $(\Pi_* \Delta)(z) = \text{span}\,\{e_1, \ldots, e_r\}$ for each $z \in \mathcal{V}$, where e_i is the ith canonical basis vector.

(A distribution of the form $\Delta_{e_1,\ldots,e_r}$ is said to be a *flat* distribution. So, the claim is that a distribution is involutive if and only if it is locally flat in appropriate coordinates.) □

Exercise 4.4.18 Suppose that $f_1, \ldots, f_r$ are smooth vector fields on an open set $\mathcal{O} \subseteq \mathbb{R}^n$, and that $f_1(x), \ldots, f_r(x)$ are linearly independent for each $x \in \mathcal{O}$. Show that the following two properties are equivalent:

- $[f_i, f_j] = 0$ for each $i, j \in \{1, \ldots, r\}$.
- For each $x^0 \in \mathcal{O}$ there is an open subset $\mathcal{O}_0 \subseteq \mathcal{O}$ which contains x^0 and a diffeomorphism $\Pi : \mathcal{O}_0 \to \mathcal{V}$ into some open subset $\mathcal{V} \subseteq \mathbb{R}^n$ such that $(\Pi_* f_i)(z) = e_i$ for each $z \in \mathcal{V}$, where e_i is the ith canonical basis vector.

(That is, the vector fields commute if and only if there is a local change of variables where they all become $f_i \equiv e_i$.) □

Exercise 4.4.19 Provide an example of a set of vector fields $f_1, \ldots, f_r$ so that (a) the distribution $\Delta = \Delta_{f_1,\ldots,f_r}$ has constant rank r and, locally about each point $x^0 \in \mathcal{O}$, transforms under a diffeomorphism into $\Delta_{e_1,\ldots,e_r}$ (in the sense of Exercise 4.4.17), but (b) there is some point x^0 such that, for no possible diffeomorphism Π defined in a neighborhood of x^0, $(\Pi_* f_i)(z) = e_i$ for all i. □

Exercise 4.4.20 Consider the following vector fields in $\mathbb{R}^2$:

$$f_1\begin{pmatrix} x_1 \\ x_2 \end{pmatrix} = \begin{pmatrix} x_1 \\ x_2 \end{pmatrix} \quad \text{and} \quad f_2\begin{pmatrix} x_1 \\ x_2 \end{pmatrix} = \begin{pmatrix} x_2 \\ -x_1 \end{pmatrix}.$$

(a) Check that $[f_1, f_2] = 0$. Explain in geometric terms why $e^{t_1 f_1} e^{t_2 f_2} x^0 = e^{t_2 f_2} e^{t_1 f_1} x^0$ for all t_1, t_2, and x^0.
(b) Find, locally about each $x^0 \neq 0$, an explicit change of coordinates Π (whose existence is assured by (a), cf. Exercise 4.4.18) so that f_1 becomes e_1 and f_2 becomes e_2 under Π. □

Exercise 4.4.21 Consider the following partial differential equation:

$$\varphi_{x_1}(x_1, x_2)\, x_1^2 + \varphi_{x_2}(x_1, x_2) = 0.$$

(a) Explain how we know that, locally about each $x^0 = (x_1^0, x_2^0) \in \mathbb{R}^2$, there exists a solution φ of this equation so that $\varphi(x^0) = 0$ and $\nabla\varphi(x) \neq 0$ everywhere in its domain.
(b) Now find one such solution, about each point. (You will need different functions φ around different points.)
(*Hint:* You may want to introduce an appropriate vector field and follow the construction in the proof of Lemma 4.4.16.) □

4.5 Necessity of Accessibility Rank Condition

We wish to prove now two partial converses of Theorem 9 (p. 156). The first one, valid for arbitrary smooth systems, says that if the sets $\mathcal{R}_{\mathcal{V}}^{\leq T}(x)$ have nonempty interior (for all $T, \mathcal{V}$) for each x in an open and dense subset of $\mathcal{X}$, then the accessibility rank condition also holds in a (possibly different) open and dense subset of $\mathcal{X}$. The second one, valid for analytic systems, drops the "possibly different". (The counterexamples in Exercise 4.3.6 show that more than smoothness is needed in order to prove this stronger statement.)

Given any system $\dot{x} = f(x, u)$, the set of points where the accessibility rank condition holds is an open set (possibly empty), because if $X_1(x), \ldots, X_n(x)$ are linearly independent, then they are also linearly independent in a neighborhood of x. Thus, asking that the rank condition holds in a dense set is equivalent to asking that it holds in an open dense set.

Theorem 11 *The following statements are equivalent:*

1. *There is an open and dense subset $\mathcal{X}_1 \subseteq \mathcal{X}$ such that* $\operatorname{int} \mathcal{R}_{\mathcal{V}}^{\leq T}(x^0) \neq \emptyset$ *for each $x^0 \in \mathcal{X}_1$, each neighborhood $\mathcal{V}$ of x^0, and each $T > 0$.*

2. *There is an open and dense subset $\mathcal{X}_2 \subseteq \mathcal{X}$ such that the accessibility rank condition holds at each $x^0 \in \mathcal{X}_2$.*

Proof. Theorem 9 (p. 156) shows that if the accessibility rank condition holds for each $x^0 \in \mathcal{X}_2$, then $\operatorname{int} \mathcal{R}_{\mathcal{V}}^{\leq T}(x^0) \neq \emptyset$ for each $x^0 \in \mathcal{X}_2$ and all T, $\mathcal{V}$. Assume now, conversely, that there is a set $\mathcal{X}_1$ as in the first statement. Suppose that the second statement is not true. In that case, there is some open subset $\mathcal{X}_0 \subseteq \mathcal{X}$ such that the accessibility rank condition fails for every $x^0 \in \mathcal{X}_0$. (Otherwise, the set where it holds is dense, and hence open dense.) Since $\mathcal{X}_0$ is open and $\mathcal{X}_1$ is open dense, the intersection $\mathcal{X}_1 \bigcap \mathcal{X}_0$ is a nonempty open set, so we may, and will, assume without loss of generality that $\mathcal{X}_0 \subseteq \mathcal{X}_1$. Let $r < n$ be defined as the maximal possible dimension of the vector spaces $\mathcal{A}_{\rm LA}(x)$, over all $x \in \mathcal{X}_0$.

Pick $X_1, \ldots, X_r \in \mathcal{A}_{\rm LA}$ and $x^0 \in \mathcal{X}_0$ so that $X_1(x^0), \ldots, X_r(x^0)$ are linearly independent. Then, $X_1(x), \ldots, X_r(x)$ are linearly independent for all x near x^0. Restricting $\mathcal{X}_0$ if needed, we may assume that $X_1(x), \ldots, X_r(x)$ are linearly independent for all $x \in \mathcal{X}_0$, and they generate $\mathcal{A}_{\rm LA}(x)$ (because of maximality of r). We introduce the constant-rank distribution $\Delta := \Delta_{X_1,\ldots,X_r}$. We have that $[X_i, X_j] \in \mathcal{A}_{\rm LA}$ for all i, j, because, by definition, $\mathcal{A}_{\rm LA}$ is a Lie algebra of vector fields. By the equivalence 3a $\Leftrightarrow$ 3b in Lemma 4.4.9, Δ is involutive.

We apply Lemma 4.4.16 (Frobenius), to conclude that there is some open subset $\mathcal{V} \subseteq \mathcal{X}_0$ containing x^0, an $\varepsilon > 0$, and a diffeomorphism $\Pi : \mathcal{V} \to (-\varepsilon, \varepsilon)^n$, so that, if $\Pi_2 : \mathcal{V} \to (-\varepsilon, \varepsilon)^{n-r}$ denotes the last $n - r$ coordinates of Π, then

$$(\Pi_2)_*(x)\, f(x, u) \;=\; 0 \tag{4.45}$$

for all $u \in \mathcal{U}$ and $x \in \mathcal{V}$ (using the fact that, by definition, $f_u \in \mathcal{A} \subseteq \mathcal{A}_{\rm LA}$, so $f_u(x) = f(x, u) \in \Delta(x)$ for all x).

We claim that $\mathcal{R}_{\mathcal{V}}^{\leq T}(x^0)$ has an empty interior, in fact for every $T > 0$, which contradicts $x^0 \in \mathcal{X}_1$. Since Π is a diffeomorphism, it will be enough to show that the image

$$\Pi\left(\mathcal{R}_{\mathcal{V}}^{\leq T}(x^0)\right)$$

has an empty interior.

Pick any control $\omega \in \mathcal{L}_{\mathcal{U}}^{\infty}(0,T)$, admissible for x^0, with the property that the path $x(t) := \phi(t,0,x^0,\omega)$ satisfies $x(t) \in \mathcal{V}$ for all $0 \leq t \leq T$. Let $z(t) := \Pi(x(t))$. This is absolutely continuous, and

$$\dot{z}(t) = \Pi_*(x(t))\, f(x(t),\omega(t)) = g(z(t),\omega(t))$$

for almost all $t \in [0,T]$, where we define $g_u := \Pi_* f_u$ (cf. Equation (4.44)):

$$g(z,u) := \Pi_*\left(\Pi^{-1}(z)\right)\, f\left(\Pi^{-1}(z),u\right)$$

for each $z \in (-\varepsilon,\varepsilon)^n$ and each $u \in \mathcal{U}$. We partition z into blocks of size $n-r$ and r respectively, so that the differential equation satisfied by z can be written as $\dot{z}_1 = g_1(z_1,z_2,u)$, $\dot{z}_2 = g_2(z_1,z_2,u)$. By (4.45), $g_2 \equiv 0$. Thus the equation is

$$\begin{aligned}\dot{z}_1 &= g_1(z_1,z_2,u)\\ \dot{z}_2 &= 0,\end{aligned}$$

in the z variables, and this shows that

$$\Pi\left(\mathcal{R}_{\mathcal{V}}^{\leq T}(x^0)\right) \subseteq (-\varepsilon,\varepsilon)^r \times \Pi_2(x^0)$$

has empty interior. ∎

Analytic Case

We now consider analytic systems $\dot{x} = f(x,u)$, meaning that $f_u = f(\cdot,u)$ is analytic, for each $u \in \mathcal{U}$. Models derived from classical mechanics are most often analytic.

Lemma 4.5.1 Suppose that the accessibility rank condition holds at $z^0 = e^{tf_u}x^0$, for some t, x^0, and $u \in \mathcal{U}$, and the system is analytic. Then the accessibility rank condition holds at x^0 as well.

Proof. Let $X = f_u$. Pick a set $Y_1,\ldots,Y_n$ of vector fields in $\mathcal{A}_{\mathrm{LA}}$ such that

$$Y_1(e^{tX}x^0),\ldots,Y_n(e^{tX}x^0)$$

are linearly independent. By Corollary 4.4.6, the vectors $\mathrm{ad}_X^k Y_j(x^0)$ span $\mathbb{R}^n$. The vector fields $\mathrm{ad}_X^k Y_j$ are in $\mathcal{A}_{\mathrm{LA}}$, because X as well as the Y_j's are. Therefore $\mathcal{A}_{\mathrm{LA}}(x^0) = \mathbb{R}^n$, that is, the accessibility rank condition holds at x^0. ∎

Theorem 12 *For analytic systems, the following two properties are equivalent:*

- *For each state x and neighborhood $\mathcal{V}$ of x, and each $T > 0$, $\operatorname{int}\mathcal{R}_{\mathcal{V}}^{\leq T}(x) \neq \emptyset$.*
- *The accessibility rank condition holds at every state x.*

Proof. Theorem 9 (p. 156) established the sufficiency of the rank condition. Suppose now that $\operatorname{int}\mathcal{R}_{\mathcal{V}}^{\leq T}(x) \neq \emptyset$ for all $x, \mathcal{V}, T$. By Theorem 11 (p. 177), the accessibility rank condition holds at every state z in some open and dense subset $\mathcal{Z}$ of $\mathcal{X}$. Pick any $x^0 \in \mathcal{X}$. Consider the set

$$\mathcal{W} := \operatorname{int}\mathcal{R}(x^0)\bigcap \mathcal{Z}.$$

This set is open (intersection of open sets) and nonempty (because $\operatorname{int}\mathcal{R}(x^0)$ is open and $\mathcal{Z}$ is dense). Pick any point $z^0 \in \mathcal{W}$; thus, there is some $T > 0$ and some $\omega \in \mathcal{L}_{\mathcal{U}}^{\infty}(0,T)$ such that $z^0 = \phi(T, 0, x^0, \omega)$. Approximating ω by piecewise constant controls, we can conclude that there is a convergent sequence of states $z_k \to z^0$, each z_k reachable from x^0 by piecewise constant controls (see Lemma 2.8.2). Pick any k large enough so that $z_k \in \mathcal{Z}$. Then, the accessibility rank condition holds at z_k, and there exist $u_1, \ldots, u_\ell \in \mathcal{U}$ and positive $t_1, \ldots, t_\ell$ such that $e^{t_1 f_{u_1}} \ldots e^{t_\ell f_{u_\ell}} x^0 = z_k$. Applying Lemma 4.5.1 ℓ times, we have that the accessibility rank condition holds at x^0. ∎

4.6 Additional Problems

Exercise 4.6.1 Let $G : \mathcal{X} \to \mathbb{R}^{n\times m}$ be a matrix function of class C^∞, defined on an open subset $\mathcal{X} \subseteq \mathbb{R}^n$. Consider the following properties:

(C) For each $x^0 \in \mathcal{X}$, there is some neighborhood $\mathcal{X}_0$ of x^0, and there are two matrix functions $C : \mathcal{X}_0 \to \mathbb{R}^{n\times n}$ and $D : \mathcal{X}_0 \to \mathbb{R}^{m\times m}$, both of class C^∞ and nonsingular for each $x \in \mathcal{X}_0$, such that

$$C(x)\,G(x)\,D(x) = \begin{pmatrix} I \\ 0 \end{pmatrix} \quad \text{for all } x \in \mathcal{X}_0\,,$$

where I is the $m \times m$ identity matrix.

(F) For each $x^0 \in \mathcal{X}$ there are $\mathcal{X}_0$, C, and D as above, and there is some diffeomorphism $\Pi : \mathcal{X}_0 \to V_0$ into an open subset of $\mathbb{R}^n$ so that $C(x) = \Pi_*(x)$ for all $x \in \mathcal{X}_0$.

(a) Show that (C) holds if and only if $\operatorname{rank} G(x) = m$ for all $x \in \mathcal{X}$.
(b) Show that (F) holds if and only if the columns $g_1, \ldots, g_m$ generate a distribution $\Delta_{g_1,\ldots,g_m}$ which is involutive and has constant rank m. (That is, $\operatorname{rank} G(x) = m$ for all $x \in \mathcal{X}$ and the n^2 Lie brackets $[g_i, g_j]$ are pointwise linear combinations of the g_i's.)

(c) Give an example where (C) holds but (F) fails.
(d) Interpret (b), for a system without drift $\dot{x} = G(x)u$, in terms of a change of variables (a feedback equivalence) $(x, u) \mapsto (z, v) := (\Pi(x), D(x)^{-1}u)$. What are the equations for $\dot{z}$ in terms of v, seen as a new input?
(*Hint:* (a) and (b) are both easy, from results already given.) □

Exercise 4.6.2 Suppose that $\Delta = \Delta_{X_1,\ldots,X_r}$ has constant rank r, is involutive, and is invariant under the vector field f. Pick any $x^0 \in \mathcal{O}$, and let $\mathcal{O}_0$ and Π be as in Frobenius' Lemma 4.4.16. Define, for $z \in (-\varepsilon, \varepsilon)^n$, $g(z) := \Pi_*(\Pi^{-1}(z))\, f(\Pi^{-1}(z))$, and partition $g = (g_1, g_2)'$ and $z = (z_1, z_2)'$ as in the proof of Theorem 11. Show that g_2 does not depend on z_1, that is to say, the differential equation $\dot{x} = f(x)$ transforms in the new coordinates $z = \Pi(x)$ into:

$$
\begin{aligned}
\dot{z}_1 &= g_1(z_1, z_2) \\
\dot{z}_2 &= g_2(z_2)\,.
\end{aligned}
$$

Explain how, for linear systems $\dot{x} = Ax$, this relates to the following fact from linear algebra: if A has an invariant subspace, then there is a change of coordinates so that A is brought into upper triangular form consistent with that subspace. (*Hint:* (For the proof that $\partial g_2/\partial z_1 = 0$.) We have that $g_2(\Pi(x)) = \Pi_{2*}(x) f(x)$. On the other hand, each row of $\Pi_{2*}(x)f(x)$ is of the form $L_f\psi_i$, where ψ_i's are the rows of Π_2. We know that $L_{X_j}\psi_i = 0$ for all i, j (this is what Lemma 4.4.16 gives), and also $L_{[f,X_j]}\psi_i = 0$ (because Δ is invariant under f), so conclude that $L_{X_j}(L_f\psi_i) = 0$. This gives that the directional derivatives of the rows of $g_{2*}(\Pi(x))$ along the directions $e_j(x) := \Pi_*(x)X_j(x)$ are all zero. Now observe that the vectors $e_i(x)$ are all of the form $(e_{i1}, 0)'$, and they are linearly independent.) □

Exercise 4.6.3 Use the conclusion of Exercise 4.6.2 to prove: for a control-affine system $\dot{x} = g_0 + \sum u_i g_i$, if $\Delta = \Delta_{X_1,\ldots,X_r}$ has constant rank r, is involutive, and is invariant under g_0, and in addition all the vector fields $g_i \in_p \Delta$ for $i = 1, \ldots, m$, then there is a local change of coordinates, about each state, in which the system can be expressed in the form

$$
\begin{aligned}
\dot{z}_1 &= g_0^1(z_1, z_2) + \sum_{i=1}^{m} u_i g_i^1(z_1, z_2) \\
\dot{z}_2 &= g_0^2(z_2)\,.
\end{aligned}
$$

Compare with the Kalman controllability decomposition for linear systems. □

Exercise 4.6.4 Show that, for analytic systems with connected state space $\mathcal{X}$, if the accessibility rank condition holds at even *one* point $x^0 \in \mathcal{X}$, then it must also hold on an open dense subset of $\mathcal{X}$. (*Hint:* Use that if an analytic function vanishes in some open set, then it must be identically zero.) □

4.7 Notes and Comments

The systematic study of controllability of nonlinear continuous-time systems started in the early 1970s. The material in this chapter is based on the early work in the papers [255], [284], [391], and [185], which in turn built upon previous PDE work in [91] and [182]. Current textbook references are [199] and [311]. An excellent exposition of Frobenius' Theorem, as well as many other classical facts about Lie analysis, can be found in [53].

Under extra conditions such as reversibility, we obtained an equivalence between controllability and accessibility. Another set of conditions that allows this equivalence is based on having suitable Hamiltonian structures, see, e.g., [51]. The advantage of considering accessibility instead of controllability is that, in effect, one is dealing with the transitivity of a *group* (as opposed to only a semigroup) action, since positive times do not play a distinguishing role in the accessibility rank condition. (Indeed, for analytic systems on connected state spaces, accessibility is exactly the same as controllability using possibly "negative time" motions in which the differential equation is solved backward in time.)

Also, for continuous-time systems evolving on Lie groups according to right-invariant vector fields, better results on controllability can be obtained; see, for instance, [126], [211], and [351]. Nonlinear *global* controllability problems are emphasized in [160], [161], and [162].

The question of providing necessary and sufficient characterizations for controllability is still open even for relatively simple classes such as bilinear continuous-time systems, but substantial and very deep work has been done to find sufficient conditions for controllability; see, for instance, [388], or [239]. It is worth remarking that it can be proved formally that accessibility is "easier" to check than controllability, in the computer-science sense of computational complexity and NP-hard problems; see [366], [240].

For discrete-time nonlinear accessibility, see for instance [206] and [208].

Example 4.3.13 is from [308], which introduced the model and pursued the analysis of controllability in Lie-algebraic terms.

Much recent work in nonlinear control has dealt with the formulation of explicit algorithms for finding controls that steer a given state to a desired target state, under the assumption that the system is controllable. This is known as the *path-planning* problem. See for instance the survey paper [251] and the many articles in the edited book [30].

Chapter 5

Feedback and Stabilization

The introductory Sections 1.2 to 1.5, which the reader is advised to review at this point, motivated the search for *feedback* laws to control systems. One is led then to the general study of the effect of feedback and more generally to questions of stability for linear and nonlinear systems. This Chapter develops basic facts about linear feedback and related topics in the algebraic theory of control systems including a proof of the Pole-Shifting Theorem described in Chapter 1, as well as an elementary introduction to Lyapunov's direct method and a proof of a "linearization principle" for stability. Some more "advanced" topics on nonlinear stabilization are also included, mostly to indicate some of the directions of current research.

5.1 Constant Linear Feedback

In this Section, $\mathbb{K}$ is an arbitrary field; unless otherwise stated, (A, B) denotes an arbitrary but fixed pair with $A \in \mathbb{K}^{n\times n}$ and $B \in \mathbb{K}^{n\times m}$, and a "continuous-time" or a "discrete-time" system means a linear and time-invariant system. In the special case $m = 1$, we often denote B as lowercase b.

It is natural to study systems under a change of basis in the state space, $x = Tz$. We already considered such transformations when dealing with the Kalman controllability decomposition (Lemma 3.3.3). Recall that $\mathcal{S}_{n,m}$ is the set of all pairs with given n, m, and $\mathcal{S}^c_{n,m}$ is the set of all such controllable pairs.

Definition 5.1.1 *Let* (A, B) *and* $(\widetilde{A}, \widetilde{B})$ *be two pairs in* $\mathcal{S}_{n,m}$. *Then* (A, B) **is similar to** $(\widetilde{A}, \widetilde{B})$, *denoted*

$$(A, B) \sim (\widetilde{A}, \widetilde{B})$$

if

$$T^{-1}AT = \widetilde{A} \quad \textit{and} \quad T^{-1}B = \widetilde{B}$$

for some $T \in GL(n)$. □

Lemma/Exercise 5.1.2 The following facts hold for the relation $\sim$:

1. It is an equivalence relation on $\mathcal{S}_{n,m}$.
2. If $(A,B) \sim (\widetilde{A},\widetilde{B})$, (A,B) is controllable if and only if $(\widetilde{A},\widetilde{B})$ is. □

Lemma 5.1.3 Let (A,b) be any pair with $m=1$, and write

$$\chi(s) = \det(sI - A) = s^n - \alpha_n s^{n-1} - \ldots - \alpha_2 s - \alpha_1$$

for the characteristic polynomial of A. With the matrices

$$A^{\dagger} := \begin{pmatrix} 0 & 0 & \cdots & 0 & \alpha_1 \\ 1 & 0 & \cdots & 0 & \alpha_2 \\ 0 & 1 & \cdots & 0 & \alpha_3 \\ \vdots & \vdots & \ddots & \vdots & \vdots \\ 0 & 0 & \cdots & 1 & \alpha_n \end{pmatrix} \qquad b^{\dagger} := \begin{pmatrix} 1 \\ 0 \\ 0 \\ \vdots \\ 0 \end{pmatrix}$$

it follows that

$$A\,\mathbf{R}(A,b) = \mathbf{R}(A,b)\,A^{\dagger} \quad \text{and} \quad b = \mathbf{R}(A,b)\,b^{\dagger},$$

where $\mathbf{R} = \mathbf{R}(A,b)$ is the controllability matrix of (A,b).

In particular, (A,b) is controllable if and only if it is similar to $(A^{\dagger},b^{\dagger})$.

Proof. The first part is an immediate consequence of the Cayley-Hamilton Theorem. Since $(A^{\dagger},b^{\dagger})$ is controllable, because

$$\mathbf{R}(A^{\dagger},b^{\dagger}) = I,$$

the sufficiency in the last statement follows from Lemma 5.1.2; for the necessity part, $T := \mathbf{R}(A,b)$ provides the required similarity. ∎

The system $(A^{\dagger},b^{\dagger})$ is sometimes called the *controllability form* of (A,b). From the above Lemma and the fact that $(A^{\dagger},b^{\dagger})$ depends only on the characteristic polynomial of A, and because $\sim$ is an equivalence relation, we conclude the following. It says that, for single-input controllable systems ($m=1$), the characteristic polynomial of A uniquely determines the pair (A,b) up to similarity.

Corollary 5.1.4 Assume that (A,b) and $(\widetilde{A},\widetilde{b})$ are two pairs in $\mathcal{S}^c_{n,1}$. Then, (A,b) is similar to $(\widetilde{A},\widetilde{b})$ if and only if the characteristic polynomials of A and $\widetilde{A}$ are the same. □

Definition 5.1.5 *The* **controller form** *associated to the pair* (A,b) *is the pair*

$$A^{\flat} := \begin{pmatrix} 0 & 1 & 0 & \ldots & 0 \\ 0 & 0 & 1 & \ldots & 0 \\ \vdots & \vdots & \vdots & \ddots & \vdots \\ 0 & 0 & 0 & \ldots & 1 \\ \alpha_1 & \alpha_2 & \alpha_3 & \ldots & \alpha_n \end{pmatrix} \qquad b^{\flat} := \begin{pmatrix} 0 \\ 0 \\ \vdots \\ 0 \\ 1 \end{pmatrix} \tag{CF}$$

where $s^n - \alpha_n s^{n-1} - \ldots - \alpha_2 s - \alpha_1$ *is the characteristic polynomial of* A. □

The symbols $A^\flat$ and $b^\flat$ are read as "A-flat" and "b-flat," respectively, the terminology motivated by the fact that the nontrivial coefficients appear horizontally, along a row, in contrast to the controllability form, where they appear vertically.

Lemma/Exercise 5.1.6 The pair $(A^\flat, b^\flat)$ is controllable, and the characteristic polynomial of $A^\flat$ is

$$s^n - \alpha_n s^{n-1} - \ldots - \alpha_2 s - \alpha_1 . \qquad \square$$

Proposition 5.1.7 The single-input ($m = 1$) pair (A, b) is controllable if and only if it is similar to its controller form.

Proof. Assume that the pair is controllable. Then, since $A^\flat$ has the same characteristic polynomial as A, and the pair $(A^\flat, b^\flat)$ is controllable, it follows from Corollary 5.1.4 that (A, b) is similar to its controller form. Conversely, if this similarity holds, then the second statement in Lemma 5.1.2 says that (A, b) is controllable, because $(A^\flat, b^\flat)$ is controllable. ∎

Thus, up to a change of variables, single-input controllable linear, time-invariant systems are *precisely* those obtained from nth order constant coefficient scalar differential equations

$$x^{(n)} - \alpha_n x^{(n-1)} - \ldots - \alpha_1 x = u$$

by the usual introduction of variables $x_i := x^{(i-1)}$, and analogously for discrete-time systems. Recall that in Chapter 1 we saw how the notion of state-space system was motivated by this construction.

We shall be interested in studying the effect of constant linear feedback. The following Lemma will be helpful in that regard:

Lemma/Exercise 5.1.8 For any pair (A, B) and any $F \in \mathbb{K}^{m \times n}$,

$$\mathcal{R}(A + BF, B) = \mathcal{R}(A, B) .$$

In particular, $(A + BF, B)$ is controllable if and only if (A, B) is. □

Definition 5.1.9 *The nth degree monic polynomial χ is* **assignable** *for the pair (A, B) if there exists a matrix F such that $\chi_{A+BF} = \chi$.* □

Lemma 5.1.10 If $(A, B) \sim (\widetilde{A}, \widetilde{B})$, then they can be assigned the same polynomials.

Proof. Given any F, it holds that

$$\chi_{A+BF} = \chi_{\widetilde{A}+\widetilde{B}\widetilde{F}}$$

provided that one chooses $\widetilde{F} := FT$. ∎

In Chapter 8 it will be proved using optimal control techniques (Corollary 8.4.3 and Exercise 8.4.5) that, if $\mathbb{K} = \mathbb{R}$ and (A, B) is controllable, then there is a feedback matrix F such that $A + BF$ is Hurwitz and another F such that $A+BF$ is convergent (discrete-time Hurwitz). As discussed in the introductory Section 1.5, in fact one can assign an arbitrary spectrum under feedback, as was illustrated there with the linearized pendulum. This is the Pole-Shifting Theorem, to be proved below. The use of the term *pole* to refer to the eigenvalues of A originates in the classical frequency-domain theory of linear systems.

Recall (Definition 3.3.5) that χ_u is the uncontrollable part of the characteristic polynomial of A, and we let $\chi_u = 1$ if the pair (A, B) is controllable.

Theorem 13 (Pole-Shifting Theorem) *For each pair (A, B), the assignable polynomials are precisely those of the form*

$$\boxed{\chi_{A+BF} = \chi_1 \chi_u}$$

where χ_1 is an arbitrary monic polynomial of degree $r = \dim \mathcal{R}(A, B)$.

In particular, the pair (A, B) is controllable if and only if every nth degree monic polynomial can be assigned to it.

Proof. Assume first that (A, B) is not controllable. By Lemmas 3.3.3 and 5.1.10, we can assume without loss of generality that (A, B) has the form that $(\widetilde{A}, \widetilde{B})$ has in (3.6). Write any $F \in \mathbb{K}^{m\times n}$ in the partitioned form (F_1, F_2), where F_1 is of size $m \times r$ and F_2 is of size $m \times (n - r)$. Since

$$\chi_{A+BF} = \chi_{A_1+B_1F_1} \chi_u ,$$

it follows that any assignable polynomial has the factored form in the statement of the Theorem.

Conversely, we want to show that the first factor can be made arbitrary by a suitable choice of F (except for the case $r = 0$, where only χ_u is there and hence there is nothing to prove). Assume given any χ_1. If we find an F_1 so that

$$\chi_{A_1+B_1F_1} = \chi_1 ,$$

then the choice $F := (F_1, 0)$ will provide the desired characteristic polynomial. By Lemma 3.3.4, the pair (A_1, B_1) is controllable. Thus, we are left to prove only that controllable systems can be arbitrarily assigned, so we take (A, B) from now on to be controllable.

When $m = 1$, we may assume, because of Lemmas 5.1.10 and 5.1.7, that (A, b) is in controller form. Given any polynomial

$$\chi = s^n - \beta_n s^{n-1} - \ldots - \beta_2 s - \beta_1$$

to be assigned, the (unique) choice

$$f := (\beta_1 - \alpha_1, \ldots, \beta_n - \alpha_n)$$

satisfies $\chi_{A+bf} = \chi$. Thus, the result has been proved in the case $m = 1$. (The fact that f can be obtained immediately from the controller form justifies the terminology "controller form.")

Now let m be arbitrary. Pick any vector $v \in \mathcal{U} = \mathbb{K}^m$ such that $Bv \neq 0$, and let $b := Bv$. We next show that there exists an $F_1 \in \mathbb{K}^{m\times n}$ such that

$$(A + BF_1, b)$$

is itself controllable. Because the result has been established for the case when $m = 1$, there will then be, for any desired χ, a (unique) $1 \times n$ matrix f such that the characteristic polynomial of

$$A + B(F_1 + vf) = (A + BF_1) + bf$$

is χ, and the Theorem will be proved using the feedback $F := F_1 + vf$.

To establish the existence of F_1, we argue as follows. Let

$$\{x_1, \dots, x_k\}$$

be any sequence of linearly independent elements of $\mathcal{X}$, with k as large as possible, having the properties that $x_1 = Bv$ and

$$x_i - Ax_{i-1} \in \mathcal{B} := \operatorname{col} B \tag{5.1}$$

for $i = 1, \dots, k$ (with $x_0 := 0$). In other words, the sequence is required to be such that

$$Bv = x_1 \underset{1}{\rightsquigarrow} x_2 \underset{1}{\rightsquigarrow} \dots \underset{1}{\rightsquigarrow} x_k$$

and k is as large as possible.

We claim that $k = n$. Consider the span $\mathcal{V}$ of $\{x_1, \dots, x_k\}$. By maximality of k,

$$Ax_k + Bu \in \mathcal{V} \quad \text{for all } \; u \in \mathcal{U}\,. \tag{5.2}$$

This implies, in particular, that

$$Ax_k \in \mathcal{V}\,.$$

Thus, equation (5.2) shows that

$$\mathcal{B} \subseteq \mathcal{V} - \{Ax_k\} \subseteq \mathcal{V}\,,$$

so equation (5.1) implies that

$$Ax_l \in \mathcal{V}, \;\; l = 1, \dots, k-1\,.$$

We conclude that $\mathcal{V}$ is an A-invariant subspace that contains $\mathcal{B}$. It follows that $\mathcal{R}(A, B) \subseteq \mathcal{V}$. By controllability, this implies that indeed $k = n$.

Finally, for each $i = 1, \dots, k-1$, we let

$$F_1 x_i := u_i\,,$$

where $u_i \in \mathcal{U}$ is any element such that

$$x_i - Ax_{i-1} = Bu_{i-1},$$

and we define $F_1 x_k$ arbitrarily. Since

$$\mathcal{R}(A + BF_1, x_1) = (x_1, \ldots, x_n),$$

the desired single-input controllability is achieved. ■

Exercise 5.1.11 ◇ Show that, if the field $\mathbb{K}$ is infinite, (A, B) is controllable, and the matrix A is cyclic, then there exists some u so that (A, Bu) is controllable. Give counterexamples to show that this fails if the hypothesis that A is cyclic is dropped, and to show that even if A is cyclic the result fails over the field of two elements. (*Hint:* Use the fact that the geometric multiplicity of each eigenvalue of the transpose A' must be one, and then solve the inequalities

$$v'Bu \neq 0$$

for each eigenvector v of A'.) □

Exercise 5.1.12 ◇ For single-input systems, it is possible to give explicit formulas for the (unique) feedback law that assigns any given polynomial. One of these, *Ackermann's formula*, is as follows:

$$f = -(0 \cdots 0\, 1)\, \mathbf{R}(A, b)^{-1} \chi(A),$$

where χ is the desired polynomial. Prove this formula. □

Remark 5.1.13 A quick numerical procedure for pole-shifting when $\mathbb{K} = \mathbb{R}$ (or $\mathbb{C}$) is based on the following argument. Given a controllable pair (A, B), we first pick an arbitrary nonzero column b of B and then generate at random (using a pseudorandom number generator) a matrix F. With probability one,

$$(A + BF, b)$$

is controllable. Indeed, the set of F for which the property is true is the complement of an algebraic set, hence open dense and of full measure if nonempty; as we know from the proof of the Theorem that there is some such F, this set is open dense. (If the random choice happens to be one of the exceptional cases, we just generate another F.) Now Ackermann's formula (Exercise 5.1.12) can be applied to $(A + BF, b)$. □

Exercise 5.1.14 Refer to Exercise 3.2.12.

(a) Find the controller form for Σ and the similarity T needed to bring the original system into this form. (In finding the latter, note that $T : (A, b) \sim (A^\flat, b^\flat)$ is given by $T = \mathbf{R}(\mathbf{R}^\flat)^{-1}$, where $\mathbf{R} = \mathbf{R}(A, b)$ and $\mathbf{R}^\flat = \mathbf{R}(A^\flat, b^\flat)$. Take all constants equal to one, for simplicity.)

(b) Find a feedback $u = fx$ so that all eigenvalues of $A + bf$ equal -1. □

Exercise 5.1.15 Repeat Exercise 5.1.14 for the system Σ in Exercise 3.2.13, taking for simplicity $M = m = F = g = l = 1$. □

5.2 Feedback Equivalence*

The controller form $(A^\flat, b^\flat)$ in Definition 5.1.5 was useful in establishing the Pole-Shifting Theorem. We proved that every controllable single-input system can be brought into (a unique) controller form under change of basis in the state space. Further, a system in this form then can be transformed easily under feedback into any other. In particular, from the form (CF) and using the transformation

$$\left(A^\flat, b^\flat\right) \to \left(A^\flat - b^\flat(\alpha_1, \alpha_2, \alpha_3, \dots, \alpha_n), b^\flat\right)$$

there results the following pair (A_n, b_n):

$$A_n = \begin{pmatrix} 0 & 1 & 0 & \dots & 0 \\ 0 & 0 & 1 & \dots & 0 \\ \vdots & \vdots & \vdots & \ddots & \vdots \\ 0 & 0 & 0 & \dots & 1 \\ 0 & 0 & 0 & \dots & 0 \end{pmatrix} \qquad b_n = \begin{pmatrix} 0 \\ 0 \\ \vdots \\ 0 \\ 1 \end{pmatrix}. \tag{5.3}$$

The above transformation can be interpreted as the application of the feedback

$$u = fx = -(\alpha_1, \alpha_2, \alpha_3, \dots, \alpha_n)x\,,$$

and essentially the previous results show that, under change of basis and feedback, any single-input continuous-time time-invariant linear controllable system can be reduced to a scalar nth order equation of the form

$$x^{(n)} = u\,,$$

which corresponds to an nth order integrator (or an n-step delay, in the case of discrete-time systems); see Figure 5.1.

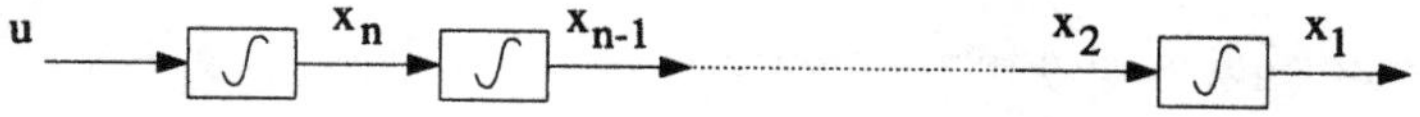

u → ∫ → x_n → ∫ → x_{n-1} ⋯ x_2 → ∫ → x_1

Figure 5.1: *Cascade of integrators.*

The purpose of this Section is to provide a generalization of this fact to the case of multi-input ($m > 1$) systems. It will state that every controllable pair (A, B) can be made equivalent, in a suitable sense to be defined below, to one in a form analogous to that obtained for the case $m = 1$; for continuous-time systems, this will result in the form

$$\begin{aligned} x_1^{(\kappa_1)} &= u_1 \\ x_2^{(\kappa_2)} &= u_2 \\ &\ \vdots \\ x_r^{(\kappa_r)} &= u_r \end{aligned}$$

* This section can be skipped with no loss of continuity.

where $(\kappa_1, \ldots, \kappa_r)$, $r = \operatorname{rank} B$, is a sequence of integers, called the *controllability indices* or *Kronecker indices* of the system. These integers satisfy

$$\kappa_1 + \ldots + \kappa_r = n$$

and are unique up to ordering. Thus, every linear continuous-time time-invariant system can be made to behave like r parallel cascades of integrators, under the action of feedback. This result is of some theoretical interest and can be used to give an alternative proof of the Pole-Shifting Theorem. In addition, because the κ_i's are unique, the result provides an effective test to determine whether two systems are equivalent under feedback transformations, in the sense defined below.

Except when explaining the interpretations for continuous-time systems, we assume that $\mathbb{K}$ is an arbitrary field.

Definition 5.2.1 *Let (A, B) and $(\widetilde{A}, \widetilde{B})$ be two pairs in $S_{n,m}$. Then (A, B) is* **feedback equivalent** *to $(\widetilde{A}, \widetilde{B})$, denoted*

$$(A, B) \equiv (\widetilde{A}, \widetilde{B})$$

if

$$T^{-1}(A + BF)T = \widetilde{A} \quad \textit{and} \quad T^{-1}BV = \widetilde{B}$$

for some $T \in GL(n)$, some $F \in \mathbb{K}^{m \times n}$, and some $V \in GL(m)$. □

Feedback equivalence corresponds to changing basis in the state and control-value spaces (invertible matrices T and V, respectively) and applying a feedback transformation $u = Fx + u'$, where u' is a new control.

Lemma/Exercise 5.2.2 The following facts hold for the relation $\equiv$:

1. It is an equivalence relation on $\mathcal{S}_{n,m}$.
2. If $(A, B) \sim (\widetilde{A}, \widetilde{B})$, then $(A, B) \equiv (\widetilde{A}, \widetilde{B})$.
3. If $(A, B) \equiv (\widetilde{A}, \widetilde{B})$, then (A, B) is controllable iff $(\widetilde{A}, \widetilde{B})$ is. □

The notion of feedback equivalence is closely related with Kronecker's work (ca. 1890) on invariants of matrix pencils. This connection is explained in the next problem; we will derive our results directly in terms of feedback equivalence, however.

Exercise 5.2.3 An expression of the type $sF + G$, where F and G are two $n \times l$ matrices over $\mathbb{K}$, is called a *pencil* of matrices. We think of this as a formal polynomial in s, with equality of $sF + G$ and $sF' + G'$ meaning pairwise equality: $F = F'$ and $G = G'$. Two pencils are called *equivalent* if there exist invertible matrices $T \in GL(n)$ and $S \in GL(l)$ such that

$$T[sF + G]S = [sF' + G'] .$$

Given a pair (A, B), consider the pencil

$$[sI - A, B] = s[I, 0] + [-A, B],$$

with $l = n + m$. *Prove:* $(A, B) \equiv (\widetilde{A}, \widetilde{B})$ if and only if the corresponding pencils are equivalent. □

A *partition* $\kappa = (\kappa_1, \ldots, \kappa_r)$ of the integer n is a sequence of positive integers

$$\kappa_1 \geq \kappa_2 \geq \ldots \geq \kappa_r$$

such that

$$\kappa_1 + \kappa_2 + \ldots + \kappa_r = n.$$

Associated to each such partition for which $r \leq m$, we consider the pair (A_κ, B_κ), defined (with the same n, m) as follows. The matrix A_κ is partitioned into r^2 blocks, and B_κ is partitioned into mr blocks:

$$A_\kappa = \begin{pmatrix} A_{\kappa_1} & 0 & \cdots & 0 \\ 0 & A_{\kappa_2} & \cdots & 0 \\ \vdots & \vdots & \ddots & \vdots \\ 0 & 0 & \cdots & A_{\kappa_r} \end{pmatrix} \quad B_\kappa = \begin{pmatrix} b_{\kappa_1} & 0 & \cdots & 0 & 0 & \cdots & 0 \\ 0 & b_{\kappa_2} & \cdots & 0 & 0 & \cdots & 0 \\ \vdots & \vdots & \ddots & \vdots & \vdots & \ddots & \vdots \\ 0 & 0 & \cdots & b_{\kappa_r} & 0 & \cdots & 0 \end{pmatrix} \tag{5.4}$$

where each A_{κ_i} has size $\kappa_i \times \kappa_i$ and is as in equation (5.3), and each b_{κ_i} is a column of length κ_i as in that equation. Note that when $m = 1$ there is only one partition, $\kappa = (n)$, and (A_κ, B_κ) is in that case just the pair (A_n, b_n) in (5.3). The above pair (A_κ, B_κ) is said to be in *Brunovsky form*. For each κ, this is a controllable pair. Note that

$$\operatorname{rank} B_\kappa = r \leq m$$

and that a continuous-time system whose matrices have this form consists of r integrator cascades, as discussed informally above.

The main Theorem to be proved in this section is as follows:

Theorem 14 *For every pair (A, B) in $S^c_{n,m}$, there is a unique partition κ such that $(A, B) \equiv (A_\kappa, B_\kappa)$.*

This implies that the partition κ is a complete invariant for controllable systems under feedback equivalence and that the number of equivalence classes is equal to the number of such partitions for which $r \leq m$. For instance, if $n = 3$ and $m = 2$, then there are exactly two equivalence classes, corresponding to the partitions (3) and $(2, 1)$.

The proof of Theorem 14 will be organized as follows. First we show how a partition κ can be associated to each pair. We show that this assignment is invariant under feedback equivalence, that is, that two feedback equivalent pairs necessarily must give rise to the same κ, and further, for the system (A_κ, B_κ)

already in Brunovsky form (5.4) the associated partition is κ. This will prove the uniqueness part of the statement. To establish existence, we show that, for the κ associated to the pair (A, B), it holds that $(A, B) \equiv (A_\kappa, B_\kappa)$; this is done by an explicit construction of the needed transformations.

Fix any controllable pair (A, B). Let b_i denote the ith column of B, and consider the sequence of vectors

$$b_1, b_2, \ldots, b_m, Ab_1, Ab_2, \ldots, Ab_m, \ldots, A^{n-1}b_1, A^{n-1}b_2, \ldots, A^{n-1}b_m\,.$$

We say that a vector in this sequence is *dependent* if it is a linear combination of the previous ones. Observe that if some $A^j b_i$ is dependent, and $j < n-1$, then $A^{j+1}b_i$ also must be dependent. If a vector is not dependent, we call it independent. We now arrange these vectors in the following pattern:

$$\begin{array}{cccc} b_1 & b_2 & \cdots & b_m \\ Ab_1 & Ab_2 & \cdots & Ab_m \\ \vdots & \vdots & \ddots & \vdots \\ A^{n-1}b_1 & A^{n-1}b_2 & \cdots & A^{n-1}b_m \end{array} \tag{5.5}$$

and we associate a preliminary partition $\lambda = (\lambda_1, \ldots, \lambda_s)$ of n as follows. The integer λ_i is defined as the number of independent vectors (in the sense explained above) in the ith row of the table (5.5), and s is the largest index so that the sth row contains at least one independent vector. From the remark about the dependence of $A^{j+1}b_i$ following from that of $A^j b_i$, we know that this is a nonincreasing sequence. Further,

$$\lambda_1 = \operatorname{rank} B \le m\,,$$

and by controllability

$$\lambda_1 + \ldots + \lambda_s = n\,,$$

so that this is a partition of n into parts none of which exceeds m. If we need to emphasize that this was done for the system (A, B), we write $\lambda(A, B)$ instead of just λ.

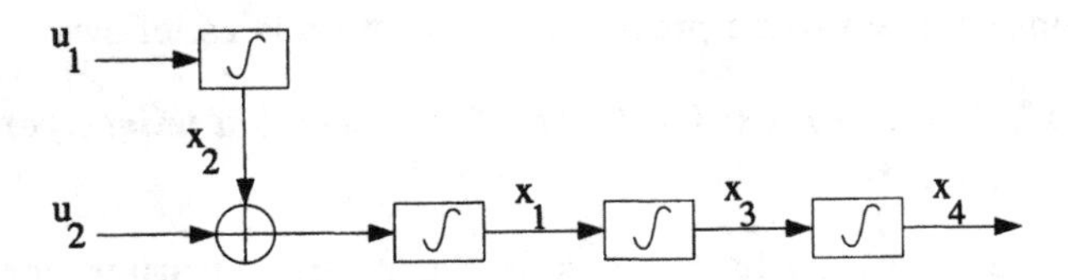

Figure 5.2: *Example leading to indices (3,1).*

As an illustration, take the following system, with $n = 4$ and $m = 2$ (see Figure 5.2):

$$A = \begin{pmatrix} 0 & 1 & 0 & 0 \\ 0 & 0 & 0 & 0 \\ 1 & 0 & 0 & 0 \\ 0 & 0 & 1 & 0 \end{pmatrix} \quad B = \begin{pmatrix} 0 & 1 \\ 1 & 0 \\ 0 & 0 \\ 0 & 0 \end{pmatrix} \tag{5.6}$$

for which, denoting by e_i the ith canonical basis element in $\mathbb{K}^4$, the above pattern is

$$\begin{matrix} e_2 & e_1 \\ e_1 & e_3 \\ e_3 & e_4 \\ e_4 & 0 \end{matrix}$$

and $\lambda(A, B) = (2, 1, 1)$.

Lemma/Exercise 5.2.4 If $(A, B) \equiv (\widetilde{A}, \widetilde{B})$, then $\lambda(A, B) = \lambda(\widetilde{A}, \widetilde{B})$. (*Hint:* Establish that for each k the rank of

$$[B, AB, \ldots, A^k B]$$

remains invariant under each of the possible transformations $B \mapsto BV$, $A \mapsto T^{-1}AT$, $A \mapsto A + BF$.) □

The controllability indices of (A, B) will be obtained from $\lambda(A, B)$ as the "conjugate" partition to λ, in the following sense:

Definition 5.2.5 *Given a partition $\lambda = (\lambda_1, \ldots, \lambda_s)$ of n, the* **conjugate partition** *to λ (also called the "dual" partition to λ) is the partition λ' given by $(\lambda'_1, \ldots, \lambda'_{s'})$, where λ'_i is the number of parts of λ that are $\geq i$.* □

For instance, the dual partition to $(2, 1, 1)$ is $(3, 1)$, since there are three λ_i's larger than or equal to 1, and just one larger than or equal to 2 (and none larger than 2).

The following concept from combinatorics is useful when studying partitions. A *Young tableaux* is a matrix of zeros and ones, with the property that there is never a one below nor to the right of any zero, and no row or column is identically zero.

Given any partition $\lambda = (\lambda_1, \ldots, \lambda_s)$, we can associate to it the Young tableaux of size $s \times \lambda_1$ that has λ_1 ones in the first row, λ_2 in the second, and so forth. For instance, associated to $(2, 1, 1)$ is the tableaux

$$\begin{matrix} 1 & 1 \\ 1 & 0 \\ 1 & 0 \end{matrix}$$

Conversely, each tableaux with n ones comes from precisely one partition of n.

Under this rule, *the tableaux of the conjugate is the transpose of the original tableaux.* From this, it follows that $\lambda'' = \lambda$ for all λ, so λ is uniquely determined by λ'. Note also that, always, $s' = \lambda_1$.

Definition 5.2.6 *The* **controllability indices** *$\kappa(A, B)$ of the controllable pair (A, B) are the elements of the sequence $\kappa_1, \ldots, \kappa_r$, where*

$$\kappa = (\kappa_1, \ldots, \kappa_r)$$

is the partition conjugate to $\lambda(A, B)$. □

For example, the system (5.6) has indices 3, 1. In general, the indices are the numbers of independent vectors in the columns of (5.5), arranged in nonincreasing order.

From Lemma 5.2.4 it follows that *controllability indices are invariant under feedback equivalence.* Observe that $r = \lambda_1 \leq m$, the rank of B. The assignment $(A, B) \to \kappa(A, B)$ is a mapping from $\mathcal{S}^c_{n,m}$ into the set of all partitions of n into $r = \operatorname{rank} B$ parts.

Lemma/Exercise 5.2.7 For each partition $\widetilde{\kappa}$ of n into r parts,

$$\kappa(A_{\widetilde{\kappa}}, B_{\widetilde{\kappa}}) = \widetilde{\kappa} \,.$$

□

The uniqueness statement in Theorem 14 now follows: If $(A, B) \equiv (A_{\kappa'}, B_{\kappa'})$ and also $(A, B) \equiv (A_{\kappa''}, B_{\kappa''})$, then because of feedback equivalence and Lemma 5.2.7,

$$\kappa' = \kappa(A_{\kappa'}, B_{\kappa'}) = \kappa(A, B) = \kappa(A_{\kappa''}, B_{\kappa''}) = \kappa'' \,,$$

and the Brunovsky form is unique.

Next we show that there is a sequence of transformations as in the definition of feedback equivalence that brings any given pair into the corresponding Brunovsky form.

We first apply a transformation consisting of a permutation of the columns of B (i.e., V is a permutation matrix, and $T = I$, $F = 0$) in such a way that the columns in (5.5) are rearranged so that the number of independent vectors in the ith column is nonincreasing. (In the above example, this means that we use

$$V = \begin{pmatrix} 0 & 1 \\ 1 & 0 \end{pmatrix}$$

and obtain the new arrangement

$$\begin{matrix} e_1 & e_2 \\ e_3 & e_1 \\ e_4 & e_3 \\ 0 & e_4 \end{matrix}$$

after permuting columns.) We denote the new matrix BV again by B and its columns by b_i. (For the above example this means that now $B = [e_1, e_2]$.)

In terms of the pattern (5.5), then, κ_i is now the number of independent vectors in the ith column, and the vectors

$$b_1, \ldots, A^{\kappa_1 - 1} b_1, b_2, \ldots, A^{\kappa_2 - 1} b_2, \ldots, b_r, \ldots, A^{\kappa_r - 1} b_r$$

form a basis of $\mathbb{K}^n$. Furthermore, from the definition of dependency it follows that for each i there is a linear combination

$$A^{\kappa_i} b_i + \sum_{j<i} \alpha_{ij0} A^{\kappa_i} b_j = - \sum_{j=1}^{r} \sum_{k=1}^{\kappa_i} \alpha_{ijk} A^{\kappa_i - k} b_j \,. \tag{5.7}$$

Next we apply yet another invertible transformation $\widetilde{B} := BV$ resulting in new columns

$$\widetilde{b}_i := b_i + \sum_{j<i} \alpha_{ij0} b_j \ , \quad i = 1, \ldots, r \, ,$$

and $\widetilde{b}_i := b_i$ for $i > r$, so that, after again calling $\widetilde{B}$ just B, we have that we can write (5.7) simply as:

$$A^{\kappa_i} b_i + \sum_{j=1}^{r} \sum_{k=1}^{\kappa_i} \alpha_{ijk} A^{\kappa_i - k} b_j = 0 \tag{5.8}$$

(with different coefficients α_{ijk}'s). For each

$$i = 1, \ldots, r \quad \text{and} \quad l := 1, \ldots, \kappa_i \, ,$$

we introduce the vectors

$$e_{il} := A^{l-1} b_i + \sum_{j=1}^{r} \sum_{k=1}^{l-1} \alpha_{ijk} A^{l-1-k} b_j$$

(with the convention that $e_{i1} = b_i$). Listing them in the order

$$e_{11}, e_{21}, \ldots ; e_{r1}, e_{12}, e_{22}, \ldots ; \ldots$$

makes it evident that they are linearly independent, since they form a triangular linear combination of the basis vectors

$$b_1, b_2, \ldots, b_r, Ab_1, Ab_2, \ldots \, ,$$

so the matrix T with columns

$$\left[e_{1\kappa_1}, e_{1(\kappa_1 - 1)}, \ldots, e_{11}, e_{2\kappa_2}, e_{2(\kappa_2 - 1)}, \ldots, e_{21}, \ldots, e_{r\kappa_r}, e_{r(\kappa_r - 1)}, \ldots, e_{r1} \right]$$

is invertible. Note that

$$Ae_{il} = e_{i(l+1)} - \sum_{j=1}^{r} \alpha_{ijl} b_j$$

for $l < \kappa_i$ and that, because of (5.8),

$$Ae_{i\kappa_i} = - \sum_{j=1}^{r} \alpha_{ij\kappa_i} b_j \, ,$$

from which it follows that the matrix on the basis given by the columns of T, that is, $\widetilde{A} := T^{-1} A T$, has the following block structure:

$$\begin{pmatrix} A_{11} & A_{12} & \cdots & A_{1r} \\ A_{21} & A_{22} & \cdots & A_{2r} \\ \vdots & \vdots & \ddots & \vdots \\ A_{r1} & A_{r2} & \cdots & A_{rr} \end{pmatrix} \tag{5.9}$$

where the off-diagonal blocks A_{ij}, $i \neq j$, have the form

$$\begin{pmatrix} 0 & 0 & 0 & \dots & 0 \\ 0 & 0 & 0 & \dots & 0 \\ \vdots & \vdots & \vdots & \ddots & \vdots \\ 0 & 0 & 0 & \dots & 0 \\ * & * & * & \dots & * \end{pmatrix}$$

and the diagonal blocks have the form

$$\begin{pmatrix} 0 & 1 & 0 & \dots & 0 \\ 0 & 0 & 1 & \dots & 0 \\ \vdots & \vdots & \vdots & \ddots & \vdots \\ 0 & 0 & 0 & \dots & 1 \\ * & * & * & \dots & * \end{pmatrix}$$

where the asterisks indicate possibly nonzero entries. Similarly, since $e_{i1} = b_i$ for each i, the matrix $T^{-1}B$ is in the desired form (5.4) except for the fact that the last $m - r$ columns may be nonzero. Applying a new change of basis V in the control-value space, these columns can be made zero (since they are linearly dependent on the first r rows, because $\operatorname{rank} B = r$).

Remark 5.2.8 At this stage, one has not yet applied any purely feedback transformations (using $F \neq 0$ in the definition of feedback equivalence) but only changes of basis on states and controls. The system has been reduced to a form analogous to the controllability form (CF). It should be noted however that, in contrast to the situation with $m = 1$, the free parameters left (the asterisks above) are not uniquely determined by the original pair, so this is not a "canonical form" in the sense of invariant theory. □

Finally, the asterisks in the blocks in (5.9) all can be zeroed by application of a feedback transformation $A \to A + BF$; in fact, from the form of B, this can be achieved simply by using as F the negative of the matrix that has those same asterisks in the same order (and arbitrary last $m - r$ rows). This completes the proof of Theorem 14. ■

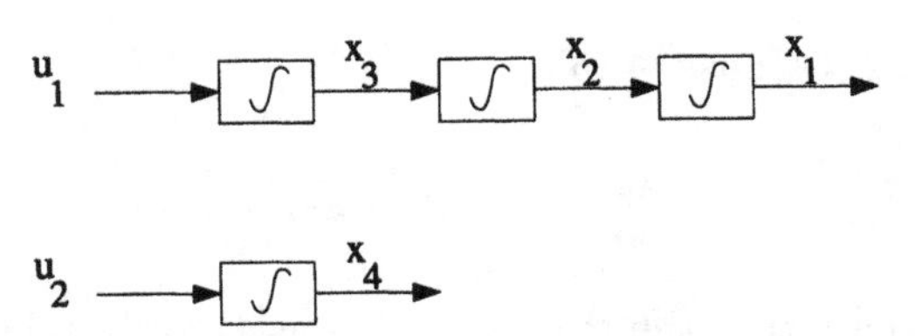

Figure 5.3: *Canonical form for example.*

Exercise 5.2.9 Write the matrix $[sI - A, B]$ for a pair in Brunovsky form. Show how to obtain from this expression, and from Exercise 5.2.3, the fact that $\operatorname{rank}[sI - A, B] = n$ for all s is necessary for controllability. □

Exercise 5.2.10 The Pole-Shifting Theorem can be proved as an easy consequence of Theorem 14. The critical step is to show that for each controllable (A, B) there is some feedback transformation F and some $u \in \mathbb{K}^m$ such that $(A+BF, Bu)$ is controllable. *Show* how to obtain such an F and u from Theorem 14. □

Exercise 5.2.11 Let $\mathbb{K} = \mathbb{R}$ or $\mathbb{C}$. For any fixed n, m, write

$$n = sm + t\,, \quad 0 \leq t < m$$

with s the integer part of n/m. Show that the set of pairs (A, B) with indices

$$\underbrace{s+1, \ldots, s+1}_{t}, \underbrace{s, \ldots, s}_{m-t}$$

forms an open dense subset of $S^c_{n,m}$. □

5.3 Feedback Linearization*

We study single-input control-affine (time-invariant, continuous-time) systems

$$\dot{x} \;=\; f(x) + ug(x)\,. \tag{5.10}$$

The state-space is, as usual, an open subset $\mathcal{X} \subseteq \mathbb{R}^n$, and the control-value space is $\mathcal{U} = \mathbb{R}$. (This is as in (4.18), with $m = 1$, but here we prefer to write f and g, instead of g_0 and g_1 respectively, for the drift and control vector fields.)

Definition 5.3.1 *Let Σ and $\widetilde{\Sigma}$ be two systems of the above form, and suppose that $\mathcal{O}$ and $\widetilde{\mathcal{O}}$ are open subsets of the respective state spaces $\mathcal{X}$ and $\widetilde{\mathcal{X}}$. We say that $(\Sigma, \mathcal{O})$ is* **feedback equivalent to** *$(\widetilde{\Sigma}, \widetilde{\mathcal{O}})$ if there exist:*

- *a diffeomorphism $T : \mathcal{O} \to \widetilde{\mathcal{O}}$, and*
- *smooth maps $\alpha, \beta : \mathcal{O} \to \mathbb{R}$, with $\beta(x) \neq 0$ for all $x \in \mathcal{O}$,*

such that, for each $x \in \mathcal{O}$:

$$T_*(x)\,\left(f(x) + \alpha(x)g(x)\right) \;=\; \widetilde{f}(T(x)) \tag{5.11}$$

and

$$\beta(x)T_*(x)g(x) \;=\; \widetilde{g}(T(x)) \tag{5.12}$$

(f, g and $\widetilde{f}, \widetilde{g}$ are the vector fields associated to the respective systems). □

* This section can be skipped with no loss of continuity.

Notice that $\mathcal{X}$ and $\widetilde{\mathcal{X}}$ must have the same dimension, because T is a diffeomorphism. One can express the two equalities (5.11)-(5.12) in the following equivalent form:

$$T_*(x)\left(f(x)+ug(x)\right) = \widetilde{f}(T(x))+\frac{1}{\beta(x)}\left(u-\alpha(x)\right)\widetilde{g}(T(x)) \quad \forall x \in \mathcal{O}\,,\ \forall u \in \mathbb{R}^m.$$

The change of variables

$$(x,u) \mapsto (z,v) := \left(T(x), \frac{1}{\beta(x)}\left[u-\alpha(x)\right]\right)$$

provides a diffeomorphism between $\mathcal{O}\times\mathbb{R}$ and $\mathcal{W}\times\mathbb{R}$, whose inverse is

$$(z,v) \mapsto (x,u) := \left(T^{-1}(z), \alpha\left(T^{-1}(z)\right) + \beta\left(T^{-1}(z)\right)v\right).$$

Solutions of $\dot{x} = f(x)+ug(x)$ are transformed into solutions of $\dot{z} = \widetilde{f}(z)+v\widetilde{g}(z)$, corresponding to the input $v = \frac{1}{\beta(x)}\left[u-\alpha(x)\right]$. We may view $u = k(x,v) = \alpha(x)+\beta(x)v$ as a feedback law which "closes the loop" about the system, with v having the role of a new input, to be used for further control purposes, see Figure 5.4. Under the change of variables $z = T(x)$, the closed-loop system

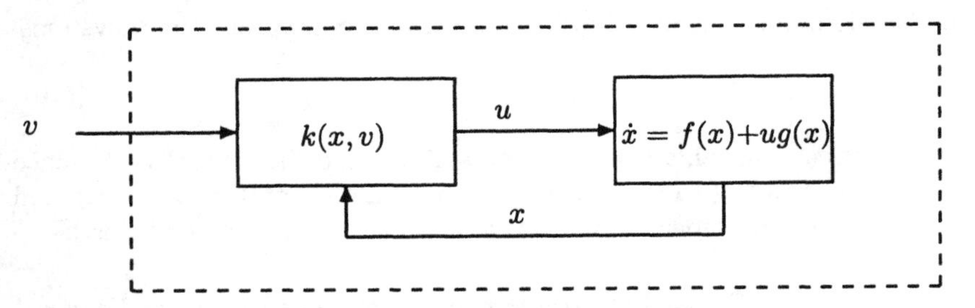

Figure 5.4: *Feedback.*

$$\dot{x} = f(x) + k(x,v)\,g(x) \tag{5.13}$$

(represented by the dashed box in Figure 5.4) transforms into $\dot{z} = \widetilde{f}(z)+v\widetilde{g}(z)$. This new system might be easier to control than the original system, especially if it is linear.

Definition 5.3.2 *The system Σ in (5.10) is* **feedback linearizable about x^0**, *where $x^0 \in \mathcal{X}$, if there exists an open subset $\mathcal{O} \subseteq \mathcal{X}$ containing x^0, a controllable n-dimensional single-input linear system $(\widetilde{\Sigma})\ \dot{z} = Az + vb$, and an open subset $\widetilde{\mathcal{O}} \subseteq \mathbb{R}^n$, such that $(\Sigma,\mathcal{O})$ is feedback equivalent to $(\widetilde{\Sigma},\widetilde{\mathcal{O}})$.* □

Given any T,α,β providing an equivalence as in this definition, T is said to be a linearizing change of variables and the map $(x,v) \mapsto \alpha(x)+\beta(x)v$ a linearizing feedback.

Suppose, as an illustration, that we wish to stabilize about $x = x^0$, and assume that we have found a feedback linearization with the extra property $T(x^0) = 0$. Since the pair (A, b) is controllable, there is some $F \in \mathbb{R}^{1\times n}$ such that $\dot{z} = (A + bF)z$ is globally asymptotically stable (Pole Shifting Theorem 13 (p. 186)). Changing back to x variables, we consider the system without inputs

$$\dot{x} \;=\; f(x) + \widetilde{k}(x)\,g(x)\,, \tag{5.14}$$

where the new feedback $\widetilde{k}(x) := \alpha(x) + \beta(x)FT(x)$ is obtained by substituting $u = k(x, v)$ and $v = Fz = FT(x)$ (this can be interpreted as adding a feedback loop around the dashed box shown in Figure 5.4). We claim that system (5.14) has x^0 as an asymptotically stable equilibrium, with a potentially large domain of attraction: *every* solution ξ with the property that $\xi(t) \in \mathcal{O}$ for all t must converge to x^0. Indeed, suppose that ξ is any solution of (5.14) which stays in $\mathcal{O}$. Let $\zeta(t) := T(\xi(t))$. Then $\dot{\zeta}(t) = (A + bF)\zeta(t)$, so $\zeta(t) \to 0$ as $t \to \infty$. Since T has a continuous inverse, this means that $\xi(t) \to x^0$ as $t \to \infty$. (In addition, if $\xi(0)$ is near x^0 then $\zeta(0)$ is near zero, so $\zeta(t)$ remains close to zero and thus $\xi(t)$ remains close to x^0 for all t.)

Lemma/Exercise 5.3.3 Prove that we have an equivalence relation, in the following sense:
(a) For each Σ and open $\mathcal{O} \subseteq \mathcal{X}$, $(\Sigma, \mathcal{O})$ is feedback equivalent to $(\Sigma, \mathcal{O})$.
(b) $(\Sigma, \mathcal{O})$ feedback equivalent to $(\widetilde{\Sigma}, \widetilde{\mathcal{O}})$ implies $(\widetilde{\Sigma}, \widetilde{\mathcal{O}})$ feedback equivalent to $(\Sigma, \mathcal{O})$.
(c) If $(\Sigma, \mathcal{O})$ is feedback equivalent to $(\widetilde{\Sigma}, \widetilde{\mathcal{O}})$ and $(\widetilde{\Sigma}, \widetilde{\mathcal{O}})$ is feedback equivalent to $(\hat{\Sigma}, \hat{\mathcal{O}})$, then $(\Sigma, \mathcal{O})$ is feedback equivalent to $(\hat{\Sigma}, \hat{\mathcal{O}})$. □

Lemma/Exercise 5.3.4 Suppose that the pairs (A, b) and $(\widetilde{A}, \widetilde{b}) \in \mathcal{S}_{n,1}$ are feedback equivalent in the sense of Definition 5.2.1. That is, there exist $T \in GL(n)$, $F \in \mathbb{R}^{1\times n}$, and a nonzero scalar V such that $T^{-1}(A + bF)T = \widetilde{A}$ and $T^{-1}bV = \widetilde{b}$. Consider the systems Σ and $\widetilde{\Sigma}$ given respectively by $\dot{x} = Ax + ub$ and $\dot{x} = \widetilde{A}x + u\widetilde{b}$. Show that, for each open subset $\mathcal{O} \subseteq \mathbb{R}^n$, $(\Sigma, \mathcal{O})$ is feedback equivalent to $(\widetilde{\Sigma}, T^{-1}\mathcal{O})$. □

The study of feedback linearization is simplified considerably by the following observation. For each positive integer n, we let Σ_n be the continuous-time linear system $\dot{x} = A_n x + ub_n$, where A_n and b_n are in the controller form given by Equation 5.3:

$$A_n \;:=\; \begin{pmatrix} 0 & 1 & 0 & \dots & 0 \\ 0 & 0 & 1 & \dots & 0 \\ \vdots & \vdots & \vdots & \ddots & \vdots \\ 0 & 0 & 0 & \dots & 1 \\ 0 & 0 & 0 & \dots & 0 \end{pmatrix} \qquad b_n \;:=\; \begin{pmatrix} 0 \\ 0 \\ \vdots \\ 0 \\ 1 \end{pmatrix}.$$

By Theorem 14 (p. 191), any single input n-dimensional controllable pair is feedback equivalent, in the sense of Definition 5.2.1, to this special pair (A_n, b_n).

Therefore, by Lemma 5.3.4, for any n-dimensional controllable single-input linear system Σ and any open subset $\mathcal{O} \subseteq \mathbb{R}^n$, $(\Sigma, \mathcal{O})$ is feedback equivalent to $(\Sigma_n, \mathcal{W})$, for some open subset $\mathcal{W} \subseteq \mathbb{R}^n$. By transitivity of equivalence, cf. Lemma 5.3.3, we conclude:

Lemma 5.3.5 The system Σ is feedback linearizable about the state x^0 if and only if there exist an open $\mathcal{O} \subseteq \mathcal{X}$ containing x^0 and an open $\mathcal{W} \subseteq \mathbb{R}^n$ such that $(\Sigma, \mathcal{O})$ is feedback equivalent to $(\Sigma_n, \mathcal{W})$. □

Observe that Equations (5.11)-(5.12) when specialized to Σ_n can be written equivalently as follows:

$$T_*(x)\, f(x) \;=\; A_n T(x) - \frac{\alpha(x)}{\beta(x)} b_n \tag{5.15}$$

and

$$T_*(x)\, g(x) \;=\; \frac{1}{\beta(x)} b_n\,. \tag{5.16}$$

The main result to be proved in this section is as follows. We employ concepts from the nonlinear controllability Chapter 4.

Theorem 15 *The system (5.10) is feedback linearizable about x^0 if and only if the distribution*

$$\Delta_{n-1} \;:=\; \Delta_{g,\mathrm{ad}_f g,\ldots,\mathrm{ad}_f^{n-2} g} \tag{5.17}$$

is involutive when restricted to some neighborhood of x^0, and the set of vectors $\left\{g(x^0), \mathrm{ad}_f g(x^0), \ldots, \mathrm{ad}_f^{n-1} g(x^0)\right\}$ is linearly independent.

(When $n = 1$, the condition is simply $g(x^0) \neq 0$.) Observe that, if the n vectors shown are linearly independent, then $\dim \Delta_{n-1}(x^0) = n - 1$. Thus, $\dim \Delta_{n-1} \equiv n - 1$ on some neighborhood $\mathcal{O}$ of x^0. On any such set $\mathcal{O}$, the involutivity condition amounts to the requirement that

$$[\mathrm{ad}_f^i g, \mathrm{ad}_f^j g](x) \in \Delta_{n-1}(x) \;\text{ for all } x \in \mathcal{O} \text{ and all } i,j \in \{0,\ldots,n-2\} \tag{5.18}$$

(cf. Lemma 4.4.93b). Equivalently, by Frobenius' Theorem 10 (p. 175), the condition is that Δ_{n-1} must be integrable (in some neighborhood of x^0).

We need to establish a preliminary general result concerning smooth functions and vector fields.

Lemma 5.3.6 Let $\mathcal{O}$ be an open subset of $\mathbb{R}^n$, $\varphi \in \mathbb{F}(\mathcal{O})$, $f, g \in \mathbb{V}(\mathcal{O})$. The following properties are equivalent, for any given nonnegative integer ℓ:

1. For each $0 \le i \le \ell$, $L_g L_f^i \varphi = 0$.
2. For each $0 \le j \le \ell$, $L_{\mathrm{ad}_f^j g} \varphi = 0$.
3. For each $i, j \ge 0$ with $i + j \le \ell$, $L_{\mathrm{ad}_f^j g} L_f^i \varphi = 0$.

Furthermore, if these properties hold, then:

$$\text{For each } i,j \geq 0 \text{ with } i+j=\ell+1\,,\; L_{\mathrm{ad}_f^{\ell+1}g}\varphi \;=\; (-1)^i L_{\mathrm{ad}_f^j g} L_f^i \varphi\,. \tag{5.19}$$

Proof. Note that statements 1 and 2 are particular cases of 3, corresponding to the choices $j=0$ and $i=0$ respectively. We will prove, separately, that each of them implies 3. The proof will be based upon a useful identity which we derive first.

Equation (4.3) says that $L_{\mathrm{ad}_f h} = L_f L_h - L_h L_f$ for any two vector fields $f,h \in \mathbb{V}(\mathcal{O})$. We may apply this formula, in particular, when h has the form $\mathrm{ad}_f^j g$, and thus derive

$$L_{\mathrm{ad}_f^{j+1}g} \;=\; L_f L_{\mathrm{ad}_f^j g} \;-\; L_{\mathrm{ad}_f^j g} L_f\,.$$

This is an equality among operators on smooth functions, so in particular we may evaluate at any function which is a directional derivative of φ, and conclude:

$$L_{\mathrm{ad}_f^{j+1}g} L_f^i \varphi \;=\; L_f L_{\mathrm{ad}_f^j g} L_f^i \varphi \;-\; L_{\mathrm{ad}_f^j g} L_f^{i+1} \varphi \quad \forall i,j \geq 0\,. \tag{5.20}$$

Suppose now that 1 holds. We prove 3, by induction on j. The case $j=0$ is just 1 (since $\mathrm{ad}_f^0 g = g$, by definition). Suppose that 3 has been shown to hold, for a given j (and for all i). We need to prove that, for any i, $L_{\mathrm{ad}_f^{j+1}g} L_f^i \varphi = 0$ if $i+(j+1) \leq \ell$. This is a consequence of identity (5.20). Indeed, if $i+(j+1)\leq \ell$ then also $i+j \leq \ell$, so, by inductive hypothesis, $L_{\mathrm{ad}_f^j g} L_f^i \varphi = 0$, and so

$$L_{\mathrm{ad}_f^{j+1}g} L_f^i \varphi \;=\; -L_{\mathrm{ad}_f^j g} L_f^{i+1} \varphi\,. \tag{5.21}$$

The right-hand side vanishes, again by inductive hypothesis since $(i+1)+j \leq \ell$, so $L_{\mathrm{ad}_f^{j+1}g} L_f^i \varphi = 0$, as claimed.

Next, supposing that 2 holds, we prove 3, this time by induction on i. The case $i=0$ is just 2. Suppose 3 has been shown to hold, for a given i (and for all j). We must show, for any j, that $L_{\mathrm{ad}_f^j g} L_f^{i+1} \varphi = 0$ if $(i+1)+j \leq \ell$. As $i+j \leq \ell$ by inductive hypothesis $L_{\mathrm{ad}_f^j g} L_f^i \varphi = 0$, so Equation (5.21) again holds. By induction, the left-hand side vanishes, so the right-hand term does too.

Finally, we assume that 3 is true and show

$$L_{\mathrm{ad}_f^{\ell+1}g}\varphi \;=\; (-1)^i L_{\mathrm{ad}_f^{\ell+1-i}g} L_f^i \varphi\,, \;\; \text{for } i=0,\ldots,\ell+1 \tag{5.22}$$

by induction on i. When $i=0$, this is clear. Assume now that (5.22) has been proved for some $i<\ell+1$. We consider Equation (5.20), applied with $j=\ell-i$. Since the first term in the right vanishes, because $i+j=\ell$ and 3 is true, we have that

$$L_{\mathrm{ad}_f^{\ell+1-i}g} L_f^i \varphi \;=\; -L_{\mathrm{ad}_f^{\ell-i}g} L_f^{i+1} \varphi \;=\; -L_{\mathrm{ad}_f^{(\ell+1)-(i+1)}g} L_f^{i+1} \varphi\,.$$

Substituting into (5.22) completes the induction. ∎

From this Lemma, the following consequence is clear:

Corollary 5.3.7 Let $\mathcal{O}$ be an open subset of $\mathbb{R}^n$, $\varphi \in \mathbb{F}(\mathcal{O})$, $f, g \in \mathbb{V}(\mathcal{O})$. The following properties are equivalent, for any given nonnegative integer ℓ:

1. For all $x \in \mathcal{O}$, $L_g L_f^i \varphi(x) = 0$ if $0 \le i \le \ell$, and $L_g L_f^{\ell+1}\varphi(x) \neq 0$.
2. For all $x \in \mathcal{O}$, $L_{\mathrm{ad}_f^j g}\varphi(x) = 0$ if $0 \le j \le \ell$, and $L_{\mathrm{ad}_f^{\ell+1} g}\varphi(x) \neq 0$.
3. For all $x \in \mathcal{O}$, and for $i, j \ge 0$, $L_{\mathrm{ad}_f^j g} L_f^i \varphi(x) = 0$ if $i + j \le \ell$, and $L_{\mathrm{ad}_f^j g} L_f^i \varphi(x) \neq 0$ if $i + j = \ell + 1$. □

Corollary 5.3.8 Let $\mathcal{O}$ be an open subset of $\mathbb{R}^n$, $\varphi \in \mathbb{F}(\mathcal{O})$, $f, g \in \mathbb{V}(\mathcal{O})$, and suppose that the equivalent properties in Corollary 5.3.7. Then, for each $x \in \mathcal{O}$ the $\ell + 2$ vectors

$$g(x), \mathrm{ad}_f g(x) \ldots, \mathrm{ad}_f^{\ell+1} g(x)$$

are linearly independent, and the $\ell + 2$ row vectors

$$\nabla\varphi(x), \nabla L_f\varphi(x), \ldots, \nabla L_f^{\ell+1}\varphi(x)$$

are linearly independent.

Proof. Consider the following matrix

$$M(x) = \left(\nabla\varphi(x), \nabla L_f\varphi(x), \ldots, \nabla L_f^{\ell+1}\varphi(x)\right)' \cdot \left(g(x), \mathrm{ad}_f g(x), \ldots, \mathrm{ad}_f^{\ell+1} g(x)\right)$$

of size $(\ell + 2) \times (\ell + 2)$, for each $x \in \mathcal{O}$. Its (i, j)th entry is:

$$M(x)_{i,j} = \nabla\left(L_f^{i-1}\varphi\right)(x) \cdot \mathrm{ad}_f^{j-1} g(x).$$

By (3), $M(x)$ has the triangular structure [0/*] for every x, where the skew-diagonal elements are nonzero. Thus $M(x)$ is invertible, which proves the desired conclusions. ∎

The main step in proving Theorem 15 is given by the following technical result.

Proposition 5.3.9 Let Σ be a system $\dot{x} = f(x) + ug(x)$, and let $\mathcal{O}$ be an open subset of the state space $\mathcal{X} \subseteq \mathbb{R}^n$. Assume that $T = (T_1, \ldots, T_n)' : \mathcal{O} \to \mathbb{R}^n$ and $\alpha, \beta : \mathcal{O} \to \mathbb{R}$ are smooth mappings, with $\beta(x) \neq 0$ for all $x \in \mathcal{O}$, and are such that Equations (5.15) and (5.16) hold for all $x \in \mathcal{O}$. Then, writing $\varphi := T_1$, the following properties hold for each $x \in \mathcal{O}$:

$$L_g L_f^i \varphi(x) = 0\,,\ i = 0, \ldots, n-2, \quad L_g L_f^{n-1}\varphi(x) \neq 0 \tag{5.23}$$

$$T_{i+1} = L_f^i\varphi, i = 1, \ldots, n-1,\ \alpha(x) = \frac{-L_f^n\varphi(x)}{L_g L_f^{n-1}\varphi(x)}\ \beta(x) = \frac{1}{L_g L_f^{n-1}\varphi(x)} \tag{5.24}$$

$$\nabla\varphi(x) \,.\, \mathrm{ad}_f^j g(x) = 0\,,\ j = 0, \ldots, n-2, \quad \nabla\varphi(x) \,.\, \mathrm{ad}_f^{n-1} g(x) \neq 0 \tag{5.25}$$

$$g(x), \mathrm{ad}_f g(x) \ldots, \mathrm{ad}_f^{n-1} g(x) \text{ are linearly independent} \tag{5.26}$$

$$T_*(x) \text{ is nonsingular.} \tag{5.27}$$

Conversely, suppose that (5.26) is true for all $x \in \mathcal{O}$, and assume that $\varphi \in \mathbb{F}(\mathcal{O})$ is so that (5.25) holds for all $x \in \mathcal{O}$. Then $L_g L_f^{n-1}\varphi(x) \neq 0$ for all x, so we may define $T_1 := \varphi$ and $T_2, \ldots, T_n, \alpha, \beta$ via (5.24); with these definitions, Equations (5.15) and (5.16) hold for all $x \in \mathcal{O}$.

Proof. Suppose that T, α, β are so that Equations (5.15) and (5.16) hold for all $x \in \mathcal{O}$, and let $\varphi := T_1$. Observe that $T_*(x)f(x) = (L_f T_1(x), \ldots, L_f T_n(x))'$ and $T_*(x)g(x) = (L_g T_1(x), \ldots, L_g T_n(x))'$, and because of the form of the matrix A_n, $A_n T(x) = (T_2(x), \ldots, T_n(x), 0)'$. So Equation (5.15) says

$$(L_f T_1(x), \ldots, L_f T_n(x))' = \left(T_2(x), \ldots, T_n(x), -\frac{\alpha(x)}{\beta(x)}\right)' .$$

Comparing the first $n-1$ entries gives $T_2 = L_f\varphi$, $T_3 = L_f T_2 = L_f^2\varphi$, and, in general $T_{i+1} = L_f^i\varphi$, $j = 1, \ldots, n-1$, as required for (5.24), while the last coordinate gives $L_f^n\varphi = -\alpha/\beta$. On the other hand, Equation (5.16) gives

$$(L_g T_1(x), \ldots, L_g T_n(x))' = \left(0, \ldots, 0, \frac{1}{\beta(x)}\right)' .$$

From the last coordinate, $L_g L_f^{n-1}\varphi = 1/\beta$, which together with $L_f^n\varphi = -\alpha/\beta$ gives the missing part of (5.24), This also says that $L_g L_f^{n-1}\varphi$ is everywhere nonzero, so the last part of (5.23) holds. On the other hand, the first $n-1$ coordinates provide $L_g T_i = L_g L_f^{i-1}\varphi = 0$ for $i = 1, \ldots, n-1$, so all of (5.23) has been proved as well.

We now show that (5.23) implies all three of (5.25), (5.26), and (5.27). The first of these is simply the equivalence of 1 and 2 in Corollary 5.3.7, when ℓ is chosen as $n-2$, and the last two follow from Corollary 5.3.8, because $T = (\varphi, L_f\varphi, \ldots, L_f^{n-1}\varphi)'$.

We conclude by establishing the last (converse) statement. Let $\varphi \in \mathbb{F}(\mathcal{O})$ be given, and assume that (5.25) and (5.26) hold for all x. By the equivalence of 1 and 2 in Corollary 5.3.7, also (5.23) holds, and in particular $L_g L_f^{n-1}\varphi(x) \neq 0$ for all x, so we may indeed define α and β as in (5.24). We also let $T_i := L_f^{i-1}\varphi$,for $i = 0, \ldots, n-1$. Comparing coordinates, both (5.15) and (5.16) are verified. ∎

Proof of Theorem 15

Suppose that Σ is feedback linearizable around x^0. Then, there exists some neighborhood $\mathcal{O}$ of x^0 and T, α, and β (nowhere vanishing) so that both (5.15) and (5.16) hold on $\mathcal{O}$. Then, by Proposition 5.3.9, there exists a $\varphi \in \mathbb{V}(\mathcal{O})$ so that (5.25) holds. This implies, in particular, that $\nabla\varphi(x) \neq 0$ for all x, and $\nabla\varphi(x).\mathrm{ad}_f^j g(x) = 0$ for each $j = 0, \ldots, n-2$, so the distribution Δ_{n-1} (which has constant rank $n-1$, by (5.26)) is integrable, by the characterizations (4.37)-(4.38). So Δ_{n-1}, seen as a distribution on $\mathcal{O}$, is involutive.

Conversely, assume that Δ_{n-1} is involutive, on some open set $\mathcal{V}$ which contains x^0, and $g(x^0), \mathrm{ad}_f g(x^0), \ldots, \mathrm{ad}_f^{n-1} g(x^0)$ are linearly linearly independent. Using if necessary a smaller $\mathcal{V}$, we may assume that $g(x), \mathrm{ad}_f g(x), \ldots, \mathrm{ad}_f^{n-1} g(x)$ (and, in particular, the first $n-1$ vectors in this list) are linearly independent for each $x \in \mathcal{V}$. So, on $\mathcal{V}$, Δ_{n-1} is constant rank and involutive. By Frobenius' Theorem 10 (p. 175), Δ_{n-1} must be integrable, so there exists a neighborhood $\mathcal{O} \subseteq \mathcal{V}$ of x^0 and a $\varphi \in \mathbb{V}(\mathcal{O})$ so that, for all $x \in \mathcal{O}$, $\nabla\varphi(x) \neq 0$ and, for each $j = 0, \ldots, n-2$, $\nabla\varphi(x).\mathrm{ad}_f^j g(x) = 0$. If it were the case that $\nabla\varphi(x).\mathrm{ad}_f^{n-1} g(x) = 0$ for some $x \in \mathcal{O}$, then $\mathrm{ad}_f^{n-1} g(x)$ would be in the span of $g(x), \mathrm{ad}_f g(x), \ldots, \mathrm{ad}_f^{n-2} g(x)$ (because these are $n-1$ linearly independent vectors orthogonal to the nonzero vector $\nabla\varphi(x)$), and this contradicts the fact that $g(x), \mathrm{ad}_f g(x), \ldots, \mathrm{ad}_f^{n-1} g(x)$ are linearly independent. Therefore (5.25) and (5.26) hold for all $x \in \mathcal{O}$, and the second part of Proposition 5.3.9 provides a T, as well as an α and a (nowhere vanishing) β so that both (5.15) and (5.16) hold. This provides a feedback linearization, except for the fact that T is not necessarily a diffeomorphism. However, the first part of Proposition 5.3.9, applied to this T, shows that $T_*(x)$ is nonsingular. Thus, by the Implicit Function Theorem, we may restrict $\mathcal{O}$ to a smaller neighborhood of x^0 in such a manner that, in this restricted set, T indeed admits a smooth inverse. ∎

Example 5.3.10 We consider a model of an arm driven by a motor through a torsional spring, see Figure 5.5; this is used as a simplified model of a single-link

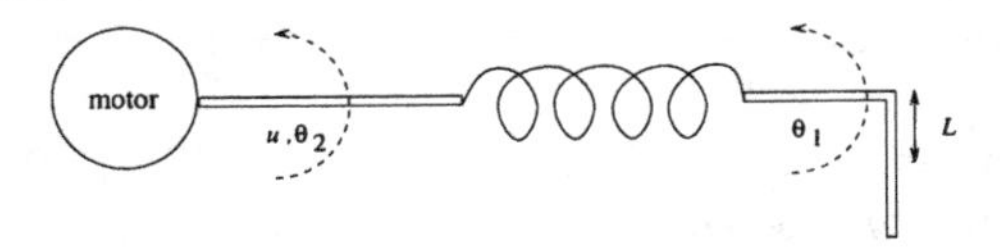

Figure 5.5: *Flexible link.*

flexible-joint robot. The equations for the angular positions θ_1 and θ_2 of the arm and shaft can be shown to be (see [378]):

$$\begin{aligned} I\ddot{\theta}_1 + MgL\sin\theta_1 + k(\theta_1 - \theta_2) &= 0 \\ J\ddot{\theta}_2 - k(\theta_1 - \theta_2) &= u\,, \end{aligned}$$

where I and J are moments of inertia, k is a spring constant (notice that the term $k(\theta_1 - \theta_2)$ stands for a restoring force proportional to the difference between the angles, that is, the torsion of the spring), M is the mass, L is the distance from the joint to the center of mass of the arm, and u, the control, is the torque being applied by a motor. We let $x_1 := \theta_1$, $x_2 := \dot{\theta}_1$, $x_3 := \theta_2$, and $x_4 := \dot{\theta}_2$ and view the system as a system with state space $\mathcal{X} = \mathbb{R}^4$. (More precisely, one could represent this as a system whose state space is a differentiable manifold, accounting for the fact that $\theta_1 - \theta_2$ only matters modulo 2π.) Letting $a := MgL/I$, $b := k/I$, $c := k/J$, and $d := 1/J$, this is a system $\dot{x} = f(x) + ug(x)$

with

$$f(x) = \begin{pmatrix} x_2 \\ -a\sin x_1 - b(x_1 - x_3) \\ x_4 \\ c(x_1 - x_3) \end{pmatrix}, \quad g(x) = \begin{pmatrix} 0 \\ 0 \\ 0 \\ d \end{pmatrix}.$$

We compute

$$\left(g(x), \mathrm{ad}_f g(x), \mathrm{ad}_f^2 g(x), \mathrm{ad}_f^3 g(x)\right) = \begin{pmatrix} 0 & 0 & 0 & -bd \\ 0 & 0 & bd & 0 \\ 0 & -d & 0 & cd \\ d & 0 & -cd & 0 \end{pmatrix}$$

(constant matrix). This shows that $g(x^0), \mathrm{ad}_f g(x^0), \mathrm{ad}_f^2 g(x^0), \mathrm{ad}_f^3 g(x^0)$ are linearly independent for each x^0 and, since $[\mathrm{ad}_f^i g, \mathrm{ad}_f^j g] = 0 \in_p \Delta_3$ for all $i, j \in \{0, 1, 2\}$, that Δ_3 is involutive. Thus, the system is feedback linearizable around every x^0. However, much more is true in this example. Let us compute φ as in Proposition 5.3.9. Property (5.25), and the explicit form found above for the $\mathrm{ad}_f^i g$'s, says that we must satisfy, for each x near any given x^0,

$$\frac{\partial \varphi}{\partial x_4} = \frac{\partial \varphi}{\partial x_3} = \frac{\partial \varphi}{\partial x_2} = 0 \text{ and } \frac{\partial \varphi}{\partial x_1} \neq 0.$$

This set of partial differential equations have an obvious solution, which happens to be globally defined on $\mathbb{R}^4$, namely, $\varphi(x) := x_1$. This means that Proposition 5.3.9 can be satisfied globally. With this choice of φ, we compute

$$\begin{aligned} L_f\varphi(x) &= x_2 \\ L_f^2\varphi(x) &= -a\sin x_1 - b(x_1 - x_3) \\ L_f^3\varphi(x) &= -ax_2\cos x_1 - b(x_2 - x_4) \\ L_f^4\varphi(x) &= ax_2^2\sin x_1 + (a\cos x_1 + b)(a\sin x_1 + b(x_1 - x_3)) + bc(x_1 - x_3) \\ L_g L_f^3\varphi(x) &= bd. \end{aligned}$$

The obtained mapping $T(x) = (\varphi(x), L_f\varphi(x), L_f^2\varphi(x), L_f^3\varphi(x))'$ happens to define a global diffeomorphism from $\mathbb{R}^4$ to $\mathbb{R}^4$: its inverse is given by

$$T^{-1}(z) = (z_1, z_2, z_1 + (1/b)(z_3 + a\sin z_1), z_2 + (1/b)(z_4 + az_2\cos z_1))'.$$

Thus, with $\beta(x) \equiv 1/bd$ and $\alpha(x) = -(1/bd)L_f^4\varphi(x)$, one obtains a global feedback linearization. □

Exercise 5.3.11 Property 5.23 in Proposition 5.3.9 says that Σ must have *relative degree n* with respect to φ on the open set $\mathcal{O}$, and in particular at the point x^0. In general, for any system $(\Sigma)\, \dot{x} = f(x) + ug(x)$, together with a smooth $\varphi : \mathcal{X} \to \mathbb{R}$ (which we think of as an "output map" for Σ), and a state x^0, one defines the relative degree of (Σ, φ) at x^0 as the smallest positive integer

r, if one exists, with the property that $L_g L_f^i \varphi(x) = 0$ for each $0 \le i \le r-2$ and each x in some neighborhood $\mathcal{O}$ of x^0 (this condition is vacuous if $r = 1$), and $L_g L_f^{r-1} \varphi(x^0) \ne 0$. If for every $i \ge 0$ it is the case that $L_g L_f^i \varphi(x) = 0$ on some neighborhood of x^0, we will say that the relative degree exists but is infinite.
(a) Show that, if (Σ, φ) has a finite relative degree r around x^0, then there are real numbers $\mu_0, \ldots, \mu_r$, and $\nu \ne 0$, with the following property: For each continuous input u defined on some interval $[0, \tau]$, $\tau > 0$, let $x(t) = \phi(t, 0, x^0, u)$ be the solution of $\dot{x}(t) = f(x(t)) + u(t)g(x(t))$, which is defined on some maximal interval $[0, \varepsilon)$, $\varepsilon > 0$, and let $y(t) = \varphi(x(t))$. Then y is class C^r, and its derivatives at $t = 0$ satisfy $y^{(i)}(0) = \mu_i$ for $i = 0, \ldots, r-1$ and $y^{(r)}(0) = \mu_r + \nu u(0)$. Note that this says that the first $r-1$ derivatives of the "output" y are independent of the input, but the rth derivative does depend on u.
(b) Show that for a controllable linear system $\dot{x} = Ax + ub$ with $\varphi(x) = cx$ linear and $c \ne 0$, a finite relative degree is always defined, it is $\le n$, and it is the smallest r so that $cA^{r-1}b \ne 0$.
(c) Give an example of Σ, φ, x^0 for which the relative degree does not exist. To make the example interesting, you are not allowed to pick one where the relative degree exists but is infinite. □

Exercise 5.3.12 Suppose that x^0 is an equilibrium state: $f(x^0) = 0$. Show that, if the system is feedback linearizable about x^0, then it is possible to choose $T(x^0) = 0$ and $\alpha(x^0) = 0$. □

Exercise 5.3.13 Suppose that x^0 is an equilibrium state: $f(x^0) = 0$. Prove that the linear independence condition in the statement of Theorem 15 is equivalent to asking that the pair (A, b) be controllable, where (A, b) is the linearization of Σ about $x = x^0$ and $u = 0$. (That is, $A = f_*(x^0)$ and $b = g(x^0)$.) □

Exercise 5.3.14 Suppose that x^0 is an equilibrium state: $f(x^0) = 0$ and the system Σ has dimension 2. Show that Σ is feedback linearizable about x^0 if and only if the linearization of Σ about $x = x^0$ and $u = 0$ is controllable. □

Exercise 5.3.15 Give an example of a Σ of dimension 3, and an equilibrium state x^0, such that the linearization of Σ about $x = x^0$ and $u = 0$ is controllable but Σ is not feedback linearizable about x^0. □

Exercise 5.3.16 Check that the system $\dot{x}_1 = \sin x_2$, $\dot{x}_2 = x_1^2 + u$ is feedback linearizable about $x^0 = (0, 0)'$. Next, find explicit expressions for T, α, and β, so that both (5.15) and (5.16) are verified for all $x \in \mathcal{O} := \mathbb{R} \times (-\pi/2, \pi/2)$. □

Exercise 5.3.17 Show that if the conditions of Theorem 15 hold, then the distribution $\Delta_k := \Delta_{g, \mathrm{ad}_f g, \ldots, \mathrm{ad}_f^{k-1} g}$ is integrable, on some neighborhood of x^0, for each $k = 1, \ldots, n-2$. (*Hint:* For each Δ_k, show that $\psi = L_f^j \varphi$, $j = 0, \ldots, n-k-1$ provide independent integrals, if φ is as in the proofs.) □

Exercise 5.3.18 ◇ Show that Σ is feedback linearizable in the restricted sense that one may pick $\beta(x) \equiv 1$ if and only if, in addition to the conditions of Theorem 15 (p. 200), one has that

$$\left[\mathrm{ad}_f^k g, \mathrm{ad}_f^j g\right] \in_p \Delta_k\,, \quad k = 0, \ldots, n-1\,,\ j = 0, \ldots, k-1$$

(on some neighborhood of x^0). □

Exercise 5.3.19 ◇ A far more restrictive problem is that of asking that Σ be linearizable by means of coordinate changes alone, i.e., that there be some diffeomorphism defined in a neighborhood of x^0, and a controllable pair (A, b), so that $T_*(x)f(x) = AT(x)$ and $T_*(x)g(x) = b$. This can be seen as feedback linearization with $\alpha \equiv 0$ and $\beta \equiv 1$. Show that such a linearization is possible if and only if $g(x^0), \mathrm{ad}_f g(x^0), \ldots, \mathrm{ad}_f^{n-1} g(x^0)$ are linearly independent and $[\mathrm{ad}_f^i g, \mathrm{ad}_f^j g] = 0$ for all $i, j \geq 0$. □

5.4 Disturbance Rejection and Invariance*

A large amount of the linear systems literature has been devoted to the study of synthesis problems in which a feedback law $u = Fx$ is sought to achieve objectives other than —or more often, in addition to— stabilization or pole assignment.

This section describes one of these objectives, the "disturbance rejection" (sometimes also called "disturbance decoupling") problem. The method of solution illustrates the application of what is sometimes called the (linear-) "geometric approach" to linear control. This approach is based on the use of certain spaces that are invariant under the A matrix, modulo the action of controls. Illustrating the use of these spaces is our main purpose here; the material will not be used in later sections.

Still $\mathbb{K}$ denotes an arbitrary field, and unless otherwise stated, (A, B) is an arbitrary pair, $A \in \mathbb{K}^{n \times n}$, $B \in \mathbb{K}^{n \times m}$. For the case of no controls, $B = 0$, the following reduces to the classical notion of invariant subspace. We let $\mathcal{B} = \mathrm{col}\, B$, i.e. $\mathrm{im}\, B$ when the latter is thought of as a map

$$B : \mathcal{U} = \mathbb{K}^m \to \mathcal{X} = \mathbb{K}^n\,.$$

Definition 5.4.1 *A subspace $\mathcal{V}$ of $\mathcal{X} = \mathbb{K}^n$ is* **A-invariant mod B** *if*

$$\boxed{A\mathcal{V} \subseteq \mathcal{V} + \mathcal{B}}$$

One says equivalently that $\mathcal{V}$ is **(A, B)-invariant.** □

Since the concept does not depend on B itself but only on its image $\mathcal{B}$, the alternative terminologies **A-invariant mod $\mathcal{B}$** and $(A, \mathcal{B})$-invariant are also used.

* This section can be skipped with no loss of continuity.

Exercise 5.4.2 If $\mathcal{V}$ is A-invariant mod B and if $\mathcal{B} \subseteq \mathcal{V}$ then $\mathcal{V}$ must also contain the controllable subspace $\mathcal{R}(A, B)$. □

By definition, $\mathcal{V}$ is (A, B)-invariant if and only if for each $x \in \mathcal{V}$ there exist a $z \in \mathcal{V}$ and a $u \in \mathcal{U}$ such that

$$Ax = z + Bu\,.$$

For a discrete-time system with matrices (A, B), this implies that for each state x in $\mathcal{V}$ there is another state z in $\mathcal{V}$ such that $x \underset{1}{\leadsto} z$ (use $-u$ as a control), and hence, by induction, the state may be kept in $\mathcal{V}$ for arbitrary long times. An analogous statement holds for continuous-time systems; the following observation is useful in establishing it:

Lemma 5.4.3 The subspace $\mathcal{V}$ of $\mathcal{X}$ is A-invariant mod B if and only if there exists a matrix $F \in \mathbb{K}^{m\times n}$ such that $\mathcal{V}$ is $(A + BF)$-invariant.

Proof. Sufficiency is trivial. To prove necessity, let $\{v_1, \ldots, v_r\}$ be a basis of the A-invariant mod B subspace $\mathcal{V}$. First choose for each i some $u_i \in \mathcal{U}$ such that

$$Av_i + Bu_i \in \mathcal{V}\,.$$

Now define on $\mathcal{V}$

$$Fv_i := u_i$$

and let F be arbitrary on a complement of $\mathcal{V}$. ■

The matrix F is interpreted as a feedback transformation $u = Fx$.

Exercise 5.4.4 Let Σ be a linear time-invariant continuous-time system (A,B). Prove that the subspace $\mathcal{V}$ is A-invariant mod B iff for each $x \in \mathcal{V}$ and each $\tau > 0$ there is some $\omega \in \mathcal{U}^{[0,\tau)}$ such that

$$\xi(t) \in \mathcal{V} \quad \text{for all} \quad t \in [0, \tau]$$

if ξ is the resulting path $\psi(x, \omega)$. □

The above concept of relative invariance arose originally when treating the following problem. For concreteness we deal with continuous-time systems, but the discrete-time case is entirely analogous.

Consider linear time-invariant continuous-time systems with outputs, over $\mathbb{R}$, with $m + l$ inputs partitioned as (u, v):

$$\dot{x} = Ax + Bu + Ev \quad y = Cx\,, \tag{5.28}$$

where v is to be interpreted as a "disturbance." This disturbance cannot be measured directly by the controller, and the objective is to design a feedback law

$$u = Fx + u'$$

such that v has no effect whatsoever on the output, no matter what u' or the initial condition x^0 of (5.28) are.

For instance, in the system

$$\begin{aligned} \dot{x}_1 &= x_2 + u \\ \dot{x}_2 &= v \\ y &= x_1 \end{aligned}$$

the disturbance v is not decoupled from the output —in particular, for $x^0 = 0$ and $u \equiv 0$ one has $\ddot{y} = v$. But using the feedback transformation

$$u = -x_2 + u'$$

there results a system for which

$$y(t) = x_1^0 + \int_0^t u'(s)ds$$

is independent of v.

In general terms, then, we wish to have

$$y(t) = Ce^{t\widetilde{A}}x^0 + C\int_0^t e^{(t-\tau)\widetilde{A}}(Bu'(\tau) + Ev(\tau))\,d\tau$$

independent of v, for some matrix of the form

$$\widetilde{A} = A + BF\,.$$

This is equivalent to the requirement that

$$C\int_0^t e^{(t-\tau)\widetilde{A}}Ev(\tau)\,d\tau \;=\; 0$$

for all (measurable and essentially bounded) $v : [0,T] \to \mathbb{R}^l$, all $T \geq 0$, and all $t \in [0,T]$. Considering this expression as the i/o behavior of the system $(\widetilde{A}, E, C)$ with zero initial state, we see from Lemma 2.7.13 that the problem becomes:

$$\textit{Find a matrix } F \textit{ so that } C(A+BF)^iE = 0 \textit{ for all } i \geq 0. \qquad \text{(DRP)}$$

In principle, the problem of deciding whether there is such an F (and finding one in that case) is highly nonlinear in terms of the data A, B, C, E; however, the following Lemma transforms it into a geometric question. Let $\mathcal{E} := \operatorname{im} E$.

Lemma 5.4.5 The DRP problem is solvable if and only if there exists a subspace $\mathcal{V} \subseteq \mathbb{R}^n$ so that

$$\mathcal{E} \subseteq \mathcal{V} \subseteq \ker C$$

and

$$\mathcal{V} \text{ is } A\text{-invariant mod } B\,.$$

Proof. Assume that there is such a subspace $\mathcal{V}$. By Lemma 5.4.3 there exists some F so that $\mathcal{V}$ is $A+BF$-invariant. Thus, for all $i \geq 0$,

$$(A+BF)^i\mathcal{E} \subseteq (A+BF)^i\mathcal{V} \subseteq \mathcal{V} \subseteq \ker C\,,$$

and therefore, $C(A+BF)^iE = 0$, as wanted. Conversely, given such an F, it is only necessary to define

$$\mathcal{V} := \mathcal{R}(A+BF, E)\,,$$

which has the desired properties. ∎

Deciding whether there is such a subspace turns out to be easy, once another concept is introduced. For any given (A,B,C), we consider the set of possible (A,B)-invariant subspaces $\mathcal{V}$ which are included in $\ker C$. There always is at least one such subspace, since $\mathcal{V} = 0$ is (A,B)-invariant. We order these subspaces by inclusion. If $\mathcal{V}_1$ and $\mathcal{V}_2$ are two maximal spaces with these properties, then their sum is again an (A,B)-invariant subspace included in $\ker C$, so, by maximality, we have that

$$\mathcal{V}_1 = \mathcal{V}_1 + \mathcal{V}_2 = \mathcal{V}_2\,.$$

In other words, there is a *unique* such maximal (A,B)-invariant subspace included in $\ker C$. This is denoted

$$\mathcal{V}^*(A,B,C)\,.$$

Furthermore, any $\mathcal{V}$ which is (A,B)-invariant and satisfies $\mathcal{V} \subseteq \ker C$ must be a subspace of $\mathcal{V}^*(A,B,C)$, by the same argument. So if any such space contains $\mathcal{E}$, $\mathcal{V}^*$ does also. Thus, we have proved the following result.

Theorem 16 *The* DRP *problem is solvable iff* $\mathcal{E} \subseteq \mathcal{V}^*(A,B,C)$. □

The space $\mathcal{V}^*$ can be calculated recursively, as shown by the next problem. On a computer, it is necessary then only to solve a sequence of linear equations in order to find this space, and a further Gaussian elimination can be used to check the property that $\mathcal{E} \subseteq \mathcal{V}^*(A,B,C)$.

Exercise 5.4.6 Consider the following algorithm:

$$V^0 \;:=\; \ker C\,,$$

and, for $i = 1,\ldots,n$,

$$V^i \;:=\; \ker C \bigcap A^{-1}(\mathcal{B} + V^{i-1})\,.$$

Prove that $V^n = \mathcal{V}^*(A,B,C)$. □

Remark 5.4.7 For motivation regarding the DRP problem, the reader might want to review the material on PID control in Chapter 1. The objective there was somewhat weaker than here, however, in that it was only required that the effect of disturbances be asymptotically damped, as opposed to completely canceled. Such a more realistic objective can also be studied with geometric techniques. We picked cancellation merely as the simplest illustration of these tools. □

5.5 Stability and Other Asymptotic Notions

In studying stability, we restrict attention to *time-invariant* systems only. Time-varying system stability is an important area, and many results that are relatively easy to establish in the time-invariant case become much more difficult and interesting in the general time-varying case. However, the main system theoretic ideas are already illustrated by the case we treat here, while many technical complications can be avoided. The objective here merely is to provide a basic introduction to these ideas in the context of control problems.

From now on in this chapter, and unless otherwise stated, *system* means *time-invariant topological system* in the sense of Definition 3.7.1, that is, we ask that trajectories depend continuously on the initial state. This provides the right generality for establishing many of the basic facts. Readers interested only in continuous-time systems may simply substitute "continuous-time system" for "topological system," but this does not simplify the proofs in any substantial manner. When dealing with stability notions for continuous-time systems and for linear systems, unless otherwise stated we will assume we are working over $\mathbb{K} = \mathbb{R}$.

The next definition is an infinite-time version of that in Definition 3.7.5.

Definition 5.5.1 *Let $y, z \in \mathcal{X}$, and assume that $\mathcal{V}$ is a subset of $\mathcal{X}$ containing both y and z. Then, z* **can be asymptotically controlled to y without leaving $\mathcal{V}$** *if there exists some control $\nu \in \mathcal{U}^{[0,\infty)}$ admissible for z so that:*

- *For the path $\zeta := \psi(z, \nu)$,*
$$\lim_{t\to\infty} \zeta(t) = y$$
- *$\zeta(t) \in \mathcal{V}$ for all $t \in \mathcal{T}_+$.*

When $\mathcal{V} = \mathcal{X}$, one just says that z can be **asymptotically controlled** *to y.* □

Definition 5.5.2 *Let Σ be a topological system and x^0 an equilibrium state. Then Σ is:*

- **Locally asymptotically controllable to x^0** *if for each neighborhood $\mathcal{V}$ of x^0 there is some neighborhood $\mathcal{W}$ of x^0 such that each $x \in \mathcal{W}$ can be asymptotically controlled to x^0 without leaving $\mathcal{V}$.*

- **Globally asymptotically controllable to x^0** *if it is locally asymptotically controllable and also every $x \in \mathcal{X}$ can be asymptotically controlled to x^0.*

For systems with no control, the more standard terminology is to say that the system Σ is (locally or globally) **asymptotically stable** *with respect to x^0, or that x^0 is an asymptotically stable state for the system.* □

For linear systems, and unless otherwise stated, we always take $x^0 = 0$ when considering asymptotic controllability (or stability).

We use also the term *asycontrollable* instead of *asymptotically controllable.* The interpretation of Σ being asymptotically controllable is that it should be possible to drive states close to the desired x^0 (or arbitrary states, in the global case) into x^0 *without large excursions.*

Asycontrollability is a necessary condition for feedback stabilizability, in the following sense:

Lemma/Exercise 5.5.3 Consider the continuous-time system Σ:

$$\dot{x} = f(x, u)$$

of class C^1 and let $x^0 \in \mathcal{X}$. Assume that there exists some function

$$k : \mathcal{X} \to \mathcal{U}$$

of class C^1 so that x^0 is a local (respectively, global) asymptotically stable state for the "closed-loop system"

$$\dot{x} = f(x, k(x)) .$$

Then Σ is locally (respectively, globally) asymptotically controllable to x^0. The analogous result holds for discrete-time systems of class C^0. □

Remark 5.5.4 One could consider also a weaker notion, dropping the large excursion part and requiring only that states close to x^0 be asymptotically controllable to x^0. For systems with no control, the standard terminology is then that x^0 is an *attractor.* For linear systems attractivity results in an equivalent definition (see below). □

Proposition 5.5.5 Consider the differential equation

$$\dot{x} = Ax$$

where $A \in \mathbb{R}^{n \times n}$, thought of as a continuous-time linear system Σ with no controls. Then, the following statements are equivalent:

- Σ is locally asymptotically stable.
- Σ is globally asymptotically stable.
- A is Hurwitz.

Analogous statements hold for the discrete-time system $x^+ = Ax$ and the discrete-time Hurwitz (convergent) property.

Proof. First we observe that the local notion implies the global one in the linear case. Indeed, if there exists some $\delta > 0$ so that each solution with initial condition $\|z\| < \delta$ converges to the origin, then for any initial condition $y \neq 0$, letting

$$z := \frac{\delta}{2\,\|y\|}y\,,$$

it follows that $e^{tA}z \to 0$, and therefore also that

$$e^{tA}y = \frac{2\,\|y\|}{\delta}e^{tA}z \;\to\; 0\,.$$

Since all solutions of $\dot{x} = Ax$ converge to zero if and only if A is Hurwitz, by Proposition C.5.1 in Appendix C.5 on stability of linear equations, it is only necessary to prove that, if A is Hurwitz, then for each $\varepsilon > 0$ there is a $\delta > 0$ such that $\|\xi(t)\| \leq \varepsilon$ for all $t > 0$ whenever $\|\xi(0)\| < \delta$. Consider the function

$$\alpha(t) := \left\|e^{tA}\right\|, t \geq 0\,.$$

Since A is Hurwitz, $\alpha(t) \to 0$ as $t \to \infty$, so by continuity of α, it must be bounded, $\alpha(t) \leq K$ for all t. Since

$$\|\xi(t)\| \leq \alpha(t)\,\|\xi(0)\| \leq K\,\|\xi(0)\|\,,$$

we may pick $\delta := \varepsilon/K$. The discrete-time case is entirely analogous. ∎

Proposition 5.5.6 Let $\Sigma = (A, B)$ be a continuous-time (respectively, discrete-time) time-invariant linear system. The following statements are then equivalent:

1. Σ is globally asymptotically controllable (to 0).
2. Σ is locally asymptotically controllable (to 0).
3. Each root of χ_u has negative real part (respectively, has magnitude less than 1).
4. There is some $F \in \mathbb{R}^{m\times n}$ such that $A + BF$ is Hurwitz (respectively, discrete-time Hurwitz).

Proof. We only prove the continuous-time case, the discrete-time case being analogous. As in the proof of the previous lemma, the local and global properties (1) and (2) are equivalent, since controlling z asymptotically to zero using the control ω is equivalent to controlling

$$\frac{2\,\|y\|}{\delta}z$$

using the scaled control

$$\frac{2\,\|y\|}{\delta}\omega\,.$$

Assume now that (1) holds. Pick any

$$z_2 \in \mathbb{R}^{n-r}$$

where the integer r is as in the Kalman controllability decomposition, Lemma 3.3.3, and consider

$$x := T\begin{pmatrix} 0 \\ z_2 \end{pmatrix}$$

where T is the similarity used in that Lemma. By assumption, there is some control ω on $[0, \infty)$ such that $\xi(t) \to 0$, where $\xi := \psi(x, \omega)$. It follows that also $\zeta(t) := T^{-1}\xi(t)$ converges to zero. But $\zeta(t)$ has the block form

$$\begin{pmatrix} \zeta_1(t) \\ \zeta_2(t) \end{pmatrix}$$

with

$$\dot{\zeta}_2(t) = A_3\zeta_2(t)\ , \quad \zeta_2(0) = z_2\,.$$

Since z_2 was arbitrary, this means that A_3 in Lemma 3.3.3 must be Hurwitz, so (3) is proved.

That (3) implies (4) is a consequence of Theorem 13 (Pole-Shifting), since the polynomial χ_1 in the statement of that Theorem can be made Hurwitz by an appropriate choice of the feedback F.

Finally, (4) implies (2) because of Lemma 5.5.3; explicitly, by Proposition 5.5.5, $\dot{x} = (A + BF)x$ is globally asymptotically stable. Given any $x \in \mathcal{X}$, consider the solution ξ of this closed-loop system with $\xi(0) = x$. The control

$$\omega(t) := F\xi(t)$$

is then as desired for global asymptotic controllability. ∎

Thus, in the linear case we may check that Σ is asymptotically controllable in two steps: First obtain the decomposition in Lemma 3.3.3, and then check that A_3 is Hurwitz. Since global and local notions coincide, we talk in the linear case just of *asymptotically controllable* or *asymptotically stable* systems.

The equivalence of (4) with Σ being global asymptotically controllable is a characteristic of linear systems. Because of this equivalence, one often refers to asymptotically controllable linear systems as *stabilizable* systems. For nonlinear systems, being asymptotically controllable is *not* in general equivalent to the existence of a stabilizing feedback law $u = k(x)$, at least if k is required to be smooth or even continuous; one exception is the local result presented later, which applies if the linearization at the origin is itself already asymptotically controllable. See Section 5.9 below for further discussion of this point.

There is an analogue of the Hautus controllability condition, Lemma 3.3.7, that applies to asycontrollability:

Exercise 5.5.7 Prove that the continuous-time (respectively, discrete-time) linear time-invariant system $\Sigma = (A, B)$ is asymptotically controllable if and only if

$$\text{rank}\,[\lambda I - A, B] = n$$

for all $\lambda \in \mathbb{C}$ with nonnegative real part (respectively, with magnitude greater than or equal to 1). □

Exercise 5.5.8 State and prove results analogous to Proposition 3.3.13 for the distance to the sets of non-asymptotically controllable discrete-time or continuous-time time-invariant linear systems. □

Exercise 5.5.9 ◇ Let $\mathcal{S}^{c,\text{dtH}}_{n,m}$, respectively $\mathcal{S}^{c,\text{H}}_{n,m}$, denote the set of all controllable pairs $(A, B) \in \mathcal{S}^{c}_{n,m}$ for which A is discrete-time Hurwitz (convergent), respectively Hurwitz, and consider the map

$$\beta:\ (A, B) \to \left((A - I)(A + I)^{-1}, (A + I)^{-1}B\right).$$

Prove that this map induces a bijection between $\mathcal{S}^{c,\text{dtH}}_{n,m}$ and $\mathcal{S}^{c,\text{H}}_{n,m}$. *Hint:* Use the Hautus criterion to check controllability of $\beta(A, B)$. □

5.6 Unstable and Stable Modes*

The conditions for a linear time-invariant system to be asymptotically controllable can be restated in various manners. In this Section we prove a result that can be read as saying that a system is asymptotically controllable if and only if its unstable modes are controllable. For simplicity, we let $\mathbb{K} = \mathbb{R}$, though many of the results could be stated in far more generality. Accordingly, "polynomial" will mean polynomial with real coefficients.

Assume that $\chi = \chi(z)$ is a polynomial that factors as

$$\chi = \chi_g \chi_b$$

in such a way that χ_g and χ_b (read "the good part" and "the bad part" of χ) have no common complex roots. By the Euclidean algorithm, we know then that there exist two polynomials p and q such that

$$p\chi_g + q\chi_b = 1,$$

and therefore, by specialization, any square matrix C satisfies

$$I = p(C)\chi_g(C) + q(C)\chi_b(C), \tag{5.29}$$

where I is the identity matrix of the same size as C. Further, the product decomposition of χ gives that $\chi(C) = \chi_g(C)\chi_b(C) = \chi_b(C)\chi_g(C)$, so the implication

$$\chi(C) = 0 \Rightarrow \chi_b(C)\chi_g(C) = 0 \tag{5.30}$$

holds.

* This section can be skipped with no loss of continuity.

Lemma 5.6.1 For any decomposition into two polynomials χ_b and χ_g as above, and any matrix C for which $\chi(C) = 0$,

$$\operatorname{im} \chi_g(C) = \ker \chi_b(C).$$

Proof. If $\chi_b(C)x = 0$, then it follows from (5.29) that

$$x = p(C)\chi_g(C)x = \chi_g(C)\widetilde{x},$$

where $\widetilde{x} := p(C)x$. Conversely, if $x = \chi_g(C)y$, then

$$\chi_b(C)x = \chi_b(C)\chi_g(C)y = 0$$

from (5.30). ∎

The Lemma applies in particular when χ is the characteristic polynomial of C.

We will say that a polynomial χ is a **Hurwitz polynomial** if all its roots have negative real part. The characteristic polynomial of C is Hurwitz if and only if C is a Hurwitz matrix. Similarly, one can define **convergent** (discrete-time Hurwitz) polynomials.

Recall that the *minimal polynomial* $m_C = m_C(z)$ of a matrix C is the (unique) monic polynomial of smallest possible degree that annihilates C, that is,

$$m_C(C) = 0,$$

and that m_C divides any polynomial which annihilates C. Since also the roots of the minimal polynomial are (in general with different multiplicities) the eigenvalues of C, the following holds:

Lemma 5.6.2 The matrix C is Hurwitz (or convergent) if and only if there is some Hurwitz (respectively, convergent) polynomial χ so that $\chi(C) = 0$. □

Given any matrix A, we now decompose its characteristic polynomial χ as $\chi_g\chi_b$ by letting χ_g collect all of the roots with negative real part. If discrete-time systems are of interest, one picks a decomposition into roots with magnitude less than one and magnitude ≥ 1. That χ_g (and hence also the quotient $\chi_b = \chi/\chi_g$) has real coefficients follows from the fact that complex conjugate roots appear in pairs.

For a linear system (A, B), we have introduced at this point two different factorizations of the characteristic polynomial of A:

$$\boxed{\chi = \chi_c\chi_u = \chi_g\chi_b}$$

one corresponding to the Kalman controllability decomposition (and depending on B) and the other to stability (independent of B, but dependent on the notion of stability, such as discrete-time or continuous-time, being used).

In terms of these polynomials, the equivalence between (1) and (3) in Proposition 5.5.6, which states that the system is asymptotically controllable if and only if each root of χ_u has negative real part, can be restated as:

$$\chi_u \text{ divides } \chi_g$$

as a polynomial. On the other hand, the following property is easy to establish for the Kalman controllability decomposition:

Lemma/Exercise 5.6.3 Let χ be any polynomial, and pick any pair (A, B). Let A_3 be any matrix as in the controllability decomposition (3.6) in Chapter 3. Then,

$$\operatorname{im} \chi(A) \subseteq \mathcal{R}(A, B) \Leftrightarrow \chi(A_3) = 0 .$$ □

We have the following characterization of asymptotic controllability:

Proposition 5.6.4 Let (A, B) be a time-invariant continuous-time or discrete-time linear system. The following properties are equivalent:

1. (A, B) is asymptotically controllable.
2. $\operatorname{im} \chi_g(A) \subseteq \mathcal{R}(A, B)$.
3. $\ker \chi_b(A) \subseteq \mathcal{R}(A, B)$.

Proof. That the last two properties are equivalent follows from Lemma 5.6.1. By Lemma 5.6.3, we must prove that the system is asymptotically controllable if and only if

$$\chi_g(A_3) = 0 \tag{5.31}$$

for some A_3 as in that Lemma. But if (5.31) holds, then Lemma 5.6.2 implies that A_3 is Hurwitz (or convergent, in the discrete-time case), and so the system is indeed asymptotically controllable. Conversely, if the system is asymptotically controllable, then χ_u, the characteristic polynomial of A_3, divides χ_g, from which (5.31) holds because of the Cayley-Hamilton Theorem. ∎

The space $\operatorname{im} \chi_g(A) = \ker \chi_b(A)$, or more precisely the kernel of the corresponding complex mapping, can be thought of as the set of unstable modes of the system, as in the next exercise. Thus, the above result can be restated roughly in the following way:

A linear system is asymptotically controllable if and only if its unstable modes are controllable

which is an intuitive fact: Stable modes decay by themselves, and unstable modes can be controlled to zero.

Exercise 5.6.5 Let v be an eigenvector of A. Prove that v corresponds to an eigenvalue with nonnegative real part if and only if it is in the kernel of $\chi_b(A)$ seen as a complex matrix. □

The following exercise makes use of the concept of (A, B)-invariant subspace introduced in Section 5.4. It deals with the question of deciding, for the time-invariant continuous-time linear system

$$\dot{x} = Ax + Bu \,, \quad y = Cx \,,$$

whether there is any feedback law $u = Fx$ so that

$$C\xi(t) \to 0 \quad \text{as } t \to \infty$$

for every solution of

$$\dot{x} = (A + BF)x \,.$$

In other words, the state may not be stabilized, but the observations should be. When $C = I$, this is just the stabilization problem, and asymptotic controllability is necessary and sufficient. More generally, the result of the exercise can be interpreted as saying that unstable modes must be either controllable or not "observable" by C.

Exercise 5.6.6 ◇ Show that the above problem is solvable if and only if

$$\ker \chi_b(A) \subseteq \mathcal{R}(A, B) + \mathcal{V}^*(A, B, C) \,. \qquad \square$$

5.7 Lyapunov and Control-Lyapunov Functions

For nonlinear systems, the best one often can do in order to establish that a system is asymptotically stable or asymptotically controllable is to use the Lyapunov function method. The intuitive idea behind this method is that, if there exists some sort of "energy" measure of states such that this energy diminishes along suitably chosen paths, then the system can be made to approach a minimal-energy configuration.

We first illustrate the basic ideas of the Lyapunov direct method with the simplest (and most classical) case. Assume that $\mathcal{X}$ is an open subset of $\mathbb{R}^n$, f is a vector field defined on $\mathcal{X}$, and $V : \mathcal{X} \to \mathbb{R}$ is a differentiable function. Consider the new function

$$L_f V : \mathcal{X} \to \mathbb{R} \,, \quad (L_f V)(x) := \nabla V(x) \, f(x)$$

(we drop the parentheses and write simply $L_f V(x)$; also, we often insert a dot and write "$\nabla V(x) \cdot f(x)$" in order to help readability). This is the directional, or Lie, derivative of V in the direction of the vector field f (also considered in Section 4.1). It is the expression that appears when taking the derivative of V along a path of the system with no controls $\dot{x} = f(x)$ and state space $\mathcal{X}$. That is, if ξ is any path of $\dot{x} = f(x)$, then

$$\frac{dV(\xi(t))}{dt} = L_f V(\xi(t)) \,.$$

The Lyapunov stability method is based on the following observation: If there is a function V such that

$$L_f V(x) < 0$$

for all nonzero x, then $V(\xi(t))$ must decrease. If in addition one can guarantee that $V(x) \to 0$ implies $x \to 0$, then this means that $\xi(t) \to 0$. The main point of this is that there is no need to compute explicitly the solutions of the differential equation (hence the term *direct method*). On the other hand, a certain amount of ingenuity and/or physical intuition is needed in order to come up with a suitable "energy" function V.

We will prove the basic results for discrete as well as continuous-time systems, and in fact, with the appropriate definitions, for arbitrary topological systems. Moreover, we give a control-theoretic version of the results. To avoid confusion with the more classical term, we use the terminology "control-Lyapunov function" for systems with controls, and "Lyapunov function" for systems without controls.

Definition 5.7.1 *A* **local control-Lyapunov function for the system** Σ *(relative to the equilibrium state x^0) is a continuous function $V : \mathcal{X} \to \mathbb{R}$ for which there is some neighborhood $\mathcal{O}$ of x^0 such that the following properties hold:*

1. *V is* **proper at** x^0, *that is,*

$$\{x \in \mathcal{X} \mid V(x) \le \varepsilon\}$$

is a compact subset of $\mathcal{O}$ for each $\varepsilon > 0$ small enough.

2. *V is* **positive definite** *on $\mathcal{O}$:*

$$V(x^0) = 0\,, \quad \text{and } V(x) > 0 \text{ for each } x \in \mathcal{O}, x \neq x^0\,.$$

3. *For each $x \neq x^0$ in $\mathcal{O}$ there is some time $\sigma \in \mathcal{T}, \sigma > 0$, and some control $\omega \in \mathcal{U}^{[0,\sigma)}$ admissible for x such that, for the path $\xi = \psi(x,\omega)$ corresponding to this control and this initial state,*

$$V(\xi(t)) \le V(x) \quad \text{for all } t \in [0,\sigma)$$

and

$$V(\xi(\sigma)) < V(x)\,.$$

A **global control-Lyapunov function for** Σ *(relative to x^0) is a continuous V which is (globally) proper, that is, the set*

$$\{x \in \mathcal{X} \mid V(x) \le L\}$$

is compact for each $L > 0$, and such that (2) and (3) are satisfied with $\mathcal{O} = \mathcal{X}$.

For systems without controls, we say simply (local or global) **Lyapunov function**. □

Remark 5.7.2 Properness amounts to the requirement that $V^{-1}(C)$ be compact whenever $C \subseteq \mathbb{R}$ is compact. When $\mathcal{X} = \mathbb{R}^n$, a continuous $V : \mathbb{R}^n \to \mathbb{R}_{\geq 0}$ is proper if and only if $V^{-1}(C)$ is bounded whenever C is bounded, or equivalently,

$$\lim_{\|x\| \to \infty} V(x) = \infty .$$

A V which satisfies this last property, equivalent to properness in the special case $\mathcal{X} = \mathbb{R}^n$, is also said to be *weakly coercive* or *radially unbounded.* □

Local control-Lyapunov functions are often only specified in a neighborhood of x^0. If the state space is well-behaved enough, they can be extended while preserving the desired properties. The next exercise illustrates this.

Exercise 5.7.3 Let $V : \mathcal{O} \to \mathbb{R}$ be a continuous and positive definite function defined on some neighborhood $\mathcal{O}$ of x^0, where $\mathcal{X} = \mathbb{R}^n$. Assume that V satisfies property (3) in Definition 5.7.1. Show that, then, there exists a local control-Lyapunov function W for the same system, with $W = V$ on some neighborhood of x^0. (*Hint:* For $\delta > 0$ small enough, let $W = V$ on the ball $\mathcal{B}_\delta(x^0)$ and $W(x) := V(\delta(x - x^0)/\left\|x - x^0\right\|)$ outside. Verify the definition of properness for ε less than $\inf\{V(x), \left\|x - x^0\right\| = \delta\}$.) □

The main result is as follows:

Theorem 17 *If there exists a local (respectively, global) control-Lyapunov function V for Σ, then Σ is locally (respectively, globally) asymptotically controllable.*

Proof. Assume that V is a local control-Lyapunov function. Choose an $\alpha_0 > 0$ so that $\{x \mid V(x) \leq \alpha_0\}$ is a compact subset of $\mathcal{O}$ (property (1) in Definition 5.7.1). We first remark that, if ξ is a path on $[0, \infty)$ such that

$$V(\xi(t)) \to 0 \quad \text{as} \quad t \to \infty ,$$

then, necessarily,

$$\xi(t) \to x^0 .$$

This fact is, in turn, a consequence of the following observation:

Claim 1: For each open neighborhood $\mathcal{W}$ of x^0 there is some $\beta > 0$ such that the set

$$\{x \mid V(x) \leq \beta\}$$

is included in $\mathcal{W}$. Indeed, if this were not the case, there would exist a sequence $\{x_n\}$ of elements of $\mathcal{X}$ with the properties that $x_n \notin \mathcal{W}$ for all n but

$$V(x_n) \to 0 \quad \text{as} \quad n \to \infty .$$

Without loss of generality, we assume that all x_n are in the set

$$K := \mathcal{W}^c \bigcap \{x \mid V(x) \leq \alpha_0\}$$

(we denote the complement of a set S as S^c), which is itself compact. So $\{x_n\}$ has a convergent subsequence in K,

$$x_{n_k} \to x_0 \,,$$

and by continuity of V, it follows that $V(x_0) = 0$, so by positive definiteness $x_0 = x^0$. But then x^0 is in K, contradicting the fact that it is in $\mathcal{W}$. This proves the claim.

For any $x \in \mathcal{X}$, we say that $z \in \mathcal{X}$ is *nicely reachable from* x if there exist a $\sigma > 0$ and a $\omega \in \mathcal{U}^{[0,\sigma)}$ such that for the trajectory $\xi := \psi(x, \omega)$ it holds that $\xi(\sigma) = z$,

$$V(\xi(t)) \leq V(x) \quad \text{for all } t \in (0, \sigma)\,,$$

and

$$V(\xi(\sigma)) < V(x)\,.$$

Property (3) in the definition of control-Lyapunov function says that for each state $y \in \mathcal{O}, y \neq x^0$ there is *some* state nicely reachable from y. Note also that, if z is nicely reachable from x, and if y is nicely reachable from z, then y is also nicely reachable from x. For each $x \in \mathcal{O}$, we let

$$B(x) := \inf\{V(z) \mid z \text{ is nicely reachable from } x\}\,.$$

By the above remarks, necessarily $B(x) < V(x)$ if $x \neq x^0$.

Claim 2:

$$V(x) < \alpha_0 \;\Rightarrow\; B(x) = 0\,.$$

Indeed, suppose that there is some x (necessarily $\neq x^0$), with $V(x) < \alpha_0$ but

$$B(x) = \alpha > 0\,.$$

Because of the choice of α_0, necessarily $x \in \mathcal{O}$. Let $\{z_n\}$ be a sequence of elements nicely reachable from x such that

$$V(z_n) \to \alpha$$

monotonically as $n \to \infty$. Thus, all the z_n are in the compact set

$$C := \{z \mid V(z) \leq V(x)\}\,,$$

and, without loss of generality, we may assume then that the sequence is convergent,

$$z_n \to z, \;\; V(z) = \alpha\,.$$

Since $\alpha \neq 0$, also $z \neq x^0$. Further, since

$$\alpha < V(x) < \alpha_0\,,$$

also $z \in \mathcal{O}$, so there is some y that is nicely reachable from z. We pick an $\varepsilon > 0$ such that both

$$V(z) < V(x) - \varepsilon \;\text{ and }\; V(y) < V(z) - \varepsilon \tag{5.32}$$

and we let $\nu \in \mathcal{U}^{[0,\sigma)}$ be admissible for z and such that the path $\zeta := \psi(z,\nu)$ is so that $\zeta(\sigma) = y$ and

$$V(\zeta(t)) \leq V(z) \quad \text{for all } t \in (0,\sigma]\,.$$

Since V is uniformly continuous on the compact set C, there is a $\delta > 0$ such that

$$|V(a) - V(b)| < \varepsilon \quad \text{whenever } d(a,b) < \delta \text{ and } a,b \in C\,.$$

Using that Σ is a topological system, we know then that there is some integer N such that ν is admissible for z_N and such that

$$d_\infty(\zeta_N, \zeta) < \delta\,,$$

where ζ_N is the path $\psi(z_N, \nu)$. So from the choice of δ, it follows that

$$|V(\zeta_N(t)) - V(\zeta(t))| < \varepsilon \tag{5.33}$$

for each $t \in [0,\sigma]$. We claim that the state $y_N := \zeta_N(\sigma)$ is nicely reachable from x; this will give a contradiction, because

$$|V(y_N) - V(y)| < \varepsilon$$

together with

$$V(y) < B(x) - \varepsilon$$

imply that $V(y_N) < B(x)$, contradicting the definition of $B(x)$. Concatenating with ν a ω that (nicely) controls x to z_N, it is only necessary to prove that

$$V(\zeta_N(t)) \leq V(x) \quad \text{for all } t\,,$$

but this is a consequence of (5.33) and (5.32). The contradiction is due to the assumption that $\alpha = B(x) > 0$. This establishes Claim 2.

We next claim that, if $V(x) < \alpha_0, x \neq x^0$, then there is a sequence of states

$$\{x_n, n \geq 0\}$$

such that $x_0 = x$, an increasing sequence of times $t_n \in \mathcal{T}_+, t_0 = 0$, and controls

$$\omega_n \in \mathcal{U}^{[t_n, t_{n+1})}$$

such that $t_n \to \infty$ as $n \to \infty$ and so that for each $n \geq 0$ the following properties hold:

1. ω_n is admissible for x_n.

2. $\phi(t_{n+1}, t_n, x_n, \omega_n) = x_{n+1}$.

3. With $\xi_n := \psi(x_n, \omega_n)$, $V(\xi_n(t)) \leq 2^{-n}V(x)$ for all $t \in [t_n, t_{n+1}]$.

Assume for a moment that the claim is proved. Then, we may define a control ω on $[0, \infty)$ as the concatenation of the ω_n's. This control is admissible for x. This is verified by checking that each restriction to an interval $[0, \tau)$ is admissible: Since $t_n \to \infty$, there is some n so that $t_n > \tau$; thus, the restriction to $[0, \tau)$ is also the restriction of a finite concatenation of the ω_n and hence is indeed admissible. Let (ξ, ω) be the corresponding trajectory; since ξ is a concatenation of the ξ_n, it follows that

$$V(\xi(t)) \to 0 \quad \text{as } t \to \infty$$

and therefore also that $\xi(t) \to x^0$.

To prove the claim, it will be sufficient by induction to show that for each $x \in \mathcal{X}$ for which

$$0 < V(x) < \alpha_0$$

there is some $\sigma > 1$, and some control ω of length σ, admissible for x, such that the path $\xi := \psi(x, \omega)$ satisfies

$$V(\xi(t)) \leq V(x) \quad \text{for all } t$$

and also

$$V(\xi(\sigma)) < (1/2)V(x)\,.$$

By continuity of V at x^0, there is some $\varepsilon > 0$ such that

$$V(z) < (1/2)V(x) \quad \text{whenever } d(z, x^0) < \varepsilon\,.$$

Let $\omega^0 \in \mathcal{U}^{[0,1)}$ be a constant control such that

$$\phi(t, 0, x^0, \omega^0) \equiv x^0\,.$$

Since Σ is topological, there is some $\delta > 0$ such that, for each $y \in \mathcal{X}$ with $d(y, x^0) < \delta$, necessarily ω^0 is admissible for y and the path $\zeta := \psi(y, \omega^0)$ satisfies $d(\zeta(t), x^0) < \varepsilon$ for all t, and hence by the choice of ε so that

$$V(\zeta(t)) < (1/2)V(x)$$

for all $t \in [0, 1]$. Because of Claim 1, there is some $\delta_0 > 0$ with the property that

$$V(y) < \delta_0 \Rightarrow d(y, x^0) < \delta\,.$$

Now use the fact that $B(x) = 0$. There is, then, some $\tau > 0$ and some control $\omega_1 \in \mathcal{U}^{[0,\tau)}$ such that the path $\xi_1 := \psi(x, \omega_1)$ satisfies that

$$V(\xi_1(t)) \leq V(x) \quad \text{for all } t \in [0, \tau]$$

and so that $V(\xi_1(\tau)) < \delta_0$; hence, because of the choice of δ, also

$$d(\xi_1(\tau), x^0) < \delta\,.$$

The concatenation of ω_1 and (a translate of) ω^0 is then as desired.

Now we prove that Σ is locally asymptotically controllable. Let $\mathcal{V}$ be any neighborhood of x^0. Choose some $\alpha_1 > 0$ such that

$$\{y \mid V(y) < \alpha_1\} \subseteq \mathcal{V},$$

and let

$$\alpha := \min\{\alpha_0, \alpha_1\}.$$

We take

$$\mathcal{W} := \{y \mid V(y) < \alpha\}$$

in the definition of locally asymptotically controllable. For any $y \in \mathcal{W}, y \neq x^0$, the previous claim applies, because $V(y) < \alpha_0$. Thus, there is some trajectory (ξ, ω) on $[0, \infty)$ such that $\xi(0) = y$,

$$\xi(t) \to x^0 \quad \text{as } t \to \infty,$$

and

$$V(\xi(t)) \leq V(y) < \alpha_1 \quad \text{for all } t.$$

In particular, $\xi(t) \in \mathcal{V}$ for all t.

Assume finally that V is a global control-Lyapunov function, and pick any $y \in \mathcal{X}$. Let $\beta := V(y)$. The only fact about $\mathcal{O}$ and α_0 used in the above proof, besides property (3) in the definition of control-Lyapunov function, was that the set

$$\{y \mid V(y) \leq \alpha_0\}$$

is compact and is contained in $\mathcal{O}$. So here we may take in particular $\mathcal{O} := \mathcal{X}$ and $\alpha_0 := \beta + 1$. Pick, in the previous paragraph, $\mathcal{V} := \mathcal{X}$. Then $\alpha_1 := \beta + 1$ can be used, and so with $\mathcal{W}$ as above we have that $y \in \mathcal{W}$ and hence that y is asymptotically controllable to x^0. ∎

The question of establishing converses to Theorem 17 is a classical one, not studied in this text; see the discussion at the end of the chapter. Property (3) in the definition of control-Lyapunov function (V decreases along appropriate trajectories) cannot be checked, for continuous-time systems, without solving differential equations. As discussed in the introduction to this section, Lyapunov analysis relies on a *direct* criterion for decrease, expressed infinitesimally in terms of directional derivatives of V along control directions, as follows.

Lemma 5.7.4 Let Σ be a continuous-time system and $V : \mathcal{X} \to \mathbb{R}$ a continuous function. Assume that $\mathcal{O} \subseteq \mathcal{X}$ is an open subset for which the restriction of V to $\mathcal{O}$ is continuously differentiable and properties (1) and (2) in the definition of control-Lyapunov function hold. Then, a sufficient condition for V to be a control-Lyapunov function is that for each $x \in \mathcal{O}, x \neq x^0$, there exist some $u \in \mathcal{U}$ such that

$$\nabla V(x).f(x,u) < 0. \tag{5.34}$$

If this property holds with $\mathcal{O} = \mathcal{X}$, V is a global control-Lyapunov function.

Proof. We need only to establish property (3) in the definition. Pick any $x \in \mathcal{O}, x \neq x^0$, and any $u \in \mathcal{U}$ such that equation (5.34) holds. For small $\sigma > 0$, the control on $[0, \sigma]$ with $\omega(t) \equiv u$ is admissible for x. We let (ξ, ω) be the corresponding trajectory. By continuity of ∇V and of $f(\cdot, u)$, we may choose σ small enough so that

$$\frac{dV(\xi(t))}{dt} = \nabla V(\xi(t)).f(x, \omega(t)) < 0$$

for all $t \in [0, \sigma]$. It follows that $V(\xi(t)) < V(x)$ for all $t > 0$. ∎

Observe that, for systems with no controls, condition (5.34) amounts to the requirement that $L_f V(x) < 0$ for all $x \in \mathcal{O}, x \neq x^0$; this is the most classical case studied for differential equations. For continuous-time systems $\dot{x} = f(x)$ with no controls, it is customary to write just "$\dot{V}(x)$" instead of $(L_f V)(x) = \nabla V(x) f(x)$.

Example 5.7.5 Consider the system with $\mathcal{X} = \mathbb{R}^2$, $\mathcal{U} = \mathbb{R}$, and equations

$$\begin{aligned} \dot{x}_1 &= x_2 \\ \dot{x}_2 &= -\sin x_1 + u \end{aligned}$$

where we take all notions with respect to the equilibrium state $x^0 = 0$ (and corresponding input $u^0 = 0$). This corresponds to a model for a pendulum where we do not identify states whose angles differ by 2π. The squared norm

$$V(x) := x_1^2 + x_2^2$$

is a positive definite proper smooth function on $\mathbb{R}^2$, so we test to see if it is a local or global control-Lyapunov function for Σ. Using the criterion in Lemma 5.7.4, we check the sign of

$$\nabla V(x).f(x, u) = 2x_1x_2 - 2x_2 \sin x_1 + 2x_2 u\,.$$

When $x_2 \neq 0$, this expression is indeed negative for an appropriate choice of u. But at any x such that $x_2 = 0$, it vanishes identically. Thus, the criterion cannot be applied. After some trial and error (typical of the use of the Lyapunov function technique), we try the alternative

$$V(x) := 2x_1^2 + x_2^2 + 2x_1x_2\,,$$

which is still smooth, positive definite, and proper (since it is also a positive definite quadratic form). Now however,

$$\nabla V(x).f(x, u) = (4x_1 + 2x_2)x_2 + (2x_1 + 2x_2)(-\sin x_1 + u)\,.$$

When $x_1 + x_2 \neq 0$, again the expression can be made negative by an appropriate choice of u. But the expression is automatically negative for any other nonzero x. Indeed, if $x_1 + x_2 = 0$ then the second term vanishes, but the first one is now $-2x_2^2$, which is negative unless $x_2 = 0$, which can only happen if $x_1 = -x_2 = 0$ too. Thus, the expression is negative except at $x = 0$, and V is a global control-Lyapunov function for Σ. So Σ is globally asymptotically controllable. □

For linear systems, one often considers quadratic Lyapunov functions. As usual, we write $P > 0$ to indicate that a matrix P is symmetric and positive definite, $P \geq 0$ to indicate positive semidefinite, and $P \leq 0, P < 0$ for negative semidefinite and definite respectively. We remind the reader that, unless otherwise stated, asymptotic stability notions and Lyapunov functions for linear systems are always meant with respect to $x^0 = 0$.

Lemma/Exercise 5.7.6 Let Σ be a linear (time-invariant) continuous-time system with no controls, $\dot{x} = Ax$, and let $P > 0$, $P \in \mathbb{R}^{n\times n}$. Prove that the condition

$$A'P + PA < 0$$

is sufficient for $V(x) := x'Px$ to be a Lyapunov function for Σ. □

Lemma/Exercise 5.7.7 Let Σ be a linear (time-invariant) discrete-time system with no controls over $\mathbb{R}$, $x^+ = Ax$, and let $P > 0$, $P \in \mathbb{R}^{n\times n}$. Prove that the condition

$$A'PA - P < 0$$

is sufficient for $V(x) := x'Px$ to be a Lyapunov function for Σ. □

Condition (5.34) in Lemma 5.7.4 is often too restrictive. The next condition is less restrictive, but applies only to systems with no controls. The strict inequality is replaced by a weak inequality, but an extra condition is imposed. The application of Theorem 17 to the type of function described in the next Lemma is a weak version of what is often called the *LaSalle invariance principle.*

Lemma 5.7.8 Assume that Σ is a continuous-time system with no controls, that $V : \mathcal{X} \to \mathbb{R}$ is a continuously differentiable map, and that the open set $\mathcal{O} \subseteq \mathcal{X}$ is such that properties (1) and (2) in the definition of Lyapunov function hold. Then, V is a (local) Lyapunov function if the following properties hold:

1. For each $x \in \mathcal{O}$,
$$\nabla V(x).f(x) \leq 0 \tag{5.35}$$

2. whenever ξ is a trajectory on the infinite interval $\mathcal{I} = [0, \infty)$ for which
$$\nabla V(\xi(t)).f(\xi(t)) \equiv 0 \text{ for all } t\,, \tag{5.36}$$
necessarily $\xi(t) \equiv x^0$.

Further, V is a global Lyapunov function provided that these conditions hold with $\mathcal{O} = \mathcal{X}$ and V is proper.

Proof. We start by picking an $\varepsilon_0 > 0$ so that property (1) in the definition of Lyapunov function holds, that is, $\{x \in \mathcal{X} \mid V(x) \leq \varepsilon_0\}$ is compact and contained in $\mathcal{O}$. Since V is continuous at x^0, there is some $\delta > 0$ such that $V(x) < \varepsilon_0$ for all x in the closed ball $\overline{\mathcal{B}}_\delta(x^0)$. We claim that the definition of Lyapunov function is satisfied with $\widetilde{\mathcal{O}} := \mathcal{B}_\delta(x^0)$.

Observe that, since $\mathcal{B}_\delta(x^0) \subseteq \{x \in \mathcal{X} \mid V(x) \leq \varepsilon_0\} \subseteq \mathcal{O}$, V is positive definite on $\widetilde{\mathcal{O}}$. Furthermore, V is proper, because the compact set $\{x \in \mathcal{X} \mid V(x) \leq \varepsilon\}$ is included in $\widetilde{\mathcal{O}}$ for all $\varepsilon \leq \varepsilon_0$ small enough (proof: otherwise, there is some sequence $\{x_k\}$, contained in the compact set $\{x \in \mathcal{X} \mid V(x) \leq \varepsilon_0\} \setminus \mathcal{B}_\delta(x^0)$, such that $V(x_k) \to 0$; without loss of generality, $x_k \to \hat{x}$, which gives a contradiction since by continuity of V it holds that $V(\hat{x}) = 0$, but $\hat{x} \notin \mathcal{B}_\delta(x^0)$).

We need only to establish property (3). For this, take any $x \in \widetilde{\mathcal{O}}$, $x \neq x^0$, and consider the maximal solution ξ with $\xi(0) = x$, which is defined on some interval of the form $\mathcal{I} = [0, \sigma)$, with $\sigma \leq \infty$.

We first show that $\sigma = \infty$. Condition (5.35) implies that $V(\xi(t))$ is nonincreasing as long as $\xi(t)$ remains in the original set $\mathcal{O}$, for which it is in turn sufficient that it remain in the set

$$S := \{x \mid V(x) \leq \varepsilon_0\}$$

because of the choice of ε_0. Assume that there would be some $T > 0$ in $\mathcal{I}$ so that $V(\xi(T)) \geq \varepsilon_0$. Since $V(\xi(\cdot))$ is continuous, there is a first such T; thus, we may assume that $V(\xi(t)) < \varepsilon_0$ for all $t \in [0, T)$. So $V(\xi(\cdot))$ is nonincreasing on $[0, T]$, which implies that $V(\xi(T)) \leq V(x) < \varepsilon_0$, a contradiction. We conclude that ξ remains in the compact set S for all $t \in \mathcal{I}$, so by Proposition C.3.6 in Appendix C, $\mathcal{I} = [0, \infty)$, as desired.

Thus, the trajectory is defined for all t, and V is nonincreasing. Assume that it were the case that

$$V(\xi(t)) = V(x) \quad \text{for all } \; t\,.$$

It follows that $dV/dt \equiv 0$ identically, and hence condition (5.36) holds there. We conclude that $\xi \equiv x^0$, contradicting the assumption that $\xi(0) = x \neq x^0$. Thus, there is some t so that $V(\xi(t)) < V(x)$, and V is a Lyapunov function as desired. ∎

Exercise 5.7.9 Use Lemma 5.7.8 to establish the following generalization of Lemma 5.7.6: Assume that Σ is as in Lemma 5.7.6, and $P > 0$, $P \in \mathbb{R}^{n \times n}$. Prove that a sufficient condition for $V(x) := x'Px$ to be a Lyapunov function for Σ is that

$$A'P + PA \;\leq\; -C'C\,,$$

(the inequality is understood in the usual sense for symmetric matrices: $Q_1 \leq Q_2$ means that $x'Q_1x \leq x'Q_2x$ for all x) where C is a $p \times n$ matrix (p some integer) for which the pair (A, C) is observable, that is, (A', C') is controllable. (*Hint:* See Corollary 3.5.9 in Section 3.5.) □

Exercise 5.7.10 State and prove a discrete-time analogue of Exercise 5.7.9. □

Example 5.7.11 Consider now a system with no controls corresponding to a damped harmonic oscillator (or a pendulum linearized about $\theta = \dot{\theta} = 0$),

$$\begin{aligned} \dot{x}_1 &= x_2 \\ \dot{x}_2 &= -x_1 - x_2 . \end{aligned}$$

We want to study its asymptotic stability (with respect to $x^0 = 0$) using Lyapunov function techniques. Again trying the squared norm $V(x) := x_1^2 + x_2^2$, this is now a global Lyapunov function. To check this, we apply Lemma 5.7.8. Note that

$$\nabla V(x).f(x) = -2x_2^2 \leq 0$$

for all x. Further,

$$\nabla V(\xi(t)).f(\xi(t)) \equiv 0$$

along a trajectory ξ implies that $\xi_2 \equiv 0$ along the trajectory, so also $\dot{\xi}_2 \equiv 0$ and hence from the second equation $\xi_1 \equiv 0$. Thus, $\xi \equiv 0$, and the Lemma applies.

Alternatively in this simple example, one can compute the eigenvalues of

$$A = \begin{pmatrix} 0 & 1 \\ -1 & -1 \end{pmatrix}$$

which have real part $-\frac{1}{2}$, so A is a Hurwitz matrix. □

Example 5.7.12 Consider the double integrator system ($\mathcal{X} = \mathbb{R}^2$ and $\mathcal{U} = \mathbb{R}$):

$$\begin{aligned} \dot{x}_1 &= x_2 \\ \dot{x}_2 &= u . \end{aligned}$$

The linear feedback law $u = -x_1 - x_2$ globally stabilizes this system, cf. Example 5.7.11. We show now how to globally stabilize this system with a simple feedback law which is *globally bounded in magnitude.* Let $\theta : \mathbb{R} \to \mathbb{R}$ be any locally Lipschitz and strictly increasing continuous function such that $\theta(0) = 0$ (for instance, we may pick $\theta(s) := \tanh s$, which is bounded). We show that the feedback law

$$u := -\theta(x_1 + x_2)$$

globally asymptotically stabilizes the given system. In order to construct a Lyapunov function, let $\Theta(s) := \int_0^s \theta(t)\, dt$. This function is clearly positive definite and proper; we claim that

$$V(x_1, x_2) := \Theta(x_1) + \Theta(x_1 + x_2) + x_2^2$$

is a Lyapunov function for the resulting closed-loop system. Indeed, a simple calculation shows that

$$\nabla V(x) \cdot f(x) = [\theta(x_1) - \theta(x_1 + x_2)]\, x_2 - [\theta(x_1 + x_2)]^2 .$$

Since θ is monotonic, the term $[\theta(x_1) - \theta(x_1 + x_2)]\, x_2$ is always nonpositive, and it is strictly negative unless $x_2 = 0$. If $x_2 = 0$, then the second term is negative, unless also $x_1 = 0$. Thus, $\nabla V(x) f(x) < 0$ for all $x \neq 0$, as required. □

Exercise 5.7.13 For the system in Example 5.7.11, find a quadratic global Lyapunov function V that satisfies the strict inequality condition in Lemma 5.7.4, that is, $\nabla V(x).f(x) < 0$ for all $x \neq 0$. □

Exercise 5.7.14 Modify Example 5.7.11 to deal with the full nonlinear model of a damped pendulum with no controls,

$$\begin{aligned}\dot{x}_1 &= x_2\\ \dot{x}_2 &= -\sin x_1 - x_2\,.\end{aligned}$$

Now all the states with $x_1 = k\pi$ and $x_2 = 0$ are equilibria, so for $x^0 = 0$ there is at most local asymptotic stability. From physical considerations, a reasonable Lyapunov function to try is

$$V(x) = (1 - \cos x_1) + (1/2)x_2^2$$

(sum of potential and kinetic energy), say for $x_1 \in [-\pi/2, \pi/2]$, and $1 + (1/2)x_2^2$ for all other x_1. Show that this is a local Lyapunov function. □

Exercise 5.7.15 Consider any system with $\mathcal{X} = \mathbb{R}^2$, $\mathcal{U} = \mathbb{R}$, and equations

$$\begin{aligned}\dot{x}_1 &= f_1(x_1, x_2) + u x_1^r\\ \dot{x}_2 &= f_2(x_1, x_2) + u x_2^r\end{aligned}$$

where f_1 and f_2 are arbitrary and $r \in \mathbb{Z}_+$. Then:

1. Show, using a control-Lyapunov function, that if r is odd then this system is globally asymptotically controllable (with respect to the origin).
2. Give examples to show that, when r is even, the system may or may not be locally asymptotically controllable. □

Exercise 5.7.16 Consider a continuous-time system $\dot{x} = f(x)$ with no controls and $\mathcal{X} = \mathbb{R}^n$. Suppose that $V : \mathbb{R}^n \to \mathbb{R}$ is proper and positive definite, and satisfies $\dot{V}(x) = L_f V(x) < 0$ for all $x \neq 0$ (this is the Lyapunov condition in Lemma 5.7.4). Show that there exists a continuous function $\alpha : [0, \infty) \to [0, \infty)$ which is positive definite (that is, $\alpha(0) = 0$ and $\alpha(r) > 0$ for all $r > 0$) such that the following differential inequality holds:

$$\nabla V(x) \cdot f(x) \;=\; \dot{V}(x) \;\leq\; -\alpha(V(x)) \quad \text{for all } x \in \mathbb{R}^n\,.$$

(*Hint:* Study the maximum of $L_f V(x)$ on the set where $V(x) = r$.) □

Exercise 5.7.17 ◇ Suppose that V is as in Exercise 5.7.16. Show that there is a proper and positive definite $W : \mathbb{R}^n \to \mathbb{R}$ so that

$$\nabla W(x) \cdot f(x) \;=\; \dot{W}(x) \;\leq\; -W(x) \quad \text{for all } x \in \mathbb{R}^n\,,$$

that is, provided we modify V, we can choose $\alpha =$ identity. (*Hint:* Try $W = \rho \circ V$, for a suitably constructed $\rho : [0, \infty) \to [0, \infty)$.) □

Linear Systems

For linear systems, Lyapunov functions can be obtained constructively; we now show that *for any asymptotically controllable linear system there is always a global quadratic control-Lyapunov function.* The material in Appendix C.5 on Stability of Linear Equations should be reviewed at this point, as well as Proposition A.4.6 and Corollary A.4.7 in Section A.4 of the Linear Algebra Appendix.

Since a continuous-time system $\Sigma = (A, B)$ is asymptotically controllable if and only if there is an F such that $\dot{x} = (A + BF)x$ is asymptotically stable (Lemma 5.5.6), and similarly for discrete-time systems and convergent matrices, it is enough to prove the claim for systems without controls.

We need first a very useful matrix Lemma. A matrix equation of the type $MX + XN = Q$ is often called a *Sylvester* equation.

Lemma 5.7.18 Let M and N be two fixed $n \times n$ matrices over $\mathbb{K} = \mathbb{R}$ or $\mathbb{C}$, and consider the linear operator

$$\mathcal{L} : \mathbb{K}^{n\times n} \to \mathbb{K}^{n\times n}, \quad \mathcal{L}(X) := MX + XN\,.$$

If both M and N are Hurwitz, then $\mathcal{L}$ is invertible.

Proof. There are two proofs one can give: one algebraic and the other analytic. The algebraic proof uses Corollary A.4.7 in Appendix A.4. This Corollary can be applied with $A = M$, $B = -N'$; since N and M are Hurwitz, M and $-N'$ have no common eigenvalues. There is also a purely analytic proof, which results in an explicit formula for the solution, as follows.

It will suffice to prove that $\mathcal{L}$ is onto. Thus, we need to show that given any matrix Q there exists some X so that

$$MX + XN = Q\,.$$

Since M and N are Hurwitz,

$$\left\|e^{tM}Qe^{tN}\right\| \leq c\,\|Q\|\,e^{2\lambda t}$$

for all $t \geq 0$, where $\lambda < 0$ is any real number such that all eigenvalues of M and N have real parts less than λ. Therefore,

$$P := -\int_0^\infty e^{tM}Qe^{tN}\,dt$$

is well-defined. Further,

$$\begin{aligned} MP + PN &= -\int_0^\infty [Me^{tM}Qe^{tN} + e^{tM}Qe^{tN}N]\,dt \\ &= -\int_0^\infty \frac{d(e^{tM}Qe^{tN})}{dt}\,dt \\ &= Q - \lim_{t\to\infty} e^{tM}Qe^{tN} \;=\; Q\,, \end{aligned}$$

as desired. ∎

Theorem 18 *Let $A \in \mathbb{R}^{n\times n}$. The following statements are equivalent:*

1. *A is Hurwitz.*
2. *For each $Q \in \mathbb{R}^{n\times n}$, there is a unique solution P of the Lyapunov matrix equation*
$$A'P + PA = Q\,,$$
and if $Q < 0$ then $P > 0$.
3. *There is some $P > 0$ such that $A'P + PA < 0$.*
4. *There is some $P > 0$ such that $V(x) := x'Px$ is a Lyapunov function for the system $\dot{x} = Ax$.*

Proof. Note that (2) trivially implies (3), and that (3) implies (4) (use Lemma 5.7.4, or refer to Exercise 5.7.6). Also, (4) implies (1) by Theorem 17 (p. 220). (An alternative and direct proof that (3) implies (1) is also given in the proof of Lemma 8.4.7 in Chapter 8.)

So it is only left to show that (1) implies (2). Pick any Q. Since A is Hurwitz, A' also is, and thus Lemma 5.7.18 shows that there is for each Q a unique solution P. The formula for P,

$$P := -\int_0^\infty e^{tA'}Qe^{tA}\,dt\,,$$

shows that $P > 0$ if $Q < 0$, since the latter implies that $-e^{tA'}Qe^{tA}$ is a positive definite matrix for each t. ∎

Observe that, for example, when $Q := -I$ the proof gives the Lyapunov function

$$V(x) = \int_0^\infty \left\|e^{tA}x\right\|^2\,dt\,.$$

Lemma/Exercise 5.7.19 Let $A \in \mathbb{R}^{n\times n}$. The following statements are then equivalent:

1. A is convergent (discrete-time Hurwitz).
2. For each $Q \in \mathbb{R}^{n\times n}$, there is a unique solution P of the discrete Lyapunov matrix equation
$$A'PA - P = Q\,,$$
and if $Q < 0$ then $P > 0$.
3. There is some $P > 0$ such that $A'PA - P < 0$.
4. There is some $P > 0$ such that $V(x) := x'Px$ is a Lyapunov function for the system $x^+ = Ax$. □

Remark 5.7.20 As pointed out in Appendix C.5, there are direct algebraic tests for determining whether a matrix is Hurwitz, *without computing eigenvalues.* The following is a simple example of one such test; it is more complex than those used in practice, but it is relatively trivial to establish. Given an $n \times n$ real matrix A, set up the equation

$$A'X + XA = -I\,,$$

and solve for symmetric X. This is a linear equation, so Gaussian elimination can be used. If no solution exists, then A is already known not to be Hurwitz. If a solution X is found, we check whether it is positive definite, for instance by evaluating all the principal minors of X. If $X > 0$, then A is Hurwitz, by part (3) of Theorem 18 (p. 231). If not, either the solution was not unique or the unique solution is not positive definite, so by part (2) of the Theorem we know that A is not Hurwitz.

There is a more abstract way to prove the existence of an algebraic test for stability. Since the Hurwitz property can be expressed in first-order logic, one also knows from logical arguments based on the Tarski-Seidenberg Theorem for real-closed fields that Hurwitz matrices are precisely those whose entries satisfy *some* expression in terms of polynomial equalities and inequalities. □

Even though we have not treated stability of time-varying systems, it is worth pointing out through an example some of the difficulties that may arise in the more general theory. Consider a differential equation

$$\dot{x}(t) = A(t)x(t)\,.$$

It is a common mistake to assume that if $A(t_0)$ is Hurwitz for each t_0 then the differential equation must have some sort of stability property. The following exercise shows that this need not be the case.

Exercise 5.7.21 For any fixed $1 < a < 2$, consider the following matrix of functions of time:

$$A(t) = \begin{pmatrix} -1 + a\cos^2 t & 1 - a\sin t\cos t \\ -1 - a\sin t\cos t & -1 + a\sin^2 t \end{pmatrix}.$$

Prove that $A(t_0)$ is Hurwitz for each fixed $t_0 \in \mathbb{R}$, but that

$$\xi(t) = e^{(a-1)t}\begin{pmatrix} \cos t \\ -\sin t \end{pmatrix}$$

satisfies $\dot{x}(t) = A(t)x(t)$ but does not converge to the origin. □

Theorem 17 (p. 220) and Exercise 5.7.9 can be used to construct an explicit feedback matrix F as needed for part 4 of Proposition 5.5.6.:

Exercise 5.7.22 Suppose that the pair (A, B) is controllable. Pick any $T > 0$ and consider the controllability Gramian (cf. Exercise 3.5.5)

$$W := \int_0^T e^{-sA} BB' e^{-sA'}\, ds\,.$$

Let $F := -B'W^{-1}$. Prove that $A + BF$ is a Hurwitz matrix, i.e., the feedback

$$u = -B'W^{-1}x$$

stabilizes $\dot{x} = Ax+Bu$. (*Hint:* Prove, equivalently, that the transpose $(A+BF)'$ is Hurwitz, by an application of Exercise 5.7.9, taking "P" there as W, "A" as $(A+BF)'$, and "C" as B'. You will need to use that $(A+BF, B)$ is controllable, and you will need to evaluate the integral appearing in the expression for $AW + WA'$.) □

5.8 Linearization Principle for Stability

Consider as an illustration the following discrete-time system with no controls and with $\mathcal{X} = \mathbb{R}$:

$$x(t+1) = \frac{1}{2}x(t) + x^2(t)\,.$$

The linearization at $x^0 = 0$ is the system

$$x(t+1) = \frac{1}{2}x(t)\,,$$

which is asymptotically stable. We may expect that the original system is itself (at least locally) asymptotically stable with respect to $x^0 = 0$. Indeed, the time-one mapping

$$\phi(t+1, t, x) = \mathcal{P}(x) = \frac{1}{2}x + x^2$$

is a contraction on some interval $[-\varepsilon, \varepsilon]$, because $\mathcal{P}'(0) = 1/2 < 1$. For instance, we may take $\varepsilon = 1/8$. On this interval, $x = 0$ is the only fixed point of $\mathcal{P}$. By the Contraction Mapping Theorem, it then follows that

$$\mathcal{P}^n(x) \to 0$$

for each initial

$$x \in [-1/8, 1/8]\,.$$

From this we may conclude that Σ is locally asymptotically stable. Note by the way that Σ is not *globally* asymptotically stable, since $x = 1/2$ is an equilibrium point.

We shall leave the general discrete-time case as an exercise and prove the continuous-time version of the above observation.

Theorem 19 *Assume that Σ is a time-invariant continuous-time system of class C^1,*

$$\dot{x} = f(x,u)$$

with $\mathcal{X} \subseteq \mathbb{R}^n$ and $\mathcal{U} \subseteq \mathbb{R}^m$ open, and let (x^0, u^0) be an equilibrium pair, i.e.,

$$f(x^0, u^0) = 0 .$$

Assume that the linearization of Σ at (x^0, u^0) is asymptotically controllable. Then Σ is locally asymptotically controllable (to x^0). Moreover, there exists in that case a matrix $F \in \mathbb{R}^{m\times n}$ such that the closed-loop system

$$\dot{x} = f_{cl}(x) := f(x, u^0 + F(x - x^0)) \tag{5.37}$$

is locally asymptotically stable.

Proof. It will be enough to prove the last statement, since this implies that Σ is locally asymptotically controllable. Let (A, B) be the linearization at (x^0, u^0). By Proposition 5.5.6, there is some F such that

$$A_{cl} := A + BF$$

is Hurwitz. Pick the (unique) $P > 0$ such that

$$A_{cl}'P + PA_{cl} = -I .$$

Let Σ_{cl} be the system with no controls having the equation (5.37).

We claim that the linearization of Σ_{cl} at $x = x^0$ is the linear system with matrix A_{cl}. To establish this fact, write first

$$f(x,u) = A(x - x^0) + B(u - u^0) + g(x - x^0, u - u^0) ,$$

where

$$\lim_{\|(\alpha,\beta)\|\to 0} \frac{\|g(\alpha,\beta)\|}{\|(\alpha,\beta)\|} = 0 .$$

So it holds that

$$f_{cl}(x) = A_{cl}(x - x^0) + \gamma(x - x^0) ,$$

where

$$\gamma(x - x^0) = g(x - x^0, F(x - x^0))$$

is of order $o(x - x^0)$, as wanted.

Consider the function

$$V(x) := (x - x^0)'P(x - x^0)$$

on $\mathbb{R}^n$. For small enough $\varepsilon > 0$, the sets

$$\{x \in \mathbb{R}^n \mid V(x) \le \varepsilon\}$$

are included in $\mathcal{X}$, so they are compact subsets of $\mathcal{X}$. We take the restriction of V to $\mathcal{X}$; thus (1) and (2) in the definition of Lyapunov function are satisfied for any open $\mathcal{O} \subseteq \mathcal{X}$. We now find an $\mathcal{O}$ so that the criterion in Lemma 5.7.4 is satisfied. Note that

$$\nabla V(x).f_{cl}(x) = -\left\|x - x^0\right\|^2 + 2(x - x^0)'P\gamma(x - x^0).$$

Since

$$\frac{\left\|2(x - x^0)'P\gamma(x - x^0)\right\|}{\left\|x - x^0\right\|^2} \leq c\frac{\left\|\gamma(x - x^0)\right\|}{\left\|x - x^0\right\|} \to 0$$

as $\left\|x - x^0\right\| \to 0$, it follows that $\nabla V(x).f_{cl}(x) < 0$ whenever $\left\|x - x^0\right\|$ is small enough. This determines an open neighborhood of x^0 as desired. ∎

Lemma/Exercise 5.8.1 Assume that Σ is a time-invariant discrete-time system of class $\mathcal{C}^1$,

$$x^+ = \mathcal{P}(x, u),$$

with $\mathcal{X} \subseteq \mathbb{R}^n$ and $\mathcal{U} \subseteq \mathbb{R}^m$ open, and let (x^0, u^0) be an equilibrium pair, i.e.

$$\mathcal{P}(x^0, u^0) = x^0.$$

Assume that the linearization of Σ at (x^0, u^0) is asymptotically controllable. Then Σ is locally asymptotically controllable (to x^0). Moreover, there exists in that case a matrix $F \in \mathbb{R}^{m \times n}$ such that the closed-loop system

$$x^+ = \mathcal{P}_{cl}(x) := \mathcal{P}(x, u^0 + F(x - x^0)) \tag{5.38}$$

is locally asymptotically stable. □

Exercise 5.8.2 ◇ Our proof of the above Theorem is based on Lyapunov stability ideas. An alternative proof could be based on a contraction argument as in the discussion at the beginning of this section, using that for linear systems with no controls the matrix A is Hurwitz (convergent in the discrete-time case) if and only if the map

$$x \mapsto e^{tA}x$$

(or, $x \mapsto A^t x$ in discrete-time) is a contraction for $t > 0$ large enough. Carry out such a proof in detail. □

Exercise 5.8.3 Use Exercise 5.7.11 and Theorem 19 (p. 234) to show that the nonlinear damped pendulum treated in Exercise 5.7.14 is locally asymptotically stable. □

Exercise 5.8.4 Find explicitly a matrix $F \in \mathbb{R}^{1 \times 2}$ as concluded in Theorem 19 (p. 234), for the system ($\mathcal{X} = \mathbb{R}^2, \mathcal{U} = \mathbb{R}$, $x^0 = 0$)

$$\begin{aligned} \dot{x}_1 &= x_1 x_2 + x_2 \\ \dot{x}_2 &= u. \end{aligned}$$

Show that Σ is not *globally* asymptotically controllable. □

First-Order Necessary Conditions for Stabilization

Consider a continuous-time time-invariant system

$$\dot{x} = f(x,u) \tag{5.39}$$

of class $\mathcal{C}^1$ and an equilibrium pair (x^0, u^0). We say that this system is (locally) $\mathcal{C}^1$ *stabilizable* (respectively, *smoothly stabilizable*) with respect to this equilibrium pair if there exists a function

$$k : \mathcal{X}_0 \to \mathcal{U}, \quad k(x^0) = u^0,$$

of class $\mathcal{C}^1$ (respectively, of class $\mathcal{C}^\infty$) defined on some neighborhood $\mathcal{X}_0$ of x^0 for which the closed-loop system (with state-space $\mathcal{X}_0$)

$$\dot{x} = f(x, k(x)) \tag{5.40}$$

is locally asymptotically stable. If $\mathcal{X}_0 = \mathcal{X}$ and (5.40) is globally asymptotically stable, we say that the system (5.39) is *globally* $\mathcal{C}^1$ (or $\mathcal{C}^\infty$) stabilizable.

Theorem 19 showed that local smooth stabilizability holds provided that the linearization about the equilibrium pair is itself asymptotically controllable. This condition is not necessary, as illustrated by

$$\dot{x} = u^3,$$

which has the linearization $A = B = 0$ but which can be smoothly stabilized by $k(x) := -x$. However, there is a partial converse to the fact that stabilizability of the linearization implies local stabilizability. In order to present it, we first need a result which complements Theorem 18 (p. 231) on Lyapunov equations.

Proposition 5.8.5 Assume that $A \in \mathbb{R}^{n\times n}$ has at least one eigenvalue with positive real part. Then there exists a symmetric $n \times n$ matrix P, a constant $c > 0$, and a positive definite matrix Q such that the following two properties hold:

1. $A'P + PA = cP + Q$ and
2. there is some x so that $x'Px > 0$.

Proof. Assume that A has eigenvalues with positive real part. Let

$$c := \min\{a \mid a > 0 \text{ and } a + ib \text{ is an eigenvalue of } A\}.$$

Consider the matrix $C := A - \frac{c}{2}I$; this has no purely imaginary eigenvalues but still has some eigenvalues with positive real part. Decomposing into its stable and unstable parts, there exists an invertible matrix T such that

$$TCT^{-1} = D = \begin{pmatrix} D_1 & 0 \\ 0 & D_2 \end{pmatrix}$$

and both D_1 and $-D_2$ are Hurwitz matrices. We will obtain a symmetric matrix P so that there is some x so that $x'Px > 0$ and so that

$$D'P + PD = I\,. \tag{5.41}$$

It will then follow that

$$A'\widetilde{P} + \widetilde{P}A = c\widetilde{P} + Q\,,$$

with the choices $\widetilde{P} = T'PT$ and $Q = T'T$. Both $\widetilde{P}$ and Q are symmetric, and because T is invertible, it holds that $Q > 0$ and there is some x so that $x'\widetilde{P}x > 0$. Thus, it is only necessary to find P as above.

Since D_1 is Hurwitz, there exists a symmetric (positive definite) P_1 of appropriate size so that $D_1'P_1 + P_1D_1 = -I$, by Theorem 18 (p. 231). Equivalently,

$$D_1'(-P_1) + (-P_1)D_1 = I\,.$$

Similarly, since $-D_2$ is Hurwitz, there is a $P_2 > 0$ so that $(-D_2)'P_2 + P_2(-D_2) = -I$, or equivalently

$$D_2'P_2 + P_2D_2 = I\,.$$

With

$$P := \begin{pmatrix} -P_1 & 0 \\ 0 & P_2 \end{pmatrix}$$

we conclude that (5.41) holds, and also $(0\ x)P\begin{pmatrix} 0 \\ x \end{pmatrix} > 0$ for all $x \neq 0$. ∎

We wish to prove that certain systems are not locally (and hence not globally) asymptotically stable. In order to establish this, we prove that the systems in question are not even "stable" in the sense of Lyapunov. We have not discussed (non-asymptotic) stability, so we define directly the opposite concept:

Definition 5.8.6 *Let Σ be a time-invariant system with no controls, and let x^0 be an equilibrium state. The system Σ is* **unstable** *if there exists some $\varepsilon > 0$ so that, for each $\delta > 0$, there is some state x in the ball $\mathcal{B}_\delta(x^0)$ of radius δ about x^0 and there is some $T \in \mathcal{T}_+$ so that $\phi(T,x)$ is defined and $\phi(T,x) \notin \mathcal{B}_\varepsilon(x^0)$.* □

Clearly, an unstable system cannot be asymptotically stable.

Theorem 20 *Let $(\Sigma) : \dot{x} = f(x)$ be a continuous-time time-invariant system with no controls, and let $x^0 \in \mathcal{X}$ be an equilibrium state. Let $V : \mathcal{B} \to \mathbb{R}$ be a C^1 function defined on the ball $\mathcal{B} = \mathcal{B}_\varepsilon(x^0)$, and assume that there exists some constant $c > 0$ such that*

$$W(x) := L_fV(x) - cV(x) \geq 0 \quad \text{for all } x \in \mathcal{B}$$

and so that

$$x^0 \in \operatorname{clos}\{x \mid V(x) > 0\}\,. \tag{5.42}$$

Then the system Σ is unstable.

Proof. We use the same ε for the definition of instability. Pick any $\delta > 0$, assumed without loss of generality to be less than ε. By the second assumption, there is some $x \in \mathcal{B}_\delta(x^0)$ for which $V(x) > 0$. Picking a smaller ε if needed, we may assume that V is bounded by β on $\mathcal{B}$. We need to show that it cannot happen that $\xi = \psi(x)$ is defined and remains in $\mathcal{B}$ for all $t > 0$. Assume then that this is the case.

Consider the function $\alpha(t) := e^{-ct}V(\xi(t))$. This is differentiable, and its derivative satisfies

$$\dot{\alpha}(t) = e^{-ct}W(\xi(t)) \geq 0 \quad \text{for all } t > 0\,.$$

Thus, $e^{-ct}\beta \geq \alpha(t) \geq \alpha(0) = V(x)$ for all t, with $c > 0$, contradicting the fact that $V(x) > 0$. ∎

Corollary 5.8.7 Let $(\Sigma) : \dot{x} = f(x)$ be a continuous-time time-invariant system with no controls, and let $x^0 \in \mathcal{X}$ be an equilibrium state. Let A be the linearization matrix at x^0, that is,

$$\dot{x} = A(x - x^0) + g(x - x^0)\,,$$

where $g(x)$ is a continuous function that has order $o(x)$ as $x \to 0$. If A has any eigenvalues with positive real part, then Σ is unstable.

Proof. Let P, Q, c be given as in Proposition 5.8.5, and let $V(x) := (x - x^0)'P(x - x^0)$ in Theorem 20, restricted to $\mathcal{X}$. Note that, if $y'Py > 0$ then $V(x^0 + \lambda y) > 0$ for all $\lambda \neq 0$, so (5.42) holds.

With the same c,

$$W(x) = L_f V(x) - cV(x) = (x - x^0)'Q(x - x^0) + 2(x - x^0)'Pg(x - x^0)\,.$$

If we prove that $W(x) > 0$ for all small enough x, the result will follow. For this, note that

$$\frac{x'Qx}{\|x\|^2} \geq \sigma \quad \text{for each } x \neq 0\,,$$

where σ is the smallest eigenvalue of Q, while

$$\frac{|2x'Pg(x)|}{\|x\|^2} \leq 2\,\|P\|\,\frac{\|g(x)\|}{\|x\|}\,,$$

which is smaller than σ for all small x. ∎

Corollary 5.8.8 Let Σ be a continuous-time time-invariant system of class $\mathcal{C}^1$. Let (x^0, u^0) be an equilibrium pair for Σ, and let (A, B) be the linearization at this pair. Assume that (A, B) has an uncontrollable eigenvalue with positive real part, that is, that χ_u (Definition 3.3.5) has a root with real part > 0. Then the system Σ is not locally $\mathcal{C}^1$ stabilizable.

Proof. Assume that the system would be C^1 stabilizable by means of the feedback law $k : \mathcal{X} \to \mathcal{U}$, and write

$$k(x) = F(x - x^0) + g_2(x - x^0),$$

where $g_2(x) = o(x)$ as $x \to 0$. Note that $A + BF$ still has an eigenvalue with positive real part, by (the straightforward part of) Theorem 13 (p. 186). If we also expand

$$f(x, u) = A(x - x^0) + B(x - x^0) + g_1(x - x^0)$$

with $g_1(x) = o(x)$, we may conclude that the closed-loop equations are

$$\dot{x} = (A + BF)(x - x^0) + g(x - x^0),$$

where $g(x) := Bg_2(x) + g_1(x, k(x))$ is again of order $o(x)$. But the closed-loop system cannot be stable, by Corollary 5.8.7. ∎

Since the zeros of χ_u are the right-hand plane numbers for which the rank of $[sI - A, B]$ drops (by Lemma 3.3.10), we may summarize the discussion as follows, for time-invariant continuous-time systems Σ and their linearizations:

$$\boxed{\text{rank}\,[sI - A, B] = n \text{ for all } \text{Re}\, s \geq 0 \;\Rightarrow\; \Sigma \text{ locally } C^1 \text{ stabilizable}}$$

$$\boxed{\text{rank}\,[sI - A, B] < n \text{ for some } \text{Re}\, s > 0 \;\Rightarrow\; \Sigma \text{ not locally } C^1 \text{ stabilizable}}$$

The critical case, when the only uncontrollable unstable eigenvalues are purely imaginary, gives rise to the more interesting theory, since the linearization does not characterize stabilizability.

It is important to emphasize that the above results refer to C^1 stabilizability. There may exist, even in the second case, a feedback that is smooth everywhere but at x^0, and even continuous everywhere, which gives an asymptotically stable system. An example of this is provided by the system

$$\dot{x} = x + u^3,$$

($x^0 = 0$ and $u^0 = 0$) for which the feedback $u := -2x^{1/3}$ stabilizes. Proposition 5.9.10 below deals with stabilization by $k(x)$ smooth away from the origin.

5.9 Introduction to Nonlinear Stabilization*

The previous section discussed first-order conditions for stabilization. One of the conclusions was that, if the linearization about the equilibrium being considered has any uncontrollable eigenvalues with positive real part, then C^1 stabilizability is impossible. Thus, from the point of view of the theory of local C^1 stabilization, the "critical case," in which there are some purely imaginary uncontrollable eigenvalues, is the one to be studied further.

---

* This section can be skipped with no loss of continuity.

In this section, we discuss several further issues regarding nonlinear stabilization. We start with three different approaches to building feedback. All are Lyapunov-function based, and the resulting techniques have found wide applicability, both for local (in critical cases) and global stabilization. The first two, "damping control" and "backstepping", allow the construction of feedback that is smooth at the origin, while the third one, the use of "universal formulas" for stabilization, typically guarantees smoothness only away from the equilibrium of interest.

We close the section with "Brockett's necessary condition" for nonlinear stabilization.

Damping Control

The first method starts with a V which insures boundedness of trajectories (typically, V is an energy function associated to a conservative system), and adds a "friction" term implemented through a feedback law.

Proposition 5.9.1 We consider smooth control-affine systems

$$\dot{x} = f(x) + \sum_{i=1}^{m} u_i g_i(x)$$

as in Equation (2.34) (but writing here f instead of g_0) with $\mathcal{X} = \mathbb{R}^n$, $\mathcal{U} = \mathbb{R}^m$, and $f(0) = 0$. Suppose that V is smooth, positive definite, and (globally) proper, and satisfies

$$L_f V(x) = \nabla V(x) \cdot f(x) \leq 0 \text{ for all } x \in \mathcal{X}. \tag{5.43}$$

Consider the following set:

$$\mathcal{Q} := \{x \mid L_f V(x) = 0 \text{ and } L_f^k L_{g_i} V(x) = 0 \text{ for all } k = 0, 1, 2, \ldots, \; i = 1, \ldots, m\}$$

and suppose that $\mathcal{Q} = \{0\}$. Then, the following smooth feedback law:

$$u = k(x) := -(\nabla V(x) \cdot G(x))' = -(L_{g_1} V(x), \ldots, L_{g_m} V(x))' \tag{5.44}$$

globally asymptotically stabilizes the system (with respect to $x^0 = 0$, $u^0 = 0$).

Proof. This is a simple application of the LaSalle invariance principle, cf. Lemma 5.7.8. We consider the closed-loop system

$$\dot{x} = F(x) = f(x) - \sum_{i=1}^{m} L_{g_i} V(x)\, g_i(x)$$

and note that

$$\nabla V(x).F(x) = L_f V(x) - \sum_{i=1}^{m} (L_{g_i} V(x))^2$$

is nonpositive for all x, by assumption. Take now any trajectory ξ for which $\nabla V(\xi(t)).F(\xi(t)) \equiv 0$; we need to show that $\xi \equiv 0$. Along such a trajectory, it must be the case that $L_f V(\xi(t)) \equiv 0$ and also $L_{g_i} V(\xi(t)) \equiv 0$ for every i. For any given i, also the time derivatives of $L_{g_i} V(\xi(t))$ must vanish identically. The first derivative of this function is $L_f L_{g_i} V(\xi(t))$ (chain rule), and, inductively, the kth derivative is $L_f^k L_{g_i} V(\xi(t))$. It follows that $\xi(t) \in \mathcal{Q}$ for all t, so indeed $\xi \equiv 0$. ∎

Exercise 5.9.2 This problem is a variation of Proposition 5.9.1. Suppose given a smooth control-affine system, and V as there, so that (5.43) holds, and consider, for each $x \in \mathcal{X}$, the following vector subspace of $\mathbb{R}^n$

$$\Delta(x) := \text{span}\left\{f(x),\ \text{ad}_f^k g_i(x),\ i = 1,\ldots,m,\ k = 0,1,2,\ldots\right\}$$

(in the differential-geometric terminology of Chapter 4, Δ defines a "distribution"). Assume that $\nabla V(x) = 0$ implies $x = 0$ and that

$$\dim \Delta(x) = n \quad \text{for all } x \neq 0\,.$$

Show that the feedback law (5.44) stabilizes the system. (*Hint:* Prove by induction on k that, if ξ is a solution of $\dot{x} = f(x)$ so that $L_f V(\xi(t)) \equiv 0$ and $L_{g_i} V(\xi(t)) \equiv 0$ for all i, then also $L_{\text{ad}_f^k g_i} V(\xi(t)) \equiv 0$, for all k and all i. You will need to use the facts that $L_{\text{ad}_f^{k+1} g_i}$ can be expressed in terms of $L_f L_{\text{ad}_f^k g_i}$ and $L_{\text{ad}_f^k g_i} L_f$, and that $\nabla (L_f V)$ vanishes on the set $\{x \mid L_f V(x) = 0\}$ (why?).) □

Exercise 5.9.3 Give a feedback law $u = k(x)$, $k(0) = 0$, that stabilizes the bilinear system

$$\dot{x} = \left[\begin{pmatrix} 0 & 1 \\ -1 & 0 \end{pmatrix} + u \begin{pmatrix} 0 & 1 \\ 1 & 0 \end{pmatrix}\right] x\,.$$

(Use the above results.) □

Exercise 5.9.4 ◇ As in Exercise 4.3.15, consider the problem of controlling the angular momentum of a rigid body by the use of one or more applied torques. For concreteness, we assume that three torques are being applied. Thus (cf. Exercise 4.3.15), we have a system with $\mathcal{X} = \mathbb{R}^3$, $\mathcal{U} = \mathbb{R}^3$, and equations given by:

$$I\dot{x} = S(x)Ix + Tu\,, \tag{5.45}$$

where here I denotes the diagonal matrix with entries I_1, I_2, I_3, $T \in \mathbb{R}^{3\times 3}$, and

$$S(x) = \begin{pmatrix} 0 & x_3 & -x_2 \\ -x_3 & 0 & x_1 \\ x_2 & -x_1 & 0 \end{pmatrix}.$$

It is assumed that $I_i \neq I_j$ whenever $i \neq j$. We may also write this system in the more standard form $\dot{x} = f(x) + G(x)u$, where $G(x) = (g_1(x), g_2(x), g_3(x)) =$

$I^{-1}T$ is constant and $f(x) = I^{-1}S(x)Ix$. Motivated by the above results, we consider the following function (the kinetic energy):

$$V(x) := \frac{1}{2}\left(I_1x_1^2 + I_2x_2^2 + I_3x_3^2\right)$$

and, associated to it, the following feedback law:

$$k(x) = -\left(\nabla V(x) \cdot G(x)\right)' = -\left(L_{g_1}V(x), \ldots, L_{g_m}V(x)\right)' = -T'x\,.$$

Prove that this feedback stabilizes the system if and only if every row of T is nonzero. (*Hint:* You need to prove that $\mathcal{Q} = \{0\}$. For this, you will need to consider the various cases $\operatorname{rank} T = 1, 2, 3$ separately. The computations in each case are geometric, using the fact that $\mathcal{Q}$ may be written as intersections of appropriate lines, cones, and ellipsoids.) □

Backstepping

The following Lemma is useful in establishing that, under certain conditions, a series connection of two systems is stabilizable provided that each system is. For simplicity, we give the smooth version, but it will be clear from the proof how to relax to less differentiability. This result, often called the "integrator backstepping lemma," allows one to recursively design controllers for certain complex systems by a step-by-step procedure, starting with a simpler system and adding at each stage an integrator "in front" (at the input) of the simpler system.

Lemma 5.9.5 Assume that the smooth system (with $\mathcal{X}$ open in $\mathbb{R}^n$ and $\mathcal{U}$ open in $\mathbb{R}^m$)

$$\dot{x} = f(x, u)$$

is smoothly locally stabilizable using $u = k(x)$ and that there is a local Lyapunov function V as in Lemma 5.7.4 for the system (5.40), which is smooth. Let

$$h : \mathcal{X} \times \mathcal{U} \to \mathbb{R}^m$$

be smooth, $h(x^0, u^0) = 0$. Then, the system

$$\dot{x} = f(x, z) \tag{5.46}$$

$$\dot{z} = h(x, z) + u \tag{5.47}$$

having control set $\mathbb{R}^m$ is again smoothly locally stabilizable (with respect to the equilibrium point (x^0, u^0) and the control 0). The same result holds if "global" is used throughout.

In particular, the system obtained by integrating controls ($h \equiv 0$) is again stabilizable.

Proof. Assume that (5.40) is asymptotically stable and that

$$\nabla V(x).f(x,k(x)) < 0 \tag{5.48}$$

for $x \neq x^0$.

For any state on the surface $S = \{(x,z) \mid z = k(x)\}$ such that $x \neq x^0$, $\nabla V(x).f(x,z) < 0$; thus, x moves towards x^0, or, more precisely, $V(x)$ decreases, as long as the trajectory stays in S. See Figure 5.6 ($x^0 = 0$ and $u^0 = 0$ there). The idea of the proof will be to construct a controller which forces every state that is not on S to approach S. The dashed lines in Figure 5.6 indicate the desired motions. One obvious way to force moves towards S is to find controls so that $\|z - k(x)\|^2$ is decreased, in addition to $V(x)$ being decreased. Thus, we consider the following (smooth) function:

$$W(x,z) \;:=\; V(x) \,+\, \frac{1}{2}\,\|z - k(x)\|^2$$

as a candidate control-Lyapunov function. Note that $W(x,z)$ will be small if both $z \approx k(x)$ and $x \approx x^0$, which implies $z \approx u^0$.

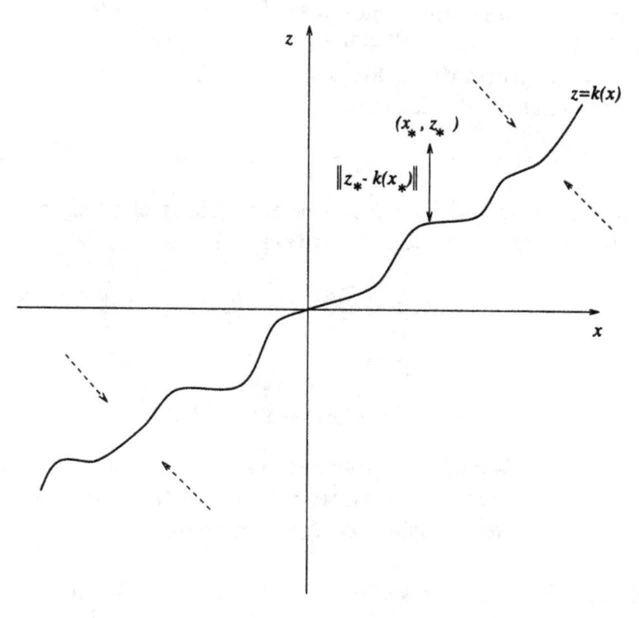

Figure 5.6: *Desired surface* $z = k(x)$.

The derivative of W, along a trajectory corresponding to a control u, is:

$$\nabla V(x).f(x,z) \,+\, (z - k(x))'\big(h(x,z) + u - k_*[x]f(x,z)\big)\,. \tag{5.49}$$

Let us write

$$u \;=\; -h(x,z) + k_*[x]f(x,z) + \hat{u}\,,$$

where $\hat{u}$ is to be chosen as a function of x and z. Then (5.49) becomes

$$\nabla V(x).f(x,z) + (z-k(x))'\hat{u}$$

and we are left with trying to make this expression negative.

Since f is smooth, there is an $n \times m$ smooth matrix G such that

$$f(x,k(x)+z) = f(x,k(x)) + G(x,z)z$$

for all (x,z), or equivalently,

$$f(x,z) = f(x,k(x)) + G(x,z-k(x))(z-k(x)) \tag{5.50}$$

for all (x,z). (For instance, if we write $\alpha(\lambda) := f(x,k(x)+\lambda z)$, then from $\alpha(1) = \alpha(0) + \int_0^1 \alpha'(\lambda)\,d\lambda$ we conclude that one choice is

$$G(x,z) := \int_0^1 f_z(x,k(x)+\lambda z)\,d\lambda\,,$$

where f_z is the Jacobian of f with respect to z.) Since already $\nabla V(x).f(x,k(x))$ is negative, for $x \neq x^0$, and since now $z-k(x)$ appears as a factor, we may simply use $\hat{u} = -[\nabla V(x).G(x,z-k(x))]' + \widetilde{u}$ to first subtract the effect of G, and then stabilize using an appropriate $\widetilde{u}$, for instance $-(z-k(x))$. Formally, then, we consider the following feedback law:

$$\widetilde{k}(x,z) := -h(x,z) + k_*[x]f(x,z) - [\nabla V(x).G(x,z-k(x))]' - z + k(x)\,.$$

Note that $\widetilde{k}(x^0,u^0) = 0$ and that $\widetilde{k}$ is smooth. Along the trajectories of (5.46)-(5.47) with this feedback law, the derivative of W is

$$\nabla V(x).f(x,z) + \left[h(x,z) + \widetilde{k}(x,z) - k_*[x]f(x,z)\right]'(z-k(x))\,,$$

which, by (5.50), reduces simply to

$$\nabla V(x).f(x,k(x)) - \|z-k(x)\|^2\,.$$

This latter expression is negative whenever $(x,z) \neq (x^0,u^0)$. Thus, W is a Lyapunov function for the composite system, which is then asymptotically stable. The arguments can be done either locally or globally. ∎

Example 5.9.6 Consider the system Σ with $\mathcal{X} = \mathcal{U} = \mathbb{R}$ and

$$\dot{x} = u^3\,.$$

This system is (globally) stabilizable (relative to $x^0 = 0$) using the feedback $u = k(x) := -x$. The Lyapunov function $V(x) = x^2$ can be used. We wish to show that the new system Σ' with equations

$$\begin{aligned}\dot{x} &= z^3\\ \dot{z} &= u\end{aligned}$$

is again smoothly stabilizable (relative to $(0,0)$). We have the expansion

$$\begin{aligned} f(x, k(x)+z) &= (k(x)+z)^3 \\ &= (k(x))^3 + [3(k(x))^2 + 3k(x)z + z^2]z \\ &= f(x, k(x)) + G(x,z)z, \end{aligned}$$

where $G(x,z) = 3x^2 - 3xz + z^2$. Equivalently,

$$\begin{aligned} f(x,z) &= f(x,k(x)) + G(x, z-k(x))(z-k(x)) \\ &= f(x,k(x)) + (z^2 - zx + x^2)(z+x)\,. \end{aligned}$$

In the above proof, we obtain

$$\begin{aligned} W(x,z) : &= V(x) + \frac{1}{2}(z-k(x))^2 \\ &= x^2 + \frac{1}{2}(z+x)^2 \end{aligned}$$

and the feedback law

$$\tilde{k}(x,z) = -z^3 - [2x(z^2 - zx + x^2)] - z - x\,.$$

Note that $\tilde{k}(0,0) = 0$ and that $\tilde{k}$ is a polynomial. Along the trajectories of Σ' with this feedback law, the derivative of W is

$$\begin{aligned} \dot{W} &= [2x + (z+x)]z^3 + (z+x)[-z^3 - 2x(x^2 - xz + z^2) - x - z] \\ &= -2x^4 - (z+x)^2\,, \end{aligned}$$

which is negative whenever $(x,z) \neq (0,0)$. This proves that W is a Lyapunov function for the composite system Σ', which is therefore (globally) asymptotically stable. □

Exercise 5.9.7 Consider the following system, with $\mathcal{X} = \mathbb{R}^3$ and $\mathcal{U} = \mathbb{R}$:

$$\begin{aligned} \dot{x}_1 &= x_2^3 \\ \dot{x}_2 &= x_3 \\ \dot{x}_3 &= u \end{aligned}$$

(and $x^0 = 0$, $u^0 = 0$). Find a smooth globally stabilizing feedback law. □

Exercise 5.9.8 The equations for the angular momentum of a rigid body controlled by two independent torques (e.g., a satellite controlled by two pairs of opposing jets, cf. also Exercises 4.3.15 and 5.9.4) can be written, for appropriate parameter values and after some simplification, as:

$$\begin{aligned} \dot{x}_1 &= a_1 x_2 x_3 + u_1 \\ \dot{x}_2 &= a_2 x_1 x_3 + u_2 \\ \dot{x}_3 &= a_3 x_1 x_2 \end{aligned}$$

where a_1, a_2, and a_3 are certain constants and $a_3 \neq 0$. We view this as a system with $\mathcal{X} = \mathbb{R}^3$ and $\mathcal{U} = \mathbb{R}^2$. *Prove* that this system can be globally smoothly stabilized (about $x = 0$ and $u = 0$), and exhibit a smooth stabilizing feedback. (*Hint:* Apply Lemma 5.9.5.) □

A Universal Formula for Feedback

Several approaches to modern feedback design start with a control-Lyapunov function for a system, and use the control-Lyapunov function in order to construct a feedback stabilizer. We now describe one technique which has proven useful in that regard, restricting attention to control-affine continuous-time time-invariant systems

$$\dot{x} = f(x) + u_1 g_1(x) + \ldots + u_m g_m(x) = f(x) + G(x)u \tag{5.51}$$

having $\mathcal{X} = \mathbb{R}^n$ and $\mathcal{U} = \mathbb{R}^m$ (as in Equations (4.18), (4.22), and (5.10), except that, here, we prefer to write f instead of g_0). We suppose that the system is smooth, and that $f(0) = 0$, and discuss stabilization with respect to the equilibrium $x^0 = 0$, $u^0 = 0$.

Given is a smooth global control-Lyapunov function V with the infinitesimal decrease property which was described in Lemma 5.7.4. That is, $V : \mathbb{R}^n \to \mathbb{R}$ is proper and positive definite, and, for each nonzero $x \in \mathbb{R}^n$, there is some $u \in \mathbb{R}^m$ so that Equation (5.34) holds. When specializing to control-affine systems, this condition says that, for each $x \neq 0$, there must be some $u \in \mathbb{R}^m$ so that

$$L_f V(x) + u_1 L_{g_1} V(x) + \ldots + u_m L_{g_m} V(x) < 0 \tag{5.52}$$

or, equivalently,

$$L_{g_1} V(x) = \ldots = L_{g_m} V(x) = 0 \text{ and } x \neq 0 \Rightarrow L_f V(x) < 0 . \tag{5.53}$$

Given such a V, one obvious way to try to control the system to zero is to use a feedback law $u = k(x)$ which selects, for each $x \neq 0$, some u so that (5.52) holds. The value of V will then decrease along trajectories, and properness and positive definiteness will assure that the trajectory itself converges to zero. There are many ways to select such an u as a function of x. For example, if we can be assured that a suitable u can be found in a compact subset $\mathcal{U}_0 \subset \mathbb{R}^m$ ($\mathcal{U}_0$ may be allowed to depend on x, if desired), and if there is a unique minimizer to

$$\min_{u \in \mathcal{U}_0} L_f V(x) + u_1 L_{g_1} V(x) + \ldots + u_m L_{g_m} V(x)$$

then one could pick $u = k(x)$ to be this minimizer. Such an approach is a useful one, and it appears often in applications. It might give rise, however, to technical difficulties. For example, the function $u = k(x)$ may not be sufficiently regular so that the closed-loop differential equation has unique solutions for each initial condition. Also, the minimization may not be trivial to perform, from a computational point of view. In addition, smoothness of $k(x)$ is often desirable, for implementation reasons. We show next that it is always possible to select a $u = k(x)$ that is a smooth function of x, at least for $x \neq 0$. (If we would be content with a class $\mathcal{C}^\ell$ feedback k, with $\ell < \infty$, it would be sufficient to assume that V, as well as the system, are merely of class $\mathcal{C}^{\ell+1}$.) That is, we look for a mapping $k : \mathbb{R}^n \to \mathbb{R}^m$, with $k(0) = 0$ and so that k restricted to $\mathbb{R}^n \setminus 0$ is smooth, so that

$$\nabla V(x).\, [f(x) + k_1(x)g_1(x) + \ldots + k_m(x)g_m(x)] \ < \ 0 \tag{5.54}$$

for all nonzero x. Note that, if any such k happens to be continuous at the origin, then the following property, the *small control property* for V, must hold as well (take $u := k(x)$):

> *For each $\varepsilon > 0$ there is a $\delta > 0$ so that, whenever $0 < \|x\| < \delta$, there is some u with $\|u\| < \varepsilon$ such that (5.52) holds.*

Exercise 5.9.9 Suppose that $F : \mathbb{R}^n \to \mathbb{R}^n$ is locally Lipschitz when restricted to $\mathbb{R}^n \setminus \{0\}$ and satisfies $F(0) = 0$, and assume that V is a proper and positive definite continuously differentiable function such that $\nabla V(x) \cdot F(x) < 0$ for all nonzero x. (For example, $F = f + \sum k_i g_i$, where k, with $k(0) = 0$, is smooth on $\mathbb{R}^n \setminus \{0\}$ and (5.54) holds for all nonzero states.) Show that there is a unique solution of $\dot{x} = F(x)$, for each initial condition, defined for all $t \geq 0$. Moreover, every solution converges to zero, and for each $\varepsilon > 0$ there is some $\delta > 0$ so that $\|x(0)\| < \delta$ implies that $\|x(t)\| < \varepsilon$ for all $t \geq 0$. That is, the system $\dot{x} = F(x)$ is globally asymptotically stable. As an illustration, compute the solutions, for each initial condition and each $t \geq 0$, of the one-dimensional system $\dot{x} = -x^{\frac{1}{3}}$ ($V(x) = x^2$ is as required for this F). (*Hint:* On the open set $\mathcal{X} := \mathbb{R}^n \setminus \{0\}$, there is local existence and uniqueness, and completeness can be proved using properness of V. Properness and positive definiteness also guarantee that 0 is an equilibrium, so behavior such as that of $\dot{x} = +x^{\frac{1}{3}}$ cannot occur.) □

Proposition 5.9.10 Let the smooth and positive definite $V : \mathbb{R}^n \to \mathbb{R}$ be so that, for each $x \neq 0$, there is an $u \in \mathbb{R}^m$ so that (5.52) holds. Then there exists a k, $k(0) = 0$, k smooth on $\mathbb{R}^n \setminus 0$, satisfying (5.54). Moreover, if V satisfies the small control property, then k can be chosen to also be continuous at 0.

Proof. We shall prove the theorem by constructing a fixed real-analytic function φ of two variables, and then designing the feedback law in closed-form by the evaluation of this function at a point determined by the directional derivatives $\nabla V(x).f(x)$ and $\nabla V(x).g_i(x)$'s.

Consider the following open subset of $\mathbb{R}^2$:

$$S := \{(a, b) \in \mathbb{R}^2 \mid b > 0 \text{ or } a < 0\}\,.$$

Pick any real analytic function $q : \mathbb{R} \to \mathbb{R}$ such that $q(0) = 0$ and $bq(b) > 0$ whenever $b \neq 0$. (Later we specialize to the particular case $q(b) = b$.) We now show that the function defined by:

$$\varphi(a,b) := \begin{cases} \dfrac{a+\sqrt{a^2+bq(b)}}{b} & \text{if } b \neq 0 \\ 0 & \text{if } b = 0 \end{cases}$$

is real-analytic on S. For this purpose, consider the following algebraic equation on p:

$$F(a,b,p) = bp^2 - 2ap - q(b) = 0 \tag{5.55}$$

which is solved by $p = \varphi(a,b)$ for each $(a,b) \in S$. We show that the derivative of F with respect to p is nonzero at each point of the form $(a,b,\varphi(a,b))$ with $(a,b) \in S$, from which it will follow by the Implicit Function Theorem that φ must be real-analytic:

$$\frac{1}{2}\frac{\partial F}{\partial p} = bp - a$$

equals $-a > 0$ when $b = 0$ and $\sqrt{a^2+bq(b)} > 0$ when $b \neq 0$.

Assume now that V satisfies (5.52). We write

$$a(x) := \nabla V(x).f(x), \quad b_i(x) := \nabla V(x).g_i(x), \; i = 1,\ldots,m,$$

and

$$B(x) := (b_1(x),\ldots,b_m(x)), \quad \beta(x) := \|B(x)\|^2 = \sum_{i=1}^m b_i^2(x)\,.$$

With these notations, condition (5.53) is equivalent to:

$$(a(x),\beta(x)) \in S \quad \text{for each } x \neq 0\,.$$

Finally, we define the feedback law $k = (k_1,\ldots,k_m)'$, whose coordinates are:

$$k_i(x) := -b_i(x)\,\varphi(a(x),\beta(x)) \tag{5.56}$$

for $x \neq 0$ and $k(0) := 0$. This is smooth as a function of $x \neq 0$. Moreover, at each nonzero x we have that (5.54) is true, because the expression there equals

$$a(x) - \varphi(a(x),\beta(x))\beta(x) = -\sqrt{a(x)^2+\beta(x)q(\beta(x))} < 0\,.$$

To conclude, we assume that V satisfies the small control property and show that, with the choice $q(b) := b$, the obtained function k is continuous at the origin. Pick any $\varepsilon > 0$. We will find a $\delta > 0$ so that $\|k(x)\| < \varepsilon$ whenever $\|x\| < \delta$. Since $k(x) = 0$ whenever $\beta(x) = 0$, we may assume that $\beta(x) \neq 0$ in what follows. Let $\varepsilon' := \varepsilon/3$.

Since V is positive definite, it has a minimum at 0, so $\nabla V(0) = 0$. Since the gradient is continuous, every $b_i(x)$ is small when x is small. Together with the small control property, this means that there is some $\delta > 0$ such that, if $x \neq 0$

satisfies $\|x\| < \delta$, then there is some u with $\|u\| < \varepsilon'$ so that $a(x) + B(x)u < 0$ and

$$\|B(x)\| < \varepsilon' . \tag{5.57}$$

Since $a(x) < -B(x)u$, the Cauchy-Schwartz inequality gives $a(x) < \varepsilon' \|B(x)\|$. Thus

$$|a(x)| < \varepsilon' \|B(x)\| \tag{5.58}$$

provided $0 < \|x\| < \delta$ and $a(x) > 0$. On the other hand, observe that

$$b\,\varphi(a, b) \;=\; a + \sqrt{a^2 + b^2} \;\leq\; 2\,|a| + b$$

for every $(a, b) \in S$ for which $b > 0$. Thus, if $0 < \|x\| < \delta$ and $a(x) > 0$, necessarily

$$\varphi(a(x), \beta(x)) \;\leq\; \frac{2\varepsilon'}{\|B(x)\|} + 1$$

and hence also, using (5.57),

$$\|k(x)\| \;=\; \varphi(a(x), \beta(x))\; \|B(x)\| \;\leq\; 3\varepsilon' = \varepsilon$$

as desired. There remains to consider those x for which $a(x) \leq 0$. In that case,

$$0 \;\leq\; a(x) + \sqrt{a(x)^2 + \beta(x)^2} \;\leq\; \beta(x)$$

so $0 \leq \varphi(a(x), \beta(x)) \leq 1$ and therefore

$$\|k(x)\| = \varphi(a(x), \beta(x))\; \|B(x)\| \leq \varepsilon' < \varepsilon$$

as desired too. ∎

It is worth writing out formula (5.56) explicitly in the special case $m = 1$:

$$k(x) \;=\; \begin{cases} -\dfrac{L_fV(x) + \sqrt{[L_fV(x)]^2 + [L_gV(x)]^4}}{L_gV(x)} & \text{if } L_gV(x) \neq 0 \\ 0 & \text{if } L_gV(x) = 0 . \end{cases}$$

As an example, consider the case of one-dimensional systems with a scalar control ($m = n = 1$),

$$\dot{x} = f(x) + ug(x) .$$

This system is stabilizable if the following assumption holds: if $g(x) = 0$ and $x \neq 0$ then $xf(x) < 0$. The feedback law given by the above construction, using the Lyapunov function $V(x) = x^2/2$, is simply

$$k(x) = -\frac{xf(x) + |x|\sqrt{f(x)^2 + x^2 g(x)^4}}{xg(x)}$$

(which is smooth, even though the absolute value sign appears, because in the one dimensional case there are two connected components of $\mathbb{R} - \{0\}$,) so that the closed-loop system becomes

$$\dot{x} = -\text{sign}\,(x)\sqrt{f(x)^2 + x^2 g(x)^4} .$$

Exercise 5.9.11 The function φ can be obtained as the solution of an optimization problem. For each *fixed* $x \neq 0$, thought of as a parameter, not as a state, we may consider the pair $(a(x), B(x))$ as a $(1, m)$ pair describing a linear system of dimension 1, with m inputs. The equations for this system are as follows, where we prefer to use "z" to denote its state, so as not to confuse with x, which is now a fixed element of $\mathbb{R}^n \setminus \{0\}$:

$$\dot{z} = az + \sum_{i=1}^{m} b_i u_i = az + Bu\,. \tag{5.59}$$

The control-Lyapunov function condition guarantees that this system is asymptotically controllable. In fact, the condition "$B = 0 \Rightarrow a < 0$" means precisely that this system is asymptotically controllable. A stabilizing feedback law $k = k(x)$ for the original nonlinear system must have, for each fixed x, the property that

$$a + \sum_{i=1}^{m} b_i k_i < 0\,.$$

This means that $u = kz$ must be a stabilizing feedback for the linear system (5.59). Consider for this system the infinite-horizon linear-quadratic problem of minimizing (cf. Theorem 41 (p. 384))

$$\int_0^\infty u^2(t) + \beta(x) z^2(t)\, dt\,.$$

(For motivation, observe that the term u^2 has greater relative weight when β is small, making controls small if x is small.)
Prove that solving this optimization problem leads to our formula (5.56). □

We next turn to other necessary conditions for stabilizability. As we will remark, both of these hold even for weaker notions than C^1 stabilizability.

Topological Obstructions to Stability

When dealing with time-invariant systems with no controls, we write just $\phi(t, x)$ instead of $\phi(t, 0, x, \omega)$.

Let Σ be any (topological) system with no controls and x^0 an equilibrium state. Assume that the system is locally asymptotically stable (with respect to x^0). The *domain of attraction* of x^0 is the set

$$D = D(x^0) := \{x \in \mathcal{X} \mid \lim_{t \to \infty} \phi(t, x) = x^0\}\,.$$

We need the following Lemma:

Lemma 5.9.12 For each $x \in D$ and each neighborhood $\mathcal{V}$ of x^0 there exists some neighborhood $\mathcal{V}'$ of x, $\mathcal{V}' \subseteq D$, and some $T \in \mathcal{T}_+$, such that for all $y \in \mathcal{V}'$ it holds that

$$\phi(t, y) \in \mathcal{V} \quad \text{for all} \quad t \geq T\,.$$

In particular, D is open.

Proof. Let $\mathcal{W}$ be an open neighborhood of x^0 included in the domain of attraction D and for which all trajectories starting in $\mathcal{W}$ remain in $\mathcal{V}$; such a neighborhood exists by the definition of asymptotic stability.

Now pick any point x in D. By definition of D, there exists some $T \geq 0$ such that $\phi(T, x) \in \mathcal{W}$. By continuity of $\phi(T, \cdot)$, there is then a neighborhood $\mathcal{V}'$ of x such that

$$\phi(T, y) \in \mathcal{W} \quad \text{for all } y \in \mathcal{V}' .$$

Since trajectories starting at $\phi(T, y)$ are in D, it follows that $\mathcal{V}' \subseteq D$. Since these stay in $\mathcal{V}$, the other conclusion follows too. ∎

Theorem 21 *If $\mathcal{T} = \mathbb{R}$ and $\phi(t, x)$ is continuous on t (as happens with continuous-time systems), then the domain of attraction is a contractible set.*

Proof. To prove contractibility, we need to provide a continuous mapping

$$H : [0, 1] \times D \to D$$

such that $H(0, x) = x$ for all $x \in D$ and (for instance) $H(1, x) = x^0$ for all x. We define H as follows, for $t < 1$:

$$H(t, x) := \phi\left(\frac{t}{1-t}, x\right)$$

and $H(1, \cdot) \equiv x^0$. By Remark 3.7.3, H is continuous in both variables, for $t < 1$.

We are left to prove only continuity at each $(1, x)$. Take any such x, and pick any neighborhood $\mathcal{V}$ of $H(1, x) = x^0$. We wish to find some $\delta > 0$ and some neighborhood $\mathcal{V}'$ of x such that

$$H(s, y) \in \mathcal{V} \quad \text{whenever } y \in \mathcal{V}' \text{ and } s > 1 - \delta .$$

By Lemma 5.9.12, there is a $\mathcal{V}'$ and a T so that this happens provided that $\frac{s}{1-s} > T$. It is only necessary then to pick $\delta := \frac{1}{T+1}$. ∎

Corollary 5.9.13 If Σ is a continuous-time system for which $\mathcal{X}$ is not contractible, then Σ is not C^1 globally stabilizable (about any x^0).

Proof. If the system were stabilizable, the resulting closed-loop system would be a continuous-time system that is asymptotically stable and for which the domain of attraction is all of $\mathcal{X}$. ∎

Example 5.9.14 Consider the continuous-time system on $\mathcal{X} = \mathbb{R}^2 - \{0\}$ and $\mathcal{U} = \mathbb{R}^2$,

$$\begin{aligned} \dot{x}_1 &= u_1 \\ \dot{x}_2 &= u_2 . \end{aligned}$$

This system is completely controllable, because any two points in the plane can be joined by a differentiable curve that does not cross the origin and any such curve is a path for the system. Furthermore, every point is an equilibrium point. About any x^0 there is local smooth stabilizability since the linearization has the same equations as the original system, but considered as a system with $\mathcal{X} = \mathbb{R}^2$. More explicitly, we may use the control law $u_i := -(x_i - x_i^0)$, which is well-defined on some neighborhood of any (nonzero) x^0. But, for no x^0 is the system globally smoothly stabilizable, because $\mathcal{X}$ is not a contractible space.

This system is not complete (for instance, the control constantly equal to $(1,0)'$ on $[0,1]$ is not admissible for $x = (-1,0)'$); a similar but complete example is as follows: Take the same state space and equations

$$\begin{aligned} \dot{\rho} &= \rho u_1 \\ \dot{\theta} &= u_2 . \end{aligned}$$

in polar coordinates or

$$\dot{x} = \begin{pmatrix} u_1 & -u_2 \\ u_2 & u_1 \end{pmatrix} x$$

in the (x_1, x_2) coordinates. □

Remark 5.9.15 The result in Corollary 5.9.13 can be extended to show that there does not exist for noncontractible $\mathcal{X}$ a globally stabilizing feedback k for which the closed-loop system is just locally Lipschitz on $\mathcal{X}$. Indeed, for any such k we may consider the closed-loop system defined by the equation

$$\dot{x} = f(x, k(x)) .$$

By the local existence Theorem, and the well-posedness Theorem 55 (p. 486) in the Appendix on differential equations, this defines a topological system, and it also satisfies the hypotheses of Theorem 21. □

Brockett's Necessary Condition

There is a powerful and easy to check necessary condition for local stabilization, which we present next. As a motivation, consider first the very special case of a continuous-time system $\dot{x} = f(x,u)$, of dimension one, with $\mathcal{X}$ being an open subset of $\mathbb{R}$ which contains zero. If there would be a feedback law $u = k(x)$ which stabilizes this system with respect to $x^0 = 0$, then clearly it must be the case that $f(x_1, k(x_1)) > 0$ for some (actually, for all) $x_1 < 0$, and that $f(x_2, k(x_2)) < 0$ for some $x_2 > 0$ (otherwise, if, for example, $f(x, k(x)) \geq 0$ for all x near zero, then no trajectories ξ of the closed-loop system with initial condition $\xi(0) > 0$ can approach 0). It follows, using continuity of $x \mapsto f(x, k(x))$ and connectedness of $[x_1, x_2]$ (i.e., the Intermediate Value Theorem), that there is some open interval V, containing zero, with the following property: for each $p \in V$, there is some $x \in \mathcal{X}$ and some $u \in \mathcal{U}$ (namely, $u = k(x)$) so that $f(x,u) = p$. In other words, $(x,u) \mapsto f(x,u)$ contains a neighborhood of zero in its image. For arbitrary dimensions but *linear* systems, the same statement is true. Indeed, asymptotic controllability (with respect to $x^0 = 0$ and $u^0 = 0$) implies, via the Hautus condition:

$$\text{rank}\,[A, B] = n\,,$$

which means that the mapping $(x, u) \mapsto Ax + Bu$ must be onto (since this is a linear map, the conclusion is equivalent to asking that there exists a neighborhood of zero included in its image). There is a common generalization of these two observations:

Theorem 22 *Assume that the $\mathcal{C}^1$ continuous-time system Σ:*

$$\dot{x} = f(x, u)$$

is locally $\mathcal{C}^1$ stabilizable with respect to x^0. Then the image of the map

$$f : \mathcal{X} \times \mathcal{U} \to \mathbb{R}^n$$

contains some neighborhood of x^0.

We prove this Theorem below, after reviewing a result from nonlinear analysis. First, we illustrate with an application.

Example 5.9.16 Consider the system on $\mathcal{X} = \mathbb{R}^3$ and $\mathcal{U} = \mathbb{R}^2$ with equations:

$$\begin{aligned} \dot{x}_1 &= u_1 \\ \dot{x}_2 &= u_2 \\ \dot{x}_3 &= x_2 u_1 - x_1 u_2 \end{aligned}$$

and equilibrium state 0. No point of the form

$$\begin{pmatrix} 0 \\ 0 \\ \varepsilon \end{pmatrix} \quad (\varepsilon \neq 0)$$

is in the image of f, so there is no $\mathcal{C}^1$ feedback stabilizing the system. This system, incidentally, is completely controllable, a fact that is easy to establish using Lie-algebraic techniques. It is also a bilinear system. Note that the linearized system has a single uncontrollable mode, $\lambda = 0$. (This is usually called "Brockett's example"; see also Exercises 4.3.14, 4.3.16, and 9.4.2, and Example 9.2.14.) □

The result from nonlinear analysis that we need is as follows; it depends on basic facts from degree theory which are cited in its proof.

Fact 5.9.17 Let $\overline{B}_\rho(0)$ be the closed ball of radius ρ centered at the origin in $\mathbb{R}^n$ and S_ρ its boundary:

$$S_\rho = \{x \mid \|x\| = \rho\}\,.$$

Assume given a continuous function

$$H : [0,1] \times \overline{\mathcal{B}}_\rho(0) \to \mathbb{R}^n$$

such that $H(1,x) = -x$ for all x and

$$H(t,x) \neq 0$$

for all $x \in S_\rho$ and all t. Let

$$F : \overline{\mathcal{B}}_\rho(0) \to \mathbb{R}^n : x \mapsto H(0,x).$$

Then there exists some $\varepsilon > 0$ such that the image of F contains the ball $\overline{\mathcal{B}}_\varepsilon(0)$.

Proof. For any continuous function $G : \overline{\mathcal{B}}_\rho(0) \to \mathbb{R}^n$ and any $p \in \mathcal{B}_\rho(0)$ for which $G(x) \neq p$ for all $x \in S_\rho$, one defines in nonlinear analysis the *degree* of G with respect to p, $\deg(G,p)$. (If G is C^1 and if $\det G_*[x] \neq 0$ for every x in the preimage $G^{-1}(p)$ – which is a finite set, by the Implicit Mapping Theorem and the compactness of $\overline{\mathcal{B}}_\rho(0)$ – then $\deg(G,p)$ is $\sum_{x\in G^{-1}(p)} \text{sign}\,(\det G_*[x])$, that is to say, the number of solutions of $G(x) = p$, where each solution is counted with sign, which depends on the orientation. If the Jacobian condition is not satisfied, or if the map G is merely continuous, small perturbations of G are used, and one proves that the definition is the same no matter which particular perturbation was taken. In any case, the index is always an integer.) See, for instance, [47], Chapter 2, for details. The degree is a homotopy invariant, as long as the homotopy H satisfies that $H(x,t) \neq p$ for boundary x (above reference, property 2-7). Applied to the above data, this gives that

$$\deg(F,0) = \deg(-x,0) = (-1)^n,$$

the last equality by exercise 2-3 in that reference. Since F never vanishes on S_ρ, there is some $\varepsilon > 0$ so that $|F(x)| > \varepsilon$ for all $x \in S_\rho$. It follows that

$$F(x) \neq y \quad \text{for all } y \in \overline{\mathcal{B}}_\varepsilon(0) \text{ and } x \in S_\rho,$$

so the degree $\deg(F,y)$ is well-defined for all such y. Since the degree depends continuously on y (property 2-8 in the above reference), and it is an integer, it follows that

$$\deg(F,y) = \deg(F,0) \neq 0$$

for all $y \in \overline{\mathcal{B}}_\varepsilon(0)$. But nonzero degree implies that $F(x) = y$ has at least one solution (property 2-15 in the reference), as required. ∎

We can now complete the proof of Theorem 22. Without loss of generality, we take $x^0 = 0$. Considering the obtained closed-loop system, we must show that, if a system

$$\dot{x} = f(x)$$

defined on some ball about the origin is asymptotically stable with respect to 0, then the image of the mapping

$$x \mapsto f(x)$$

must contain a neighborhood of the origin. Pick a ball $\overline{\mathcal{B}}_\rho(0)$ which is included in the domain of attraction $D \subseteq \mathcal{X} \subseteq \mathbb{R}^n$.,

We apply the above fact, using the following homotopy, for $x \in \overline{\mathcal{B}}_\rho(0)$ and $t \in [0,1]$:

$$H(t,x) := \begin{cases} f(x) & \text{if } t = 0 \\ -x & \text{if } t = 1 \\ \frac{1}{t}\left[\phi\left(\frac{t}{1-t}, x\right) - x\right] & \text{if } 0 < t < 1 \end{cases}$$

which connects f and $-x$. Note that $H(t,x) \neq 0$ for all $x \neq 0$ and all t, since there cannot be nonzero equilibria or periodic points.

At points of the form (t,x) with $t > 0$, H is the product of the continuous function $1/t$ with

$$H_0(t,x) - x\,,$$

where H_0 is the homotopy used in the proof of Theorem 21, so it is continuous there. It only remains to show continuity at $t = 0$. That is, we need to establish that, for each x and each $\varepsilon > 0$, there is some neighborhood $\mathcal{W}$ of x and some $\delta > 0$ such that

$$|H(t,y) - f(x)| < \varepsilon \text{ whenever } y \in \mathcal{W} \text{ and } t < \delta\,.$$

Since for every s and y it holds that $\frac{1}{s}\left(\phi(s,y) - y\right) = \frac{1}{s}\int_0^s f(\phi(\tau,y))\,d\tau$,

$$\frac{1+s}{s}\left(\phi(s,y) - y\right) - f(x) = \frac{1}{s}\int_0^s \{f(\phi(\tau,y)) - f(x)\}\,d\tau + \int_0^s f(\phi(\tau,y))\,d\tau\,.$$

By continuity of $f(\phi(\cdot,\cdot))$, there are a neighborhood $\mathcal{W}$ of x and $\delta > 0$ so that

$$\|f(\phi(\tau,y)) - f(x)\| < \varepsilon/2$$

whenever $y \in \mathcal{W}$ and $\tau < \delta$. Let M be a bound on the values of $\|f(\phi(\tau,y))\|$ for $y \in \mathcal{W}$ and $\tau \leq \delta$. Thus, if $s < \delta$ it holds that, with $t = \frac{s}{1+s}$,

$$|H(y,t) - f(x)| < \varepsilon/2 + M\frac{t}{1-t} \tag{5.60}$$

provided $y \in \mathcal{W}$ and $s = \frac{t}{1-t} < \delta$. The proof of Theorem 22 is completed by picking a smaller δ so that $t < \delta$ implies that the second term in (5.60) is less than $\varepsilon/2$. ∎

Remark 5.9.18 The proof uses only the existence of k so that the closed-loop system is Lipschitz. Note also that, for any given neighborhood $\mathcal{V}$ of x^0, the image $f(\mathcal{V} \times \mathcal{U})$ must contain a neighborhood of x^0: this is clear from the proof, or applying the Theorem to the system restricted to the state-space $\mathcal{V}$. □

Exercise 5.9.19 Consider the following system, with $\mathcal{X} = \mathbb{R}^3$ and $\mathcal{U} = \mathbb{R}^3$:

$$\begin{aligned}\dot{x}_1 &= u_2 u_3\\ \dot{x}_2 &= u_1 u_3\\ \dot{x}_3 &= u_1 u_2\end{aligned}$$

and its equilibrium pair $x^0 = 0$, $u^0 = 0$.
(1) Show that $V(x) = \|x\|^2$ is a control-Lyapunov function for this system, so the system is asymptotically controllable.
(2) Prove that there is no possible local C^1 stabilizer for this system. □

Exercise 5.9.20 Consider a system without drift

$$\dot{x} = G(x)u$$

that is, a control-affine system for which $f = 0$. (Recall, cf. Equation (4.22) and material after that, that controllability of such systems is equivalent to the accessibility rank condition.) We assume that $m < n$ and G is smooth. Show that (with $x^0 = 0$, $u^0 = 0$):
(1) If $\operatorname{rank} G(0) = m$ then there is no possible local C^1 stabilizer for this system.
(2) Give an example where there is a global C^1 stabilizer. □

5.10 Notes and Comments

Constant Linear Feedback

See Chapter 2 of [212] for an extensive discussion of controller and controllability forms, as well as related topics such as their use in analog computer simulations.

The Pole-Shifting Theorem has a long history. The proof for single-input systems, simple once the controller form is obtained, appears to have been discovered independently during the late 1950s. Kalman credits J. Bertram, ca. 1959, and Kailath credits also [329]. (In 1962, Harvey and Lee reported in [174] a weak version of the Theorem, asserting only stabilizability, as a consequence of the single-input case.) The general, multivariable, case, over the complexes, was first proved by Langenhop in [265] and independently by Popov ca. 1964, who discussed it in [321]. Over the real field, a complete proof was given by Wonham in [431]. The simple proof for general fields, by reduction to the single-input case, was given in [191]. The result that there is some F_1 so that $(A + BF_1, b)$ is controllable is known as *Heymann's Lemma*; our proof is from [177]. Pole-placement problems for "singular" or "descriptor" systems $E\dot{x} = Ax + Bu$ are discussed for instance in [94].

For a solution of Exercise 5.1.11, see [430]; for Exercise 5.1.12, see [212], Section 3.2.

For any given pair (A, B) and desired characteristic polynomial χ, there are (when $m > 1$) many feedback matrices F such that $\chi_{A+BF} = \chi$. As a set of equations on the nm entries of F, assignability introduces only n constraints, one for each of the coefficients of the desired characteristic polynomial. It is possible to exploit this nonuniqueness in various ways, for instance in order to impose other design requirements like the assignment of some eigenvectors, or the optimization of some criterion, such as the minimization of $\|F\|$. Furthermore, the Pole-Shifting Theorem does not say anything about the possible Jordan form of $A+BF$, that is, the possible similarity class of the closed loop A matrix. The more precise result is sometimes called *Rosenbrock's Theorem*, and is given in [331]; a purely state-space proof of this Theorem was given in [117] (see also [119]).

Design issues in control theory depend on the "zeros" of a system as well as its poles (the eigenvalues of A). The book [212] discusses various notions of "zero" for linear systems, a topic that will not be discussed here. Recently there has been work characterizing system zeros using techniques of commutative algebra; see [436] and references therein.

Feedback Equivalence

The Brunovsky form originated with [77]; see also [224], on which our discussion is based (but note that some of the formulas in the original paper are incorrect).

Feedback Linearization

The original formulation of the feedback linearization problem was given in [71], for the restricted case treated in Exercise 5.3.18. The problem in the general form, including the multiple-input case ($m > 1$) is from [207] and [197], and is covered in detail in textbooks such as [199] and [311]. We restricted attention to the single-input case because this is sufficient for illustrating the techniques but is technically much simpler than the general case. (In general, instead of A_n and b_n in the controller form 5.3, one must base the proof on the Brunovsky form and Theorem 14. Frobenius' Theorem must also be generalized, to deal with nested families of distributions $\Delta^1 \subseteq \Delta^2 \subseteq \ldots$ rather than single distributions, as part of the proof of the multiinput case.) Exercise 5.3.19 can be solved directly, not using feedback linearization ideas; see for instance [311], Theorem 5.3. Example 5.3.10 is from [378].

Disturbance Rejection and Relatively Invariant Subspaces

The notion of (A, B)-invariant subspace and the applications to solving systems problems, were proposed independently by [39] and [432]. It forms the basis of the approach described in [430], which should be consulted for a serious introduction to the topic of relatively invariant subspaces and many control applications. See [178] for a different manner of presenting these concepts and relations with other formalisms.

There has been much research during the last few years attempting to generalize these techniques to nonlinear systems. See, for instance, the textbooks

[199] and [311] for a treatment of that topic. The basic idea is to think of the space $\mathcal{V}$ not as a subset of the state space, but as what is sometimes called a "distribution" of subspaces in the tangent space to $\mathcal{X}$. That is to say, one thinks of $\mathcal{V}$ as an assignment of a (possibly different) space $\mathcal{V}(x)$ at each $x \in \mathcal{X}$. (See Definition 4.4.8.) Invariance then is defined with respect to the possible directions of movement, and an analogue of Lemma 5.4.3 is obtained, under strong extra assumptions, from the classical Frobenius' Theorem on partial differential equations (cf. Theorem 10 (p. 175).) See also [166], [167], [194], [200], [257], [302], and [303] in this context, including discrete-time nonlinear versions.

Stability and Other Asymptotic Notions

There are a large number of references on stability, most of them in the classical dynamical systems literature. The paper [63] gives an excellent bibliographical survey of those results on stability that are of major importance for control. Related to stability questions in feedback systems, it is worth pointing out that recently there have been studies of control laws in relation to chaotic and other highly irregular behavior; see, for instance, [29] and [341], as well as the relevant papers in the book [273] and the very early work already discussed in [213].

Attractivity is, in general, not equivalent to asymptotic stability, contrary to what happens for linear systems (Proposition 5.5.5); for a counterexample, see, for instance, [169], pages 191-194. The weaker notion of attractivity is less interesting, partly because positive results usually will establish the stronger property anyway, and also because of considerations that arise in input/output design. It is also possible to introduce notions associated to simply "stability" (as opposed to *asymptotic stability*). For differential equations (no controls) this is defined as the property that states near x^0 should not give rise to trajectories that go far from x^0. Except in the context of proving an instability result, we did not discuss this notion.

Stable and Unstable Modes

For a solution of Exercise 5.6.6 see [430], Chapter 4. That book also has many applications of the concepts discussed in this section.

Lyapunov and Control-Lyapunov Functions

There are various results which provide converses to Theorem 17, at least for systems with no controls. The book [49] covers abstract dynamical systems. For continuous-time systems $\dot{x} = f(x)$, with $f(0) = 0$ and evolving in $\mathbb{R}^n$, and under weak assumptions on f (locally Lipschitz is sufficient), Massera's Theorem, given in [296], asserts that, if the system is asymptotically stable with respect to 0 then there is a *smooth* Lyapunov function V, which satisfies the infinitesimal decrease condition $\dot{V}(x) = L_f V(x) < 0$ for all $x \neq 0$; if the system is globally asymptotically stable, then a proper V exists. In other words, Lemma 5.7.4 is necessary as well as sufficient in the case of systems with no controls. For other proofs see also [169], [262] (allows continuous f; even though solutions may not be unique, a notion of asymptotic stability can be defined in that case), and [428]. The reader may also find a proof in [279], which extends Massera's Theorem to smooth Lyapunov functions for systems subject to disturbances (robust stability), and is presented in the more general context of stability with respect to sets rather than merely equilibria.

For control systems, the situation with converses of Lemma 5.7.4 is far more delicate. The existence of V with the infinitesimal decrease property is closely tied to the existence of continuous fedback stabilizers; see [23], [372]. If differentiability is relaxed, however, one may always find control-Lyapunov functions for asymptotically controllable systems: see the result in [363], and the recent applications to discontinuous feedback design in [92].

In any case, all converse Lyapunov results are purely existential, and are of no use in guiding the search for a Lyapunov function. The search for such functions is more of an art than a science, and good physical insight into a given system plus a good amount of trial and error is typically the only way to proceed. There are, however, many heuristics that help, as described in differential equations texts; see, for instance, [334]. Another possibility is to build control-Lyapunov functions recursively, via "backstepping" as discussed in Lemma 5.9.5 and pursued systematically in [259].

For Exercise 5.7.17, see e.g. [323].

Example 5.7.12 implies that there exists a *bounded* feedback of the type $u = \theta(Fx)$ which stabilizes a double integrator. The result might not be surprising, since we know (cf. Exercise 3.6.8) that the double integrator (A has all eigenvalues at zero) is null-controllable using bounded controls. It is a bit surprising, on the other hand, that, for the *triple* integrator $\dot{x}_1 = x_2$, $\dot{x}_2 = x_3$, $\dot{x}_3 = u$, while still true that every state can be controlled to zero using bounded controls, it is in general impossible to find any globally stabilizing feedback of the form $u = \theta(Fx)$, with θ a saturation function; see [150]. On the positive side, however, slightly more complicated feedback laws do exist in general, for stabilizing linear systems with saturating inputs; see [392].

The example in Exercise 5.7.21 is due to [295]. In the adaptive control literature one finds extensive discussions of results guaranteeing, for slow-enough time variations, that a time-varying system is stable if the "frozen" systems are.

Linearization Principle for Stability

Even though, as discussed in Section 5.9, the linearized condition in Theorem 19 is sufficient but not necessary, this condition does become necessary if one imposes *exponential stability* rather than just asymptotic stability; see, for instance, [372] and references therein, as well as [157].

Theorem 20 is basically Lyapunov's Second Theorem on Instability, proved in 1892; it is often referred to (in generalized form) as "Chaetaev's Theorem."

Introduction to Nonlinear Stabilization

The question of existence of smooth and more generally continuous feedback laws in the nonlinear case is the subject of much current research; it is closely related to the existence of Lyapunov functions and to the existence of smooth solutions to optimal control problems. See, for instance, the survey papers [372] and [101], as well as [1], [4], [23], [27], [54], [72], [92], [99], [100], [109], [186], [187], [210], [238], [268], [294], [339], [342], [355], [403], [407], and [443]. The recent textbook [259] covers many questions of nonlinear feedback design. Discontinuous feedback of various types is, of course, also of interest, and is widespread in practice, but less has been done in the direction of obtaining basic mathematical results; see, for instance, [385] for *piecewise analytic* feedback, [362] for *piecewise linear* sampled feedback, [353] for *pulse-width modulated control* (a type of variable rate sampling), and [92] for a notion of feedback based upon the idea of sampling at arbitrarily fast rates. A major recent result in [99] showed that *time-varying* smooth feedback stabilizers exist for a large class of systems (see also the related work, and alternative proof, in [374]). There is also an extensive engineering literature, e.g., [406], on *sliding mode control*, in which discontinuity surfaces play a central role.

The line of work illustrated by Proposition 5.9.1 and Exercise 5.9.2 is based on [202], [210], and [268]. Exercise 5.9.4 is from [8].

The assumption in Lemma 5.9.5 about the existence of a suitable Lyapunov function is redundant, since stabilizability implies it (Massera's Theorem, quoted above). The proof of this Lemma was given independently in [80], [249], and [403]. An alternative proof of the same result about stability of cascade systems is given in [368], using only elementary definitions. A local version of the result has been known for a long time; see, for instance, [415]. For closely related work, see also [250].

Proposition 5.9.10 is essentially from [23]. The proof that we gave, based on the use of the "universal" formula (5.56), is from [369]. The result in [23] is far more general than stated here, however, especially in that it includes consideration of control-value sets $\mathcal{U}$ which are proper subsets of $\mathbb{R}^m$. For universal formulas that apply to more general sets $\mathcal{U}$, and for applications, see e.g. [199], [259], [278]. The material that centers around Proposition 5.9.10 can be stated in necessary and sufficient form, because the existence of a stabilizing feedback implies, via the converse Lyapunov theorem cited earlier, the existence of appropriate control-Lyapunov functions; see [23].

Theorem 22 was stated originally in [72]; see [443]. The proof given here is close to that for vector fields in [254]. See also [338].

There is also a wide literature regarding the stabilization of infinite dimensional linear systems; see, for instance, the survey article [324], as well as [113] and the general references cited for infinite dimensional linear systems theory.

Chapter 6

Outputs

Except for the basic definitions given in Chapter 2, the other major ingredient in control theory, taking into account the constraints imposed by the impossibility of measuring all state variables, has not been examined yet. These constraints were illustrated in Chapter 1 through the example of proportional-only (as opposed to proportional-derivative) control. Section 1.6, in particular, should be reviewed at this point, for a motivational introduction to observability, observers, and dynamic feedback, topics which will be developed next.

As discussed in Chapter 1, a natural set of questions to consider when measurements are restricted revolves around the concept of observability. This deals, in some sense, with the inversion of the relation between states and outputs to recover states, a problem dual to that of controllability, which deals instead with the correspondence between controls and states. This duality will not be made explicit in general —there has been work framing it in the language of category theory— but the discussion in the first sections of this Chapter closely parallels the beginning of Chapter 3. Moreover, in the special but most important case of *linear* systems, the duality between control and observation takes a particularly elegant form, and it allows the application of results from previous chapters to problems of observability.

6.1 Basic Observability Notions

In all of the definitions that follow, $\Sigma = (\mathcal{T}, \mathcal{X}, \mathcal{U}, \phi, \mathcal{Y}, h)$ is an arbitrary system with outputs.

Recall from Definition 2.2.4 that for each initial state x^0 the i/o response of the initialized system (Σ, x^0) is the map λ_{Σ,x^0} (denoted λ_{x^0}, or even just λ, if Σ and/or x^0 are clear from the context), which assigns to each triple (τ, σ, ω) with ω admissible for x^0 the function value

$$h(\tau, \phi(\tau, \sigma, x^0, \omega))$$

representing the last output, at time τ, due to applying the control $\omega \in \mathcal{U}^{[\sigma,\tau)}$ when starting at the initial state x^0. Especially when σ, τ are fixed, we also denote $\lambda_{x^0}(\tau, \sigma, \omega)$ as $\lambda_{x^0}^{\sigma,\tau}(\omega)$ or just $\lambda(\omega)$.

For instance, for the linear continuous-time time-varying system

$$
\begin{aligned}
\dot{x} &= A(t)x + B(t)u \\
y &= C(t)x
\end{aligned}
\tag{6.1}
$$

one has that

$$
\lambda_{x^0}^{\sigma,\tau}(\omega) = \eta(\tau) = C(\tau)\Phi(\tau,\sigma)x^0 + C(\tau)\int_\sigma^\tau \Phi(\tau,s)B(s)\omega(s)\,ds\,,
$$

where Φ denotes as usual the fundamental solution of $\dot{x} = A(t)x$.

Definition 6.1.1 *Let $x, z \in \mathcal{X}$ and $\sigma, \tau \in \mathcal{T}$, with $\sigma \le \tau$.*

- *The control $\omega \in \mathcal{U}^{[\sigma,\tau)}$* **distinguishes between the events** *(x, σ) and (z, σ) if it is admissible for both and*

$$
\lambda_x^{\sigma,\tau}(\omega) \neq \lambda_z^{\sigma,\tau}(\omega)\,.
$$

- *The events (x, σ) and (z, σ) are* **distinguishable on the interval** *$[\sigma, \tau]$, or* **in time** *(at most) $T = \tau - \sigma$, if there is some $t \in [\sigma, \tau]$ and some $\omega \in \mathcal{U}^{[\sigma,t)}$ that distinguishes them. If (x, σ) and (z, σ) are distinguishable on at least one interval, then they are just called distinguishable.*

- *The states x and z are distinguishable by a given control, or in time T, or just distinguishable, if there is some $\sigma \in \mathcal{T}$ so that the respective property holds for the events (x, σ) and (z, σ).* □

Distinguishable events (x, σ) and (z, σ) are therefore those that can be separated by applying *some* control, in the sense that if the system is initially in state x at time σ it will behave differently than if it started in state z at time σ. They are indistinguishable (i.e., not distinguishable) if every control leads to the same output. If one thinks of a system as a "black box" into which inputs can be applied and from which outputs emerge, distinguishability of internal states —or events, if time is of importance— is precisely the possibility of differentiating among them on the basis of input/output experiments.

Note that distinguishability in time T implies distinguishability in any larger time $T' > T$, by definition. Also,

$$
h(\sigma, x) \neq h(\sigma, z)
$$

is equivalent to distinguishability in time 0 because

$$
\lambda_x^{\sigma,\sigma}(\diamond) = h(\sigma, x) \neq h(\sigma, z) = \lambda_z^{\sigma,\sigma}(\diamond)\,,
$$

where $\diamond$ is the control of length zero. One may call this "instantaneous distinguishability."

Example 6.1.2 Consider the states $x = 1$ and $z = -1$ for the continuous-time system

$$\dot{x} = x^2 \;, \quad y = x^2$$

having $\mathcal{X} = \mathbb{R}$ and $\mathcal{U}$ a one-point set. On any interval $[0, t]$, $0 < t < 1$, the outputs are

$$\lambda_{-1}^{0,t} = \left(\frac{1}{1+t}\right)^2$$

and

$$\lambda_{1}^{0,t} = \left(\frac{1}{1-t}\right)^2 ,$$

which are distinct; thus, $1, -1$ are distinguishable in time t. (We omit "ω" from the notation for λ since there are no controls.) On any interval $[0, t]$ with $t \geq 1$, on the other hand, $\lambda_1^{0,t}$ is undefined, so $1, -1$ cannot be distinguished by a control of length larger than 1, but they are still distinguishable in any time larger than 1 according to our definition. Note that each pair of states x, z so that $x + z \neq 0$ is instantaneously distinguishable by the measurement $y = x^2$, but how the dynamical behavior is essential in separating any other pair of states x and $-x$. □

We use the notation $(x, \sigma) \underset{\tau}{\sim} (z, \sigma)$ to indicate that the events (x, σ) and (z, σ) are indistinguishable on the interval $[\sigma, \tau]$ and just $(x, \sigma) \sim (z, \sigma)$ if they are indistinguishable. For states, we write $x \underset{T}{\sim} z$ if x cannot be distinguished from z in time T and just $x \sim z$ to indicate that they are indistinguishable. Observe that

$$(x, \sigma) \sim (z, \sigma) \quad \text{iff} \quad (x, \sigma) \underset{\tau}{\sim} (z, \sigma) \text{ for all } \tau \geq \sigma$$

as well as

$$x \underset{T}{\sim} z \quad \text{iff} \quad (x, \sigma) \underset{\tau}{\sim} (z, \sigma) \text{ for all } \sigma, \tau \text{ so that } \tau - \sigma = T$$

and

$$x \sim z \;\text{ iff }\; x \underset{T}{\sim} z \text{ for all } T \in \mathcal{T}_+ .$$

Lemma/Exercise 6.1.3 Assume that Σ is a complete system. Show that in this case, on events and states, $\sim$ and $\underset{\tau}{\sim}$ are equivalence relations. □

Definition 6.1.4 *The system* Σ *is* **observable on the interval** $[\sigma, \tau]$ *if for every pair of distinct states* x *and* z *the events* (x, σ) *and* (z, σ) *are distinguishable on the interval* $[\sigma, \tau]$. *It is* **observable in time** T *if any two distinct states are distinguishable in time* T *and just* **observable** *if any two distinct states are distinguishable.* □

As in Chapter 3, these notions simplify considerably for linear systems. Recall that a **linear** system is one that is either as in Definition 2.4.1 or as in

Definition 2.7.2. By linearity of $\phi(\tau,\sigma,\cdot,\cdot)$ (see Lemmas 2.4.2 and 2.7.4) and of $h(t,\cdot)$, it follows that

$$\lambda_x^{\sigma,\tau}(\omega) = h(\tau,\phi(\tau,\sigma,x,\omega))$$

is linear on (x,ω). From $(x,\omega) = (x,\mathbf{0}) + (0,\omega)$, it follows that

$$\boxed{\lambda_x^{\sigma,\tau}(\omega) = \lambda_x^{\sigma,\tau}(\mathbf{0}) + \lambda_0^{\sigma,\tau}(\omega)} \tag{6.2}$$

for all pairs (x,ω), where $\mathbf{0}$ denotes the control $\omega \equiv 0$ of length $\tau-\sigma$. Equation (6.2) is sometimes called the **superposition principle**, because it expresses the fact that for linear systems the output from any initial state is the superposition (sum) of the zero-state response and the response to the zero input.

In particular, for any two states x, z, if

$$\lambda_x^{\sigma,\tau}(\omega) \neq \lambda_z^{\sigma,\tau}(\omega)$$

for some ω, then also

$$\lambda_x^{\sigma,\tau}(\mathbf{0}) \neq \lambda_z^{\sigma,\tau}(\mathbf{0})$$

and so

$$\lambda_x^{\sigma,\tau}(\omega) \neq \lambda_z^{\sigma,\tau}(\omega)$$

for *all* ω. In particular, the events (x,σ) and (z,σ) are distinguishable on $[\sigma,\tau]$ if and only if $\mathbf{0}$ distinguishes them on some subinterval $[\sigma,t]$. Similar remarks apply to the distinguishability of states. The equality $\lambda_{x-z}^{\sigma,\tau}(\mathbf{0}) = \lambda_x^{\sigma,\tau}(\omega) - \lambda_z^{\sigma,\tau}(\omega)$ implies:

Lemma 6.1.5 For linear systems, (x,σ) and (z,σ) are indistinguishable on $[\sigma,\tau]$ if and only if

$$\lambda_{x-z}^{\sigma,t}(\mathbf{0}) = 0$$

for all $t \in [\sigma,\tau]$; that is, $(x-z,\sigma)$ is indistinguishable from $(0,\sigma)$ using zero controls on $[\sigma,\tau)$. Similarly, any two states x, z are indistinguishable if and only if $x-z$ is indistinguishable from 0. □

Corollary 6.1.6 For linear systems, observability on the interval $[\sigma,\tau]$ is equivalent to

$$\bigcap_{t=\sigma}^{\tau} \ker\left(x \mapsto \lambda_x^{\sigma,t}(\mathbf{0})\right) \;=\; 0\,.$$ □

For linear systems observable on $[\sigma,\tau]$ a *single experiment*, namely applying the zero control (on increasing intervals), serves to separate all possible pairs. This is in general false for nonlinear systems, as illustrated by the following example using a finite discrete-time system.

Example 6.1.7 Consider the discrete-time time-invariant finite system with

$$\mathcal{X} = \{a,b,c,d\}\,,\quad \mathcal{U} = \{u,v\}\,,\quad \mathcal{Y} = \{\alpha,\beta\}\,,$$

local-in-time transition function $\mathcal{P}$ given by the following table:

$\mathcal{P}$	u	v
a	c	c
b	c	b
c	a	a
d	d	a

and output function $h(a) = h(b) = \alpha$ and $h(c) = h(d) = \beta$. The pairs

$$\{a, c\}, \{a, d\}, \{b, c\}, \{b, d\}$$

are instantaneously distinguishable. To distinguish

$$\{a, b\}$$

one may use any sequence starting with v and to distinguish

$$\{c, d\}$$

any sequence starting with u. In summary, every pair of distinct states can be distinguished in time 1, and the system is observable in time 1.

There is however no single "universal" sequence $\omega \in \mathcal{U}^{[\sigma,\tau)}$ so that for every pair $x \neq z$ it would hold that

$$\lambda_x^{\sigma,t}(\omega|_{[\sigma,t)}) \neq \lambda_z^{\sigma,t}(\omega|_{[\sigma,t)})$$

for some t (that is allowed to depend on x and z). This is because if ω starts with u, then all outputs when starting at a or b coincide, since the state trajectories coincide after the first instant, but if ω starts with v, then ω cannot separate c from d. □

This Example illustrates the "hypothesis testing" character of observability: For each pair of states there is some experiment that allows one to distinguish between them, but there is in general no "universal" experiment that allows a diagnosis of the initial state. The notion introduced next is somewhat better behaved in that sense.

Final-State Observability

Note that in Example 6.1.7, even though the initial-state information may have been lost, it is still the case that after applying a suitable input sequence and measuring the outputs, the final state may be determined completely. If the ultimate objective is to control the system to a given state, this knowledge is sufficient. As an illustration, assume that it is desired to bring the system to $x = a$. This is a reasonable objective, since every state can be controlled to a, but the goal now is to choose the controls based only on output measurements.

Consider the control on $[0,2)$ with $\omega(0)=u$ and $\omega(1)=v$. The outputs at times $0,1,2$ constitute one of the sequences

$$\alpha\beta\alpha \quad \beta\beta\alpha \quad \beta\alpha\beta\,,$$

and if either of the first two happened —corresponding to the possible initial states a, b, or d— then we know that the final state, the state at time 2, is already a, which is as desired. If the last happens —initial state was c— the final state is known to be c. In either case, all future behavior of the system is completely known. Control to a is achieved in the last case simply by applying one further input, either u or v.

This discussion motivates the next definition, totally parallel to that of distinguishability. The terminology that we use is not standard. Since as proved later this concept does not give anything essentially new for continuous-time systems, this subsection can be skipped by the reader interested only in continuous-time.

As in Chapter 2, Remark 2.2.2, $\bar{\lambda}$ denotes the entire output corresponding to a control ω, that is,

$$\bar{\lambda}_x^{\sigma,\tau}(\omega)(t) := \lambda_x^{\sigma,t}(\omega|_{[\sigma,t)}) \tag{6.3}$$

for each $t \in [\sigma,\tau]$ and each state x, and we drop x or σ,τ when they are clear from the context.

Definition 6.1.8 *The control $\omega \in \mathcal{U}^{[\sigma,\tau)}$* **final-state distinguishes between the events** *(x,σ) and (z,σ) if ω is admissible for both x and z and either*

- $\bar{\lambda}_x^{\sigma,\tau}(\omega) \neq \bar{\lambda}_z^{\sigma,\tau}(\omega)$, *or*
- $\phi(\tau,\sigma,x,\omega) = \phi(\tau,\sigma,z,\omega)$. □

As with distinguishability, if there exists one such ω on some interval $[\sigma,t)$ so that $t \leq \tau$, then (x,σ) and (z,σ) are *final-state distinguishable on the interval* $[\sigma,\tau]$ or *in time* $\tau-\sigma$. If (x,σ) and (z,σ) are final-state distinguishable on some interval $[\sigma,\tau]$, then they are just called final-state distinguishable. For states x and z, they are final-state distinguishable by a given control, or in time T, or just final-state distinguishable if there is some $\sigma \in \mathcal{T}$ for which the respective property holds for the events (x,σ) and (z,σ). A system Σ is final-state observable on an interval $[\sigma,\tau]$ if for every pair of states x and z the events (x,σ) and (z,σ) are final-state distinguishable on the interval $[\sigma,\tau]$. It is final-state observable in time T if any two states are final-state distinguishable in time T, and just final-state observable if any two states are final-state distinguishable.

It is immediate from the definition that distinguishable states are also final-state distinguishable. Thus, observable systems are always final-state observable. The converse is not true, as illustrated by the discrete-time system

$$x^+ = 0, \quad y = 0\,,$$

with $\mathcal{X} = \mathcal{U} = \mathcal{Y} = \mathbb{R}$. No two states are distinguishable, and hence the system is in particular unobservable, but every control (of length at least one) final-state distinguishes, since the state is zero after one instant, so the system is final-state observable. However, for continuous-time systems the concepts are equivalent:

Proposition 6.1.9 A continuous-time system is observable on $[\sigma, \tau]$ (or observable in time T, or observable) if and only if it is final-state observable on $[\sigma, \tau]$ (respectively, in time T, final-state observable).

Proof. Necessity always holds, so pick a final-state observable system and any two distinct events $(x, \sigma) \neq (z, \sigma)$. Let ω be so that it final-state distinguishes the events (x, σ) and (z, σ); without loss of generality, assume $\omega \in \mathcal{U}^{[\sigma,\tau)}$. If $\bar{\lambda}_x^{\sigma,\tau}(\omega) \neq \bar{\lambda}_z^{\sigma,\tau}(\omega)$, then some restriction $\omega_{[\sigma,t)}$ distinguishes between them, and there is nothing left to prove. So assume that

$$\phi(\tau, \sigma, x, \omega) = \phi(\tau, \sigma, z, \omega) .$$

Let $\xi(t) = \phi(t, \sigma, x, \omega)$ and $\zeta(t) = \phi(t, \sigma, z, \omega)$. Then ξ and ζ are two solutions of $\dot{x} = f(x, \omega)$ defined on all of $[\sigma, \tau]$ and having the same value at time $t = \tau$. By uniqueness of solutions, $\xi \equiv \zeta$, contradicting the assumption that $\xi(\sigma) = x \neq z = \zeta(\sigma)$. This gives the implication on any fixed interval $[\sigma, \tau]$, and the other implications follow from here. ∎

As discussed earlier, observability requires multiple experiments, except for linear systems and some other restricted classes of systems. But if a system is final-state observable, it is often possible to prove that it is final-state distinguishable using single (long-enough) experiments, as in the following exercise.

Exercise 6.1.10 Assume that Σ is a final-state observable time-invariant discrete-time complete system for which $\operatorname{card} \mathcal{X} < \infty$. Show that there exists some $T \geq 0$ and some fixed control ω of length T so that ω final-state distinguishes *every* pair of states. (*Hint:* Consider any control for which the set of pairs of states (x, z) that are final-state indistinguishable by ω is of minimal possible cardinality.) □

Continuous-Time Systems

There is an analogue of the discussion in Section 3.9, dealing with the sufficiency of various classes of controls in testing observability for continuous-time systems.

For any fixed σ, τ, let $\mathcal{A}$ be any subset of $\mathcal{L}_{\mathcal{U}}^{\infty}(\sigma, \tau)$ so that for each $\omega \in \mathcal{L}_{\mathcal{U}}^{\infty}$ there exists a sequence of elements

$$\omega^{(j)} \in \mathcal{A}, \quad \omega^{(j)} \to \omega \quad \text{pointwise a.e.} \; ,$$

which is equibounded (all values in a fixed compact).

Examples of sets $\mathcal{A}$ that satisfy the above conditions are discussed in Remark C.1.2 in Appendix C. They include analytic controls when $\mathcal{U}$ is an open convex subset of $\mathbb{R}^m$.

The following observation then is very useful.

Proposition 6.1.11 If (x_1, σ) and (x_2, σ) are distinguishable, then they can be distinguished by a control in $\mathcal{A}$.

Proof. By Theorem 1 (p. 57), Part 2(i), and the continuity of h, it follows that, if $\omega \in \mathcal{U}^{[\sigma,\tau)}$ is admissible for some x, and if $\omega^{(j)} \to \omega$ pointwise almost everywhere and the sequence is equibounded, then $\omega^{(j)}$ is also admissible for x for large j, and

$$\lambda_x^{\sigma,\tau}(\omega^{(j)}) \to \lambda_x^{\sigma,\tau}(\omega).$$

Thus, if ω distinguishes two events and $\omega^{(j)} \to \omega$ for some equibounded sequence in $\mathcal{A}$, then for large enough j one of the controls in the approximating sequence already distinguishes. ∎

An interesting consequence of Proposition 6.1.11 is the fact that, for analytic systems, observability can be achieved almost instantaneously. The analogous controllability result is of course false, except for linear systems (cf. Remark 3.7.9).

Proposition 6.1.12 Let Σ be a continuous-time system over $\mathbb{K} = \mathbb{R}$,

$$\dot{x} = f(t, x, u), \quad y = h(t, x),$$

for which the maps f and h are real-analytic in all their arguments, $\mathcal{U}$ and $\mathcal{Y}$ being assumed to be open subsets of Euclidean spaces $\mathbb{R}^m$ and $\mathbb{R}^p$, respectively, with $\mathcal{U}$ convex. Assume that (x, σ) and (z, σ) are distinguishable on $[\sigma, \tau]$ and pick any $\delta > 0$. Then, these events are distinguishable in time δ.

In particular, if Σ is observable on $[\sigma, \tau]$, then it is observable in arbitrarily small time $\delta > 0$.

Proof. The two states x and z can be distinguished by an analytic control ω on $[\sigma, \tau]$, by Proposition 6.1.11. Let $\xi = \psi(\tau, \sigma, x, \omega)$ and $\zeta = \psi(\tau, \sigma, z, \omega)$ be the ensuing paths and let

$$\alpha(t) := \bar{\lambda}_x^{\sigma,\tau}(\omega)(t) = h(t, \xi(t)), \quad \beta(t) := \bar{\lambda}_z^{\sigma,\tau}(\omega)(t) = h(t, \zeta(t)).$$

These are analytic functions of time, because h as well as each of ξ and ζ are (cf. Proposition C.3.12 in Appendix C). Thus, if these functions were to coincide on $[0, \delta]$, then they would be identically the same, contradicting the fact that they differ at $t = \tau$. ∎

Exercise 6.1.13 Prove that if Σ is a continuous-time system as in Proposition 6.1.12, then indistinguishability is an equivalence relation. □

6.2 Time-Invariant Systems

Throughout this section, unless otherwise stated, Σ is an arbitrary *time-invariant* system with outputs.

For each fixed $t \in \mathcal{T}_+$ we denote

$$I^t := \{(x,z) \in \mathcal{X} \times \mathcal{X} \mid x \underset{t}{\sim} z\},$$

that is, the relation $\underset{t}{\sim}$ seen as a set of ordered pairs. Note that

$$I^t \subseteq I^s$$

for each pair $s \leq t$. We also let

$$I := \bigcap_{t \in \mathcal{T}_+} I^t$$

denote the set of indistinguishable pairs. The following can be seen as a dual statement to the one in Lemma 3.2.6:

Lemma 6.2.1 If there are any $S, T \in \mathcal{T}_+, T > 0$, such that

$$I^S = I^{S+T},$$

then necessarily $I^S = I$.

Proof. It will be enough to establish that

$$I^S = I^{S+kT} \tag{6.4}$$

for all positive integers k, since then for each $t \in \mathcal{T}_+$ there is some k so that $S + kT > t$ and therefore $I^S = I^{S+kT} \subseteq I^t$. We prove (6.4) by induction on k, the case $k = 1$ holding by hypothesis. Assume then that (6.4) holds for $k - 1$ and pick any

$$(x,z) \in I^S = I^{S+T}. \tag{6.5}$$

We want to prove that $(x,z) \in I^{S+kT}$. If this were not to be the case, then there would be some $\omega \in \mathcal{U}^{[0,t)}$ that distinguishes between them, with $t \leq S + kT$. It cannot be the case that $t < T$, since that would contradict the fact that $(x,z) \in I^{S+T}$. So we may consider the control

$$\omega_1 := \omega|_{[0,T)},$$

which must be admissible for both x and z because ω was. Let

$$x_1 := \phi(T, 0, x, \omega_1), \quad z_1 := \phi(T, 0, z, \omega_1).$$

The control

$$\omega_2 := \omega|_{[T,t)}$$

distinguishes x_1 and z_1, in time $t-T \leq S+(k-1)T$, so by inductive hypothesis these states can also be distinguished by some control ω_3 of length at most S. Finally, introduce the concatenation

$$\omega' := \omega_1\omega_3,$$

which is a control of length at most $S+T$ distinguishing x and z. The existence of such a control contradicts (6.5). ∎

The following two results are analogues of Lemma 3.2.4 and Corollary 3.2.7, respectively.

Corollary 6.2.2 Assume that Σ is a complete discrete-time system and $1 < \operatorname{card}(\mathcal{X}) = n < \infty$. Then

$$I = I^{n-2},$$

and in particular Σ is observable if and only if it is observable in time $n-2$.

Proof. Consider the sequence of sets

$$I^{n-1} \subseteq I^{n-2} \subseteq \ldots \subseteq I^0 \subseteq \mathcal{X} \times \mathcal{X}.$$

The inclusions cannot be all strict. Indeed, recall (Lemma 6.1.3) that, since Σ is complete, indistinguishability in time i is an equivalence relation. Let l_i be the number of equivalence classes under $\underset{i}{\sim}$; strict inclusions would imply

$$l_{n-1} > l_{n-2} > \ldots > l_0 > 1,$$

which is impossible since all $l_i \leq n$. Thus, either $I^0 = \mathcal{X} \times \mathcal{X}$, i.e., h is constant and therefore $I = \mathcal{X} \times \mathcal{X}$, or we are in the situation of Lemma 6.2.1. ∎

Exercise 6.2.3 Show that, for arbitrary finite discrete-time systems, not necessarily complete,

$$I = I^{\frac{n(n-1)}{2}}.$$

Give an example to illustrate that the formula $I = I^{n-2}$ may not hold if the system is not complete. □

If Σ is a linear system and $T \in \mathcal{T}_+$, we let

$$\mathcal{O}_T := \bigcap_{t=0}^{T} \{x \mid \lambda_x^{0,t}(\mathbf{0}) = 0\}$$

so that $x \underset{T}{\sim} z$ if and only if $x - z \in \mathcal{O}_T$. In this case, a strict inclusion

$$I^T \subset I^S$$

is equivalent to

$$\dim \mathcal{O}_T < \dim \mathcal{O}_S,$$

so there cannot be any proper chain

$$I^{t_n} \subset I^{t_{n-1}} \subset \ldots \subset I^{t_0} \subset \mathcal{X} \times \mathcal{X},$$

and this implies, by the same argument as that used to prove Corollary 3.2.7:

Corollary 6.2.4 Assume that Σ is a linear system of dimension $n < \infty$. Then:

(a) If Σ is discrete-time, $I = I^{n-1}$.

(b) If Σ is continuous-time, $I = I^{\delta}$ for all $\delta > 0$.

In particular, observable finite dimensional discrete-time systems are observable in $n-1$ steps, and observable finite dimensional continuous-time systems are observable in arbitrarily small time. □

The continuous-time part is of course also a consequence of Proposition 6.1.12. Observe that for a discrete-time linear system over a finite field of q elements, Corollary 6.2.2 gives an estimate of $q^n - 2$ steps, while using linearity gives the far smaller estimate $n-1$.

For discrete-time linear systems, $\mathcal{O}_{n-1} = \ker \mathbf{O}_n(A, C)$, where for each k we are denoting

$$\mathbf{O}_k(A, C) = \begin{pmatrix} C \\ CA \\ \vdots \\ CA^{k-1} \end{pmatrix} \tag{6.6}$$

and this kernel is zero precisely when the transposed matrix has full row rank, from which the following fact follows (denoting $\mathbf{O}_n$ simply as $\mathbf{O}$):

Theorem 23 *The n-dimensional discrete-time linear system Σ is observable if and only if*

$$\operatorname{rank} \mathbf{O}(A, C) = n\,,$$

or equivalently, if and only if (A', C') is controllable. □

For continuous-time linear systems,

$$\mathcal{O}_{\varepsilon} = \{x \mid Ce^{tA}x = 0 \text{ for all } t \in [0, \varepsilon]\}\,,$$

which is equivalent, by analyticity of $Ce^{tA}x$, to the set of conditions

$$CA^i x = 0,\ i = 0, 1, 2, \ldots\,,$$

or equivalently, because of the Cayley-Hamilton Theorem, to these same conditions for $i = 0, \ldots, n-1$. Thus, again observability is equivalent to the kernel of the matrix in (6.6) being trivial for $k = n$, and we conclude as follows.

Theorem 24 *The n-dimensional continuous-time linear system Σ is observable if and only if*

$$\operatorname{rank} \mathbf{O}(A, C) = n\,,$$

or equivalently, if and only if (A', C') is controllable. □

By the two Theorems, there is no ambiguity in calling a pair of matrices (A, C) with $A \in \mathbb{K}^{n\times n}$ and $C \in \mathbb{K}^{p\times n}$ an **observable pair** if $\mathbf{O}(A, C)$ has rank n. In general, all the material in Section 3.3 dualizes to results about observable pairs. For instance, the Hautus controllability condition becomes:

$$(A, C) \text{ is observable iff rank} \begin{bmatrix} \lambda I - A \\ C \end{bmatrix} = n \text{ for all } \lambda \in \widetilde{\mathbb{K}} \tag{6.7}$$

and the set of all observable pairs is generic, in the sense of Proposition 3.3.12, among all pairs.

The Kalman controllability decomposition in Lemma 3.3.3 dualizes to an *observability decomposition*: If $\operatorname{rank} \mathbf{O}(A, C) = r \leq n$, then there exists a $T \in GL(n)$ such that the matrices $\widetilde{A} := T^{-1}AT$ and $\widetilde{C} := CT$ have the block structure

$$\boxed{\widetilde{A} = \begin{pmatrix} A_1 & 0 \\ A_2 & A_3 \end{pmatrix} \qquad \widetilde{C} = \begin{pmatrix} C_1 & 0 \end{pmatrix}} \tag{6.8}$$

where A_1 is $r \times r$, C_1 is $p \times r$, and (Lemma 3.3.4) the pair (A_1, C_1) is observable. As before, if $r = 0$, the decomposition is trivial, and $A_1, 0, A_2, C_1$ are not there.

As in the controllability case, two pairs (A, C) and $(\widetilde{A}, \widetilde{C})$ are said to be *similar* if such a T exists.

This decomposition corresponds, using continuous-time to illustrate, to a change of variables

$$x(t) := Tz(t)$$

under which the equations $\dot{x} = Ax, y = Cx$ are transformed into

$$\begin{aligned} \dot{z}_1 &= A_1 z_1 \\ \dot{z}_2 &= A_2 z_1 + A_3 z_2 \\ y &= C_1 z_1 , \end{aligned}$$

where z_1, z_2 are r- and $(n-r)$-dimensional respectively. Thus, the z_2 coordinate does not affect the output, even indirectly.

Exercise 6.2.5 Refer to Exercise 3.2.12. Show that using the altitude h as the measurement map, the system Σ is observable. Show that using instead any other coordinate as an output (for instance, ϕ, which can be measured with a gyroscope) would not result in observability. □

Exercise 6.2.6 Refer to Exercise 3.2.13.

(a) Show that the system is not observable if one uses the pendulum angle $x_3 = \phi$ as the measurement map. Provide an example of two indistinguishable states, and explain intuitively why they are indistinguishable.

(b) Show that using the cart position $x_1 = \delta$ as the measurement map, the system Σ is observable, as long as $m > 0$. However, when $m = 0$, which models the case where the pendulum mass is negligible compared to that of the cart, the system is not observable. (Why is this difference more or less clear intuitively?)

(c) Prove that if both δ and ϕ are measured, i.e.

$$C = \begin{pmatrix} 1 & 0 & 0 & 0 \\ 0 & 0 & 1 & 0 \end{pmatrix},$$

then the system is observable even if $m = 0$. □

Observability Under Additional Structure

Many possible nonlinear generalizations of Corollaries 6.2.2 and 6.2.4 are available when additional algebraic assumptions are made on the system structure. The following discussion is meant to illustrate this fact through some examples.

Consider complete discrete-time (still time-invariant) *polynomial systems.* These are systems with $\mathcal{X} = \mathbb{R}^n$, $\mathcal{U} = \mathbb{R}^m$, $\mathcal{Y} = \mathbb{R}^p$, and equations

$$x^+ = \mathcal{P}(x, u), \quad y = h(x),$$

for which the maps $\mathcal{P} : \mathbb{R}^{n+m} \to \mathbb{R}^n$ and $h : \mathbb{R}^n \to \mathbb{R}^p$ are polynomial.

Proposition 6.2.7 Let Σ be as above. There exists then a finite set of controls $\omega_1, \ldots, \omega_k$ such that, for any pair of states x and z, these states are distinguishable if and only if one of the controls ω_i distinguishes them. In particular, if Σ is observable, then there is some finite time T so that Σ is observable in time T.

Proof. For each possible control ω let Δ_ω be the polynomial in $2n$ variables corresponding to the function

$$\Delta_\omega(x, z) := \lambda_x(\omega) - \lambda_z(\omega).$$

In particular, $\Delta_\diamond(x, z) = h(x) - h(z)$. Observe that ω does not distinguish between the two states x and z precisely when $\Delta_\omega(x, z) = 0$.

From the *Hilbert Basis Theorem,* we know that there exists a finite set of controls $\omega_1, \ldots, \omega_k$ so that for each other ω there are polynomials $p_1, \ldots, p_k$ with

$$\Delta_\omega(x, z) = \sum_{i=1}^{k} p_i(x, z)\Delta_{\omega_i}(x, z).$$

This implies that, if for any given pair (x^0, z^0) of states it holds that

$$\Delta_{\omega_i}(x^0, z^0) = 0 \ \text{ for all } i = 1, \ldots, k,$$

then, for the same pair, necessarily $\Delta_\omega(x^0, z^0) = 0$ for every other control, and the conclusion is established. (Recall that the Hilbert Basis Theorem asserts that every ideal in the polynomial ring $\mathbb{R}[x_1, \ldots, x_n, z_1, \ldots, z_n]$ is finitely generated. We apply this to the ideal generated by the Δ_ω's. Each element of this ideal is a finite linear combination of the elements Δ_ω, so each member of a finite generating set can in turn be written in terms of a finite number of the Δ_ω's.) ■

Exercise 6.2.8 Show that the result does not extend to analytic systems by considering the system

$$x^+ = \frac{1}{2}x \,, \quad y = \sin x \,,$$

which is observable but not in finite time. □

Exercise 6.2.9 There is a continuous-time version of the above result. Consider continuous-time polynomial systems

$$\dot{x} = f(x,u) \,, \quad y = h(x) \,,$$

where $\mathcal{X}, \mathcal{U}, \mathcal{Y}, h$ are as above and f is polynomial. We assume completeness. Show that, again, there exists then a finite set of controls $\omega_1, \ldots, \omega_k$ such that, for any pair of states x and z, these states are distinguishable if and only if one of the controls ω_i distinguishes them. (*Hint:* Use analytic controls and argue in terms of the corresponding derivatives $y(0), \dot{y}(0), \ldots$ that there must exist for each such control a polynomial Δ_ω so that x and z are indistinguishable by ω if and only if $\Delta_\omega(x,z) = 0$. The Hilbert Basis Theorem is needed for that. Now apply the Basis Theorem again.) □

Abstract Duality

The duality between controllability and observability can be formalized in various manners for nonlinear systems. More precisely, a form of duality holds between reachability from a fixed initial state and observability. For linear systems, reachability from the origin is equivalent to complete controllability, and this explains the use of "controllability" in the linear case. The following exercise illustrates one abstract —and trivial— version of this duality.

Exercise 6.2.10 Let Σ be a complete time-invariant system with outputs and $\operatorname{card} \mathcal{Y} \geq 2$. Consider the system

$$\widetilde{\Sigma} := (\mathcal{T}, \widetilde{\mathcal{X}}, \mathcal{U}, \widetilde{\phi})$$

defined as follows. Its state space $\widetilde{\mathcal{X}}$ is the set of all maps $\mathcal{X} \to \mathcal{Y}$, and for each $\omega \in \mathcal{U}^{[\sigma,\tau)}$ and each state $\alpha : \mathcal{X} \to \mathcal{Y}$ of $\widetilde{\mathcal{X}}$,

$$\widetilde{\phi}(\tau, \sigma, \alpha, \omega)$$

is the map

$$x \mapsto \alpha(\phi(\tau, \sigma, x, \widetilde{\omega})) \,,$$

where $\widetilde{\omega}$ is the time-reversed control $\widetilde{\omega}(t) := \omega(\sigma + \tau - t)$. Note that the output map h can be seen as an element of $\widetilde{\mathfrak{X}}$. *Prove* that if $\widetilde{\Sigma}$ is reachable from h, i.e.,

$$\mathcal{R}(h) = \widetilde{\mathfrak{X}},$$

then Σ is observable. □

Sampled Observability

Let $\Sigma = (A, B, C)$ be a linear time-invariant continuous-time system with outputs over $\mathbb{K} = \mathbb{R}$. Consider the situation where measurements of the state of the system are only possible at discrete instants $t = k\delta$, for integers k, where $\delta > 0$ is a fixed number, called the sampling interval. This situation was discussed in Section 1.3 on digital control, and a discrete-time system with outputs, $\Sigma_{[\delta]}$, which models the system obtained by disregarding the intermediate observations, was introduced in Definition 2.10.1. In analogy with the discussion in Section 3.4 we ask when is Σ ***δ-sampled observable***, that is, when is $\Sigma_{[\delta]}$ observable as a discrete-time time-invariant linear system? Note that

$$\Sigma_{[\delta]} = (e^{\delta A}, A^{(\delta)}B, C),$$

so observability is equivalent to controllability of

$$(e^{\delta A'}, C')$$

and therefore δ-sampled observability holds provided that (A', C') is δ-sampled-controllable (Lemma 3.4.1). In particular, if δ satisfies the spectrum conditions in Theorem 4 (p. 102), then observability of (A, C) implies δ-sampled observability:

Proposition 6.2.11 Assume that Σ is observable and that

$$\delta(\lambda - \mu) \neq 2k\pi i, \quad k = \pm 1, \pm 2, \ldots, \tag{6.9}$$

for every two eigenvalues λ, μ of A. Then Σ is also δ-sampled observable. □

Thus, again using the terminology in Section 3.4:

> *Observability is preserved if the sampling frequency is larger than twice the largest frequency of A*

(cf. Lemma 3.4.4).

Exercise 6.2.12 Let $\omega_i, i = 1, \ldots, k$ be k different positive real numbers. Show that there is some continuous-time time-invariant linear system with outputs and no inputs $\Sigma = (A, 0, C)$ such that:

- Σ is observable, and

- for each set of $2k$ real numbers $a_i, \varphi_i, i = 1, \ldots, k$, there is some initial state x so that

$$\lambda_x^{0,t} = \eta(t) = \sum_{i=1}^{k} a_i \sin(2\pi\omega_i t + \varphi_i)$$

for all $t \geq 0$.

Conclude from the above discussion that, if

$$\frac{1}{\delta} > 2 \max_{i=1,\ldots,m} |\omega_i| ,$$

then the complete function $\eta(t)$ can be recovered from the values

$$\eta(0), \eta(\delta), \eta(2\delta), \ldots$$

for every set of a_i's and φ_i's. □

6.3 Continuous-Time Linear Systems

Pick any finite dimensional continuous-time linear system with outputs but no inputs $\Sigma = (A(t), 0, C(t))$. Its **adjoint** Σ^* is the system

$$\dot{x}(t) = -A(t)^* x(t) + C(t)^* u(t) . \tag{6.10}$$

Corollary 3.5.9 asserts that controllability of Σ^* on an interval $[\sigma, \tau]$ is equivalent to there not being any nonzero solution ξ of the equation $\dot{x} = -(-A^*)^* x = Ax$ on $[\sigma, \tau]$ for which $C(t)\xi(t) \equiv 0$. But then Corollary 6.1.6 gives:

$$\boxed{\Sigma \text{ is observable on } [\sigma, \tau] \iff \Sigma^* \text{ is controllable on } [\sigma, \tau]}$$

Another way to state this adjointness is in terms of the operator

$$L : \ \Omega = \mathcal{L}_p^2(\sigma, \tau) \to \mathcal{X} = \mathbb{K}^n : \ \omega \mapsto \int_\sigma^\tau k(t)^* \omega(t)\, dt \tag{6.11}$$

which was considered in Section 3.5, where

$$k(s) = C(s)\Phi(s, \tau) , \tag{6.12}$$

that is, the operator corresponding to the input-to-state map for the adjoint system. It was proved in Section 3.5 that L is onto (Σ^* is controllable on the interval $[\sigma, \tau]$) if and only if L^* is one-to-one, where

$$(L^* x)(t) = C(t)\Phi(t, \tau)x = C(t)\Phi(t, \sigma)\Phi(\sigma, \tau)x .$$

But this mapping is one-to-one precisely when there is no nonzero x so that the trajectory with initial state $\Phi(\sigma,\tau)x$ gives a zero output, that is, if the original system is observable.

This adjointness permits applying the results of Section 3.5 to observability problems for continuous-time linear systems. For example, observability on $[\sigma,\tau]$ is equivalent to the positive definiteness of

$$W = \int_{\sigma}^{\tau} \Phi(s,\tau)^* C(s)^* C(s) \Phi(s,\tau)\, ds\,, \tag{6.13}$$

which in principle can be checked numerically.

Least-Squares Observation

The optimal-norm property of the pseudoinverse of the input-to-state operator in Section 3.5 dualizes into an optimal-observation property. This is discussed next, after some general remarks about pseudoinverses.

In general, consider a bounded linear operator L from a Hilbert space Ω into a finite dimensional space $\mathcal{X}$, as in Section 3.5. Assume from now on that L is onto, so that the conclusions of Corollary 3.5.3 hold: The adjoint L^* is one-to-one, and

$$W : \mathcal{X} \to \mathcal{X},\ W := LL^*$$

is positive definite. The pseudoinverse

$$L^{\#} := L^* W^{-1} : \mathcal{X} \to \Omega$$

was defined in Equation (3.20), and it was shown that it provides the minimum-norm solution of $L\omega = x$. The next result proves that its adjoint

$$(L^{\#})^* = W^{-1} L : \Omega \to \mathcal{X}$$

gives the least-squares solution of $L^* z = \omega$. (For operators between finite dimensional spaces, both of these are particular cases of the pseudoinverse construction in Appendix A.2 on singular values.)

Lemma 6.3.1 For each $\omega \in \Omega$, let $z := W^{-1} L\omega$. Then

$$\|L^* z - \omega\| \le \|L^* \zeta - \omega\|$$

for all $\zeta \in \mathcal{X}$, with equality only if $\zeta = z$.

Proof. Pick any $\omega \in \Omega$. Observe that for all $x \in \mathcal{X}$ it holds that

$$\langle L^* x, L^{\#} L\omega - \omega \rangle = 0 \tag{6.14}$$

because, as $LL^{\#}L = L$, this expression equals $\langle x, L\omega - L\omega \rangle$. In particular, given any $\zeta \in \mathcal{X}$ we may apply (6.14) to the element $x = \zeta - z$ to obtain

$$\langle L^* \zeta - L^{\#} L\omega, L^{\#} L\omega - \omega \rangle = 0$$

and therefore that

$$\begin{aligned}\|L^*\zeta - \omega\|^2 &= \left\|L^*\zeta - L^\# L\omega\right\|^2 + \left\|L^\# L\omega - \omega\right\|^2 \\ &= \left\|L^*\zeta - L^\# L\omega\right\|^2 + \|L^* z - \omega\|^2 ,\end{aligned}$$

which establishes the desired inequality. Equality can hold only if

$$L^*\zeta = L^\# L\omega = L^* z ,$$

which, since L^* is one-to-one, is equivalent to $\zeta = z$. ∎

Now consider the system $\Sigma = (A(t), 0, C(t))$ and an interval $[\sigma, \tau]$. We introduce the reachability operator associated to the adjoint system Σ^*, that is, L as in (6.11), so that k is as in (6.12), W is as in (6.13),

$$L\eta = \int_\sigma^\tau \Phi(s,\tau)^* C^*(s)\eta(s)\, ds ,$$

and

$$(L^*\zeta)(t) = C(t)\Phi(t,\tau)\zeta = C(t)\Phi(t,\sigma)\Phi(\sigma,\tau)\zeta .$$

We now let the operator

$$M_0 : \mathcal{L}_p^2(\sigma,\tau) \to \mathcal{X}$$

be defined by the formula

$$\boxed{M_0\eta := \Phi(\sigma,\tau)W^{-1} \int_\sigma^\tau \Phi(s,\tau)^* C(s)^* \eta(s)\, ds}$$

which is obtained as the composition $\Phi(\sigma,\tau)W^{-1}L$.

Using the "$\bar{\lambda}$" notation for the entire output function as in Equation (6.3), Lemma 6.3.1 gives that $x = M_0\eta$ is the least-squares solution of

$$\bar{\lambda}_x(\mathbf{0}) = \eta$$

that is, for any given output function η, $M_0\eta$ is the state that minimizes the difference between this function and the one that would correspond to initial state x and control $\equiv 0$.

In general, for the system (6.1), let

$$M(\omega,\eta) := M_0\left(\eta - \bar{\lambda}_0(\omega)\right)$$

be the operator on $\mathcal{L}_m^2(\sigma,\tau) \times \mathcal{L}_p^2(\sigma,\tau)$ that subtracts the effect of any given control ω on the output and then provides the best least-squares solution for this new control.

The main result on optimal observation follows immediately from the preceding discussion:

Theorem 25 *Assume that* $\Sigma = (A, B, C)$ *is observable. Take any* $\omega \in \mathcal{L}^2_m(\sigma, \tau)$ *and any* $\eta \in \mathcal{L}^2_p(\sigma, \tau)$, *and let* $\hat{x} := M(\omega, \eta)$. *Then,*

$$\|\bar{\lambda}_{\hat{x}}(\omega) - \eta\| \leq \|\bar{\lambda}_x(\omega) - \eta\|$$

for all $x \in \mathcal{X}$, *with equality only if* $x = \hat{x}$. □

This can be interpreted as follows: If the actual output is

$$y(t) = C(t)x(t) + d(t),$$

where $d(\cdot)$ represents an unknown disturbance, then the estimate $\hat{x}$ provides the state that would have produced this output under the assumption that d was as small as possible.

Exercise 6.3.2 Let Σ be a continuous-time linear system with outputs, and pick $\sigma < \tau \in \mathbb{R}$. Consider the *observability Gramian*

$$W_o(\sigma, \tau) := \int_\sigma^\tau \Phi(s, \sigma)^* C(s)^* C(s) \Phi(s, \sigma)\, ds .$$

Show: Σ is observable in $[\sigma, \tau]$ if and only if $W_o(\sigma, \tau)$ has rank n, and, in that case, the operator that computes the initial state x from the observation $\eta(t) = C(t)\Phi(t, \sigma)x$ is given by:

$$x = M_0\eta = W_o(\sigma, \tau)^{-1} \int_\sigma^\tau \Phi(s, \sigma)^* C(s)^* \eta(s)\, ds .$$

(This is a dual to Exercise 3.5.5.) □

Analytically-Varying Systems

If (A, B, C) is so that A and C have real-analytic entries on some interval $\mathcal{I} = [\sigma, \tau]$, and if $t_0 \in \mathcal{I}$ is an arbitrary point in that interval, then observability on $[\sigma, \tau]$ is equivalent to

$$\operatorname{rank} \begin{bmatrix} C_0(t_0) \\ C_1(t_0) \\ \vdots \\ C_k(t_0) \end{bmatrix} = n \quad \text{for some } k, \tag{6.15}$$

where, inductively on i,

$$C_0(t) := C(t)$$

and

$$C_{i+1}(t) := -C_i(t)A(t) - \frac{d}{dt}C_i(t)$$

for $t \in \mathcal{I}$. This is simply what results from applying the criterion in Corollary 3.5.18 to the adjoint system. The negative sign in the first term on the right hand side comes from the adjoint equation. Equivalently, multiplying C_i by $(-1)^i$, one could take

$$C_{i+1}(t) = C_i(t)A(t) + \frac{d}{dt}C_i(t)$$

for $i \geq 1$.

Remark 6.3.3 Of course, the above rank condition can also be obtained directly, simply by observing that, if $y = Cx$ and $\dot{x} = Ax$, then in the analytic case $y \equiv 0$ if and only if all its derivatives vanish at t_0, but these derivatives are

$$y^{(i)}(t_0) = C_i(t_0)x(t_0),$$

and unobservability is equivalent to the existence of some $x(t_0) \neq 0$ such that this happens. □

Remark 6.3.4 More generally, if Σ is a linear system with outputs and the entries of B are continuous functions of time, one may define the adjoint as a system with outputs

$$(-A(t)^*, C(t)^*, B(t)^*)$$

with the same state space, control-value set $\mathcal{Y}$, and output-value set $\mathcal{U}$. Since $(\Sigma^*)^* = \Sigma$, one can also state the adjointness result as: Σ is controllable if and only if Σ^* is observable. □

6.4 Linearization Principle for Observability

This section provides a result showing that, if a linearization is observable, then the original system is locally observable in a suitable sense.

Definition 6.4.1 *The topological system Σ is* **locally observable** *about the state $x^0 \in \mathcal{X}$ if there exists some neighborhood V of x^0 such that every state in V different from x^0 is distinguishable from x^0.* □

Theorem 26 *Assume that Σ is a continuous-time system over $\mathbb{R}$ of class $\mathcal{C}^1$, and let $\Gamma = (\xi, \omega)$ be a trajectory for Σ on an interval $\mathcal{I} = [\sigma, \tau]$. Then a sufficient condition for Σ to be locally observable about $x^0 = \xi(\sigma)$ is that $\Sigma_*[\Gamma]$ be observable on $[\sigma, \tau]$.*

Proof. We first remark a general fact about continuous-time linear systems. We claim that, if Σ is a linear system which is observable on $[\sigma, \tau]$, then there exist $t_1, \ldots, t_n \in [\sigma, \tau]$ and integers $i_1, \ldots, i_n$ such that the mapping

$$x \mapsto (C_{i_1}(t_1)\Phi(t_1, \sigma)x, \ldots, C_{i_n}(t_n)\Phi(t_n, \sigma)x)$$

is one-to-one, where C_j is the jth column of C. This can be shown as follows. Assume inductively that we have constructed subspaces $\mathcal{X} = K_0, \ldots, K_r$, times $t_1, \ldots, t_r \in [\sigma, \tau]$, and integers $i_1, \ldots, i_r$, such that for $r > 0$

$$K_r = \{x \mid C_{i_1}(t_1)\Phi(t_1, \sigma)x = \ldots = C_{i_r}(t_r)\Phi(t_r, \sigma)x = 0\}$$

and so that K_i has dimension $n - i$ for each i. If $r = n$, then we have already obtained the desired t_i's. Otherwise, pick any nonzero $x \in K_r$. By observability, there is some $t \in [\sigma, \tau]$ so that $C(t)\Phi(t, \sigma)x \neq 0$ and hence some column, say the jth, of this which is nonzero. Letting $t_{r+1} := t$ and $i_{r+1} := j$ provides the induction step. For $r = n$ the claim follows. Apply this remark to the system $\Sigma_*[\Gamma]$ to obtain $t_1, \ldots, t_n$ and $i_1, \ldots, i_n$.

Consider now the mapping

$$\beta : \mathcal{X} \to \mathbb{R}^n : x \mapsto \left(\lambda_{x,i_1}^{\sigma,t_1}(\omega_1), \ldots, \lambda_{x,i_n}^{\sigma,t_n}(\omega_n)\right) ,$$

where ω_j is the restriction of ω to the interval $[\sigma, t_j)$ and the subscript i_j indicates the i_jth coordinate of λ_x^{σ,t_j}. The differential of this mapping, evaluated at x^0, can be obtained by looking at each component. Consider any mapping of the type

$$\mathcal{X} \to \mathbb{R} : x \mapsto \lambda_{x,i}^{\sigma,t}(\omega) ,$$

for any control ω. Its differential at x^0 is the composition of the partial differential of the mapping α in Theorem 1 (p. 57) with respect to x, evaluated at (x^0, ω), with the differential of the ith coordinate of h, that is, the mapping

$$x \mapsto C_i(t)\Phi(t, \sigma)x$$

for the linearization.

We conclude that the differential of β is one-to-one, by construction of the t_j's and the i_j's. Thus, from the Inverse Function Theorem, the map β is one-to-one in a neighborhood of x^0, as wanted. ∎

Of course this condition is far from necessary, as the example

$$\dot{x} = 0, \quad y = x^3$$

(and $x^0 = 0$) shows.

Remark 6.4.2 There is a higher order test for observability as well. Here we only present a brief outline of this test. Assume given a continuous-time system affine in controls,

$$\dot{x} = g_0(x) + \sum_{i=1}^{m} g_i(x)u_i ,$$

where we assume that all vector fields (that is, vector functions) g_i are of class $\mathcal{C}^\infty$. Consider the vector space spanned by the set of all functions of the type

$$L_{g_{i_1}} \ldots L_{g_{i_k}} h_j(x) \tag{6.16}$$

over all possible sequences $i_1, \ldots, i_k$, $k \geq 0$, out of $\{0, \ldots, m\}$ and all $j = 1, \ldots, p$, where $L_g\alpha = \nabla\alpha.g$ for any function α and any vector field g. This is called the *observation space* $\mathcal{O}$ associated to the system. We say that two states x_1 and x_2 are *separated* by $\mathcal{O}$ if there exists some $\alpha \in \mathcal{O}$ such that $\alpha(x_1) \neq \alpha(x_2)$. One can prove that if two states are separated by $\mathcal{O}$ then they are distinguishable. A sketch of the argument is as follows. Assume that x_1 is indistinguishable from x_2 and consider a piecewise constant control which is equal to u^1 on $[0, t_1)$, equal to u^2 on $[t_1, t_1 + t_2)$, ..., and equal to u^k on $[t_1 + \ldots + t_{k-1}, t_1 + \ldots + t_k)$. For small enough t_i's this control is admissible for both x_1 and x_2, and by indistinguishability we know that the resulting output at time $t = t_1 + \ldots + t_k$ is equal for both. In general, we denote the jth coordinate of this output value by

$$h_j(t_1, t_2, \ldots, t_k, u^1, u^2, \ldots, u^k, x) \tag{6.17}$$

if the initial state is x. It follows that the derivatives with respect to the t_i's of this output are also equal, for x_1 and x_2, for every such piecewise constant control. One may prove by induction that

$$\left.\frac{\partial^k}{\partial t_1 \ldots \partial t_k}\right|_{t_1=t_2=\ldots=0} h_j\left(t_1, t_2, \ldots, t_k, u^1, u^2, \ldots, u^k\right) = L_{X_1} L_{X_2} \ldots L_{X_k} h_j(x)$$

where $X_l(x) = g_0(x) + \sum_{i=1}^m u_i^l g_i(x)$. This expression is a multilinear function of the u_i^l's, and a further derivation with respect to these control value coordinates shows that the generators in (6.16) must coincide at x_1 and x_2. In the analytic case, separability by $\mathcal{O}$ is necessary as well as sufficient, because (6.17) can be expressed as a power series in terms of the generators (6.16).

The *observability rank condition* at a state $x_0 \in \mathcal{X}$ is the condition that the dimension of the span of

$$\{\nabla L_{g_{i_1}} \ldots L_{g_{i_k}} h_j(x_0) \mid i_1, \ldots, i_k \in \{0, \ldots, m\}, j = 1, \ldots p\}$$

be n. An application of the Implicit Function Theorem shows that this is sufficient for the distinguishability of states near x_0. For more details, see, for instance, [185], [199], and [311]. □

Exercise 6.4.3 Give an example of a system of class $\mathcal{C}^\infty$ with $\mathcal{X} = \mathcal{U} = \mathcal{Y} = \mathbb{R}$ which is observable but for which $\mathcal{O}$ does not separate. □

Exercise 6.4.4 Let $\dot{x} = f(x)$, $y = h(x)$ be a smooth continuous-time system with no controls. Assume that $f(0) = 0$ and $h(0) = 0$. Relate the observability rank condition to the observability of the linearization about $(0, 0)$. □

6.5 Realization Theory for Linear Systems

Motivated by the discussion in Sections 3.3 and 6.2, it becomes of interest to study *triples of matrices* (A, B, C) over a field $\mathbb{K}$,

$$A \in \mathbb{K}^{n\times n}\,, \quad B \in \mathbb{K}^{n\times m}\,, \quad C \in \mathbb{K}^{p\times n}\,,$$

corresponding to linear time-invariant systems with outputs, and it is unambiguous to call such a triple "controllable" or "observable," meaning that the corresponding discrete-time and (for $\mathbb{K} = \mathbb{R}$ or $\mathbb{C}$) continuous-time systems have that property. The *dimension* of the triple is the number n, and m and p are the *number of inputs* and *outputs*, respectively. We denote by

$$S_{n,m,p}$$

the set of all such (A, B, C)'s, and we use

$$S_{n,m,p}^{c,o}$$

for the set of controllable and observable ones. Note that for $\mathbb{K} = \mathbb{R}$ or $\mathbb{C}$, $S_{n,m,p}^{c,o}$ is an open dense subset of $S_{n,m,p}$, in the topology for

$$\mathbb{K}^{n\times n} \times \mathbb{K}^{n\times m} \times \mathbb{K}^{p\times n}$$

obtained when identifying this space with $\mathbb{K}^{n(n+m+p)}$.

Assume that (A, B, C) is so that $\operatorname{rank}\mathbf{O}(A, C) = r$ and pick a decomposition as in Equation (6.8), where $T \in GL(n)$. Write

$$\widetilde{B} := T^{-1}B = \begin{pmatrix} B_1 \\ B_2 \end{pmatrix}$$

and consider the pair (A_1, B_1). If this is not controllable, we can decompose it via Lemma 3.3.3. Let $S_1 \in GL(r)$ be so that

$$S_1^{-1}A_1S_1 = \begin{pmatrix} A_{11} & A_{12} \\ 0 & A_{22} \end{pmatrix} \quad \text{and} \quad S_1^{-1}B_1 = \begin{pmatrix} B_{11} \\ 0 \end{pmatrix}$$

and pick $S := \begin{pmatrix} S_1 & 0 \\ 0 & I \end{pmatrix} \in GL(n)$. Letting $Q := TS$, the following useful decomposition results:

Lemma 6.5.1 For any triple (A, B, C) there exists some $Q \in GL(n)$ so that

$$Q^{-1}AQ = \begin{pmatrix} A_{11} & A_{12} & 0 \\ 0 & A_{22} & 0 \\ A_{31} & A_{32} & A_{33} \end{pmatrix}$$

as well as

$$Q^{-1}B = \begin{pmatrix} B_{11} \\ 0 \\ B_{31} \end{pmatrix} \quad \text{and} \quad CQ = (\,C_{11} \quad C_{12} \quad 0\,)$$

for some matrices $A_{11}, \ldots$, and the triple

$$(A_{11}, B_{11}, C_{11})$$

is controllable and observable.

Proof. The pair (A_{11}, B_{11}) is controllable by construction, so it only remains to establish observability. Since

$$\mathbf{O}_r\left(\begin{pmatrix} A_{11} & A_{12} \\ 0 & A_{22} \end{pmatrix}, \begin{pmatrix} C_{11} & C_{12} \end{pmatrix}\right) = \begin{pmatrix} \mathbf{O}_r(A_{11}, C_{11}) & * \end{pmatrix},$$

one only needs to show that the matrix on the left has rank r. By construction, however, the pair there is similar to the pair (A_1, C_1), which was already known to be observable, so the result follows. ∎

Lemma/Exercise 6.5.2 With the notations in Lemma 6.5.1,

$$CA^iB = C_{11}A_{11}^iB_{11}$$

for all $i = 0, 1, 2, \ldots$. □

Definition 6.5.3 *The triple* (A, B, C) *is* **canonical** *if it is controllable and observable.* □

For instance, the triple (A_{11}, B_{11}, C_{11}) in the above construction is canonical. One also calls a finite dimensional time-invariant discrete-time or continuous-time system Σ a *canonical system* if the corresponding triple is.

A sequence

$$\mathcal{A} = \{\mathcal{A}_i, i = 1, 2, \ldots\},$$

where $\mathcal{A}_i \in \mathbb{K}^{p\times m}, i = 1, 2, \ldots$ for some fixed p and m, is a **Markov sequence**, and the $\mathcal{A}_i$'s are its **Markov parameters**. From Lemmas 2.4.6 and 2.7.13, the problem of realizing a discrete-time time-invariant linear behavior or a continuous-time time-invariant integral behavior is equivalent to the algebraic problem of factoring a Markov sequence as in (2.18), that is

$$\mathcal{A}_i = CA^{i-1}B \text{ for all } i > 0.$$

We say that a triple (A, B, C) for which this factorization holds is a *realization* of $\mathcal{A}$; if there is a realization, $\mathcal{A}$ is *realizable.*

From Lemmas 6.5.1 and 6.5.2 it follows that:

Corollary 6.5.4 The Markov sequence $\mathcal{A}$ is realizable if and only if there exists a canonical triple realizing it. □

Thus, for any (continuous-time or discrete-time finite dimensional) time-invariant linear system Σ, there exists a canonical system Σ_c having the same behavior as Σ. Next we study when a given sequence $\mathcal{A}$ is realizable, and we show that canonical realizations are essentially unique.

Realizability

Given a Markov sequence $\mathcal{A}$ and any pair of positive integers s, t, the (s,t)-th block **Hankel Matrix** associated to $\mathcal{A}$ is the matrix over $\mathbb{K}$:

$$\mathcal{H}_{s,t}(\mathcal{A}) := \begin{pmatrix} \mathcal{A}_1 & \mathcal{A}_2 & \cdots & \mathcal{A}_t \\ \mathcal{A}_2 & \mathcal{A}_3 & \cdots & \mathcal{A}_{t+1} \\ \vdots & \vdots & \cdots & \vdots \\ \mathcal{A}_s & \mathcal{A}_{s+1} & \cdots & \mathcal{A}_{s+t-1} \end{pmatrix} \tag{6.18}$$

of size $ps \times mt$ made up of st blocks, whose i,jth block is $\mathcal{A}_{i+j-1}$.

For any triple (A, B, C) and each k we consider the matrices $\mathbf{O}_k$ as in equation (6.6) as well as the reachability matrices

$$\mathbf{R}_k(A, B) = [B, AB, \ldots, A^{k-1}B]$$

(so that $\mathbf{R}_n = \mathbf{R}$). The main relations between these matrices and Hankel matrices are as follows:

Lemma 6.5.5 The triple (A, B, C) realizes $\mathcal{A}$ if and only if

$$\mathbf{O}_s(A, C)\, \mathbf{R}_t(A, B) \;=\; \mathcal{H}_{s,t}(\mathcal{A}) \tag{6.19}$$

for all s and t. □

Corollary 6.5.6 If (A, B, C) realizes $\mathcal{A}$, then

$$\operatorname{rank} \mathcal{H}_{s,t}(\mathcal{A}) \;\leq\; \max\{\operatorname{rank} \mathbf{O}_s(A, C), \operatorname{rank} \mathbf{R}_t(A, B)\} \leq n$$

for all s, t. □

Note that if the sequence $\mathcal{A}$ is realizable by the triple (A, B, C), then also

$$\mathbf{O}_s(A, C)\, A\, \mathbf{R}_t(A, B) \;=\; \mathcal{H}^1_{s+1,t}\,, \tag{6.20}$$

the submatrix of $\mathcal{H}_{s+1,t}$ formed by dropping the first block row.

A triple is canonical precisely if both $\mathbf{O}(A, C) = \mathbf{O}_n(A, C)$ and $\mathbf{R}(A, B)$ have rank n. In that case, there exist two matrices $\mathbf{O}^{\#}(A, C)$ and $\mathbf{R}^{\#}(A, B)$ such that

$$\mathbf{O}^{\#}(A, C)\, \mathbf{O}(A, C) \;=\; \mathbf{R}(A, B)\, \mathbf{R}^{\#}(A, B) \;=\; I\,. \tag{6.21}$$

We use the notation "#" to emphasize that in the particular case of $\mathbb{K} = \mathbb{R}$ or $\mathbb{C}$ one could use the respective pseudoinverses, though many other one-sided inverses exist.

Corollary 6.5.7 If (A, B, C) realizes $\mathcal{A}$ and (A, B, C) is a canonical triple, then

$$\operatorname{rank} \mathcal{H}_{s,t}(\mathcal{A}) = n$$

whenever $s, t \geq n$.

Proof. Using one-sided inverses satisfying (6.21), $\mathbf{O}(A,C)\mathbf{R}(A,B) = \mathcal{H}_{n,n}(\mathcal{A})$ implies that

$$I = \mathbf{O}^{\#}(A,C)\mathcal{H}_{n,n}(\mathcal{A})\mathbf{R}^{\#}(A,B)\,,$$

so $\mathcal{H}_{n,n}(\mathcal{A})$ must have rank $\geq n$, and the conclusion follows. ∎

Definition 6.5.8 *Let (A,B,C) and $(\widetilde{A},\widetilde{B},\widetilde{C})$ be two triples in $S_{n,m,p}$. Then (A,B,C)* **is similar to** *$(\widetilde{A},\widetilde{B},\widetilde{C})$, denoted*

$$(A,B,C) \sim (\widetilde{A},\widetilde{B},\widetilde{C})$$

if

$$T^{-1}AT = \widetilde{A}\,,\;\; T^{-1}B = \widetilde{B}\,,\;\; \text{and}\;\; CT = \widetilde{C}$$

for some $T \in GL(n)$. □

This is an equivalence relation, and the Markov sequences realized by any two similar triples are the same. The main result given below implies that, under a minimality assumption, the converse also holds; namely, if two minimal triples realize the same Markov sequence, then they must be similar.

Equivalence corresponds to a change of variables in the state space. Note that Lemma 6.5.1 provides a particularly important decomposition under similarity.

Definition 6.5.9 *The n-dimensional triple (A,B,C) is* **minimal** *if any other triple $(\widetilde{A},\widetilde{B},\widetilde{C})$ realizing the same Markov sequence $\mathcal{A}$ must have dimension at least n.* □

Minimality is in principle hard to check, since it involves comparisons with all other possible realizations of the same Markov sequence. The next result shows that this property is equivalent to being canonical and hence can be checked directly in terms of the data describing the system; we use the statement to summarize all relevant properties of realizations.

Theorem 27 *Assume that $\mathcal{A}$ is a Markov sequence. Then the following properties hold:*

1. *If there is a realization of $\mathcal{A}$, then there is also a canonical realization.*
2. *A realization of $\mathcal{A}$ is minimal if and only if it is canonical.*
3. *Any two minimal realizations of $\mathcal{A}$ must be similar.*

Proof. Statement 1 was given earlier as Corollary 6.5.4. We now prove statement 2.

Let (A,B,C) be a canonical triple of dimension n. By Corollary 6.5.7, $\operatorname{rank} \mathcal{H}_{n,n}(\mathcal{A}) = n$. It follows from Corollary 6.5.6 that every other realization has dimension at least n, so the triple is minimal.

On the other hand, if (A, B, C) is minimal, then it must be canonical, since otherwise from Lemma 6.5.2 and the construction in Lemma 6.5.1 it would follow that the triple (A_{11}, B_{11}, C_{11}) has lower dimension and realizes the same Markov sequence.

Next we establish statement 3. Assume that (A, B, C) and $(\widetilde{A}, \widetilde{B}, \widetilde{C})$ both realize $\mathcal{A}$ and are minimal, so in particular they must have the same dimension. Denote

$$\mathbf{R} = \mathbf{R}(A, B)\,,\ \widetilde{\mathbf{R}} = \mathbf{R}(\widetilde{A}, \widetilde{B})\,,\ \mathbf{O} = \mathbf{O}(A, C)\,,\ \widetilde{\mathbf{O}} = \mathbf{O}(\widetilde{A}, \widetilde{C})$$

and note that

$$\mathbf{OR} = \widetilde{\mathbf{O}}\widetilde{\mathbf{R}} \quad \text{and} \quad \mathbf{O}A\mathbf{R} = \widetilde{\mathbf{O}}\widetilde{A}\widetilde{\mathbf{R}}$$

because of Equations (6.19) and (6.20) applied with $s = t = n$. Let $\widetilde{\mathbf{R}}^{\#}, \ldots,$ be one-sided inverses as in Equation (6.21), for each of the triples. Then with

$$T := \mathbf{R}\widetilde{\mathbf{R}}^{\#} = \left(\widetilde{\mathbf{O}}^{\#}\mathbf{O}\right)^{-1}$$

it follows that $T^{-1}AT = \widetilde{A}$. Applying Equation (6.19) with $s = n, t = 1$, and observing that $\mathbf{R}_1 = B, \widetilde{\mathbf{R}}_1 = \widetilde{B}$, we also obtain that $T^{-1}B = \widetilde{B}$. Finally, applying Equation (6.19) with $s = 1, t = n$ we obtain the remaining equality $CT = \widetilde{C}$. ∎

In terms of realizations of discrete-time time-invariant linear behaviors and continuous-time time-invariant integral behaviors, the Theorem asserts that canonical realizations exist if the behavior is realizable, and they are unique up to a change of basis in the state space.

Remark 6.5.10 If two controllable triples are similar, then the similarity must be given by the formulas obtained in the proof of Theorem 27, so in particular similarities between canonical systems are unique. This is because if T is as in Definition 6.5.8, then necessarily

$$T^{-1}A^iB = \widetilde{A}^i\widetilde{B}$$

for all i, so it must hold that $T^{-1}\mathbf{R} = \widetilde{\mathbf{R}}$, and therefore

$$T = \mathbf{R}\widetilde{\mathbf{R}}^{\#}\,.$$

In particular, this means that the only similarity between a canonical system and itself is the identity. In the terminology of group theory, this says that the action of $GL(n)$ on triples is free. □

Theorem 27 leaves open the question of deciding *when* a realization exists; this is addressed next.

Definition 6.5.11 *The* **rank** *of the Markov sequence* $\mathcal{A}$ *is*

$$\sup_{s,t} \operatorname{rank} \mathcal{H}_{s,t}(\mathcal{A})\,.$$

□

Remark 6.5.12 In terms of the *infinite Hankel matrix* which is expressed in block form as

$$\mathcal{H}(\mathcal{A}) = \begin{pmatrix} \mathcal{A}_1 & \mathcal{A}_2 & \cdots & \mathcal{A}_t & \cdots \\ \mathcal{A}_2 & \mathcal{A}_3 & \cdots & \mathcal{A}_{t+1} & \cdots \\ \vdots & \vdots & \cdots & \vdots & \vdots \\ \mathcal{A}_s & \mathcal{A}_{s+1} & \cdots & \mathcal{A}_{s+t-1} & \cdots \\ \vdots & \vdots & \cdots & \vdots & \vdots \end{pmatrix}$$

one may restate the definition of rank as follows.

The *rank* of an infinite matrix such as $\mathcal{H}(\mathcal{A})$ is, by definition, the dimension of the column space of $\mathcal{H}(\mathcal{A})$, which is seen as a subspace of the space $\mathbb{K}^\infty$ consisting of infinite column vectors $(x_1, x_2, x_3, \ldots)'$ with entries over $\mathbb{K}$, with pointwise operations.

When this rank is finite and less than n, all columns are linear combinations of at most $n-1$ columns, and therefore all submatrices of $\mathcal{H}(\mathcal{A})$ must have rank less than n; this implies that $\mathcal{A}$ must have rank less than n. Conversely, we claim that if $\mathcal{A}$ has rank less than n, then the rank of the infinite matrix $\mathcal{H}(\mathcal{A})$ is less than n. If this were not to be the case, there would be a set of n independent columns $c_1, \ldots, c_n$ of $\mathcal{H}(\mathcal{A})$. If so, let M_i be the $i \times n$ matrix obtained by truncating each column c_j at the first i rows. Then some M_i has rank n: Consider the nonincreasing sequence of subspaces

$$Q_i := \ker M_i \subseteq \mathbb{K}^n$$

and let $Q := \cap Q_i$; by dimensionality, there is some k so that $Q_k = Q$, and if $x \in Q$ then $M_i x = 0$ for all i means that x is a linear relation between the c_j's, so $x = 0$. This means that $Q_k = 0$ and so M_k has rank n as desired. Let s, t be such that M_k is a submatrix of $\mathcal{H}_{s,t}$. Then

$$\operatorname{rank} \mathcal{A} \geq \operatorname{rank} \mathcal{H}_{s,t} \geq \operatorname{rank} M_k = n\,,$$

contradicting $\operatorname{rank} \mathcal{A} < n$.

The conclusion is that the rank of $\mathcal{A}$ is equal to the rank (possibly infinite) of $\mathcal{H}(\mathcal{A})$. □

Theorem 28 *The Markov sequence $\mathcal{A}$ is realizable if and only if it has finite rank n. In addition, if this holds, then:*

(a) There is a canonical realization of dimension n.

(b) $\operatorname{rank} \mathcal{H}_{n,n} = n$.

Proof. If there is a realization of dimension k, then it follows from Corollary 6.5.6 that $\mathcal{A}$ has rank at most k, and in particular the necessity statement is obtained. We show next that if $\mathcal{A}$ has rank n, then there is some rank n canonical realization; this will establish property (a) as well as sufficiency. Property (b) is then a consequence of Corollary 6.5.7.

For the construction of the canonical realization, it is convenient to first generalize the notion of realizability to allow for infinite dimensional triples. In general, we consider objects $(\mathcal{X}, A, B, C)$ consisting of a vector space $\mathcal{X}$ and linear maps $A : \mathcal{X} \to \mathcal{X}$, $B : \mathbb{K}^m \to \mathcal{X}$, and $C : \mathcal{X} \to \mathbb{K}^p$, and say that $\mathcal{A}$ is realizable by $(\mathcal{X}, A, B, C)$ if

$$C \circ A^{i-1} \circ B = \mathcal{A}_i$$

for all i. These objects can be identified to discrete-time time-invariant linear systems; when $\mathcal{X}$ is finite dimensional, choosing a basis on $\mathcal{X}$ provides a triple realizing $\mathcal{A}$.

Now given any Markov sequence $\mathcal{A}$, we let $\mathcal{X}_0$ denote the space $\mathbb{K}^\infty$ introduced in Remark 6.5.12. Note that the *shift operator*

$$\sigma : \begin{pmatrix} x_1 \\ x_2 \\ x_3 \\ \vdots \end{pmatrix} \mapsto \begin{pmatrix} x_2 \\ x_3 \\ x_4 \\ \vdots \end{pmatrix}$$

is linear on $\mathcal{X}_0$. Let A be defined as σ^p, the shift by p positions. Let $B : \mathbb{K}^m \to \mathcal{X}_0$ be defined on the natural basis as follows:

$$Be_j := \; j\text{th column of } \mathcal{H}(\mathcal{A})$$

and let $C : \mathcal{X}_0 \to \mathbb{K}^p$ be the projection on the first p coordinates.

We claim that $(\mathcal{X}_0, A, B, C)$ is a realization of $\mathcal{A}$. Consider any $i \geq 0$ and any $j = 1, \ldots, m$. Then CA^iBe_j is the vector consisting of the entries in positions $ip+1, \ldots, ip+p$ of the jth column of $\mathcal{H}(\mathcal{A})$, that is, the jth column of $\mathcal{A}_{i+1}$, as wanted.

The abstract realization just obtained is observable, in the sense that

$$\bigcap_{i=0}^{\infty} \ker CA^i \;=\; 0\,,$$

since for each i the elements of $\ker CA^i$ are precisely those vectors whose entries in positions $ip+1, \ldots, ip+p$ all vanish. The desired canonical realization will be obtained by restricting to an A-invariant subspace of $\mathcal{X}_0$; this restriction will also be observable, since the kernels of the restrictions still must intersect at zero.

Let $\mathcal{X}$ be the subspace of $\mathcal{X}_0$ spanned by all the iterates A^iBe_j, over all i, j. This is A-invariant by definition; we denote the restriction of A to $\mathcal{X}$ again as A. Moreover, $\mathcal{X}$ contains the image of B, so we may consider B as a map from $\mathbb{K}^m$ into $\mathcal{X}$; similarly, we restrict C to $\mathcal{X}$ and denote it also by C. We claim next that this space equals the column space of $\mathcal{H}(\mathcal{A})$. Here the Hankel pattern becomes essential: Every column of $\mathcal{H}(\mathcal{A})$ is of the form A^iBe_j for some i and j; namely, the $im+j$th column ($1 \leq j \leq m$) is the same as A^iBe_j, for each $i \geq 0$.

We conclude that $(\mathcal{X}, A, B, C)$ realizes $\mathcal{A}$, where $\mathcal{X}$ is the column space of $\mathcal{H}(\mathcal{A})$ and therefore is n-dimensional. ∎

The criterion in Theorem 28 is of interest because it allows us to check realizability, in the sense that one may look at Hankel matrices of increasing size and if a realization exists it will eventually be found. If a realization does not exist, however, there is no way of knowing so by means of this technique.

If one is interested merely in investigating the existence of realizations, as opposed to studying minimality, the problem can be reduced to the scalar case ($m = p = 1$), which in turn belongs to the classical theory of linear difference equations:

Exercise 6.5.13 Given any Markov sequence $\mathcal{A}$, introduce the pm sequences corresponding to each coordinate, that is,

$$\mathcal{A}^{ij} := (\mathcal{A}_1)_{ij}, (\mathcal{A}_2)_{ij}, (\mathcal{A}_3)_{ij}, \ldots$$

for each $i = 1, \ldots, p$ and $j = 1, \ldots, m$. Show, not using any of the results just developed, that $\mathcal{A}$ is realizable if and only if each $\mathcal{A}^{ij}$ is. □

Exercise 6.5.14 Calculate a canonical realization, and separately calculate the rank of the Hankel matrix, for each of these examples with $m = p = 1$:

1. The sequence of natural numbers $1, 2, 3, \ldots$.

2. The Fibonacci sequence $0, 1, 1, 2, 3, 5, 8, \ldots$. □

6.6 Recursion and Partial Realization

An alternative characterization of realizability is through the concept of recursive sequences. We shall say that $\mathcal{A}$ is a *recursive Markov sequence* if there exist a positive integer n and scalars $\alpha_1, \ldots, \alpha_n$ such that

$$\mathcal{A}_{k+n+1} = \alpha_n \mathcal{A}_{k+n} + \alpha_{n-1}\mathcal{A}_{k+n-1} + \ldots + \alpha_2 \mathcal{A}_{k+2} + \alpha_1 \mathcal{A}_{k+1} \qquad (6.22)$$

for all $k \geq 0$; in this case $\mathcal{A}$ is said to satisfy a recursion of order n.

Proposition 6.6.1 The Markov sequence $\mathcal{A}$ is realizable if and only if it is recursive.

Proof. If (A, B, C) realizes $\mathcal{A}$, let

$$\chi_A(s) = s^n - \alpha_n s^{n-1} - \ldots - \alpha_1$$

so that by the Cayley-Hamilton Theorem

$$A^n = \alpha_n A^{n-1} + \ldots + \alpha_1 I .$$

Multiplying this last equation by CA^k on the left and by B on the right, the recursion (6.22) results. Conversely, if there is a recursion of order n, all columns of the infinite Hankel matrix must be linear combinations of the first nm columns (that is, the columns appearing in the first n blocks), so the matrix has finite rank and therefore $\mathcal{A}$ is realizable. ∎

Corollary 6.6.2 If $\mathcal{A}$ is a Markov sequence of finite rank n, then $\mathcal{A} \equiv 0$ if and only if

$$\mathcal{A}_1 = \mathcal{A}_2 = \ldots = \mathcal{A}_n = 0 .$$

Proof. By Theorem 28 there exists a realization of dimension n, and therefore, by the proof of the Proposition, there is a recursion of order n. Recursively, all $\mathcal{A}_i = 0$. ∎

Corollary 6.6.3 If $\mathcal{A}^1$ and $\mathcal{A}^2$ are two Markov sequences of finite ranks n_1 and n_2, respectively, then $\mathcal{A}^1 = \mathcal{A}^2$ if and only if

$$\mathcal{A}_i^1 = \mathcal{A}_i^2 \,,\; i = 1, \ldots, n_1 + n_2 .$$

Proof. The sequence $\mathcal{A} := \mathcal{A}^1 - \mathcal{A}^2 = \{\mathcal{A}_i^1 - \mathcal{A}_i^2\}$ has rank at most $n := n_1 + n_2$, as follows by considering its realization:

$$\begin{pmatrix} A_1 & 0 \\ 0 & A_2 \end{pmatrix} \begin{pmatrix} B_1 \\ B_2 \end{pmatrix} \quad (C_1 \quad -C_2)$$

where (A_i, B_i, C_i) is a realization of $\mathcal{A}^i$ of dimension n_i. Then Corollary 6.6.2 gives the result. ∎

Thus, sequences of rank $\leq n$ are uniquely determined by their first $2n$ Markov parameters.

There is a direct construction of a realization from the recursion (6.22) and the Markov sequence. This is as follows. Let $\mathcal{H}_{n+k,n}^k$ be the matrix obtained by dropping the first k blocks of rows from $\mathcal{H}_{n+k,n}(\mathcal{A})$, that is, the matrix

$$\begin{pmatrix} \mathcal{A}_{k+1} & \cdots & \mathcal{A}_{k+n} \\ \vdots & \cdots & \vdots \\ \mathcal{A}_{k+n} & \cdots & \mathcal{A}_{k+2n-1} \end{pmatrix}$$

including the case $k = 0$, where this is just $\mathcal{H}_{n,n}(\mathcal{A})$. Then (6.22) implies

$$A\,\mathcal{H}_{n+k,n}^k = \mathcal{H}_{n+k+1,n}^{k+1} \tag{6.23}$$

for all $k \geq 0$, where

$$A := \begin{pmatrix} 0 & I & 0 & \dots & 0 \\ 0 & 0 & I & \dots & 0 \\ \vdots & \vdots & \vdots & \ddots & \vdots \\ 0 & 0 & 0 & \dots & I \\ \alpha_1 I & \alpha_2 I & \alpha_3 I & \dots & \alpha_n I \end{pmatrix} \tag{6.24}$$

analogously to the controller form in Definition 5.1.5, except that this is now a block matrix. Each block has size $p \times p$, and the matrix A is of size $np \times np$. Equation (6.23) implies that

$$A^i \mathcal{H}_{n,n}(\mathcal{A}) = \mathcal{H}^i_{n+i,n}$$

for all $i \geq 0$, and so

$$(I \quad 0 \quad \cdots \quad 0)\, A^i \mathcal{H}_{n,n}(\mathcal{A}) \begin{pmatrix} I \\ 0 \\ \vdots \\ 0 \end{pmatrix} = \mathcal{A}_{i+1} \tag{6.25}$$

for all $i \geq 0$. Letting

$$C := (I \quad 0 \quad \cdots \quad 0) \tag{6.26}$$

and

$$B := \mathcal{H}_{n,n}(\mathcal{A}) \begin{pmatrix} I \\ 0 \\ \vdots \\ 0 \end{pmatrix} = \begin{pmatrix} \mathcal{A}_1 \\ \mathcal{A}_2 \\ \vdots \\ \mathcal{A}_n \end{pmatrix}, \tag{6.27}$$

there results a realization of $\mathcal{A}$ of dimension np. Since $\mathbf{O}_n(A, C) = I$:

Lemma 6.6.4 The system (A, B, C) given by (6.24), (6.27), and (6.26) is an observable realization of $\mathcal{A}$. □

This is often called an **observability form realization** of $\mathcal{A}$.

Remark 6.6.5 Let $\mathcal{A}$ be any recursive Markov sequence. Consider the transposed sequence

$$\mathcal{A}' := \mathcal{A}_1', \mathcal{A}_2', \ldots .$$

This satisfies a recursion with the same coefficients α_i's. For these coefficients we let (A, B, C) be the observability form realization of $\mathcal{A}'$. Since $CA^{i-1}B = \mathcal{A}_i'$ for all i, also $B'(A')^{i-1}C' = \mathcal{A}_i$ for all i, and the system (A', C', B') is a controllable realization of $\mathcal{A}$. We have obtained the **controllability form realization** of $\mathcal{A}$:

$$A = \begin{pmatrix} 0 & 0 & \cdots & 0 & \alpha_1 I \\ I & 0 & \cdots & 0 & \alpha_2 I \\ 0 & I & \cdots & 0 & \alpha_3 I \\ \vdots & \vdots & \ddots & \vdots & \vdots \\ 0 & 0 & \cdots & I & \alpha_n I \end{pmatrix} \quad B = \begin{pmatrix} I \\ 0 \\ 0 \\ \vdots \\ 0 \end{pmatrix} \quad C = (\mathcal{A}_1 \quad \mathcal{A}_2 \quad \cdots \quad \mathcal{A}_n)$$

of dimension mn. □

Corollary 6.6.6 If $p = 1$ or $m = 1$, the minimal possible order of a recursion equals the rank of $\mathcal{A}$. Furthermore, there is a unique such minimal order recursion, and its coefficients are those of the (common) characteristic polynomial of canonical realizations of $\mathcal{A}$.

Proof. The proof of Proposition 6.6.1 shows that, for arbitary m, p, if there is a realization of dimension n, then there is a recursion of order n. Conversely, in the cases $p = 1$ or $m = 1$ there is always a realization of dimension equal to the order of any given recursion, namely the observability form or the controllability form realization, respectively. The second assertion follows from the constructions. ∎

Remark 6.6.7 Observability form realizations are in general not controllable, except if $p = 1$. In this case, if $(\alpha_1, \ldots, \alpha_n)$ give a minimal recursion then by Corollary 6.6.6 this realization must be minimal. In fact,

$$\mathbf{R}_n(A, B) = \mathcal{H}_{n,n} ,$$

which has rank n. Of course, for $p > 1$ it is possible to reduce any such realization to a controllable one using the Kalman decomposition. □

Exercise 6.6.8 Show by providing a counterexample that the hypothesis that either $p = 1$ or $m = 1$ cannot be dropped in the first assertion on Corollary 6.6.6. □

Given any finite sequence $\mathcal{A}_1, \mathcal{A}_2, \ldots, \mathcal{A}_r$, this is always part of a realizable Markov sequence, since

$$\mathcal{A}_1, \mathcal{A}_2, \ldots, \mathcal{A}_r, 0, 0, \ldots, 0, \ldots$$

has finite rank. More interesting is the following fact. Given any finite sequence $\mathcal{A}_1, \mathcal{A}_2, \ldots, \mathcal{A}_r$, whenever $s+t \leq r+1$ we can consider the Hankel matrices $\mathcal{H}_{s,t}$ obtained using formula (6.18).

Lemma 6.6.9 Assume that $\mathcal{A}_1, \mathcal{A}_2, \ldots, \mathcal{A}_{2n}$ are $2n$ matrices in $\mathbb{K}^{p \times m}$ for which

$$\operatorname{rank} \mathcal{H}_{n,n} = \operatorname{rank} \mathcal{H}_{n+1,n} = \operatorname{rank} \mathcal{H}_{n,n+1} = n . \tag{6.28}$$

Then there exists a (unique) Markov sequence $\mathcal{A}$ of rank n whose first $2n$ Markov parameters are $\mathcal{A}_1, \mathcal{A}_2, \ldots, \mathcal{A}_{2n}$.

Proof. From the equality $\operatorname{rank} \mathcal{H}_{n,n} = \operatorname{rank} \mathcal{H}_{n+1,n}$ it follows that the last p rows of $\mathcal{H}_{n+1,n}$ must be linear combinations of the rows of $\mathcal{H}_{n,n}$. This means that there must exist $p \times p$ matrices

$$C_i \, , \; i = 1, \ldots, n$$

such that

$$\mathcal{A}_j = C_1\mathcal{A}_{j-1} + \ldots + C_n\mathcal{A}_{j-n} \tag{6.29}$$

for each $j = n+1, \ldots, 2n$. Similarly, $\operatorname{rank}\mathcal{H}_{n,n} = \operatorname{rank}\mathcal{H}_{n,n+1}$ implies that the last m columns of $\mathcal{H}_{n,n+1}$ must be linear combinations of the columns of $\mathcal{H}_{n,n}$. So there must also exist $m \times m$ matrices

$$D_i\,,\; i = 1, \ldots, n$$

such that

$$\mathcal{A}_j = \mathcal{A}_{j-1}D_1 + \ldots + \mathcal{A}_{j-n}D_n \tag{6.30}$$

for each $j = n+1, \ldots, 2n$.

We now *define* $\mathcal{A}_j$ for $j > 2n$ recursively using the formula (6.29), and let $\mathcal{A}$ be the Markov sequence so obtained. It follows from this definition that all rows of $\mathcal{H}(\mathcal{A})$ are linearly dependent on the first pn rows. So $\mathcal{A}$ has rank at most pn; we next show that its rank is in fact just n. For this, it is enough to establish that (6.30) holds for all $j > 2n$ as well, since this will then imply that all columns depend linearly on the first nm columns, and therefore that the rank of $\mathcal{H}(\mathcal{A})$ is the same as the rank of $\mathcal{H}_{n,n}$, which is n. By induction on j,

$$\begin{aligned}
\mathcal{A}_{j+1} &= \sum_{i=1}^{n} C_i\mathcal{A}_{j+1-i} \\
&= \sum_{i=1}^{n} C_i \sum_{l=1}^{n} \mathcal{A}_{j+1-i-l}D_l \\
&= \sum_{l=1}^{n}\left(\sum_{i=1}^{n} C_i\mathcal{A}_{j+1-i-l}\right) D_l \\
&= \sum_{l=1}^{n} \mathcal{A}_{j+1-l}D_l\,,
\end{aligned}$$

as desired.

Uniqueness follows from the fact that any other extension would agree in its first $2n$ parameters and hence would have to be the same, because of Corollary 6.6.3. ∎

The previous result can be used to give an explicit description of the quotient space obtained when identifying triples up to similarity. Let $\mathcal{M}_{n,m,p}$ denote the set

$$\{(\mathcal{A}_1, \ldots, \mathcal{A}_{2n}) \mid \operatorname{rank}\mathcal{H}_{n,n} = \operatorname{rank}\mathcal{H}_{n+1,n} = \operatorname{rank}\mathcal{H}_{n,n+1} = n\}\,.$$

Note that this can be thought of as a subset of $\mathbb{K}^{2nmp}$, and as such it is a set defined by polynomial equalities and inequalities. Let

$$\beta : S^{c,o}_{n,m,p} \to \mathcal{M}_{n,m,p}$$

be the "behavior" function

$$(A,B,C) \mapsto (CB,\ldots,CA^{2n-1}B).$$

That β indeed maps into $\mathcal{M}_{n,m,p}$ follows from Theorem 28 (p. 288). Moreover, from Lemma 6.6.9 we know that β is onto. From the uniqueness result, we also know that, if

$$\beta(A,B,C) = \beta(\widetilde{A},\widetilde{B},\widetilde{C}),$$

then (A,B,C) and $(\widetilde{A},\widetilde{B},\widetilde{C})$ are two canonical triples realizing the same Markov sequence and so, by Theorem 27 (p. 286), Part 3, that these two triples are similar. This discussion can be summarized as follows.

Corollary 6.6.10 The map β induces a bijection between the quotient space $S^{c,o}_{n,m,p}/\sim$ and $\mathcal{M}_{n,m,p}$. □

Exercise 6.6.11 Let $\mathbb{K} = \mathbb{R}$ or $\mathbb{C}$. Let $\mathcal{M}_{n,m,p}$ have the topology induced by $\mathbb{K}^{2nmp}$, and let $S^{c,o}_{n,m,p}/\sim$ be given the quotient topology, when $S^{c,o}_{n,m,p}$ is thought of as a subspace of $\mathbb{K}^{n(n+p+m)}$. Show that β induces a homeomorphism on the quotient space. (*Hint:* Establish first that the realization given in the proof of Theorem 28 can be made to be locally continuous on the sequence. This is done by observing that, by Cramer's rule after choosing a set of n linearly independent columns of $\mathcal{H}_{n,n}$, the entries of the operator A can be taken to be rational on the coefficients of the Markov parameters. This provides a covering of $\mathcal{M}_{n,m,p}$ by open subsets V and continuous mappings $\rho : V \to S^{c,o}_{n,m,p}$ such that $\beta\rho$ is the identity.) □

Remark 6.6.12 (This remark uses some concepts from differential geometry.) In the case when $\mathbb{K} = \mathbb{R}$ (or the complex case) one can show that the quotient space $S^{c,o}_{n,m,p}/\sim$ has a differentiable manifold structure under which the natural projection is smooth. This fact can be established constructively, by exhibiting an explicit set of charts for this quotient manifold, or one can use a general theorem on group actions, the quicker path which we choose here. Consider the action of $GL(n)$ on $S^{c,o}_{n,m,p}$,

$$T.(A,B,C) := (T^{-1}AT, T^{-1}B, CT),$$

seen as a smooth action of a Lie group on a manifold ($S^{c,o}_{n,m,p}$ is an open subset of $\mathbb{R}^{n(n+m+p)}$). This action is free (see Remark 6.5.10). According to Proposition 4.1.23 in [2], the quotient will have a differentiable manifold structure for which the quotient mapping $S^{c,o}_{n,m,p} \to S^{c,o}_{n,m,p}/\sim$ is a smooth submersion, provided that the graph of the similarity relation is closed and that the action is proper. Moreover, in this case the natural map

$$S^{c,o}_{n,m,p} \to S^{c,o}_{n,m,p}/\sim$$

defines a principal fibre bundle (same reference, Exercise 4.1M). Properness of the action means that the following property must hold: Whenever $\{\Sigma_i\}$ is a convergent sequence of triples and $\{T_i\}$ is a sequence of elements of $GL(n)$ for which $\{T_i.\Sigma_i\}$ is a convergent sequence of triples, the sequence $\{T_i\}$ must have a convergent subsequence. So we must prove that the action is closed and proper.

Assume that the sequences $\{\Sigma_i\}$ and $\{T_i\}$ are as in the above paragraph, and let $\Sigma_i' := T_i.\Sigma_i$ for each i. By assumption, there are systems Σ and Σ' so that

$$\Sigma_i \to \Sigma \quad \text{and} \quad \Sigma_i' \to \Sigma' .$$

We use primes and subscripts to denote the matrices A, B, C for the various triples. The triples Σ and Σ' must realize the same Markov sequence, since this is true of the corresponding pairs (Σ_i, Σ_i') and the sequence's elements depend continuously on the triple. Therefore the matrix

$$T = \mathbf{R}(\mathbf{R}')^{\#}$$

provides a similarity between these two triples. Observe that $\mathbf{R}_{(i)}$ (the n-step reachability matrix for the triple Σ_i) converges to $\mathbf{R}$. Moreover, and this is the critical observation, one may also assume that $(\mathbf{R}'_{(i)})^{\#}$, the one-sided inverse of the n-step reachability matrix for the triple Σ_i', also converges to $(\mathbf{R}')^{\#}$. The reason for this latter fact is that one may *pick* such a one-sided inverse continuously about any given system: Just use Cramer's rule after choosing a set of n linearly independent columns of $\mathbf{R}'$ (these columns remain linearly independent for triples near the triple Σ'). We conclude that

$$T_i = \mathbf{R}_{(i)}(\mathbf{R}'_{(i)})^{\#} \to T$$

because of uniqueness of the similarity between two minimal systems (Remark 6.5.10). This establishes properness. (In fact, we proved that the sequence T_i itself is convergent, rather than merely a subsequence.)

The proof of closeness is even easier. We need to see that, if

$$\Sigma_i \sim \Sigma_i' \quad \text{for all } i$$

and

$$\Sigma_i \to \Sigma \, , \quad \Sigma_i' \to \Sigma' \, ,$$

then necessarily Σ and Σ' are similar. This is immediate from the uniqueness Theorem, because by continuity these two triples must give rise to the same Markov sequence. □

Corollary 6.6.10, Exercise 6.6.11, and Remark 6.6.12 provide only a very brief introduction to the topic of *families of systems* and more specifically to *moduli problems* for triples. Much more detailed results are known; For instance, $S^{c,o}_{n,m,p}/\sim$ has a natural structure of nonsingular algebraic variety consistent with the differentiable manifold structure given above, for $\mathbb{K} = \mathbb{R}, \mathbb{C}$.

6.7 Rationality and Realizability

We now characterize realizability of a Markov sequence in terms of the rationality of an associated power series.

Consider the set of all semi-infinite sequences

$$a_k, a_{k+1}, \ldots, a_0, a_1, a_2, a_3, \ldots, a_l, \ldots$$

formed out of elements of $\mathbb{K}$. The starting index k may be negative or positive but it is finite; however, the sequence may be infinite to the right. We think of these sequences as formal *Laurent series* in a variable s^{-1}:

$$\sum_{i=k}^{\infty} a_i s^{-i} ,$$

where $a_i \in \mathbb{K}$ for each i. If the sequence is not identically zero and k is the smallest integer so that $a_k \neq 0$, then the series has *order* k. If $k \geq 0$, this is just a formal power series in the variable s^{-1}. If all coefficients $a_i, i > 0$ vanish, this is simply a polynomial on s. Let $\mathcal{K}((s^{-1}))$ denote the set of all such Laurent series.

The set $\mathcal{K}((s^{-1}))$ has a natural vector space structure over $\mathbb{K}$, corresponding to coefficientwise operations:

$$\left(\sum a_i s^{-i}\right) + \left(\sum b_i s^{-i}\right) := \sum (a_i + b_i) s^{-i} , \tag{6.31}$$

but the use of the power series notation also suggests the convolution product

$$\left(\sum a_i s^{-i}\right) \cdot \left(\sum b_j s^{-j}\right) := \sum c_l s^{-l} , \tag{6.32}$$

where for each $l \in \mathbb{Z}$

$$c_l := \sum_{i+j=l} a_i b_j$$

is a finite sum because there are at most finitely many nonzero coefficients with negative i and j. With these operations, $\mathcal{K}((s^{-1}))$ forms a ring. It is an integral domain, that is, the product of two nonzero sequences cannot be zero, because if k_1 and k_2 are the orders of these sequences, then the product has order $k_1 + k_2$. (In fact, since $\mathbb{K}$ is a field, it can be shown that $\mathcal{K}((s^{-1}))$ is also a field, but this is not needed in what follows.)

More generally, we consider series whose coefficients are matrices of a fixed size:

$$\sum_{i=k}^{\infty} A_i s^{-i} ,$$

where $A_i \in \mathbb{K}^{p\times m}$ for all i. The sum and product given by formulas (6.31) and (6.32) are still well defined (assuming that sizes match), where now each A_i,

B_j and C_l is a matrix. The usual distributivity and associativity properties of matrix product hold. We let $\mathcal{K}^{p\times m}((s^{-1}))$ denote the set of all matrix Laurent series with fixed p and m.

A series $W(s) \in \mathcal{K}^{p\times m}((s^{-1}))$ will be said to be **rational** if there exist a monic polynomial

$$q(s) = s^n - \alpha_n s^{n-1} - \ldots - \alpha_1 \tag{6.33}$$

and a matrix polynomial

$$P(s) = P_0 + P_1 s + \ldots + P_h s^h \tag{6.34}$$

in $\mathbb{K}^{p\times m}[s]$ such that

$$qW = P$$

(which is also written as $W = q^{-1}P$, or P/q). For instance, any series having only finitely many terms is rational, since

$$A_k s^k + \ldots + A_0 + A_1 s^{-1} + \ldots + A_l s^{-l}$$

can be written as

$$\frac{1}{s^l}(A_k s^{k+l} + \ldots + A_0 s^l + A_1 s^{l-1} + \ldots + A_l)\,.$$

For another example consider the scalar series

$$1 + s^{-1} + s^{-2} + \ldots + s^{-k} + \ldots\,,$$

which is rational since it equals $s/(s-1)$.

We associate to each Markov sequence $\mathcal{A} = (\mathcal{A}_1, \mathcal{A}_2, \mathcal{A}_3, \ldots)$ its *generating series*

$$W_{\mathcal{A}}(s) := \sum_{i=1}^{\infty} \mathcal{A}_i s^{-i}\,.$$

In terms of this we may state another criterion for realizability:

Proposition 6.7.1 A Markov sequence is realizable if and only if its generating series is rational.

Proof. Because of Proposition 6.6.1, we must simply show that $W_{\mathcal{A}}$ is rational if and only if $\mathcal{A}$ is recursive.

Note first that, in general, if $W = q^{-1}P$ is rational and has order ≥ 1 and q is as in (6.33), then the polynomial P in (6.34) can be taken to be of degree at most $n-1$, since all terms in qW corresponding to s^h, $h \geq n$, must vanish. So rationality of $W_{\mathcal{A}}$ is equivalent to the existence of elements $\alpha_1, \ldots, \alpha_n$ in $\mathbb{K}$ and matrices $P_0, \ldots, P_{n-1}$ over $\mathbb{K}$ such that

$$(s^n - \alpha_n s^{n-1} - \ldots - \alpha_1)\left(\sum_{i=1}^{\infty} \mathcal{A}_i s^{-i}\right) = P_0 + P_1 s + \ldots + P_{n-1} s^{n-1}\,. \tag{6.35}$$

If such an equation holds, then comparing coefficients of s^{-k-1} results in

$$\mathcal{A}_{k+n+1} = \alpha_n \mathcal{A}_{k+n} + \alpha_{n-1}\mathcal{A}_{k+n-1} + \ldots + \alpha_2 \mathcal{A}_{k+2} + \alpha_1 \mathcal{A}_{k+1}$$

for each $k \geq 0$, so the sequence is recursive. Conversely, if this equation holds one may just define the matrices P_i by

$$P_j := \mathcal{A}_{n-j} - \sum_{i=1}^{n-j-1} \alpha_{i+j+1}\mathcal{A}_i \quad j = 0, \ldots, n-1, \tag{6.36}$$

i.e., the equations that are derived from (6.35) by comparing the coefficients of $1, s, \ldots, s^{n-1}$. ∎

The proof shows that the minimal degree of a possible denominator q equals the minimal order of a recursion satisfied by $\mathcal{A}$.

Corollary 6.7.2 If $m = 1$ or $p = 1$ and $W_{\mathcal{A}} = P/q$ with q monic of degree equal to the rank of $\mathcal{A}$, then q is the (common) characteristic polynomial of the canonical realizations of $\mathcal{A}$.

Proof. We know from Corollary 6.6.6 that for m or p equal to 1 the coefficients of a minimal recursion are those of the characteristic polynomial. The q in the statement must be a denominator of minimal degree, since a lower degree polynomial would give rise to a lower order recursion and therefore to a lower dimensional realization. From the above construction, this polynomial then corresponds to a minimal recursion. ∎

Exercise 6.7.3 Show that, if $m = p = 1$ and if $W_{\mathcal{A}} = P/q$ with P of degree $\leq n-1$ and q of degree n, then $\mathcal{A}$ has rank n if and only if P and q are relatively prime. (*Hint:* Use Corollary 6.6.6 and the fact that $\mathcal{K}((s^{-1}))$ forms an integral domain.) □

Remark 6.7.4 One can show directly that realizability implies rationality, as follows. Elements of $\mathcal{K}^{p\times m}((s^{-1}))$ can be identified naturally with $p \times m$ matrices over the ring $\mathcal{K}((s^{-1}))$, and this identification preserves the convolution structure. Under this identification, rational elements are precisely those that correspond to matrices all whose entries are rational. In particular, the matrix $(sI - A)^{-1}$, the inverse over $\mathcal{K}((s^{-1}))^{n\times n}$, is rational. On the other hand, it holds that

$$(sI - A)^{-1} = \sum_{i=1}^{\infty} A^{i-1}s^{-i}$$

from which it follows that the series

$$C(sI - A)^{-1}B = \sum_{i=1}^{\infty} CA^{i-1}Bs^{-i} = \sum_{i=1}^{\infty} \mathcal{A}_i s^{-i}$$

is rational if realizability holds. The minimal dimension of a realization is also called the *McMillan degree* of the corresponding rational matrix. □

Remark 6.7.5 When $\mathbb{K} = \mathbb{R}$ or $\mathbb{C}$, one may use complex variables techniques in order to study realizability. Take for simplicity the case $m = p = 1$ (otherwise one argues with each entry). If we can write $W_{\mathcal{A}}(s) = P(s)/q(s)$ with $\deg P < \deg q$, pick any positive real number λ that is greater than the magnitudes of all zeros of q. Then, $W_{\mathcal{A}}$ must be the Laurent expansion of the rational function P/q on the annulus $|s| > \lambda$. (This can be proved as follows: The formal equality $qW_{\mathcal{A}} = P$ implies that the Taylor series of P/q about $s = \infty$ equals $W_{\mathcal{A}}$, and the coefficients of this Taylor series are those of the Laurent series on $|s| > \lambda$. Equivalently, one could substitute $z := 1/s$ and let

$$\widetilde{q}(z) := z^d q(1/z)\,,\ \widetilde{P}(z) := z^d P(1/z)\,,$$

with $d := \deg q(s)$; there results the equality $\widetilde{q}(z)W(1/z) = \widetilde{P}(z)$ of power series, with $\widetilde{q}(0) \neq 0$, and this implies that $W(1/z)$ is the Taylor series of $\widetilde{P}/\widetilde{q}$ about 0. Observe that on any other annulus $\lambda_1 < |s| < \lambda_2$ where q has no roots the Laurent expansion will in general have terms in s^k, with $k > 0$, and will therefore be different from $W_{\mathcal{A}}$.) Thus, if there is any function g which is analytic on $|s| > \mu$ for some μ and is so that $W_{\mathcal{A}}$ is its Laurent expansion about $s = \infty$, realizability of $\mathcal{A}$ implies that g must be rational, since the Taylor expansion at infinity uniquely determines the function. Arguing in this manner it is easy to construct examples of nonrealizable Markov sequences. For instance,

$$\mathcal{A} = 1, \frac{1}{2}, \frac{1}{3!}, \frac{1}{4!}, \ldots$$

is unrealizable, since $\mathcal{W}_{\mathcal{A}} = e^{1/s} - 1$ on $s \neq 0$. As another example, the sequence

$$\mathcal{A} = 1, \frac{1}{2}, \frac{1}{3}, \frac{1}{4}, \ldots$$

cannot be realized because $\mathcal{W}_{\mathcal{A}} = -\ln(1 - s^{-1})$ on $|s| > 1$. □

Input/Output Equations

Rationality and realizability can also be interpreted in terms of high-order differential or difference equations satisfied by input/output pairs corresponding to a given behavior.

The *input/output pairs* of the behavior Λ are by definition the possible pairs $(\omega, \bar{\lambda}(\omega))$. For each $\omega \in \mathcal{U}^{[\sigma,\tau)}$ in the domain of the behavior, this is a pair consisting of a function in $\mathcal{U}^{[\sigma,\tau)}$ and one in $\mathcal{Y}^{[\sigma,\tau]}$.

Assume from now on that Λ is a time-invariant continuous-time integral behavior with analytic kernel

$$K(t) = \sum_{k=0}^{\infty} \mathcal{A}_{k+1} \frac{t^k}{k!} \tag{6.37}$$

(assume K is entire, that is, the series converges for all t), where $\mathcal{A} = \mathcal{A}_1, \mathcal{A}_2, \ldots$ is a Markov sequence over $\mathbb{K} = \mathbb{R}$ or $\mathbb{C}$. Observe that whenever ω is $r-1$-times (continuously) differentiable the corresponding output function $\bar{\lambda}(\omega)$ is r-times differentiable; such a pair $(\omega, \bar{\lambda}(\omega))$ is *of class* $\mathcal{C}^r$.

The behavior Λ is said to satisfy the *i/o equation*

$$y^{(n)}(t) = \sum_{i=0}^{n-1} \alpha_{i+1} y^{(i)}(t) + \sum_{i=0}^{n-1} P_i u^{(i)}(t), \tag{6.38}$$

where $\alpha_1, \ldots, \alpha_n \in \mathbb{K}$ and $P_0, \ldots, P_{n-1} \in \mathbb{K}^{p\times m}$, if this equation is satisfied by every i/o pair $u = \omega, y = \bar{\lambda}(\omega)$ of Λ of class $\mathcal{C}^{n-1}$, for all $t \in [\sigma, \tau)$. The nonnegative integer n is called the *order* of the i/o equation. (For $n = 0$ the equation is $y(t) = 0$.)

It is a basic fact in linear system theory that the existence of an i/o equation is equivalent to rationality, and hence to realizability by a finite dimensional linear time-invariant system.

Proposition 6.7.6 The behavior Λ satisfies (6.38) if and only if $W_{\mathcal{A}} = q^{-1}P$, where q and P are as in Equations (6.33) and (6.34), with the same α_i's and P_i's.

Proof. We first make some general observations. If K is as in (6.37), then

$$K^{(r)}(t) = \sum_{k=0}^{\infty} \mathcal{A}_{k+r+1} \frac{t^k}{k!}$$

for each $r = 0, 1, \ldots$. If, in addition, $\mathcal{A}$ is known to satisfy the order-n recursion

$$\mathcal{A}_{k+n+1} = \sum_{j=1}^{n} \alpha_j \mathcal{A}_{k+j} \tag{6.39}$$

for all $k \geq 0$, then

$$\begin{aligned} K^{(n)}(t) &= \sum_{k=0}^{\infty} \sum_{j=1}^{n} \alpha_j \mathcal{A}_{k+j} \frac{t^k}{k!} \\ &= \sum_{j=1}^{n} \alpha_j \sum_{k=0}^{\infty} \mathcal{A}_{k+j} \frac{t^k}{k!} \\ &= \sum_{j=1}^{n} \alpha_j K^{(j-1)}(t) \end{aligned}$$

for all t.

On the other hand, for each i/o pair (ω, η) of class $\mathcal{C}^{n-1}$ one concludes by induction from

$$\eta(t) = \int_{\sigma}^{t} K(t-\mu)\omega(\mu)\, d\mu$$

that

$$\eta^{(r)}(t) = \sum_{i=1}^{r} \mathcal{A}_i \omega^{(r-i)}(t) + \int_\sigma^t K^{(r)}(t-\mu)\omega(\mu)\,d\mu \tag{6.40}$$

for each $r = 0, \ldots, n$. In particular,

$$\eta^{(n)}(t) = \sum_{i=1}^{n} \mathcal{A}_i \omega^{(n-i)}(t) + \sum_{j=1}^{n} \alpha_j \int_\sigma^t K^{(j-1)}(t-\mu)\omega(\mu)\,d\mu \tag{6.41}$$

if $\mathcal{A}$ satisfies the above recursion. Since also

$$\sum_{i=0}^{n-1} \alpha_{i+1}\eta^{(i)}(t) = \sum_{i=0}^{n-1} \alpha_{i+1} \left[\sum_{l=1}^{i} \mathcal{A}_l \omega^{(i-l)}(t) + \int_\sigma^t K^{(i)}(t-\mu)\omega(\mu)\,d\mu \right],$$

it follows that when a recursion exists

$$\begin{aligned}
\eta^{(n)}(t) \;-\; & \sum_{i=0}^{n-1} \alpha_{i+1}\eta^{(i)}(t) \\
= \;& \sum_{i=1}^{n} \mathcal{A}_i \omega^{(n-i)}(t) - \sum_{i=0}^{n-1} \alpha_{i+1} \sum_{l=1}^{i} \mathcal{A}_l \omega^{(i-l)}(t) \\
= \;& \sum_{j=0}^{n-1} \left(\mathcal{A}_{n-j} - \sum_{i=1}^{n-j-1} \alpha_{i+j+1}\mathcal{A}_i \right) \omega^{(j)}(t)
\end{aligned} \tag{6.42}$$

for all t.

We conclude from Equation (6.42) that if $qW_{\mathcal{A}} = P$, which implies that both Equations (6.39) and (6.36) hold, then also the i/o equation (6.38) is valid.

Conversely, assume that (6.38) is true for all i/o pairs of class $\mathcal{C}^{n-1}$. Pick any arbitrary vectors $v_0, \ldots, v_{n-1} \in \mathbb{K}^m$, and any interval $[\sigma, \tau)$, and for these let ω be the control on $[\sigma, \tau)$ defined by

$$\omega(t) := \sum_{i=0}^{n-1} v_i \frac{(t-\sigma)^i}{i!}.$$

Using again (6.40), the i/o equation at time $t = \sigma$ gives that

$$\mathcal{A}_1 v_{n-1} + \ldots + \mathcal{A}_n v_0 = \sum_{i=0}^{n-1} \alpha_{i+1}(\mathcal{A}_1 v_{i-1} + \ldots + \mathcal{A}_i v_0) + \sum_{i=0}^{n-1} P_i v_i$$

must hold. Since these vectors were arbitrary, one may compare coefficients and equation (6.36) results. It also follows by differentiating (6.38) that

$$y^{(n+k+1)}(t) = \sum_{i=0}^{n-1} \alpha_{i+1} y^{(i+k+1)}(t) + \sum_{i=0}^{n-1} P_i u^{(i+k+1)}(t)$$

for all $k \geq 0$ and pairs of order $\mathcal{C}^{n+k}$. Applied in particular to the constant controls $\omega \equiv v_0$, v_0 arbitrary, and evaluating at $t = \sigma$, the recursion (6.39) results, and we know then that $qW_{\mathcal{A}} = P$, as desired. ∎

A similar result holds for discrete-time systems. A linear time-invariant discrete-time behavior is said to satisfy the i/o equation of order n

$$y(t+n) = \sum_{i=0}^{n-1} \alpha_{i+1} y(t+i) + \sum_{i=0}^{n-1} P_i u(t+i) \tag{6.43}$$

if this equation holds for each i/o pair with $\omega \in \mathcal{U}^{[\sigma,\tau)}$ and each $\sigma \leq t \leq \tau - n$.

Lemma/Exercise 6.7.7 The discrete-time time-invariant behavior Λ satisfies (6.43) if and only if $W_{\mathcal{A}} = q^{-1}P$, where q and P are as in Equations (6.33) and (6.34), with the same α_i's and P_i's. □

Exercise 6.7.8 Refer to Exercise 3.2.12. (Take all constants equal to one, for simplicity.)

(a) Using $y = h$ as the output, find an i/o equation of order 4 and the transfer function of the input/output behavior of Σ.

(b) Repeat the computation taking instead $x_3 = \dot{\phi}$ as the output; why is a transfer function with denominator of degree two obtained this time? Find a two-dimensional realization of the new i/o behavior. □

Exercise 6.7.9 Refer to Exercise 3.2.13, and take for simplicity $M = m = F = g = l = 1$.

(a) Using $y = x_1 = \delta$ as the output, find an i/o equation of order 4 and the transfer function of the input/output behavior of Σ.

(b) Repeat the computation taking instead $x_3 = \phi$ as the output; show that now there is an i/o equation of order three. Find a three-dimensional realization of this i/o behavior. □

6.8 Abstract Realization Theory*

There is an abstract theory of realization, which is of interest especially in the case of realizability by finite systems. We restrict attention here to *complete time-invariant* behaviors; unless otherwise stated, throughout this section Λ denotes a fixed such behavior. Accordingly, we wish to study realizations of Λ by time-invariant initialized complete systems (Σ, x^0). The realizability condition is that

$$\lambda^{0,t}_{\Sigma,x^0}(\omega) = \lambda^{0,t}(\omega)$$

* This section can be skipped with no loss of continuity.

for all $t \in \mathcal{T}_+$ and each $\omega \in \mathcal{U}^{[0,t)}$. We drop the superscripts $0, t$ and/or the subscripts Σ and x^0 on the left when they are clear from the context.

The first observation is that, at this level of abstraction, *every* behavior is realizable. To see this, consider the set

$$\Omega := \bigcup_{T \in \mathcal{T}_+} \mathcal{U}^{[0,T)}$$

and the map

$$\phi_\Omega(\tau, \sigma, \nu, \omega) := \nu\omega^{T-\sigma}$$

corresponding to concatenation. That is, if

$$\omega \in \mathcal{U}^{[\sigma,\tau)} \text{ and } \nu \in \mathcal{U}^{[0,T)},$$

then this is the control equal to ν on $[0, T)$ and equal to

$$\omega(t - T + \sigma)$$

if $t \in [T, T + \tau - \sigma)$.

Lemma/Exercise 6.8.1 The data $(\mathcal{T}, \Omega, \mathcal{U}, \phi)$ define a complete time-invariant system. □

We add the output function

$$h_\lambda : \Omega \to \mathcal{Y} : \omega \mapsto \lambda(\omega)$$

and the initial state $x^0 = \diamond$. Since $\phi_\Omega(t, 0, \diamond, \omega) = \omega$ for all ω defined on intervals of the form $[0, t)$, it follows that

$$h_\lambda(\phi_\Omega(t, 0, \diamond, \omega)) = h_\lambda(\omega) = \lambda(\omega),$$

and the following is established:

Lemma 6.8.2 The initialized system with outputs

$$\Sigma_{\Omega,\Lambda} := (\mathcal{T}, \Omega, \mathcal{U}, \phi_\Omega, \mathcal{Y}, h_\lambda, \diamond)$$

is a realization of λ. This system is complete, time-invariant, and reachable from the state $x^0 = \diamond$. □

Of course, the size of the state space of this system is huge, since it merely memorizes all inputs. A much more interesting realization results when one identifies indistinguishable states. The construction needed is more general, and it applies to arbitrary complete time-invariant systems, as follows.

Let $\Sigma = (\mathcal{T}, \mathcal{X}, \mathcal{U}, \phi, \mathcal{Y}, h)$ be any time-invariant complete system with outputs. Consider the space

$$\widetilde{\mathcal{X}} := \mathcal{X}/\sim$$

consisting of all equivalence classes under indistinguishability. For each $\sigma \leq \tau$ in $\mathcal{T}_+$, each equivalence class $[x]$, and each $\omega \in \mathcal{U}^{[\sigma,\tau)}$ let

$$\widetilde{\phi}(\tau,\sigma,[x],\omega) := [\phi(\tau,\sigma,x,\omega)],$$

that is, pick an arbitrary element in the equivalence class, apply the transitions of Σ, and then see in which equivalence class one lands. This map is well defined in the sense that the result is independent of the particular element $[x]$. In other words, for every $x, z \in \mathcal{X}$ so that $x \sim z$ and every ω, also $x_1 := \phi(\tau,\sigma,x,\omega)$ and $z_1 := \phi(\tau,\sigma,z,\omega)$ are indistinguishable. This is clear from the fact that, if ν is a control distinguishing x_1 and z_1, then the concatenation $\omega\nu$ distinguishes x and z. Similarly, the map

$$\widetilde{h}([x]) := h(x)$$

is well defined because indistinguishable states give rise in particular to identical instantaneous outputs.

This allows us to define a system

$$\widetilde{\Sigma} := (\mathcal{T},\widetilde{\mathcal{X}},\mathcal{U},\widetilde{\phi},\mathcal{Y},\widetilde{h})$$

called the *observable reduction* of Σ. It satisfies that

$$\widetilde{h}(\widetilde{\phi}(\tau,\sigma,[x],\omega)) = \widetilde{h}([\phi(\tau,\sigma,x,\omega)]) = h(\phi(\tau,\sigma,x,\omega))$$

for all ω and x, so in particular with initial state

$$\widetilde{x^0} := [x^0]$$

it realizes the same behavior as (Σ, x^0), for each fixed $x^0 \in \mathcal{X}$. It is indeed observable, since $[x] \sim [z]$ implies

$$h(\phi(\tau,\sigma,x,\omega)) = \widetilde{h}(\widetilde{\phi}(\tau,\sigma,[x],\omega)) = \widetilde{h}(\widetilde{\phi}(\tau,\sigma,[z],\omega)) = h(\phi(\tau,\sigma,z,\omega))$$

for all ω, and therefore $[x] = [z]$. Finally, note that, if the original system Σ is reachable from a state x^0, then the observable reduction is reachable from $[x^0]$. Summarizing:

Lemma 6.8.3 For each system Σ and each $x^0 \in \mathcal{X}$:

1. $\widetilde{\Sigma}$ is observable.

2. If Σ is reachable from x^0, then $\widetilde{\Sigma}$ is reachable from $\widetilde{x^0}$.

3. The i/o behaviors of (Σ, x^0) and $(\widetilde{\Sigma}, \widetilde{x^0})$ coincide. □

An initialized system (Σ, x^0) is a **canonical system** if it is reachable from x^0 and observable. This terminology is consistent with that used for linear systems, when applied with initial state $x^0 = 0$, since reachability from the origin is equivalent to complete controllability in that case.

Given any behavior Λ, let Σ_Λ be the observability reduction of the system $\Sigma_{\Omega,\Lambda}$ in Lemma 6.8.2. This is again reachable (from $\widetilde{\diamond}$), and by construction is also observable (as well as time-invariant and complete). We conclude as follows:

Theorem 29 *The initialized system with outputs* $(\Sigma_\Lambda, \widetilde{\diamond})$ *is a complete, time-invariant, canonical realization of* λ. □

Our next objective is to show that canonical realizations are unique up to a relabeling of states. Assume given two initialized systems (Σ_i, x_i^0). A *system morphism*

$$T : (\Sigma_1, x_1^0) \to (\Sigma_2, x_2^0)$$

is by definition a map

$$T : \mathcal{X}_1 \to \mathcal{X}_2$$

such that $Tx_1^0 = x_2^0$,

$$T(\phi_1(\tau, \sigma, x, \omega)) = \phi_2(\tau, \sigma, T(x), \omega)$$

for all $\omega \in \mathcal{U}^{[\sigma,\tau)}$ and all $x \in \mathcal{X}$, and $h(x) = h(T(x))$ for all $x \in \mathcal{X}$. It is a *system isomorphism* if there exists another system morphism $S : (\Sigma_2, x_2^0) \to (\Sigma_1, x_1^0)$ such that the compositions $T \circ S$ and $S \circ T$ are the identity on $\mathcal{X}_2$ and $\mathcal{X}_1$, respectively. Since at this set-theoretic level a map is invertible if and only if it is one-to-one and onto, it follows easily that T is an isomorphism if and only if the underlying map $T : \mathcal{X}_1 \to \mathcal{X}_2$ is bijective.

Theorem 30 *Let* (Σ_1, x_1^0) *and* (Σ_2, x_2^0) *be two (complete, time-invariant) initialized systems with output. Assume that* Σ_1 *is reachable from* x_1^0, Σ_2 *is observable, and they realize the same behavior. Then there exists a system morphism* $T : (\Sigma_1, x_1^0) \to (\Sigma_2, x_2^0)$. *Furthermore:*

1. *There is a unique such morphism.*

2. *If* Σ_2 *is reachable from* x_2^0, *then* T *is onto.*

3. *If* Σ_1 *is observable, then* T *is one-to-one.*

4. *If both systems are canonical, then* T *is an isomorphism.*

Proof. Consider the set G consisting of all pairs $(x_1, x_2) \in \mathcal{X}_1 \times \mathcal{X}_2$ for which

$$h_1(\phi_1(\tau, \sigma, x_1, \omega)) = h_2(\phi_2(\tau, \sigma, x_2, \omega)) \;\; \forall \sigma \leq \tau, \forall \omega \in \mathcal{U}^{[\sigma,\tau)}$$

(indistinguishable but in different state spaces). We claim that G is the graph of a system morphism.

First note that, if (x_1, x_2) and (x_1, x_2') are both in G, then the definition of G forces x_2 and x_2' to be indistinguishable, so by observability it follows that $x_2 = x_2'$, which means that G is the graph of a partially defined map. Its domain is all of $\mathcal{X}_1$: by reachability, given any $x_1 \in \mathcal{X}_1$ there exists some control ω' so that

$$\phi_1(\sigma, 0, x_1^0, \omega') = x_1$$

and we define $x_2 := \phi_2(\sigma, 0, x_2^0, \omega')$; then for all $\omega \in \mathcal{U}^{[\sigma,\tau)}$,

$$h_1(\phi_1(\tau, \sigma, x_1, \omega)) = \lambda(\omega'\omega) = h_2(\phi_2(\tau, \sigma, x_2, \omega)),$$

which shows that $(x_1, x_2) \in G$. We have proved that G is the graph of a map $T : \mathcal{X}_1 \to \mathcal{X}_2$.

To see that T defines a morphism, observe that $(x_1^0, x_2^0) \in G$, by definition of G and the fact that the behaviors coincide, and also that for each $(x_1, x_2) \in G$ necessarily $h_1(x_1) = h_2(x_2)$ (from the case $\sigma = \tau$ in the definition of G). Finally, consider any $\omega \in \mathcal{U}^{[\sigma,\tau)}$ and let for $i = 1, 2$

$$\hat{x}_i := \phi_i(\tau, \sigma, x_i, \omega)$$

so we need to show that $(\hat{x}_1, \hat{x}_2) \in G$, too. This follows trivially from the definition of G and the semigroup property.

For any morphism T_0, if (x_1, x_2) is in the graph of T_0, then also

$$(\phi_1(\tau, \sigma, x_1, \omega), \phi_2(\tau, \sigma, x_2, \omega))$$

is in the graph, for all ω; it follows that also

$$h_1(\phi_1(\tau, \sigma, x_1, \omega)) = h_2(\phi_2(\tau, \sigma, x_2, \omega))$$

for all ω, so $(x_1, x_2) \in G$. This means that the graph of T_0 is included in G, which implies $T = T_0$, establishing the uniqueness claim.

Finally, if Σ_2 is reachable, then an argument totally analogous to that used to prove that T is defined everywhere shows that T must be onto, and an argument like the one used to show that T is single-valued gives that T is one-to-one if Σ_1 is observable. When both systems are canonical, the previous conclusions show that T must be bijective. Alternatively, the Theorem can be applied twice, resulting in a $T : \Sigma_1 \to \Sigma_2$ and an $S : \Sigma_2 \to \Sigma_1$; the compositions $T \circ S$ and $S \circ T$ are system morphisms, and since the identities $I : \Sigma_1 \to \Sigma_1$ and $I : \Sigma_2 \to \Sigma_2$ are morphisms, too, the uniqueness statements imply that the compositions equal the identity. ∎

Thus, just as in the linear case, one may conclude that every behavior is realizable by a canonical system, and canonical realizations are unique. No finiteness statements have been made yet, however.

Remark 6.8.4 One can in fact obtain many of the results for the linear theory as consequences of the above abstract considerations. For instance, take the proof of the fact that canonical realizations of linear behaviors must be unique up to a linear isomorphism. Assume that two systems are given, with the same spaces $\mathcal{U}, \mathcal{Y}$, which are assumed to be vector spaces over a field $\mathbb{K}$, that the state spaces $\mathcal{X}_1, \mathcal{X}_2$ are also vector spaces, and that the maps ϕ_i and h_i are linear in (x, ω) and x, respectively. Furthermore, we assume that the initial states are zero. Then we claim that the set G is a linear subspace, which implies that the unique morphism T must correspond to a linear map $\mathcal{X}_1 \to \mathcal{X}_2$. Indeed, if (x_1, x_2) and (z_1, z_2) are in G and if $k \in \mathbb{K}$, then linearity gives that

$$\begin{aligned} h(\phi_1(\tau, \sigma, x_1 + kz_1, \omega)) &= h(\phi_1(\tau, \sigma, x_1, \mathbf{0})) + kh(\phi_1(\tau, \sigma, z_1, \omega)) \\ &= h(\phi_2(\tau, \sigma, x_2, \mathbf{0})) + kh(\phi_2(\tau, \sigma, z_2, \omega)) \\ &= h(\phi_2(\tau, \sigma, x_2 + kz_2, \omega)), \end{aligned}$$

which implies that $(x_1 + kz_1, x_2 + kz_2) \in G$. In particular, if both systems are canonical, then G is the graph of a linear isomorphism.

Arbitrary linear isomorphisms are not very interesting when considering infinite dimensional linear systems. For example, in the context of systems whose state spaces $\mathcal{X}_i$ are Banach spaces, the output value set $\mathcal{Y}$ is a normed space, and the maps ϕ_i and h_i are assumed to be bounded (continuous) linear operators, one may want to conclude that the morphism T given by the Theorem is bounded, too. This is an easy consequence of the above proof: It is only necessary to notice that G must be closed (because of the assumed continuities), so by the Closed Graph Theorem (see, for instance, [399], Theorem 4.2-I) the operator T must indeed be continuous. If both systems are canonical, T^{-1} is also bounded, by the same argument. □

We now turn to minimality. A *minimal system* will be one for which

$$\operatorname{card} X = n < \infty$$

and with the property that any other system realizing the same behavior must have a state space of cardinality at least n.

Lemma 6.8.5 A (time-invariant, complete, initialized) system is minimal if and only if it is canonical.

Proof. Assume that Σ is minimal. If it is not observable, then its observable reduction (cf. Lemma 6.8.3) realizes the same behavior and has fewer states, a contradiction. If it is not reachable from its initial state x^0, then the restriction to $\mathcal{R}(x^0)$ gives a system with the same behavior and fewer states.

Conversely, assume that Σ is canonical. If it is not minimal, then there exists another realization Σ' of lower cardinality. Reducing if necessary by observability or reachability as in the above paragraph, we may assume that Σ' is canonical. By the isomorphism Theorem given above, Σ and Σ' must be isomorphic, contradicting the cardinality assertion. ∎

Note that the proof also shows that, if there exists any realization with finite cardinality, then there is a canonical one of finite cardinality.

Example 6.8.6 We consider again the parity check example discussed in Example 2.3.3. In particular, we shall see how to prove, using the above results, the last two claims in Exercise 2.3.4. The behavior to be realized is $\lambda(\tau, 0, \omega) =$

$$\begin{cases} 1 & \text{if } \omega(\tau-3)+\omega(\tau-2)+\omega(\tau-1) \text{ is odd and 3 divides } \tau > 0 \\ 0 & \text{otherwise} \end{cases}$$

and we take the system with

$$\mathcal{X} := \{0,1,2\} \times \{0,1\}$$

and transitions

$$\mathcal{P}((i,j),l) := (i+1 \bmod 3, j+l \bmod 2)$$

for $i = 1, 2$ and

$$\mathcal{P}((0,j),l) := (1,l)\,.$$

The initial state is taken to be $(0,0)$, and the output map has $h(i,j) = 1$ if $i = 0$ and $j = 1$ and zero otherwise. (The interpretation is that $(k,0)$ stands for the state "t is of the form $3s+k$ and the sum until now is even," while states of the type $(k,1)$ correspond to odd sums.) This is clearly a realization, with 6 states. To prove that there is no possible (time-invariant, complete) realization with less states, it is sufficient to show that it is reachable and observable.

Reachability follows from the fact that any state of the form $(0,j)$ can be obtained with an input sequence $j00$, while states of the type $(1,j)$ are reached from x^0 using input j (of length one) and states $(2,j)$ using input $j0$.

Observability can be shown through consideration of the following controls ω_{ij}, for each (i,j):

$$\omega_{01} := \diamond,\ \omega_{00} := 100,\ \omega_{10} := 10,\ \omega_{11} := 00,\ \omega_{21} := 0,\ \omega_{20} := 0\,.$$

Then, ω_{01} separates $(0,1)$ from every other state, while for all other pairs $(i,j) \neq (0,1)$,

$$\lambda_{(i,j)}(\omega_{\alpha\beta}) = 1$$

if and only if $(i,j) = (\alpha,\beta)$. □

Exercise 6.8.7 Let Λ be any time-invariant complete behavior and let $\sim$ be the following equivalence relation on Ω:

$$\omega \sim \omega' \Leftrightarrow \lambda(\omega\nu) = \lambda(\omega'\nu) \quad \forall \nu$$

(to be more precise, one should write the translated version of ν). This is the *Nerode equivalence relation.* Prove that λ admits a finite-cardinality realization if and only if there are only finitely many equivalence classes under the Nerode relation. (*Hint:* It only takes a couple of lines, using previous results.) □

The result in the Exercise can be interpreted in terms of generalized Hankel matrices. Consider the matrix with rows and columns indexed by elements of Ω, and with $\lambda(\omega\nu)$ in position (ω, ν). Then finite realizability is equivalent to this matrix having only finitely many rows, and the number of different rows is equal to the cardinality of the state space of a minimal realization. This is in complete analogy to the linear case.

6.9 Notes and Comments

Basic Observability Notions

The concept of observability originates both from classical automata theory and from basic linear systems. The material on final-state observability can be generalized to certain types of infinite systems, which in the continuous-time case give, because of Proposition 6.1.9, results about observability (see, for instance, [358] and [384]).

The text [52] has an extensive discussion of the topic of multiple experiments, final-state determination, and control for finite systems. Also, [97] addresses the observability question for such systems. In the context of VLSI design a central problem is that of testing circuits for defects, and notions of observability appear; see, for instance, [148] and the references therein.

Observability of Time-Invariant Systems

The concept of observability, its duality with controllability for linear systems, and the notion of canonical realization, all arose during the early 1960s. Some early references are [155], [215], and [320].

In fact, Kalman's original filtering paper [214] explicitly mentions the duality between the optimization versions of feedback control (linear-quadratic problem) and observer construction (Kalman filtering). The term "canonical" was used already in [217], where one can find reachability and observability decompositions. For related results on duality of time-varying linear systems, see [435].

For more on the number of experiments needed for observability, and in particular Exercise 6.2.3, see, for instance, [97], [125], and especially [156]. See also [6] for continuous-time questions related to this problem. Results on observability of recurrent "neural" nets (the systems studied in Section 3.8, with linear output $y = Cx$) can be found in [14].

In the form of Exercise 6.2.12 (which, because of the simplifying assumption that the numbers ω_i are distinct, could also have been easily proved directly), one refers to the eigenvalue criterion for sampling as *Shannon's Theorem* or the *Sampling Theorem.* It gives a sufficient condition to allow reconstruction of the signal η and is one of the cornerstones of digital signal processing. The result can be generalized, using Fourier transform techniques, to more general signals η, containing an infinite number of frequency components.

It is possible to develop a large amount of the foundations of time-invariant systems based on the notion of *observables* associated to a system; see [373].

Linearization Principle for Observability

Often one refines Definition 6.4.1 to require that states be distinguishable without large excursions. One such possibility is to ask that for each neighborhood W of x^0 there be a neighborhood V so that every state in V is distinguishable from x^0 using a control that makes the resulting trajectory stay in W. This is more natural in the context of Lyapunov stability, and can be characterized elegantly for continuous-time smooth systems; see, for instance, [185] and [365]. Other nonlinear observability references are, for instance, [5], [7], [28], [139], and [310].

Realization Theory for Linear Systems

There are papers on realization of time-varying linear systems that use techniques close to those used for the time-invariant case. See, for instance, [235] and [434].

Recursion and Partial Realization

See, for instance, [180], [181], [192], [395], and the many references therein, as well as the early paper [225], for further results on families of systems.

Small perturbations of the Markov parameters will result in a nonrealizable sequence, since all of the determinants of the submatrices of $\mathcal{H}$ are generically nonzero. In this context it is of interest to look for *partial realizations*, in which only a finite part of the sequence is matched; this problem is closely related to classical mathematical problems of *Padé approximation*. Lemma 6.6.9 is one result along these lines; see, for instance, [19] and the references therein, for a detailed treatment of partial realization questions and relations to problems of rational interpolation. The problem can also be posed as one of optimal approximation of Hankel operators; see the by now classical paper [3]. In addition, the procedures that we described are numerically unstable, but various modifications render them stable; a reference in that regard is [110]. A recursive realization procedure is given in [18], which permits realizations for additional data to make use of previous realizations.

Since a Markov sequence is specified in terms of an infinite amount of data, one cannot expect to solve completely the question of realizability unless some sort of finite description is first imposed. It is known however that, for arbitrary "computable" descriptions, the problem of deciding realizability is undecidable in the sense of logic, that is, there is no possible computer program that will always correctly determine, given a description of a Markov sequence, whether a realization exists; see, for instance, [356].

Rationality and Realizability

The ideas of realization and relations to input/output equations go back at least to the nineteenth century, in the context of using integrators to solve algebraic differential equations; see [400] as well as the extensive discussion in [212], Chapter 2. The relations between rationality and finite Hankel rank are also classical (for the scalar case) and go back to the work of Kronecker (see [152]).

Algebraic techniques for studying realizations of linear systems were emphasized by Kalman; see, for instance, [220], [221], and [223], as well as [228] for relations to econometrics.

Proposition 6.7.6 could also be proved using Laplace transform techniques; we used a direct technique that in principle can be generalized to certain nonlinear systems. For these generalizations, see, for instance, [360] and [421], which establish that certain types of i/o behaviors satisfy polynomial equations

$$E(y(t+n), y(t+n-1), \ldots, y(t), u(t+n-1), \ldots, u(t)) = 0 \tag{6.44}$$

(or $E(y^{(n)}(t), y^{(n-1)}(t), \ldots, y(t), u^{(n-1)}(t), \ldots, u(t)) = 0$) if and only if they are realizable, for discrete-time and continuous-time systems, respectively. In the nonlinear case, however, not every "causal" i/o equation necessarily gives rise to an i/o behavior, and this in turn motivates a large amount of research on such equations; see, for instance, [408]. Some authors, motivated by differential-algebraic techniques, have suggested that realizability should be *defined* in terms of i/o equations; see especially [141] and the references therein.

The reference [272] discusses i/o equations for nonlinear systems in the context of identification problems. Related material is also presented in [307], which uses i/o equations (6.44) in which E is not linear, nor polynomial, but instead is given by iterated compositions of a fixed scalar nonlinear function with linear maps. Numerical experience seems to suggest that such combinations are particularly easy to estimate using gradient descent techniques, and they are in principle implementable in parallel processors. The name "neural network" is used for this type of function E, because of the analogy with neural systems: the linear combinations correspond to dendritic integrations of signals, and the scalar nonlinear function corresponds to the "firing" response of each neuron, depending on the weighted input to it.

Realization and i/o equations can also be studied in a stochastic context, when $u(0), u(1), \ldots$ in (6.43) are random variables. In that case, especially if these variables are independent and identically distributed, (6.43) describes what is called an *ARMA* or *autoregressive moving average* model for the stochastic process, in the stochastic systems and statistics literature. (The outputs form a *time series*, and the realization is a Markov model for the series.) See, for instance, the "weak" and "strong" Gaussian stochastic realization problems studied, respectively, in [131], [132], [222], [409], and [10], [11], [280], and the related [344].

Abstract Realization Theory

It is possible to develop a unified theory of realization for discrete-time finite systems and linear finite dimensional systems, in the language of category theory; see, for instance, [22]. There are also many other approaches to realization theory, for instance for nonlinear systems evolving on manifolds in continuous-time ([199], [383], and [389]) or in discrete-time ([203]), systems on finite groups ([75]), polynomial discrete-time ([360]) and continuous-time ([38]) systems, bilinear systems ([66], [198], [359], and [382]), or infinite-dimensional linear systems ([316], [437], and [438]). The goal in these cases is to study realizations having particular structures —such as linear realizations when dealing with linear behaviors— and to relate minimality (in appropriate senses) to various notions of controllability and observability. It is also possible to base a realization theory on generalizations of Hankel matrix ideas to bilinear and other systems; see, for instance, [137] and [138], as well as applications of these ideas in [96]. One also may impose physical constraints such as in the study of Hamiltonian control systems in [106] and [204].

In principle, the complete i/o behavior is needed in order to obtain a realization. However, if the system is "mixing" enough, so that trajectories tend to visit the entire state space, one may expect that a single long-enough record will be sufficient in order to charaterize the complete behavior. In this sense one can view the recent work identifying dynamics of chaotic systems from observation data (see, e.g., [85] and [130]) as closely related to realization theory (for systems with no controls), but no technical connections between the two areas have been developed yet.

Chapter 7

Observers and Dynamic Feedback

Section 1.6 in Chapter 1 discussed the advantages of using integration in order to average out noise when obtaining state estimates. This leads us to the topic of dynamic observers. In this chapter, we deal with observers and with the design of controllers for linear systems using only output measurements.

7.1 Observers and Detectability

In this section we first discuss briefly the construction of observers for continuous-time linear systems and then study the more basic questions raised by this procedure. The next subsection can be skipped if desired, as all results are proved later, but it provides the intuitive ideas underlying everything else.

Motivational Discussion

Consider the time-invariant continuous-time system

$$\begin{aligned} \dot{x} &= Ax + Bu \\ y &= Cx \end{aligned}$$

and the problem of reconstructing the state $x(t)$ from the measurements $y(t)$. One solution to this problem when (A, C) is observable is to apply the optimal integral operator M used in Theorem 25 (p. 279). For each fixed interval, let us say of the form $[0, T]$, this provides the state $x(0)$ that produced a given output function $y(\cdot)$ on that interval. From the knowledge of $x(0)$ and the control $u(\cdot)$, one may then recover the entire state trajectory.

In a sense, this is an "open-loop" procedure, and it has the same disadvantages that open-loop control has. For instance, if there is a possibility that the state could change due to unmeasured disturbances acting on the system, it is necessary to repeat the estimation procedure. Just as in the discussion about the advantages of feedback control, one can argue that a recursive estimation procedure —in which the estimates are updated continuously, as more observations are received— is often more useful. Again, as in that case, the advantages of recursive estimation could be quantified if a model incorporating disturbances is used. In fact, in analogy to the solution to the linear-quadratic problem (Chapter 8), it turns out that this type of estimation is indeed optimal with respect to certain optimization criteria. But again, as with feedback control, we take here the direct route of simply posing recursive estimation as a problem to be studied in itself.

One obvious way to obtain an estimate for the state of the above linear system is to start with a model

$$\dot{z}(t) = Az(t) + Bu(t)$$

of the system, initialized at any state, to which the same controls are applied. If we are lucky enough that $x(0) = z(0)$, then it will follow that the solutions coincide for all t, and a perfect estimate will be maintained. But of course the problem is that $x(0)$ is unknown; the only information available about x is the measurement y. We argue then as follows. The quantity

$$d(t) := Cz(t) - y(t)$$

measures the discrepancy between the actual measured value and the output value that would result if our estimate z were indeed the correct state. This error can be computed from the available data, so it can be used to influence the dynamics of z. To keep matters simple, we restrict attention to *linear* functions of the error. That is, we propose an estimator

$$\dot{z}(t) = Az(t) + Bu(t) + Ld(t)\,,$$

where L is a $p \times n$ matrix still to be decided upon. Let $e(t) := z(t) - x(t)$ be the difference between the actual and estimated state trajectories. Immediately from the equations for x and z we conclude that the differential equation

$$\dot{e}(t) = (A + LC)e(t)$$

holds for e. If it is possible to choose a matrix L so that $A + LC$ is Hurwitz, then e converges to zero for all initial conditions. In terms of x and z, this means that, whatever the initial state and the initial estimate, the estimation error decreases asymptotically, in fact at an exponential rate. Since the matrix $A+LC$ is Hurwitz if and only if $A' + C'L'$ is, such an L can be found whenever the pair (A', C') is asymptotically controllable, and in particular if (A, C) is observable. The rest of this section will only expand on this simple construction, showing that if the pair (A', C') is not asymptotically controllable then nonlinear estimators cannot work either, and will pose the definitions in a somewhat more abstract context, applying to general systems. Relatively little is known as yet, however, about estimators for nonlinear systems.

Asymptotic Observability

Let Σ be a time-invariant complete system with outputs and assume that Σ is topological (Definition 3.7.1), that is, $\mathcal{X}$ is a metric space and paths depend continuously on initial states. Assume that $u^0 \in \mathcal{U}$, $y^0 \in \mathcal{Y}$, and the state x^0, are such that (x^0, u^0) is an equilibrium pair and $h(x^0) = y^0$. (Recall that by equilibrium pair we mean that for every $t \in \mathcal{T}_+$, $\phi(t, 0, x^0, \omega^0) = x^0$ when $\omega^0 \equiv u^0$.) Consider the set

$$\mathcal{X}^0 := \{x \in \mathcal{X} \mid \lambda_x^{0,t}(\omega^0) = y^0 \text{ for all } t \in \mathcal{T}_+\},$$

the set of states indistinguishable from x^0 when using identically u^0 controls. This set is invariant under such controls $\equiv u^0$, so there is a well defined system Σ^0 (which depends on the particular u^0, y^0 that was picked) with state space $\mathcal{X}^0$, control-value space $\{u^0\}$, and whose transition map is the restriction of ϕ. This is a classical dynamical system (no inputs nor outputs).

Definition 7.1.1 *The time-invariant complete system Σ is* **asymptotically observable** *(with respect to x^0, u^0, y^0) if Σ^0 is globally asymptotically stable (with respect to x^0).* □

In informal terms, this is the property that, if a state gives rise to an output constantly equal to y^0 when the constant control u^0 is applied, then the resulting trajectory must converge to x^0 (and if the initial state started near x^0, it never goes very far). The more standard terminology (at least for linear systems) is to say that the system is **detectable**.

Let Σ be a linear (finite dimensional) system over $\mathbb{K} = \mathbb{R}$ or $\mathbb{C}$. We say simply that Σ is asymptotically observable (or equivalently, detectable) if it is so with respect to

$$x^0 = 0 \quad u^0 = 0 \quad y^0 = 0 .$$

In this case Σ^0 is the linear system

$$\dot{x} = Ax$$

or in discrete-time $x^+ = Ax$, evolving on the state space $\mathcal{O}$, where $\mathcal{O}$ is the set of states indistinguishable from 0, the same as the kernel of $\mathbf{O}(A, C)$. Thus, detectability means precisely that

$$C\xi(t) \equiv 0 \Rightarrow \xi(t) \to 0$$

for all solutions of $\dot{x}$ (or x^+) $= Ax$. In terms of the observability decomposition in Equation (6.8), $C\xi \equiv 0$ implies that the first r components of such a solution ξ must be identically zero (since the pair (A_1, C_1) is observable), and therefore detectability is simply the requirement that A_3 be a Hurwitz (or convergent) matrix. We conclude:

Lemma 7.1.2 For linear time-invariant continuous-time or discrete-time systems, (A, C) is asymptotically observable if and only if (A', C') is asymptotically controllable. □

Note that the Hautus condition for asymptotic controllability becomes that (A, C) is asymptotically observable if and only if

$$\operatorname{rank} \begin{bmatrix} \lambda I - A \\ C \end{bmatrix} = n \quad \forall \lambda \in \mathbb{C}, \operatorname{Re} \lambda \geq 0$$

for continuous-time systems, and similarly (with the condition now being for $|\lambda| \geq 1$) in discrete-time.

Observers

Still, Σ denotes an arbitrary complete time-invariant system with outputs. An observer $\hat{\Sigma}$ for Σ (more precisely, this should be called a "final-state asymptotic observer") is a system that produces an estimate of the current state of Σ based on past observations. At any instant, the inputs to $\hat{\Sigma}$ are the current inputs and outputs of Σ, and the estimate is obtained as a function of its state and the current observation.

Definition 7.1.3 *A (strong)* **observer** *for Σ consists of a (time-invariant and complete) system $\hat{\Sigma}$ having input value space*

$$\mathcal{U} \times \mathcal{Y}$$

together with a map

$$\theta : \mathcal{Z} \times \mathcal{Y} \to \mathcal{X}$$

(where $\mathcal{Z}$ is the state space of $\hat{\Sigma}$) so that the following property holds. For each $x \in \mathcal{X}$, each $z \in \mathcal{Z}$, and each $\omega \in \mathcal{U}^{[0,\infty)}$, we let ξ be the infinite path resulting from initial state x and control ω, that is,

$$\xi(t) = \phi(t, 0, x, \omega|_{[0,t)})$$

for all $t \in \mathcal{T}_+$, and write $\eta(t) = h(\xi(t))$; we also let ζ be the infinite path resulting from initial state z and control (ω, η) for the system $\hat{\Sigma}$,

$$\zeta(t) = \hat{\phi}(t, 0, z, (\omega|_{[0,t)}, \eta|_{[0,t)}))\,,$$

and finally we write

$$\hat{\xi}(t) := \theta(\zeta(t), \eta(t))\,.$$

Then it is required that, for every x, z and every ω,

$$\mathrm{d}\left(\xi(t), \hat{\xi}(t)\right) \to 0 \text{ as } t \to \infty \tag{7.1}$$

(global convergence of estimate) as well as that for each $\varepsilon > 0$ there be some $\delta > 0$ so that for every x, z, ω

$$\mathrm{d}\left(x, \theta(z, h(x))\right) < \delta \;\Rightarrow\; \mathrm{d}\left(\xi(t), \hat{\xi}(t)\right) < \varepsilon \text{ for all } t \geq 0 \tag{7.2}$$

(small initial estimate results in small future errors). □

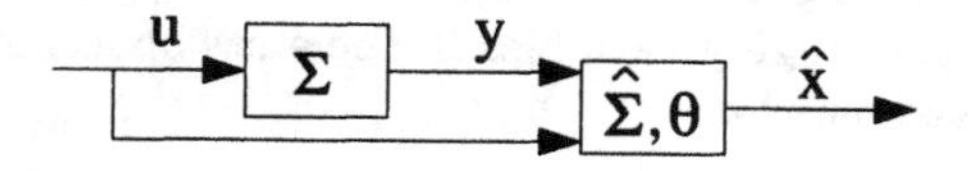

Figure 7.1: *Observer.*

By a *linear* observer for a linear (time-invariant, finite dimensional) system Σ we mean a linear system $\hat{\Sigma}$ together with a linear map θ.

Theorem 31 *Assume that Σ is a linear system (finite dimensional, time-invariant, over $\mathbb{R}$ or $\mathbb{C}$). Then the following properties are equivalent:*

1. *There is an observer for Σ.*
2. *There is a linear observer for Σ.*
3. *Σ is asymptotically observable (i.e., detectable).*

Proof. Note that Property 2 implies Property 1. We prove that 3 implies 2. For this, it is only necessary to take $\hat{\Sigma}$ of the form

$$\dot{z}\,[\text{or } z^+] \;=\; (A + LC)z + Bu - Ly\,, \tag{7.3}$$

as discussed earlier, and $\theta(z, y) := z$. By Proposition 5.5.6 and Lemma 7.1.2, there is always a choice of L that insures that $A+LC$ is a Hurwitz or convergent matrix, respectively. Since for every initial x, z and control ω one has that $e(t) := z(t) - x(t)$ satisfies

$$\dot{e}\,[\text{or } e^+] = (A + LC)e\,,$$

an observer results.

Conversely, assume that Property 1 holds. Pick an observer and let z be any initial state. We claim first that, if

$$\zeta(t) \;=\; \hat{\phi}(t, 0, z, (\mathbf{0}, \mathbf{0}))\,,$$

then $\theta(\zeta(t), \mathbf{0}) \to 0$. This is because of Property (7.1) applied to the trajectory $\omega \equiv 0, \xi \equiv 0$ of Σ. Now let ξ be *any* trajectory corresponding to the control $\omega \equiv 0$ for which $C\xi(t) \equiv 0$, as in the definition of detectability. Since ζ is the same as when $\xi \equiv 0$, it follows that again $\theta(\zeta(t), \mathbf{0}) \to 0$, so (7.1) now implies that $\xi(t) \to 0$. ∎

Remark 7.1.4 When the system Σ is discrete-time and observable, one may pick a matrix L so that $A + LC$ is nilpotent, by the Pole-Shifting Theorem. Thus, in that case a *deadbeat observer* results: The estimate becomes exactly equal to the state after n steps. □

Note that the above construction results in an observer of dimension equal to that of the original system. In general, one may also construct *reduced order observers*, in the following sense. Assume for simplicity that C has rank p. Then, —after, if necessary, changing variables in the state space and the output-value space— one may assume that

$$C = (I \quad 0)\,. \tag{7.4}$$

This means that p components of x can be measured instantaneously, which leads one to expect that an observer of dimension $n-p$ will exist. The following exercise establishes that fact.

Exercise 7.1.5 Let Σ be a detectable continuous-time time-invariant system, with C of the form in Equation (7.4). Show that there exists a linear observer of dimension $n-p$. (*Hint:* Argue as follows. First write the equations of Σ in block form

$$\begin{aligned}\dot{x}_1 &= A_{11}x_1 + A_{12}x_2 + B_1 u \\ \dot{x}_2 &= A_{21}x_1 + A_{22}x_2 + B_2 u\,,\end{aligned}$$

so that $y = x_1$ and note that the pair (A_{22}, A_{12}) is again detectable —this takes one line to prove. Next think of the second equation as defining a system whose input is $A_{21}x_1 + B_2 u$, and consider an observer

$$\dot{q} = (A_{22} + LA_{12})q - LA_{12}x_2 + A_{21}x_1 + B_2 u$$

for it. The reason that this does not yet provide a solution is that it involves the term $LA_{12}x_2$, which is not a measured quantity. However, an equation for $z := q + Lx_1$ can be derived, that does not contain any such term. Fill in the details and write explicitly the form of the observer $(\hat{\Sigma}, \theta)$.) □

The form of the observer in the proof of Theorem 31 is the most general possible if one insists that $\theta(z, y) = z$:

Exercise 7.1.6 Assume that $\Sigma = (A, B, C)$ is a continuous-time linear system admitting a linear observer

$$\dot{z} = A_0 z + B_0 u + Ny$$

with $\theta(z, y) = z$. Show that it must be the case that $B_0 = B$, $A_0 = A - NC$, and that A_0 must be a Hurwitz matrix. □

Remark 7.1.7 Consider the observer constructed in the proof of Theorem 31, say in continuous-time. The rate of convergence to zero of the error $e(t)$ will be controlled by how negative the real parts of the eigenvalues of $A+LC$ are. In order for this convergence to be fast, L must be large in magnitude. If the observation y has been corrupted by noise, say $y = Cx + d$, where d cannot be measured by the observer, then the error equation becomes

$$\dot{e}(t) = (A+LC)e(t) - Ld(t),$$

which tends to remain large if d is persistent and L is large. What this means is that there is a trade-off in achieving small error: large L may be needed for fast convergence if there is no noise, but small L amplifies less the effect of noise. This trade-off can be formulated precisely as an optimization problem. A *Kalman filter* is an observer that has been optimized with respect to uncertainty in observations as well as, if desired, state disturbances. Mathematically, the problem of designing Kalman filters turns out to be reducible, by duality, to a linear quadratic optimal control problem, and has been one of the most successfully applied parts of control theory. A deterministic version is provided in Section 8.3. □

7.2 Dynamic Feedback

We turn now to the question of stabilizing systems based only on output measurements. As the discussion in Section 1.2 regarding proportional-only control illustrated, direct output feedback alone in general is not sufficient.

Example 7.2.1 As an example of the insufficiency of output feedback, take the system

$$\begin{aligned}\dot{x}_1 &= x_2\\ \dot{x}_2 &= u\end{aligned}$$

with output map $y = x_1$. Even though this system is controllable and observable, we claim that there exists no possible continuous function

$$k : \mathbb{R} \to \mathbb{R}$$

that depends only on y and stabilizes in the sense that, for each $x = (x_1, x_2)$ and every solution $\xi = (\xi_1, \xi_2)$ of

$$\begin{aligned}\dot{x}_1 &= x_2\\ \dot{x}_2 &= k(x_1)\end{aligned}$$

with $\xi(0) = x$, the solution is defined for all $t \geq 0$ and $\xi(t) \to 0$. If any such k existed, we could consider the smooth ("energy") function

$$V(x) := x_2^2 - 2\int_0^{x_1} k(a)\,da,$$

which is constant along all trajectories. Since all trajectories converge to the origin by assumption, it follows that V must be constant and hence equal everywhere to $V(0) = 0$, contradicting $V(0,1) = 1$. □

Exercise 7.2.2 Show that, if A has zero trace and if $CB = 0$, then $A + BFC$ (the matrix obtained by feeding $u = Fy$ to $\dot{x} = Ax + Bu, y = Cx$) cannot be Hurwitz for any F. □

This section deals with the use of dynamic feedback, that is, the control of systems using other systems (as opposed to static feedback laws). As discussed in Chapter 1, this can be interpreted, in the continuous-time case when continuous-time systems are used as controllers, as the use of feedback laws that involve integrals of the observations. In order to avoid technical problems that arise only for nonlinear systems, and because the results to be presented apply mainly in the linear case, we again restrict attention to complete and time-invariant systems Σ.

We start by defining the interconnection of two systems with output Σ_1 and Σ_2. It is understood that both time sets must be the same. Subscripts i are used to indicate objects associated with system Σ_i.

Definition 7.2.3 *Let Σ_1 and Σ_2 be two systems, and let*

$$k_i : \mathcal{Y}_1 \times \mathcal{Y}_2 \to \mathcal{U}_i , \quad i = 1,2$$

be two maps. We say that **the interconnection of Σ_1 and Σ_2 through k_1 and k_2 is well-posed** *if for each $(x_1, x_2) \in \mathcal{X}_1 \times \mathcal{X}_2$ there exist unique functions*

$$\xi_i : [0,\infty) \to \mathcal{X}_i , \quad i = 1,2$$

such that, with the notations

$$\begin{aligned} \eta_i(t) &:= h_i(\xi_i(t)) \\ \omega_i(t) &:= k_i(\eta_1(t), \eta_2(t)) \end{aligned}$$

it holds that $\xi_i = \psi_i(x_i, \omega_i)$ for $i = 1,2$. □

Recall that $\xi_i = \psi_i(x_i, \omega_i)$ means that $\xi_i(t) = \phi_i(t, 0, x_i, \omega_i|_{[0,t)})$ for all $t \in \mathcal{T}_+$.

Lemma/Exercise 7.2.4 If the above interconnection is well-posed, then defining $\mathcal{X} := \mathcal{X}_1 \times \mathcal{X}_2$ and

$$\phi(t, 0, (x_1, x_2)) = (\xi_1(t), \xi_2(t))$$

results in a complete time-invariant system Σ with no controls. □

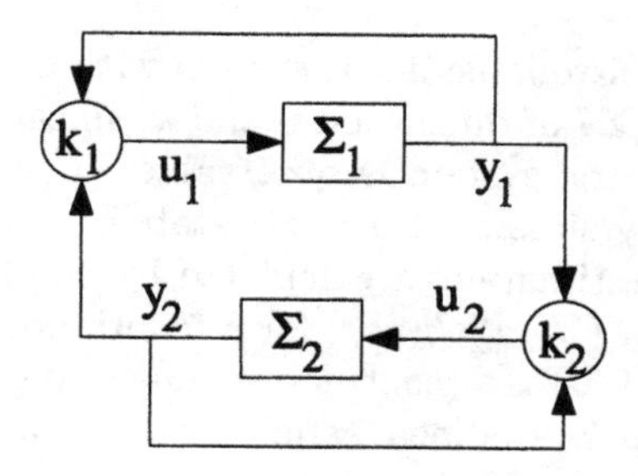

Figure 7.2: *Interconnection of two systems.*

We call the system in Lemma 7.2.4 the *interconnection of* Σ_1 *and* Σ_2 *with connecting maps* k_1, k_2. When Σ_1 and Σ_2 are topological, this interconnection is a topological system (with the product metric on $\mathcal{X}_1 \times \mathcal{X}_2$) if the functions ξ_i in Definition 7.2.3 depend continuously on (x_1, x_2) uniformly on compact time intervals. In this case, we say that the interconnection is *topologically* well-posed.

For continuous-time systems Σ_1 and Σ_2 and any maps k_1, k_2, well-posedness means that solutions of

$$\begin{aligned}\dot{x}_1 &= f_1(x_1, k_1(h_1(x_1), h_2(x_2)))\\ \dot{x}_2 &= f_2(x_2, k_2(h_1(x_1), h_2(x_2)))\end{aligned}$$

must exist and be unique for all initial states and all t. If the systems as well as k_1 and k_2 are of class $\mathcal{C}^1$, then uniqueness always holds, and the only issue is completeness. Topological well-posedness is also automatic in that case.

When the system Σ_2 has just one state, we can ignore the argument η_2 in k_1, since it is constant, as well as the function ξ_2 (see Figure 7.3). Then a well-posed interconnection is nothing more than an output feedback law $u = k(y)$ for which the closed-loop system is complete.

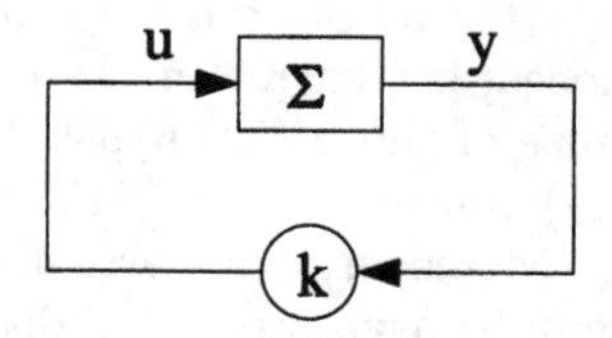

Figure 7.3: *Static feedback case of interconnection.*

If Σ_1 and Σ_2 are discrete-time systems, the interconnection is always well defined, as the corresponding difference equations always have a solution.

Definition 7.2.5 *Let* Σ_1 *be a topological system, and pick any* $x_1^0 \in \mathcal{X}$. *The system* Σ_1 *is* **dynamically stabilizable** *(with respect to* x_1^0*)* **by** $(\Sigma_2, x_2^0, k_1, k_2)$, *where* Σ_2 *is a topological system, if the interconnection of* Σ_1 *and* Σ_2 *through* k_1, k_2 *is topologically well-posed and is globally asymptotically stable with respect to* (x_1^0, x_2^0). □

For instance, any controllable linear system with $C = I$ can be dynamically stabilized with a linear Σ_2 of dimension 0 and k_1 linear. The next result shows that, when $C \neq I$ but the system is observable, one still can obtain stabilizability. The idea is very simple: If $u = Fx$ stabilizes the system, then we feed $u = Fz$, where z is an estimate of x calculated by an observer.

By a *linear* stabilizer $(\Sigma_2, x_2^0, k_1, k_2)$ for a (continuous-time or discrete-time) system Σ_1 we mean that Σ_2 is a (continuous-time or discrete-time, respectively) system, that both maps k_i are linear, and that $x_2^0 = 0$.

Theorem 32 *Assume that Σ_1 is a linear system (finite dimensional, time-invariant, over $\mathbb{R}$ or $\mathbb{C}$). Then the following properties are equivalent:*

1. *There is a dynamic stabilizer for Σ_1 (with respect to $x^0 = 0$).*

2. *There is a dynamic linear stabilizer for Σ_1 (with respect to $x^0 = 0$).*

3. *Σ_1 is asymptotically controllable and asymptotically observable.*

Proof. We first prove that Property 1 implies Property 3 for any system (not necessarily linear). More precisely, assume that the interconnection of Σ_1 and Σ_2 through k_1, k_2 is topologically well-posed and asystable with respect to (x_1^0, x_2^0), and let

$$y_i^0 := h_i(x_i^0) \quad \text{and} \quad u_i^0 := k_i(y_1^0, y_2^0) \qquad i = 1, 2\,.$$

Since (x_1^0, x_2^0) is an equilibrium point for the interconnection, it follows that (x_i^0, u_i^0) is an equilibrium pair for the system Σ_i. Then the claim is that Σ_1 must be asymptotically controllable to x_1^0 and detectable with respect to x_1^0, u_1^0, y_1^0. (An analogous statement holds for Σ_2, of course.) Asymptotic controllability is immediate from the definitions, so we prove the detectability statement.

Consider any state $x \in \mathcal{X}_1$ indistinguishable from x_1^0 when using identically u_1^0 controls, and let $\xi_1 := \psi(x, \omega_1^0)$ be the resulting infinite time path. Let $\xi_2 \equiv x_2^0$. We claim that (ξ_1, ξ_2) must be a trajectory of the interconnection. This is true by the uniqueness statement in the definition of well-posedness; observe that $\eta_1 \equiv y_1^0$ because of the indistinguishability assumption on x and that $\eta_2 \equiv y_2^0$. Thus, (ξ_1, ξ_2) converges to (x_1^0, x_2^0), which implies that also ξ_1 converges to x_1^0, as wanted. Moreover, if x starts near x_1^0, then (ξ_1, ξ_2) remains uniformly near (x_1^0, x_2^0), again by asymptotic stability of the interconnection, so ξ_1 remains near x_1^0.

We have proved that the system Σ is detectable with respect to x_1^0, u_1^0, y_1^0. When, as in the statement of the Theorem, Σ_1 is a linear system and $x^0 = 0$, and using that $y_1^0 = k_1(0) = 0$, we have then that the following property holds (using for definiteness the continuous-time case, but the discrete-time argument is exactly the same): for every state trajectory $\xi(t)$ so that $\dot{\xi} = A\xi + Bu_1^0$ and $C\xi(t) \equiv 0$, necessarily $\xi(t) \to 0$ as $t \to \infty$. On the other hand, $(0, u_1^0)$ is an equilibrium pair, so $0 \equiv A0 + Bu_1^0$, which implies that $Bu_1^0 = 0$. Thus it is also true that for every state trajectory $\xi(t)$ so that $\dot{\xi} = A\xi$ and $C\xi(t) \equiv 0$, necessarily $\xi(t) \to 0$ as $t \to \infty$, and this means that the system is detectable with respect to 0, 0, 0, which is what we mean when we say that the linear system Σ_1 is asymptotically controllable.

It is only left to show that, for linear systems, Property 2 follows from Property 3. Let (A, B, C) be the triple associated to Σ_1 and pick matrices F, L such that $A + BF$ and $A + LC$ are both Hurwitz (in continuous-time) or convergent (in discrete-time). Pick $k_2(y_1, y_2) = y_1$ and $k_1(y_1, y_2) = y_2$, and the system Σ_2 defined by

$$(A + LC + BF, -L, F)$$

so that the resulting interconnection is the linear system with no controls and "A" matrix equal to

$$\begin{pmatrix} A & BF \\ -LC & A + LC + BF \end{pmatrix} = T^{-1} \begin{pmatrix} A + BF & BF \\ 0 & A + LC \end{pmatrix} T$$

where

$$T = \begin{pmatrix} I & 0 \\ -I & I \end{pmatrix}.$$

It follows that the interconnection is asymptotically stable. ∎

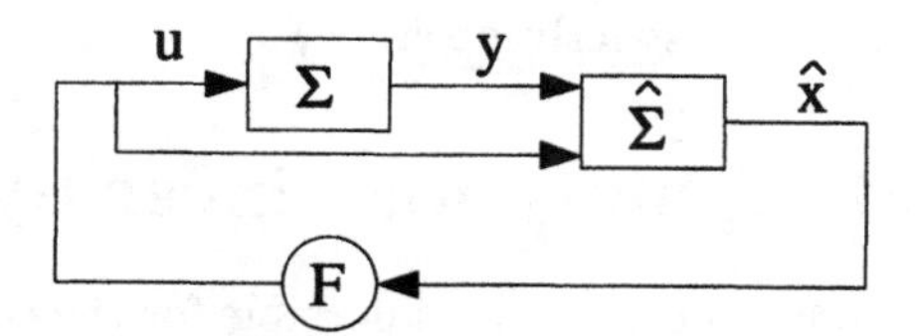

Figure 7.4: *Observer/controller configuration in proof of Theorem 32.*

Exercise 7.2.6 Let Σ_1 be a continuous-time system of class C^1,

$$\dot{x} = f(x, u), \quad y = h(x),$$

and assume that $(0, 0)$ is an equilibrium pair and $h(0) = 0$. Let Σ_* be the linearization about the zero trajectory, and assume that Σ_* is asymptotically controllable and asymptotically observable. Show that then there exist matrices F, D, E (of suitable sizes) such that the origin of

$$\begin{aligned} \dot{x} &= f(x, Fz) \\ \dot{z} &= Dz + Eh(x) \end{aligned}$$

is locally asymptotically stable. This corresponds to a notion of local dynamic stabilizability. □

Exercise 7.2.7 Refer to Exercise 3.2.12. Take all constants equal to one, for simplicity.

(a) Using $y = h$ as the output, find a dynamic feedback stabilizer for Σ.

(b) Show that there is no static output feedback $u = Fy$ which stabilizes the system.

(c) (Optional) Show that it is possible to find a stabilizing feedback law for this example which depends only on h and ϕ, i.e., there are coefficients μ, ν so that with $u = \mu h + \nu \phi$ the closed-loop system is stable. (This is of interest in practice, because h and ϕ can easily be measured with an altimeter and gyro, respectively.) □

Exercise 7.2.8 Refer to Exercise 3.2.13, and take for simplicity $M = m = F = g = l = 1$.

(a) Using $y = x_1 = \delta$ as the output, find a dynamic feedback stabilizer for Σ.

(b) Using both x_1 and x_3 as measured (see Exercise 6.2.6, part (c)) and assuming instead $m = 0$, find a dynamic feedback stabilizer.

(c) (Optional) Show that for no possible "proportional feedback" control law $u = fx = \alpha x_1 + \beta x_3$ is the closed-loop system stable. (You may want to use the fact, easily proved, that a monic polynomial can be Hurwitz only if all its coefficients are strictly positive.) □

7.3 External Stability for Linear Systems

There are many nonequivalent definitions possible for the concept of stability of i/o behaviors. Rather than dealing with these in a general setting, however, we study only the linear (finite dimensional) case, in which all definitions coincide.

Let Λ be either a linear time-invariant discrete-time behavior or a linear time-invariant continuous-time integral behavior (Definitions 2.4.3 and 2.7.8, respectively), with $\mathbb{K} = \mathbb{R}$ or $\mathbb{C}$. Let K be the impulse response of Λ. In the discrete-time case, this is a Markov sequence; in continuous-time it is a matrix of functions $K \in \mathcal{L}^{\infty,loc}_{p\times m}(0,\infty)$. For each $\omega \in \mathcal{U}^{[0,\infty)}$ (locally essentially bounded in the continuous-time case), we can consider the infinite time output $\eta \in \mathcal{Y}^{[0,\infty)}$ produced by ω,

$$\eta(t) := \lambda(t, 0, \omega|_{[0,t)}) .$$

Consistently with the discussion in Remark 2.2.2, we will denote this function η as $\bar{\lambda}(\omega)$. Formulas (2.17) and (2.27) say that

$$\bar{\lambda}(\omega) = K * \omega ,$$

where $*$ denotes convolution in discrete-time or continuous-time. Note that in continuous-time, $\bar{\lambda}(\omega) \in \mathcal{L}^{\infty,loc}_{p}(0,\infty)$.

For notational simplicity when giving unified statements for continuous-time and discrete-time, for each integer r we let $\mathcal{L}^{\infty}_{r}$ denote not just the set of essentially bounded measurable functions $[0,\infty) \to \mathbb{K}^r$ ($\mathbb{K} = \mathbb{R}$ or $\mathbb{C}$, depending on whether we are looking at real or complex systems), but also the set $(l^\infty)^r$ of all sequences

$$\eta = y_0, y_1, y_2, y_2, \ldots$$

of vectors in $\mathbb{K}^r$ satisfying

$$\|\eta\|_\infty := \sup_{i\geq 0} \|y_i\| < \infty .$$

In general, $\bar{\lambda}(\omega)$ is not necessarily in $\mathcal{L}_p^\infty$ even if $\omega \in \mathcal{L}_m^\infty$. When it is, we may view $\bar{\lambda}$ as an operator

$$\bar{\lambda} : \mathcal{L}_m^\infty \to \mathcal{L}_p^\infty .$$

This operator is continuous with respect to the essential supremum norms if and only if it is bounded, that is, if there is some constant γ so that

$$\left\|\bar{\lambda}(\omega)\right\|_\infty \leq \gamma \left\|\omega\right\|_\infty$$

for all ω. In that case, one says that the behavior Λ is **uniformly bounded-input bounded-output**, or **UBIBO** for short.

For an impulse response K, we let $\|K\|_1$ denote its (possibly infinite) $\mathcal{L}^1$ norm. By this we mean

$$\int_0^\infty \|K(t)\| \, dt$$

in continuous-time, or

$$\sum_{i=1}^{\infty} \|K_i\|$$

in discrete-time, where $\|\cdot\|$ denotes the operator norm in $\mathbb{K}^{p\times m}$ induced by Euclidean norm. We say that the impulse response is *integrable* if $\|K\|_1 < \infty$; this is equivalent to the requirement that each entry $K_{ij}(\cdot)$ of K be an integrable function.

Lemma 7.3.1 The behavior Λ is UBIBO if and only if its impulse response K is integrable.

Proof. Sufficiency follows from the fact that

$$\left\|\bar{\lambda}(\omega)\right\|_\infty \leq \|K\|_1 \, \|\omega\|_\infty$$

for all ω. To establish necessity, take any fixed i, j; we will prove that K_{ij} is integrable. Pick any $T > 0$ and let ω be defined as follows. The coordinates different from j are identically zero, and the jth coordinate is

$$\omega_j(t) := \operatorname{sign} K_{ij}(T - t)$$

for all $t \in [0, T]$ and zero otherwise (with the convention $\operatorname{sign} 0 = 0$). Then the ith coordinate of $\eta = \bar{\lambda}(\omega)$ satisfies (in continuous-time)

$$\eta_i(T) = \int_0^T |K_{ij}(t)| \, dt$$

and similarly in discrete-time. Since the behavior is UBIBO, also

$$|\eta_i(T)| \leq \|\eta\|_\infty \leq \gamma \|\omega\|_\infty \leq \gamma,$$

which implies that $\|K_{ij}\|_1 \leq \gamma < \infty$, as wanted. ∎

One of the most useful facts about stable systems is that decaying inputs produce decaying outputs:

Lemma 7.3.2 If Λ is UBIBO, then $\lim_{t\to\infty} \omega(t) = 0$ implies that

$$\lim_{t\to\infty} \bar{\lambda}(\omega)(t) = 0.$$

Proof. Assume that $\omega \to 0$, let $\eta := \bar{\lambda}(\omega)$, and pick any $\varepsilon > 0$. We wish to find a T such that $\|\eta(t)\| < \varepsilon$ for all $t > T$. The calculations will be done for continuous-time, but the discrete-time case is entirely analogous. Denoting $\alpha := \|\omega\|_\infty$ and $\gamma := \|K\|_1$, there is some T_0 so that

$$\|\omega(t)\| \leq \frac{\varepsilon}{2\gamma}$$

for all $t \geq T_0$ and

$$\int_{T_0}^{\infty} \|K(t)\| \, dt \leq \frac{\varepsilon}{2\alpha}.$$

We pick $T := 2T_0$; with this choice, for each $t > T$ it holds that

$$\begin{aligned}
\|\eta(t)\| &\leq \alpha \int_0^{T_0} \|K(t-\tau)\| \, d\tau + \frac{\varepsilon}{2\gamma} \int_{T_0}^{t} \|K(t-\tau)\| \, d\tau \\
&= \alpha \int_{t-T_0}^{t} \|K(\tau)\| \, d\tau + \frac{\varepsilon}{2\gamma} \int_0^{t-T_0} \|K(\tau)\| \, d\tau \leq \varepsilon,
\end{aligned}$$

as desired. ∎

From now on, let Σ be any (finite dimensional, time-invariant) discrete-time or continuous-time system $\Sigma = (A, B, C)$ over $\mathbb{K} = \mathbb{R}$ or $\mathbb{C}$. We shall say that Σ is *internally stable* if A is a convergent or Hurwitz matrix, respectively. Given any such system, we let Λ_Σ be its i/o behavior, and W_Σ the generating series associated to its Markov sequence.

Proposition 7.3.3 If Σ is internally stable, then Λ_Σ is UBIBO.

Proof. In continuous-time, each entry of the matrix $K(t) = Ce^{tA}B$ is a sum of terms of the form $t^l e^{\lambda t}$, with all λ having negative real parts, and this implies that K is integrable. In discrete-time, $K_i = CA^{i-1}B$, and terms of the form $t^l \lambda^i$ appear, $|\lambda| < 1$, so the same is true. ∎

The following characterization of asymptotic observability will be useful below.

Lemma 7.3.4 The system Σ is detectable if and only if the following property holds: For each solution ξ of $\dot{x} = Ax$ (or $x^+ = Ax$),

$$C\xi(t) \underset{t\to\infty}{\longrightarrow} 0 \implies \xi(t) \underset{t\to\infty}{\longrightarrow} 0\,.$$

Proof. The property clearly implies detectability ($C\xi(t) \equiv 0 \Rightarrow \xi(t) \to 0$). Conversely, assume that Σ is detectable and let

$$\dot{z}\;[\text{or}\; x^+] \;=\; (A + LC)z + Bu - Ly \tag{7.5}$$

be an observer constructed as in the proof of Theorem 31 (p. 319), where $A+LC$ is a Hurwitz or convergent matrix, respectively. Assume that $C\xi(t) \to 0$. We consider the system (7.5) with output z, and apply, to this system, Proposition 7.3.3 and Lemma 7.3.2. We conclude that the solution ζ of (7.5) with

$$y(t) = C\xi(t)\,,\; u \equiv 0\,, \;\text{ and }\; \zeta(0) = 0$$

must satisfy $\zeta(t) \to 0$. Since, by definition of observer, $\xi(t) - \zeta(t) \to 0$, the conclusion follows. ∎

A converse of Proposition 7.3.3 is provided by the following result:

Theorem 33 *If Σ is canonical and Λ_Σ is UBIBO, then Σ is internally stable.*

Proof. We provide details only for continuous-time, as the discrete-time case is similar. Pick any solution ξ of $\dot{x} = Ax$, and let $x^0 := \xi(1)$. We need to see that $\xi(t) \to 0$. By controllability of Σ, there is some control ω on $[0,1]$ steering 0 to x^0. We extend ω by $\omega(t) = 0$ for $t > 1$; note that

$$C\xi(t) \;=\; \bar{\lambda}(\omega)(t)$$

for all $t > 1$. Since Λ_Σ is UBIBO and $\omega(t) \to 0$, necessarily $C\xi(t) \to 0$. By Lemma 7.3.4 it follows that $\xi \to 0$, using detectability of Σ. ∎

Finally, one can characterize these properties in terms of rational representations:

Proposition 7.3.5 The behavior Λ_Σ is UBIBO if and only if there exists a rational representation $W_\Sigma = q^{-1}P$ in which q is a Hurwitz (or convergent, in discrete-time) polynomial.

Proof. Without loss of generality, we may assume that Σ is canonical (otherwise pick a canonical realization). If Λ_Σ is UBIBO, then Σ must be internally stable; since q may be chosen as the characteristic polynomial of A, it can be chosen to be Hurwitz or convergent, respectively. Conversely, if there is such a representation, then there is at least one realization that is internally stable. For instance, the observability form realization has a characteristic polynomial equal to q^p, and hence this is an internally stable realization. ∎

Exercise 7.3.6 Prove that "Σ is asymptotically controllable and asymptotically observable" is sufficient (instead of assuming that Σ is canonical) in Theorem 33. □

Exercise 7.3.7 Give an example of a linear system that is not internally stable but whose behavior is UBIBO. □

Exercise 7.3.8 Consider the following "non-uniform BIBO" property: For each bounded ω, $\bar{\lambda}(\omega)$ is also bounded (that is, $\bar{\lambda}$ induces an operator $\mathcal{L}_m^\infty \to \mathcal{L}_p^\infty$, but this is not necessarily continuous). Show that for every Σ as above, Λ_Σ is UBIBO if and only if it satisfies this property. (*Hint:* Just read the above proofs.) □

Exercise 7.3.9 Refer to Exercise 3.2.12. Take all constants equal to one, for simplicity. Using $y = h$ as an output, show that the i/o behavior of Σ is not UBIBO. Find an explicit input that produces a diverging output (when Σ starts at the zero state). □

Exercise 7.3.10 Repeat problem 7.3.9 for the example in Exercise 3.2.13, taking $M = m = F = g = l = 1$. Use $y = x_1 = \delta$ as the output. □

Exercise 7.3.11 Denote by $\mathcal{K}$ the class of functions $[0,\infty) \to [0,\infty)$ which are zero at zero, strictly increasing, and continuous, by $\mathcal{K}_\infty$ the set of unbounded $\mathcal{K}$ functions, and by $\mathcal{KL}$ the class of functions $[0,\infty)^2 \to [0,\infty)$ which are of class $\mathcal{K}$ on the first argument and decrease to zero on the second argument. A continuous-time time-invariant system $\dot{x} = f(x,u)$ with state space $\mathcal{X} = \mathbb{R}^n$ and control-value space $\mathcal{U} = \mathbb{R}^m$ is said to be *input-to-state stable (ISS)* if there exist a function $\beta \in \mathcal{KL}$ and a function $\gamma \in \mathcal{K}_\infty$ so that the following property holds: For each $T > 0$, if ω is a control on the interval $[0,T]$ and ξ is a solution of $\dot{\xi} = f(\xi,\omega)$ on the interval $[0,T]$, then

$$\|\xi(T)\| \leq \max\left\{\beta\Big(\|\xi(0)\|, T\Big), \gamma\left(\|\omega\|_\infty\right)\right\}. \tag{7.6}$$

Show:

1. Every ISS system is complete (i.e., every input is admissible for every state). (*Hint:* Use Proposition C.3.6.)

2. For ISS systems, if ω is any input on $[0,\infty)$ which is bounded (respectively, converges to zero as $t \to \infty$), and if ξ solves $\dot{\xi} = f(\xi,\omega)$, then ξ is bounded (respectively, converges to zero as $t \to \infty$).

3. If a system is ISS, then it is *internally stable*, in the sense that the system with no inputs $\dot{x} = f(x,0)$ is globally asymptotically stable (with respect to the origin).

4. A continuous-time time-invariant *linear* system is ISS if and only if it is internally stable. Moreover, for such systems one may always find β of the form $\beta(r,t) = c_1 e^{-\alpha t} r$ and γ of the form $\gamma(r) = c_2 r$, for some positive constants c_1, c_2, and α.

5. Show that internal stability does not imply ISS for nonlinear systems, by giving an example of a system for which the origin is globally asymptotically stable for $\dot{x} = f(x,0)$ but for which there exists a bounded control ω and an unbounded solution of $\dot{\xi} = f(\xi,\omega)$.

6. If a system is ISS, then it has *finite nonlinear gain*, meaning that there exists a function $\gamma \in \mathcal{K}_\infty$ so that the following property holds: If ω is any input on $[0,\infty)$ which is bounded, and if ξ solves $\dot{\xi} = f(\xi,\omega)$, then

$$\overline{\lim_{t\to\infty}} \|\xi(t)\| \;\leq\; \gamma\left(\|\omega\|_\infty\right). \tag{7.7}$$

(The function γ is then called a nonlinear gain.) □

Exercise 7.3.12 A continuous-time time-invariant system with outputs (but no inputs) $\dot{x} = f(x)$, $y = h(x)$, having state space $\mathcal{X} = \mathbb{R}^n$, and output-value space $\mathcal{Y} = \mathbb{R}^p$, is said to be *output-to-state stable (OSS)* if there exist a function $\beta \in \mathcal{KL}$ and a function $\gamma \in \mathcal{K}_\infty$ (terminology is as in Exercise 7.3.11) so that the following property holds: For each $T > 0$, if ξ is a solution of $\dot{\xi} = f(\xi)$ on the interval $[0,T]$, and $\eta(t) = h(\xi(t))$ for $t \in [0,T]$, then

$$\|\xi(T)\| \;\leq\; \max\left\{\beta\Big(\,\|\xi(0)\|\,,T\Big)\,,\,\gamma\left(\|\eta\|_\infty\right)\right\}. \tag{7.8}$$

Show:

1. A continuous-time time-invariant linear system is OSS if and only if it is detectable.

2. In general, a system may be detectable (with respect to $x^0 = 0$ and $y^0 = 0$) yet not be OSS. □

7.4 Frequency-Domain Considerations

The classical way to treat linear time-invariant finite dimensional systems is via frequency-domain techniques. In this section we briefly sketch some of the basic concepts of that approach, mainly in order to make the reader familiar with terminology and elementary facts. No attempt is made to cover this vast area in any detail. We restrict attention to continuous-time systems, though similar results can be derived for the discrete-time case.

Let Λ be a realizable linear time-invariant behavior of integral type over $\mathbb{K} = \mathbb{R}$, and let K be its impulse response. Realizability means that there are matrices A, B, C so that

$$K(t) = Ce^{tA}B = \sum_{j=0}^{\infty} CA^jB\frac{t^j}{j!}$$

for all $t \geq 0$. Write $\mathcal{A}_j := CA^{j-1}B$, $j = 1,2,\ldots$ for the associated Markov sequence and W for its associated generating series. Pick any real number σ larger than the norm of the matrix A; then

$$\sum_{j=0}^{\infty} \|\mathcal{A}_{j+1}\| \frac{t^j}{j!}e^{-\sigma t} \leq \|C\| \|B\| e^{(\|A\|-\sigma)t} \tag{7.9}$$

is integrable, and in particular

$$\int_0^\infty K(t)e^{-\sigma t}\, dt < \infty$$

holds. Moreover, because of (7.9), the convergence

$$\sum_{j=0}^{N} \mathcal{A}_{j+1}\frac{t^j}{j!}e^{-st} \rightarrow \sum_{j=0}^{\infty} \mathcal{A}_{j+1}\frac{t^j}{j!}e^{-st}$$

is dominated, for each fixed complex number s with $\operatorname{Re} s > \sigma$, so we conclude using the formula

$$\int_0^\infty t^j e^{-st}\, dt = \frac{j!}{s^{j+1}}$$

that

$$\int_0^\infty K(t)e^{-st}\, dt = \sum_{j=1}^{\infty} \mathcal{A}_j s^{-j} = W(s) \tag{7.10}$$

for all such s, when W is seen as a Taylor expansion about infinity.

In general, consider any measurable matrix function F defined on $[0, \infty)$ and of exponential growth, by which we mean that $\|F\|$ is locally integrable and

$$\|F(t)\|\, e^{-\sigma t} \rightarrow 0 \text{ as } t \rightarrow +\infty$$

for some $\sigma \in \mathbb{R}$. For any such F, one defines its *Laplace transform* as the (matrix) function of a complex variable s,

$$\hat{F}(s) := \int_0^\infty F(t)e^{-st}\, dt\,.$$

This is well defined, and is analytic, for all s with $\operatorname{Re} s > \sigma$, if σ is as above. For instance, $\hat{K}$ is defined whenever σ is larger than the real part of every eigenvalue of A. Observe that, writing $W = q^{-1}P$ in rational form, we conclude from Equation (7.10) and Remark 6.7.5 that the matrix of rational functions

$$H(s) := q(s)^{-1}P(s)$$

and $\hat{K}(s)$ coincide wherever they are both defined, and W is the Taylor expansion at infinity of this rational matrix. From now on we will not distinguish between H, W, and $\hat{K}(s)$.

Since, for any two scalar functions of exponential growth,

$$\widehat{k * w} = \hat{k}\hat{\omega},$$

it follows, applying this formula to each coordinate, that for each input function ω of exponential growth the corresponding output η satisfies

$$\hat{\eta}(s) = W(s)\hat{\omega}(s).$$

For this reason one calls W the *transfer matrix* (or, in the particular case $m = p = 1$, the *transfer function*) of the system (A, B, C) (or of the behavior Λ): Its value at each "complex frequency" s characterizes the i/o behavior of the system at that frequency. In classical electrical engineering design, one often starts with a desired frequency behavior (for instance, the requirement that large frequencies be attenuated but low frequencies be relatively unaffected) and one builds a system in such a manner that the transfer matrix satisfies the desired properties. The following easy remark shows how each frequency is affected.

Lemma 7.4.1 Assume that Λ is UBIBO. Pick any real number ω and any two integers $l = 1, \ldots, p$ and $j = 1, \ldots, m$. Consider the control

$$u(t) = \sin \omega t \; v_j$$

(v_j= canonical jth basis vector in $\mathbb{R}^m$) on $[0, \infty)$. Let $\eta(t)$ be the lth coordinate of the corresponding output on $[0, \infty)$, and write

$$W(i\omega)_{lj} = re^{i\varphi}$$

in polar form. Then,

$$\eta(t) - r\sin(\omega t + \varphi) \to 0$$

as $t \to \infty$.

Proof. Since the behavior is stable, $K(t)$ is integrable, so

$$\left\| \int_t^\infty K(\tau)v_j e^{i\omega(t-\tau)}\, d\tau \right\| \le \int_t^\infty \|K(\tau)\|\, d\tau \to 0$$

as $t \to \infty$. Therefore,

$$\int_0^t K(\tau)v_j e^{i\omega(t-\tau)}\, d\tau \; - \; e^{i\omega t}\int_0^\infty K(\tau)v_j e^{-i\omega\tau}\, d\tau \; \to \; 0.$$

Thus, the imaginary part of this difference goes to zero, too, as desired. ∎

Remark 7.4.2 Introducing more terminology from electrical engineering, for the case $m = p = 1$, one calls the modulus r of $W(i\omega)$ the *gain at frequency* ω and its argument φ, taken in the interval $[-\pi, \pi)$, the *phase shift at frequency* ω. Inputs at the corresponding frequencies are shifted and amplified according to these parameters. The system exhibits *phase lag* at a frequency if $\varphi < 0$ and *phase lead* if $\varphi > 0$. The *steady-state output* corresponding to $\sin \omega t$ is $r \sin(\omega t + \varphi)$. The *Bode plot* or Bode diagram consists of graphs of gain and phase as functions of $\omega \geq 0$ on a logarithmic scale. A large number of qualitative properties of the rational function $W(s)$ can be deduced immediately from such plots, even on the basis of partial information. □

Exercise 7.4.3 Let $m = p = 1$. Assume that $W(s) = \frac{b}{s+a}$ is an asymptotically stable but unknown transfer function (a, b real > 0). Suppose that the steady-state output corresponding to the input $\sin t$ is $\sqrt{2}\sin(t - \pi/4)$. (That is, there is a $-45°$ phase shift and an amplification of $\sqrt{2}$.) Find $W(s)$. □

Exercise 7.4.4 Let $m = p = 1$. Show that if Λ is UBIBO, then the output corresponding to $u \equiv 1$ converges to $W(0)$. (The quantity $W(0)$ is accordingly called the *dc gain* of the system, the gain corresponding to a nonalternating or "direct current" input.) □

Observe that, in general, if a control ω has exponential growth, then the solution of

$$\dot{\xi} = A\xi + B\omega$$

again has exponential growth, as is clear from the variation of parameters formula. Since $\dot{\xi}$ is a linear combination of $A\xi$ and $B\omega$, it also has exponential growth, and an integration by parts shows that its transform is $s\hat{\xi}(s) - \xi(0)$. If $\xi(0) = 0$, we conclude that

$$s\hat{\xi}(s) \;=\; A\hat{\xi}(s) + B\hat{\omega}$$

and therefore that

$$\hat{\eta} \;=\; C(sI - A)^{-1}B\hat{\omega}$$

for the output, consistent with the equality $W = \hat{K}$. Laplace transform techniques are useful in developing a "calculus" for systems over $\mathbb{R}$ or $\mathbb{C}$, though one could also use the formal power series approach of Section 6.7 to obtain analogous results (with the advantage that they then are valid over arbitrary fields of coefficients).

Interconnections

Consider two systems $\Sigma_1 = (A_1, B_1, C_1)$ and $\Sigma_2 = (A_2, B_2, C_2)$ and a linear interconnection as in Definition 7.2.3:

$$\begin{aligned}
\dot{x}_1 &= A_1x_1 + B_1(F_{11}C_1x_1 + F_{12}C_2x_2)\\
\dot{x}_2 &= A_2x_2 + B_2(F_{21}C_1x_1 + F_{22}C_2x_2)
\end{aligned}$$

seen as a system of dimension n_1+n_2 and no inputs or outputs. (The connecting maps are $k_1 = (F_{11}, F_{12})$ and $k_2 = (F_{21}, F_{22})$.)

We assume that Σ_1 is controllable and observable, and we wish to study the possible linear systems Σ_2 providing dynamic stabilization, that is, making the interconnection asymptotically stable. For each such possible system Σ_2, we may consider a new interconnection in which $F_{22} = 0$, $F_{21} = I$, and $F_{12} = I$, since one can redefine A_2 as $A_2 + B_2F_{22}C_2$, B_2 as B_2F_{21}, and C_2 as $F_{12}C_2$. Since a noncanonical system can always be replaced by a canonical one, we will look for canonical systems Σ_2. Therefore, the problem becomes one of finding, for the given system Σ_1, a canonical system Σ_2 and a matrix F so that

$$\begin{aligned} \dot{x}_1 &= A_1x_1 + B_1(FC_1x_1 + C_2x_2) \\ \dot{x}_2 &= A_2x_2 + B_2C_1x_1 \end{aligned}$$

is asymptotically stable. For any such interconnection, we introduce external inputs as follows:

$$\dot{x}_1 = A_1x_1 + B_1(F(C_1x_1 + u_2) + C_2x_2 + u_1) \tag{7.11}$$

$$\dot{x}_2 = A_2x_2 + B_2(C_1x_1 + u_2) \tag{7.12}$$

seen as a system of dimension $n_1 + n_2$, input space $\mathbb{R}^{m_1+m_2}$, and outputs

$$y = \begin{pmatrix} C_1x_1 \\ C_2x_2 \end{pmatrix}.$$

When the inputs are kept at zero and the outputs are ignored, this is the same as the above interconnection.

Lemma/Exercise 7.4.5 If each of the systems Σ_i is canonical, then the system (7.11-7.12) is also canonical. □

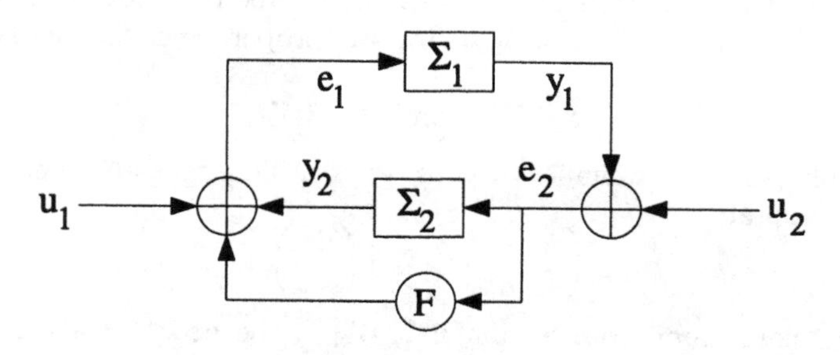

Figure 7.5: *Interconnection, after simplifications.*

From this and the equivalences between the UBIBO property and internal stability, one concludes that the problem of characterizing the linear dynamic stabilizers of a controllable and observable system Σ_1 is equivalent to the problem of finding systems Σ_2 and matrices F so that the interconection in the present input/output sense is UBIBO. One then may work with the corresponding transfer matrices W_1 of Σ_1 (of size $p \times m$) and W_2 of Σ_2 (of size $m \times p$), as illustrated next.

Consider the initial state $(0,0)$ of (7.11-7.12) and take any two inputs ω_1, ω_2 of exponential growth. Let ξ_1, ξ_2 be the resulting trajectories, and let $\eta_1 = C\xi_1$ and $\eta_2 = C\xi_2$ be the corresponding outputs. Observe that

$$\hat{\eta}_2 = W_2 \hat{e}_2 ,$$

where

$$e_2 := \eta_1 + \omega_2 ,$$

and

$$\hat{\eta}_1 = W_1 \hat{e}_1 ,$$

where

$$e_1 := F e_2 + \eta_2 + \omega_1 .$$

Note that boundedness of η_1 and η_2 is equivalent to boundedness of e_1 and e_2, when ω_1 and ω_2 are bounded. It is more convenient to work with e_1, e_2 rather than η_1, η_2. We write W instead of W_1, and we let

$$V(s) := F + W_2(s) ,$$

thought of as a matrix of rational functions. Thus, the following equations are satisfied by the Laplace transforms of all signals:

$$\begin{aligned} \hat{e}_1 &= \hat{\omega}_1 + V \hat{e}_2 \\ \hat{e}_2 &= \hat{\omega}_2 + W \hat{e}_1 . \end{aligned}$$

The matrix W is *strictly proper*, that is, its entries are rational functions vanishing at infinity. On the other hand, V is in general only *proper*, that is, each of its entries can be written as p/q with the degree of p not greater than the degree of q. The product VW is again strictly proper, so as matrices of rational functions,

$$I - VW \text{ and } I - WV$$

are invertible (the determinant is 1 as $|s| \to \infty$). Thus, the above equations can be solved to obtain

$$\begin{pmatrix} \hat{e}_1 \\ \hat{e}_2 \end{pmatrix} = W_{\text{cl}} \begin{pmatrix} \hat{\omega}_1 \\ \hat{\omega}_2 \end{pmatrix} ,$$

where the "closed loop transfer matrix" W_{cl} is the proper matrix of rational functions (of size $(m+p) \times (m+p)$)

$$\begin{pmatrix} (I - VW)^{-1} & (I - VW)^{-1} V \\ (I - WV)^{-1} W & (I - WV)^{-1} \end{pmatrix} . \tag{7.13}$$

We say that W_{cl} is *stable* if all of its entries are rational functions with no poles on $\operatorname{Re} s \geq 0$. We have then:

Proposition 7.4.6 The canonical system Σ_2 and feedback matrix F stabilize the above interconnection if and only if $W_{\rm cl}$ is stable.

Proof. Stability of the interconnection is equivalent to the requirement that e_1, e_2 be bounded for all bounded ω_1, ω_2; we need to show that this is equivalent to stability of $W_{\rm cl}$. If the latter were strictly proper, the equivalence would follow from Proposition 7.3.5. We need only to see then that the proper case can be reduced to the strictly proper one. In general, every proper matrix $U(s)$ can be written as $U_1(s) + U_0$, where U_0 is constant and U_1 is strictly proper with the same poles as U. (To see this, one writes each entry as the sum of a constant and a proper rational function, after performing a polynomial division.) Let K be so that U_1 is the transform of K. What we need to show is that the i/o behavior with impulse response K is UBIBO if and only if the operator norm of $\omega \mapsto K * \omega + U_0\omega$ is finite. But this norm differs from the one induced by convolution by K at most by the finite amount $\|U_0\|$. ∎

Since every proper V can be written, after division, in the form $F + W_2$, with W_2 strictly proper, and in turn any such W_2 can be realized as the transfer matrix of a canonical linear system, the problem of characterizing all stable interconnections can be restated simply as:

Find all proper V so that $W_{\rm cl}$ is stable.

The next section studies this problem for the case $m = p = 1$.

7.5 Parametrization of Stabilizers

We now specialize to the scalar case, $m = p = 1$, in order to avoid having to deal with the theory of stable factorizations of rational matrices. Most of what follows, however, can be developed in far more generality.

To summarize, the algebraic problem that we need to solve is as follows: Given a strictly proper rational function $w(s)$, find all those proper rational functions $v(s)$ such that all three rational functions

$$(1 - vw)^{-1} \quad (1 - vw)^{-1}v \quad (1 - vw)^{-1}w \tag{7.14}$$

are stable. Stability of a rational function w means that w is analytic on

$$\mathbb{C}_+ := \{s \mid \operatorname{Re} s \geq 0\},$$

while properness is equivalent to the property that the limit

$$w(\infty) := \lim_{|s|\to\infty} w(s)$$

exists (and is finite) and strict properness to $w(\infty) = 0$. Our objective is to parametrize the set of all v stabilizing a given w. The most convenient way to do this is through the study of the basic properties of the set $\mathbf{RH}_\infty$ of proper stable rational functions (the notation stands for real-rational *Hardy space*). We think of this set as a ring, when the pointwise operations for functions are used. The main technical fact that we need about this space is given in the following Lemma; it is a consequence of the somewhat more general fact —which we do not prove— that $\mathbf{RH}_\infty$ is a Euclidean domain.

Two elements $f, g \in \mathbf{RH}_\infty$ are said to be *coprime* if they have no common zeros in $\mathbb{C}_+$ and at least one of them is not strictly proper (in other words, f and g have no common zeros in $\mathbb{C}_+ \cup \{\infty\}$). As with polynomials, coprimeness is equivalent to a "Bezout" type identity:

Lemma 7.5.1 The functions $f, g \in \mathbf{RH}_\infty$ are coprime if and only if there exist $\tilde{\alpha}, \tilde{\beta} \in \mathbf{RH}_\infty$ such that

$$\tilde{\alpha} f + \tilde{\beta} g = 1 . \tag{7.15}$$

Proof. Since all functions are analytic on $\mathbb{C}_+$ and bounded at infinity, Equation (7.15) implies that f and g can have no common zeros in $\mathbb{C}_+ \cup \{\infty\}$. Conversely, assume that they are coprime. It is convenient to use the bilinear mapping $z := (s-1)/(s+1)$ sending $\mathbb{C}_+$ into the unit complex disk D and its inverse $s := (1+z)/(1-z)$. Given any rational function w, we consider the rational functions

$$\begin{aligned} \hat{w}(z) &:= w\left(\frac{1+z}{1-z}\right) \\ \tilde{w}(s) &:= w\left(\frac{s-1}{s+1}\right) . \end{aligned}$$

Note that if $w = w_1 w_2 + w_3$ then $\widehat{w} = \hat{w}_1 \hat{w}_2 + \hat{w}_3$, and similarly for $\tilde{}$. Furthermore, $\tilde{\hat{w}} = \hat{\tilde{w}} = w$ for all w, and $w \in \mathbf{RH}_\infty \iff \hat{w} \in S$, where S is the set of all rational functions that have no poles in D. Coprimeness of f, g is equivalent to the requirement that $\hat{f}$ and $\hat{g}$ have no common zeros in D. The problem is then reduced to showing that, given any two $a, b \in S$ with no common zeros in D, there must exist $\alpha, \beta \in S$ so that $\alpha a + \beta b = 1$.

Write $a = p_1/q_1$ and $b = p_2/q_2$ as quotients of relatively prime polynomials so that p_1 and p_2 have no common zeros in D. Let d be the GCD of the polynomials p_i and find a linear combination $d = \alpha_0 p_1 + \beta_0 p_2$ with polynomial coefficients. Note that d has no zeros in D because its zeros are the common zeros of p_1 and p_2. Therefore

$$\alpha := \frac{\alpha_0 q_1}{d} \quad \text{and} \quad \beta := \frac{\beta_0 q_2}{d}$$

are as desired. ∎

By a *coprime fractional representation* for a rational function w we mean a pair of coprime $f, g \in \mathbf{RH}_\infty$ such that $w = f/g$. Any w admits such a representation: If $w = p/q$ as a quotient of relatively prime polynomials and if k is the maximum of the degrees of p and q, then we may write

$$w = \frac{p/(s+1)^k}{q/(s+1)^k}$$

in coprime form. Observe that whenever w is strictly proper and $w = f/g$ with $f, g \in \mathbf{RH}_\infty$ necessarily $f = gw$ is strictly proper, too.

If $w = f/g$ and $v = n/d$, then the conditions for stabilizability involving (7.14) can be restated as the requirement that all three elements

$$\frac{gd}{r} \quad \frac{gn}{r} \quad \frac{fd}{r} \tag{7.16}$$

be in $\mathbf{RH}_\infty$, where $r := gd - fn$. Observe that $fn/r = gd/r - 1$ is in $\mathbf{RH}_\infty$ if and only if gd/r is. Thus, fn/r can be added to the list (7.16).

The main result is as follows. Given any w, we write it first in coprime form and find α, β as in Lemma 7.5.1.

Theorem 34 *Let $w = f/g$ be strictly proper, and let $\alpha, \beta \in \mathbf{RH}_\infty$ be so that $\alpha f + \beta g = 1$. Then the rational functions v stabilizing w are precisely the elements of the form*

$$\boxed{v = \frac{\mu g - \alpha}{\mu f + \beta}, \quad \mu \in \mathbf{RH}_\infty} \tag{7.17}$$

Proof. Assume first that v is of this form. Since w is strictly proper, f is, too; thus, $\beta g = 1 - \alpha f$ implies that $\beta(\infty) \neq 0$. We conclude that the denominator $\mu f + \beta$ in (7.17) is nonzero, so the expression is well defined as a rational function, and in particular it is not zero at infinity so that v is proper. (Observe that v is in general not stable, however.) Consider the elements

$$n := \mu g - \alpha, \quad d := \mu f + \beta$$

of $\mathbf{RH}_\infty$, and note that

$$r = gd - fn = g(\mu f + \beta) - f(\mu g - \alpha) = \alpha f + \beta g = 1 .$$

Since their numerators are in $\mathbf{RH}_\infty$, it follows that all of the elements of the list (7.16) are in $\mathbf{RH}_\infty$, which proves that v defined by formula (7.17) indeed stabilizes.

Conversely, assume that v stabilizes, and write $v = n/d$ in coprime form. Let $r := gd - fn$; we claim that r is a unit in $\mathbf{RH}_\infty$, i.e., $1/r \in \mathbf{RH}_\infty$. Indeed, if γ, δ are so that $\gamma n + \delta d = 1$, then

$$\frac{1}{r} = \frac{1}{r}(\alpha f + \beta g)(\gamma n + \delta d) = \frac{\alpha\gamma f n}{r} + \frac{\alpha\delta f d}{r} + \frac{\beta\gamma g n}{r} + \frac{\beta\delta g d}{r}$$

is in $\mathbf{RH}_\infty$, because each term is (by the assumption that v stabilizes). We may then redefine $n := n/r$ and $d := d/r$, so that without loss of generality we assume from now on that $gd - fn = 1$. We then have the equation

$$\begin{pmatrix} \alpha & \beta \\ -n & d \end{pmatrix} \begin{pmatrix} f \\ g \end{pmatrix} = \begin{pmatrix} 1 \\ 1 \end{pmatrix} \tag{7.18}$$

from which it follows, multiplying by the cofactor matrix, that

$$\mu \begin{pmatrix} f \\ g \end{pmatrix} = \begin{pmatrix} d & -\beta \\ n & \alpha \end{pmatrix} \begin{pmatrix} 1 \\ 1 \end{pmatrix} = \begin{pmatrix} d-\beta \\ n+\alpha \end{pmatrix},$$

where $\mu = \alpha d + \beta n \in \mathbf{RH}_\infty$ is the determinant of the matrix in (7.18). Therefore $d = \mu f + \beta$ and $n = \mu g - \alpha$, as wanted. ∎

Example 7.5.2 As a trivial illustration, take the system $\dot{x} = u$, $y = x$. Its transfer function is $w = 1/s$, which admits the coprime fractional representation

$$w = \frac{1/(s+1)}{s/(s+1)},$$

and we may pick $\alpha = \beta = 1$. According to the Theorem, every possible stabilizer is of the form

$$v = \frac{\mu s/(s+1) - 1}{\mu/(s+1) + 1} = \frac{\mu s - s - 1}{\mu + s + 1},$$

with μ ranging over all possible $\mu \in \mathbf{RH}_\infty$. Such v's are guaranteed to be proper and to stabilize. For instance, for $\mu = 0$ one obtains $v = -1$, which corresponds to the obvious stabilizing law $u = -y$. □

The usefulness of Theorem 34 is in the fact that it characterized all stabilizers in terms of a free parameter. Once such a parametrization is obtained, further design objectives can be translated into requirements on the parameter. In this sense it is useful to note that the closed-loop transfer matrix can also be computed easily; with the notations of the Theorem, one has

$$W_{\mathrm{cl}} = \begin{pmatrix} gd & gn \\ fd & gd \end{pmatrix} = \mu \begin{pmatrix} fg & g^2 \\ f^2 & fg \end{pmatrix} + \begin{pmatrix} g\beta & -\alpha g \\ f\beta & g\beta \end{pmatrix},$$

which is an affine expression $\mu W_1 + W_0$ on μ.

Exercise 7.5.3 With the above notations, show that, for any desired closed-loop transfer matrix $W = W_{\mathrm{cl}}$, there will be a solution if and only if one can write

$$\begin{pmatrix} 1 & 0 \\ -w & 1 \end{pmatrix} (W - W_0) = \begin{pmatrix} a & b \\ 0 & 0 \end{pmatrix}$$

with $ag^2 = bfg$, and both of these rational functions are in $\mathbf{RH}_\infty$. □

Exercise 7.5.4 Prove that, in general, μ is uniquely determined by v. That is, each possible v appears exactly once in the parametrization. □

Exercise 7.5.5 It is not totally trivial to find the coefficients α and β as in Equation (7.15). Show how to obtain these coefficients starting with any particular stabilizer. (A stabilizer can be itself obtained from state space techniques, for instance.) □

Exercise 7.5.6 Find all stabilizers for $w = 1/s^2$. Are there any that are constant? □

Nyquist Criterion

There is a graphical test for stability that is standard in engineering practice and which helps in finding stabilizers. Again for simplicity of exposition we restrict to the case $m = p = 1$ and to stabilization by constant feedback. Before presenting the main result, we need to review some basic complex analysis facts.

Assume that f is a rational function having no poles or zeros on a simple closed curve C. Assume in addition that C is oriented *clockwise* (to respect engineering conventions). Then

$$\frac{1}{2\pi i}\int_C \frac{f'(s)}{f(s)}\,ds \;=\; P_C - Z_C\,, \tag{7.19}$$

where Z_C (respectively, P_C) is the number of zeros (respectively, poles) of f in the region bounded by C (counted with multiplicities). Consider also the curve $\Gamma := f(C)$, with the induced orientation. Changing variables, the integral (7.19) is the same as

$$\frac{1}{2\pi i}\int_\Gamma \frac{d\xi}{\xi}\,.$$

In turn, this last integral is the index of Γ with respect to the origin, that is, the number of counterclockwise encirclements of 0 by the oriented curve Γ. We let $\mathrm{cw}\,(\Gamma, p)$ denote the number of *clockwise* encirclements of the point p by the curve Γ. Therefore, $\mathrm{cw}\,(\Gamma, 0) = Z_C - P_C$.

Now assume that f is proper and has no poles or zeros on the imaginary axis $i\mathbb{R}$. Let Γ be the image of the curve $i\mathbb{R} \cup \{\infty\}$, seen as an oriented curve in the Riemann sphere (transversed on the imaginary axis from negative to positive imaginary part). Let P_+ and Z_+ be the numbers of poles and of zeros of f on the half-plane $\mathbb{C}_+$; then

$$\mathrm{cw}\,(\Gamma, 0) \;=\; Z_+ - P_+\,. \tag{7.20}$$

(We could also argue directly in terms of the complex plane: The curve Γ is the closure of $f(i\mathbb{R})$, which is the same as the union of $f(i\mathbb{R})$ and the point $f(\infty)$, seen as an oriented curve, with the orientation induced by transversing $i\mathbb{R}$ from $-i\infty$ to $+i\infty$. The numbers P_+ and Z_+ are equal to the respective numbers in the region bounded by the curve C_r, for all large enough r, where C_r consists of the interval $[-ir, ir]$ of the imaginary axis and the semicircle of radius r centered at the origin and on the right-hand plane (see Figure 7.6); on the other hand, the image $f(C_r)$ approaches the image of $i\mathbb{R}$, since all values on the semicircle approach $f(\infty)$, so the index is the same.)

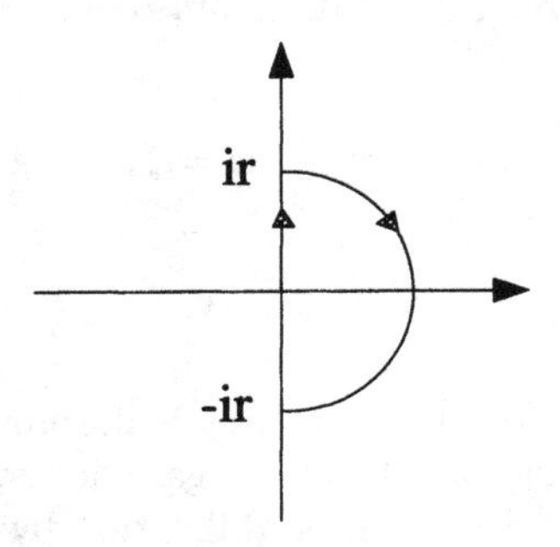

Figure 7.6: *The path C_r.*

We are ready to apply these considerations to the stabilization problem. We will deal with stabilization by constant feedback, that is, $v = k$ is a constant rational function.

Lemma/Exercise 7.5.7 Assume that w is rational and that $v = k \neq 0$ is a constant. Then the poles of the entries of W_{cl} (that is, the elements in the list (7.14),) are precisely the zeros of $w - \frac{1}{k}$. □

We call the zeros of $w - \frac{1}{k}$ that are in $\mathbb{C}_+$ the *unstable closed-loop poles* (for the system that results when using $v = k$ as a stabilizer) and the poles of w in $\mathbb{C}_+$ the *unstable open-loop poles.* Thus, $v = k$ stabilizes w if and only if there are no unstable closed-loop poles.

Let w be a strictly proper rational function. We assume that w has no poles in $i\mathbb{R}$ (the general case can be treated similarly, simply by modifying the curve using a detour about the poles). Denote by Γ the closure of the image of $i\mathbb{R}$ (transversed from $-i\infty$ to $+i\infty$) under w. The following result is often referred to as the *Nyquist criterion:*

Theorem 35 *For each $k \neq 0$ such that $\frac{1}{k} \notin \Gamma$,*

$$\begin{aligned} \operatorname{cw}(\Gamma, \frac{1}{k}) &= \text{number of unstable closed-loop poles} \\ &\quad - \text{number of unstable open-loop poles}\ . \end{aligned}$$

Proof. First note that $\operatorname{cw}(\Gamma, \frac{1}{k}) = \operatorname{cw}(\Gamma_k, 0)$, where Γ_k is the image of $i\mathbb{R}$ under $f(s) = w - \frac{1}{k}$ (since Γ_k is just a translate of Γ). Note that f has no zeros or poles on $i\mathbb{R}$, by the assumptions on w and k. Since f is proper, Equation (7.20) applies to Γ_k. ∎

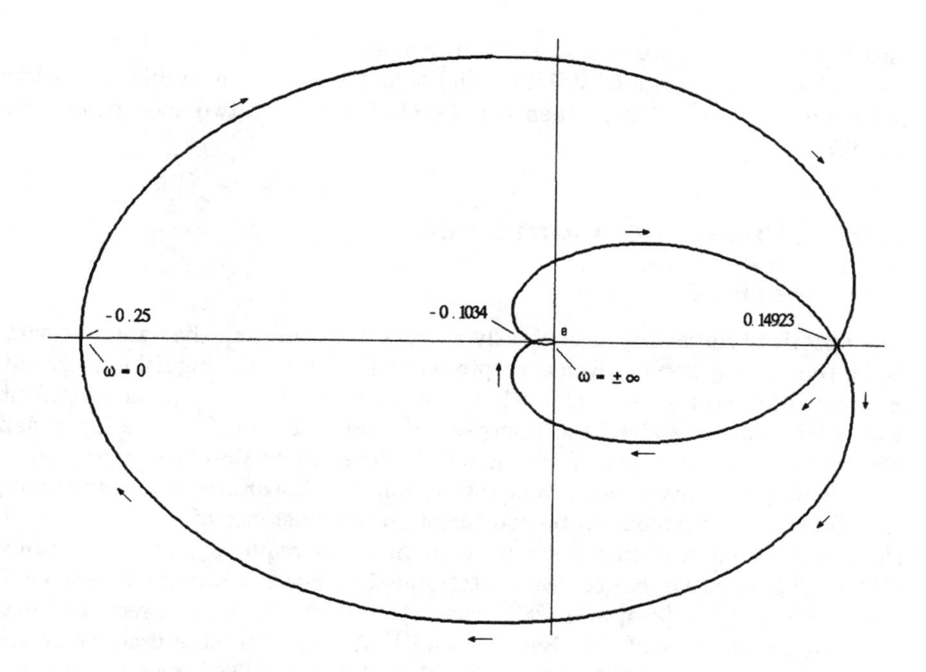

Figure 7.7: *Nyquist diagram for Example 7.5.9.*

Corollary 7.5.8 If w is stable, then for each $k \neq 0$ such that $\frac{1}{k} \notin \Gamma$, the closed-loop system with $v = k$ is stable if and only if $\mathrm{cw}\,(\Gamma, \frac{1}{k}) = 0$. □

Observe that when $\frac{1}{k} \in \Gamma$ necessarily $w - \frac{1}{k}$ has a zero in $i\mathbb{R}$, so the closed-loop system is unstable. And for $k = 0$ one has the same stability as w. Thus, the above criterion and its corollary completely determine the range of k's for which stability holds. To apply these results, one plots $w(i\omega), \omega \in \mathbb{R}_+$ and then one obtains Γ by symmetry (since w is real, $w(-i\omega) = \overline{w(i\omega)}$). Unless $w \equiv 0$, there will be a finite number of points where the curve Γ crosses the real axis. These points determine a finite set of open intervals in each of which the index is constant. For each such interval (a, b), the condition $\frac{1}{k} \in (a, b)$ provides a characterization of the number of closed-loop poles when using the feedback law $u = ky$.

Example 7.5.9 Take the transfer function

$$w(s) = \frac{s-1}{(s+1)^2(s+2)^2},$$

which is clearly stable. A computer plot reveals that the index is constant on each of the intervals (approximately)

$$(-\infty, -1/4) \quad (-1/4, -0.1034) \quad (-0.1034, 0) \quad (0, 0.14923) \quad (0.14923, +\infty)$$

(see Figure 7.7), where it is $0, 1, 3, 2, 0$, respectively. We conclude that $-4 < k < 6.7$ gives the range of stability, and that there are 3 unstable closed-loop poles when $k < -9.671$, 1 when $k \in (-9.671, -4)$, and two such poles when $k > 6.7$. □

7.6 Notes and Comments

Observers and Detectability

Asymptotic observability is obviously a necessary condition that must be satisfied if there is any hope of even asymptotically final-state distinguishing the state x^0 from other states. Any algorithm that uses only the output measurements will not be able to differentiate between x^0 and states in $\mathcal{X}^0$ if the input happens to be constanly equal to u^0. So unless these states already converge to x^0, there is no possibility of consistent estimation. For linear systems, detectability as defined here turns out to be equivalent to the existence of observers, but in the general nonlinear case a stronger definition is required (see, for instance, [361]). Observers can be seen as a deterministic version of Kalman filters, introduced originally in [214] and [229]; the deterministic treatment seems to have originated with work of J.E. Bertram and R.W. Bass, but the first systematic treatment was due to Luenberger in the early 1960s; see [287] for references and more details. The construction of reduced order observers in Exercise 7.1.5 is taken from [287], who followed [159].

The terminology "strong" in Definition 7.1.3 is meant to emphasize that the observer must provide estimates for all possible controls, and starting at arbitrary initial states. It is possible —and even desirable— to weaken this definition in various ways, by dropping these requirements. For linear systems, all reasonable definitions (even requiring that there be *some* initial state for the observer that results in converging estimates when the *zero* control is applied to the original system) are equivalent in the sense that the existence of an observer of one type implies the existence of observers of the other types. For nonlinear systems, these different notions are of course not equivalent; see, for instance, [87]. Some references on observer construction for nonlinear systems are [87], [164], [209], and [253]. There are technical reasons for not making the definition slightly more general and allowing θ to depend on $\omega(t)$: This would mean that the instantaneous value of $\hat{\xi}(t)$ may not be well defined when dealing with continuous-time systems, as there ω is in general an arbitrary measurable function.

Dynamic Feedback

The design of dynamic controllers using a combination of an observer and a state feedback law is a classical approach. It turns out that even optimal solutions, when the (linear quadratic) problem is posed in the context of stochastic control, also have this separation property. For nonlinear stabilization using such techniques, see, for instance, [416] and [419]. For linear systems over rings, see, for instance, [129] and [246]; for infinite dimensional linear systems, see [201] and the references given therein.

External Stability for Linear Systems

The topic of external stability for linear infinite dimensional and for nonlinear systems has been the subject of much research. The books [115], [414], and [422] provide further results and references, and [410] was an influential early paper.

The notion of input-to-state stability (ISS) has recently become central to the analysis of nonlinear systems in the presence of disturbances. The definition is from [370]; for recent results see [376]. In the latter reference, it is shown that a system is ISS if and only if it is internally stable, complete, and has a finite nonlinear gain. The textbook [259] discusses at length applications of ISS and related properties to practical nonlinear feedback design problems. The "dual" notion of OSS is given in [377], where a generalization to systems with inputs as well as outputs ("input/output-to-state stability" or IOSS) is also considered.

For questions of i/o stability versus stability in a state-space setting for infinite dimensional linear systems see, e.g., [285], [324], and [439].

Frequency Domain Considerations

There is a very large number of textbooks dealing with complex variables techniques for the analysis of linear systems. A modern presentation of some of these facts, together with references to a bibliography too extensive to even attempt to partially list, can be found in [144].

Parametrization of Stabilizers

See, for instance, [417] and [145] for detailed developments of the theory of parametrizations of stabilizing compensators. Theorem 34 is due to Youla and coworkers, and was given in [441]. In its version for arbitrary m, p it is the basis of many of the recent developments in multivariable linear control. Recently there have been attempts at generalizing such parametrizations to nonlinear systems as well as obtaining the associated coprime factorizations; see, for instance, [116], [171], [172], [258], [370], [398], and [412].

Chapter 8

Optimality: Value Function

Chapter 3 dealt with the abstract property of controllability, the possibility of using inputs in order to force a system to change from one state to another. Often, however, one is interested not merely in effecting such a transition, but in doing so in a "best" possible manner. One example of this was provided in that chapter, when we calculated the control of minimal norm that achieved a given transfer, in a fixed time interval.

The general *optimal control* problem concerns the minimization of some function

$$F(\xi, \omega)$$

of trajectories, the *cost function* or *performance index*. (Or, one may want to maximize instead a "utility" function F, but this amounts to minimizing the cost $-F$.) The performance index F might incorporate energy costs, or the length of time that a control takes to perform a desired objective, and the optimization may be subject as well to constraints such as maximum allowed magnitudes of controls. Of course, ξ and ω must be related by the system's equation, for instance

$$\dot{\xi} = f(\xi, \omega)$$

in continuous-time.

In broad terms, it is possible to classify the theoretical study of optimal control into two major approaches: dynamic programming, and variational methods. (We include in this second category the Pontryagin Maximum, or "Minimum", Principle.) Both approaches rely upon a transformation of a dynamic optimization into a set of static optimization problems, one for each time instant, together with a differential (or difference) equation. This equation involves the state, as well as an auxiliary quantity which serves to link the static problems. In dynamic programming, the auxiliary information is encoded into an instantaneous cost associated to a current state. In variational approaches, one works, instead, with a Lagrange multiplier.

In *dynamic programming*, the central role is played by the *value function* or *Bellman function* (named after Richard Bellman, who pioneered the use of the technique). This is the function $V(t,x)$ that provides, for any given state x at any given time t, the smallest possible cost among all possible trajectories starting at this event. The main observation is that along any optimal trajectory it must hold that the initial cost plus the cost of a single transition add up to the remaining optimal cost. In discrete-time, there results a recursive, easily programmed method for optimization, with a multi-stage optimization reduced to a sequence of single-stage minimizations. In continuous-time, the recursion becomes a differential equation in time, that is to say, a partial differential equation for $V(t,x)$, called the *Hamilton-Jacobi-Bellman* equation (HJB). In general, this PDE is very hard to solve, and no solutions exist unless one allows a generalized notion of solution. In the case of linear systems with quadratic costs, the value function is quadratic in x, and because of this, the HJB equation reduces to an ordinary differential equation, the Riccati matrix differential equation, which can be readily solved numerically.

In *variational* or *multiplier methods* (in modern control theory, *Pontryagin's Maximum, or Minimum, Principle* (PMP)), one introduces a function $p = p(t)$, as a "Lagrange multiplier" for the optimization problem, and, roughly speaking, studies the critical points of (in continuous-time):

$$F(\xi,\omega) + \langle p, \dot{\xi} - f(\xi,\omega)\rangle ,$$

seen as a function of the state $\xi(\cdot)$, the control $\omega(\cdot)$, and the multiplier $p(\cdot)$. (In the Hamiltonian formulation of classical mechanics, p is the vector of generalized momenta.) Setting derivatives to zero, one may in principle solve for $\omega(t)$ as a function of $x(t)$ and $p(t)$, and in addition one obtains a differential equation for the pair of vector functions (ξ, p).

Each of the two approaches has its advantages and disadvantages. Dynamic programming-based techniques, when applicable, often result in solutions in feedback form, which is highly desirable in applications, but the most general theorems tend to state only sufficiency conditions for optimality of a proposed solution. On the other hand, variational techniques tend to require less assumptions about the system and are thus more generally applicable, but they lead only to necessary conditions. Variational methods are the subject of Chapter 9.

The main objective of this chapter is to discuss the dynamic programming, or value function, approach, and in particular the application to linear systems with quadratic costs ("LQ problems"). For LQ problems, both methods lead to the same solution, but the discussion in terms of dynamic programming is somewhat easier. The first section discusses dynamic programming in general abstract terms. That section may be skipped, and the reader may proceed directly to Section 8.2, which discusses linear systems. However, the optional section explains the general context in which to understand the linear result, in particular the Riccati matrix differential equation. After that, we treat the infinite-time (steady-state) LQ problem. The chapter closes with an infinite-horizon problem stated in nonlinear terms.

As a final introductory remark, we note that the study of optimal control theory is important for reasons other than optimality itself. For nonlinear systems, it is often difficult to find any controls whatsoever that help achieve a desired transfer. Optimal control offers a systematic approach to search for these controls, an approach that can be used as the basis of numerical methods.

8.1 Dynamic Programming*

We first make some general remarks about abstract optimization problems. Given a system Σ and a function taking nonnegative real values

$$\mathcal{Q} : \mathcal{D}_\phi \to \mathbb{R}_+ ,$$

which we shall call the *trajectory cost* function, we read

$$\mathcal{Q}(\tau, \sigma, x, \omega)$$

as "the cost of the trajectory produced by the control $\omega \in \mathcal{U}^{[\sigma,\tau)}$ if we start at state x." The following additivity property is assumed to hold for $\mathcal{Q}$, for each triple of times $t_1 \leq t_2 \leq t_3$, and each $(x, \omega) \in \mathcal{D}_{t_1,t_3}$:

$$\mathcal{Q}(t_3, t_1, x, \omega) = \mathcal{Q}(t_2, t_1, x, \omega|_{[t_1,t_2)}) + \mathcal{Q}(t_3, t_2, \phi(t_2, t_1, x, \omega|_{[t_1,t_2)}), \omega|_{[t_2,t_3)}) \tag{8.1}$$

so that in particular it holds for the empty control that

$$\mathcal{Q}(t, t, x, \diamond) = 0$$

for all t and x.

The main examples are as follows. If Σ is a discrete-time system

$$x^+ = \mathcal{P}(t, x, u) ,$$

then for any given function

$$q : \mathcal{T} \times \mathcal{X} \times \mathcal{U} \to \mathbb{R}_+ \tag{8.2}$$

(with $\mathcal{T} = \mathbb{Z}$) we associate the trajectory cost $\mathcal{Q}$ defined by $\mathcal{Q}(\tau, \tau, \cdot, \diamond) \equiv 0$ and

$$\mathcal{Q}(\tau, \sigma, x, \omega) := \sum_{i=\sigma}^{\tau-1} q(i, \xi(i), \omega(i)) ,$$

where $\xi := \psi(x, \omega)$. If instead Σ is a continuous-time system

$$\dot{x} = f(x, u)$$

---

* This section can be skipped with no loss of continuity.

and there is given a function (8.2) (with $\mathcal{T} = \mathbb{R}$), we may consider the trajectory cost

$$\mathcal{Q}(\tau, \sigma, x, \omega) := \int_{\sigma}^{\tau} q(s, \xi(s), \omega(s))\, ds\,,$$

where ξ is as above. In order for this expression to be well defined we may assume for instance that $q(t, x, u)$ is continuous on all of its arguments. We refer to the function q, in both the continuous-time and discrete-time cases, as the *instantaneous cost.*

Assume that in addition to the trajectory cost $\mathcal{Q}$ there is also given a function

$$p : \mathcal{X} \to \mathbb{R}_+\,,$$

which we will call the *terminal cost* function. Then we define on $\mathcal{D}_\phi$,

$$\mathcal{J}(\tau, \sigma, x, \omega) := \mathcal{Q}(\tau, \sigma, x, \omega) + p(\xi(\tau))\,,$$

where $\xi = \psi(x, \omega)$, the *total cost* function.

The problem we wish to study is:

> *Given a system Σ, a trajectory cost function $\mathcal{Q}$ and a final cost function p, a pair of times $\sigma < \tau$, and a state x^0, find a control $\omega \in \mathcal{U}^{[\sigma,\tau)}$ admissible for x^0 which minimizes $\mathcal{J}(\tau, \sigma, x^0, \omega)$.*

More precisely, we want to study existence and uniqueness of, as well as algorithms for constructing, such controls. We call the data

$$\boxed{\Sigma, \mathcal{Q}, p, \sigma, \tau \qquad \text{(OCD)}}$$

a set of *optimal control data.* When an initial state x^0 is given together with (OCD), one has an *optimal control problem.* (As discussed earlier, more general problems can also be studied, for instance by adding final state constraints. In this chapter, we only deal with problems of the above type, however.)

Dynamic Programming

The dynamic programming approach is easier to understand in discrete-time, even though it applies equally well for continuous-time systems. In particular, it is worth describing first how it arises in the purely combinatorial case of automata theory.

Finite Systems

Assume that both $\mathcal{X}$ and $\mathcal{U}$ are finite in problem (OCD), and that Σ is a discrete-time system. Denote by

$$N := \text{card}\,(\mathcal{X}), \quad M := \text{card}\,(\mathcal{U})$$

the cardinalities of $\mathcal{X}$ and $\mathcal{U}$, and let

$$k := \tau - \sigma\,.$$

For simplicity of notations, we assume that Σ is complete, that is, the local transitions $\mathcal{P}(t,x,u)$ are defined for all possible (t,x,u). If this is not the case, one can modify the arguments to follow in an obvious manner in order to avoid undefined transitions.

The most naive attack on the optimization problem would consist of an enumeration of all possible controls $\omega \in \mathcal{U}^{[\sigma,\tau)}$, thought of as sequences

$$u_1, \ldots, u_k$$

of elements of $\mathcal{U}$, followed by the calculation of the complete trajectory (k steps) corresponding to such a control starting at initial state x^0, and the computation of its associated cost. Since there are M^k such sequences, this results in the order of

$$kM^k$$

evaluations of $\mathcal{P}$ and of the instantaneous cost. (Actually, slightly more computation is needed for the calculation of the final cost and for the comparisons needed to find the minimum among all resulting trajectories.)

Alternatively, one could argue as follows. We shall construct inductively in time, backward from τ toward σ, two functions

$$V : [\sigma,\tau] \times \mathcal{X} \to \mathbb{R}_+$$

and

$$K : [\sigma,\tau) \times \mathcal{X} \to \mathcal{U}$$

called, respectively, the *Bellman function* and an *optimal feedback law*, such that the following properties hold:

1. For each $s \in [\sigma,\tau]$ and each $x \in \mathcal{X}$,

$$V(s,x) = \min_{\omega} \mathcal{J}(\tau,s,x,\omega)\,. \tag{8.3}$$

2. For each $s \in [\sigma,\tau)$ and each $x \in \mathcal{X}$, if ξ is the solution on $[s,\tau]$ of the difference equation

$$\begin{aligned} \xi(l+1) &= \mathcal{P}(l,\xi(l),K(l,\xi(l))) \quad l = s, s+1, \ldots, \tau-1 \\ \xi(s) &= x \end{aligned} \tag{8.4}$$

(the *closed-loop equation*) and we define

$$\nu(l) := K(l,\xi(l))\,, \tag{8.5}$$

then this control solves the optimization problem starting at (s,x), that is,

$$V(s,x) = \mathcal{J}(\tau,s,x,\nu)\,. \tag{8.6}$$

Observe that once these functions are obtained, the original problem

$$\min_{\omega} \mathcal{J}(\tau, \sigma, x^0, \omega)$$

is also solved, but in the process one also obtained solutions for all other initial states as well, a characteristic that is useful in applications, as discussed later.

We now show how to inductively obtain V and K. We start by defining the boundary value

$$V(\tau, x) := p(x),$$

which obviously satisfies (8.3). Assume now that V already has been defined for all (s, x) such that

$$t < s \leq \tau,$$

and K has been defined for $t < s < \tau$, in such a way that the above properties hold. We now extend the definition to all pairs of the form (t, x), as follows.

Given any such (t, x) compute the minimum

$$V(t, x) := \min_{u \in \mathcal{U}} [q(t, x, u) + V(t+1, \mathcal{P}(t, x, u))] \tag{8.7}$$

and let $K(t, x)$ be defined as any u that minimizes this expression.

We must establish that with these definitions V and K satisfy the above properties. Let $\omega_0 \in \mathcal{U}^{[t,\tau)}$ be optimal from (t, x),

$$\mathcal{J}(\tau, t, x, \omega_0) = \min_{\omega} \mathcal{J}(\tau, t, x, \omega). \tag{8.8}$$

By cost additivity,

$$\mathcal{J}(\tau, t, x, \omega_0) = q(t, x, u_0) + \mathcal{J}(\tau, t+1, \mathcal{P}(t, x, u_0), \omega_1), \tag{8.9}$$

where we are denoting $u_0 := \omega_0(t)$ and $\omega_1 :=$ restriction of ω_0 to the interval $[t+1, \tau)$. The *dynamic programming principle* states that ω_1 is optimal for the problem after time $t+1$, i.e.,

$$\mathcal{J}(\tau, t+1, \mathcal{P}(t, x, u_0), \omega_1) = V(t+1, \mathcal{P}(t, x, u_0)). \tag{8.10}$$

This is proved by contradiction: If false, then there is some ν so that

$$\mathcal{J}(\tau, t+1, \mathcal{P}(t, x, u_0), \nu) < \mathcal{J}(\tau, t+1, \mathcal{P}(t, x, u_0), \omega_1)$$

from which it follows that the concatenation $\omega \in \mathcal{U}^{[t,\tau)}$ of u_0 and ν has cost

$$\mathcal{J}(\tau, t, x, \omega) = q(t, x, u_0) + \mathcal{J}(\tau, t+1, \mathcal{P}(t, x, u_0), \nu) < \mathcal{J}(\tau, t, x, \omega_0),$$

which contradicts (8.8). We conclude from (8.7), (8.9), and (8.10) that

$$V(t, x) \leq \min_{\omega} \mathcal{J}(\tau, t, x, \omega).$$

To prove equality it will be sufficient to prove that, if we solve (8.4)-(8.5) with initial time t, then (8.6) holds.

Let ν be as in (8.5); we write it as the concatenation of $u_0 := K(t,x)$ and some $\omega_1 \in \mathcal{U}^{[t+1,\tau)}$. By inductive hypothesis, since ω_1 is defined by the closed-loop equation starting at $\mathcal{P}(t,x,u_0)$, it follows that ω_1 is optimal, that is, (8.10) holds. Therefore

$$\mathcal{J}(\tau,t,x,\nu) = q(t,x,u_0) + V(t+1,\mathcal{P}(t,x,u_0)).$$

On the other hand, $u_0 = K(t,x)$ was defined by the property

$$\begin{aligned} q(t,x,u_0) + V(t+1,\mathcal{P}(t,x,u_0)) &= \min_{u\in\mathcal{U}}\left[q(t,x,u) + V(t+1,\mathcal{P}(t,x,u))\right] \\ &= V(t,x), \end{aligned}$$

so (8.6) holds, as desired. ∎

It is instructive to compare the computational effort involved in this solution method with that in the naive approach. A larger amount of storage may be required for intermediate results since, unless a "general formula" is obtained for V and K, all values must be tabulated. We ignore this (extremely important) aspect and concentrate on a rough count of the number of operations needed.

Basically, one must perform for each $t \in [\sigma,\tau)$ and each $x \in \mathcal{X}$ the minimization in equation (8.7). This takes of the order of $M = \operatorname{card}(\mathcal{U})$ operations, and it must be done kN times. Finally, a forward pass to solve the closed-loop equation is necessary, which takes an additional $O(k)$ operations. In all, the number of operations is

$$O(kNM)$$

operations as opposed to the

$$O(kM^k)$$

needed in the enumerative approach. (If the problem had been to solve for all initial states x, then this expression would instead be $O(kNM^k)$.) One typical situation is that in which continuous state and input spaces $\mathbb{R}^n$ and $\mathbb{R}^m$ are quantized by using h values in each coordinate, and the transition map is defined on the discretized values by evaluating $\mathcal{P}$ at a midpoint and then truncating the result. In this case $N = h^n$ and $M = h^m$, so the estimates are

$$O(kh^{mk}) \text{ and } O(kh^{m+n}),$$

Often m and n may be small but k and h are large, in which case the advantage of the second (dynamic programming) approach is evident. However, since the second expression is exponential in n and m, it is clear that for large n or m the computational effort required even by dynamic programming may be prohibitive (the "curse of dimensionality").

Example 8.1.1 Sometimes the minimum in equation (8.7) can be computed in closed form, in which case dynamic programming does provide an especially useful technique. For instance, consider the problem of minimizing the quadratic cost

$$\sum_{l=\sigma}^{\tau-1} \omega(l)^2 \ + \ \xi(\tau)^2$$

for the one-dimensional linear time-invariant system

$$x^+ = ax + u\,.$$

Such linear systems with quadratic cost are the subject of this chapter, and a general solution will be given later. But it is useful to compute explicitly with one example. Here $q(t,x,u) = u^2$ and $p(x) = x^2$, so we must solve the partial difference equation

$$V(t,x) = \min_u\{u^2 + V(t+1, ax+u)\} \tag{8.11}$$

with boundary condition $V(\tau, x) = x^2$. At $t = \tau$ the function V is quadratic in x, and inductively it is clear from (8.11) that, if $V(t+1,\cdot)$ is quadratic in x, then $V(t,\cdot)$ also is. So we look for a sequence of numbers π_t such that

$$V(t,x) = \pi_t x^2$$

satisfies (8.11). Minimizing

$$u^2 + \pi(ax+u)^2$$

as a function of u gives a unique minimum of

$$\frac{\pi}{1+\pi}a^2x^2$$

at

$$u = -\frac{\pi}{1+\pi}ax\,.$$

Thus, there is a simple recursion

$$\pi_t := \frac{\pi_{t+1}}{1+\pi_{t+1}}a^2\,, \qquad \pi_\tau := 1\,,$$

and the feedback law

$$K(t,x) := -\frac{\pi_{t+1}}{1+\pi_{t+1}}ax$$

provides the desired optimal control. □

One advantage of dynamic programming over other optimization techniques is that it provides automatically a feedback law. In actual control implementations, this feedback law is stored, rather than the optimal control sequence obtained in the forward pass of the algorithm (solving (8.4)). The practical benefits of feedback control were discussed in Chapter 1. Another advantage of dynamic programming is that it is easy to modify it to account for state and control constraints.

Exercise 8.1.2 Consider the discrete-time time-invariant one dimensional system with $\mathcal{X} = \mathbb{R}$, transitions

$$x^+ = x + u\,,$$

and control value space the *nonnegative* reals:

$$\mathcal{U} = \mathbb{R}_+\,.$$

With $q(t,x,u) := u^2$, $p(x) := x^2$, $\sigma = 0$, and $\tau = 3$, find $V(t,x)$ for all x and $t = 0,1,2$ using the dynamic programming technique. Next guess a general formula for arbitrary t, σ, τ and establish its validity. □

Exercise 8.1.3 Develop an approach to the general discrete-time problem, say for simplicity with finite $\mathcal{X}$ and $\mathcal{U}$, which relies on a *forward* rather than backward first pass. To each event (t,x) you will have to associate a "cost until now" together with a pointer to a triple $(t-1,z,u)$ for which $\mathcal{P}(t-1,z,u) = x$. Observe that this solution technique does not provide a feedback law for the system. □

The General Principle

The dynamic programming ideas in principle can be applied to arbitrary systems. We develop here the basic facts.

Definition 8.1.4 *Let $\Sigma, \mathcal{Q}, p, \sigma, \tau$ be a set of optimal control data* (OCD). *The* **Bellman function** *(or* **value function,** *or* **optimal cost-to-go**) *associated to this data is the function*

$$V(t,x) := \inf\{\mathcal{J}(\tau,t,x,\omega) \mid (\tau,t,x,\omega) \in \mathcal{D}_\phi\} \tag{8.12}$$

defined on $[\sigma,\tau] \times \mathcal{X}$ and with values in $\mathbb{R}_+ \cup \{+\infty\}$. □

We make the convention that $V(t,x) = +\infty$ when the set of controls on $[t,\tau)$ admissible for x is empty. Note that $V(\tau,x) = p(x)$ for all x.

The finite problem studied earlier had the following desirable property:

For each (t,x) there is a control ω so that $V(t,x) = \mathcal{J}(\tau,t,x,\omega)$

In other words, the "inf" in (8.12) is in fact achieved as a minimum. In order not to have to repeat the statement of this property, we shall say that the Bellman function of (OCD) **is achieved** if this happens. A control ω for which the above property holds will be called **optimal** (for the event (t,x)). The corresponding $\xi = \psi(x,\omega)$ is an **optimal path** and (ξ,ω) is an **optimal trajectory**.

Lemma 8.1.5 The following properties hold for any data (OCD), for each s,t such that $\sigma \le t \le s \le \tau$, and each $x \in \mathcal{X}$:

1. For all ω so that $(s,t,x,\omega) \in \mathcal{D}_\phi$,

$$V(t,x) \le \mathcal{Q}(s,t,x,\omega) + V(s,\phi(s,t,x,\omega))\,. \tag{8.13}$$

2. If the Bellman function is achieved, then

$$V(t,x) = \mathcal{Q}(s,t,x,\omega|_{[t,s)}) + V(s,\phi(s,t,x,\omega|_{[t,s)})) \tag{8.14}$$

for any ω optimal for (t,x).

It follows that

$$V(t,x) = \min\{\mathcal{Q}(s,t,x,\omega) + V(s,\phi(s,t,x,\omega))\}\,, \tag{8.15}$$

where the minimum is taken over all ω so that $(s,t,x,\omega) \in \mathcal{D}_\phi$.

Proof. Assume that s,t,x,ω are given. Pick any $\varepsilon > 0$. We prove that

$$V(t,x) \leq \mathcal{Q}(s,t,x,\omega) + V(s,\phi(s,t,x,\omega)) + \varepsilon\,. \tag{8.16}$$

Since ε is arbitrary, (8.13) will follow. By definition of $V(s,\phi(s,t,x,\omega))$, there exists some control ν on $[s,\tau)$ so that

$$\mathcal{J}(\tau,s,\phi(s,t,x,\omega),\nu) < V(s,\phi(s,t,x,\omega)) + \varepsilon$$

(if $V(s,\phi(s,t,x,\omega)) = \infty$, then there is nothing to prove, so we may assume that this is finite). Let ω' be the concatenation of ω and ν. Since ω' is admissible for the state x, it follows by definition of $V(t,x)$ that

$$V(t,x) \leq \mathcal{J}(\tau,t,x,\omega') = \mathcal{Q}(s,t,x,\omega) + \mathcal{J}(\tau,s,x,\nu)\,,$$

and this gives the inequality (8.16).

We now consider the second part of the Lemma. Let ω be optimal for (t,x). The equality (8.14) holds, by additivity of the cost, if the last term is instead

$$\mathcal{J}(\tau,s,\phi(s,t,x,\omega|_{[t,s)}),\omega|_{[s,\tau)})\,,$$

so it is only necessary to see that the restriction of ω to $[s,\tau)$ must be optimal. This is precisely the conclusion of the dynamic programming principle, as in the discrete case, since if this restriction were not to be optimal, then the concatenation of some other control on $[s,\tau)$ with its first part $\omega|_{[t,s)}$ would provide a control for (t,x) with less cost. ∎

The next Lemma states essentially that the above properties characterize the Bellman function; it is a *verification principle* in the sense that, if a function V and controls ω are shown to satisfy the Lemma, then one has verified optimality.

Lemma 8.1.6 Given a set of optimization data (OCD), assume that

$$\widetilde{V} : [\sigma,\tau] \times \mathcal{X} \to \mathbb{R}$$

is any function so that the following properties hold:

1. For each s, t such that $\sigma \leq t \leq s \leq \tau$ and each $x \in \mathcal{X}$,

$$\widetilde{V}(t, x) \leq \mathcal{Q}(s, t, x, \omega) + \widetilde{V}(s, \phi(s, t, x, \omega)) \tag{8.17}$$

for all ω on $[t, s)$ admissible for x;

2. for each s, t such that $\sigma \leq t \leq s \leq \tau$ and each $x \in \mathcal{X}$ there exists some control $\widetilde{\omega}$ such that

$$\widetilde{V}(t, x) = \mathcal{Q}(s, t, x, \widetilde{\omega}) + \widetilde{V}(s, \phi(s, t, x, \widetilde{\omega}))\,; \tag{8.18}$$

and

3. the boundary condition

$$\widetilde{V}(\tau, x) = p(x) \tag{8.19}$$

holds for all x.

Then it must hold that the Bellman function is achieved and $V = \widetilde{V}$.

Proof. Fix any t, x. As a particular case of the first property and using the boundary condition one concludes for any admissible ω on $[t, \tau)$ that

$$\widetilde{V}(t, x) \leq \mathcal{Q}(\tau, t, x, \omega) + p(\phi(\tau, t, x, \omega)) = \mathcal{J}(\tau, t, x, \omega)\,.$$

By the second property and an analogous argument there is at least some $\widetilde{\omega}$ so that

$$\widetilde{V}(t, x) = \mathcal{J}(\tau, t, x, \widetilde{\omega})\,,$$

so the function $\widetilde{V}$ must be the Bellman function for the problem, and the existence of the controls $\widetilde{\omega}$ implies that this is achieved (with the respective $\widetilde{\omega}$'s being optimal). ∎

The Continuous-Time Case

Let us turn now to a continuous-time system Σ (Definition 2.6.4 in Chapter 2)

$$\dot{x} = f(t, x, u)$$

over $\mathbb{K} = \mathbb{R}$ for which f is assumed to be continuous on t. First we will outline the derivations; rigorous proofs are given later.

We assume given optimal control data (OCD) for this system, where

$$q : \mathbb{R} \times \mathcal{X} \times \mathcal{U} \to \mathbb{R}_+ \quad \text{and} \quad p : \mathcal{X} \to \mathbb{R}_+$$

are both continuously differentiable. These give rise, as above, to the performance criterion or cost

$$\mathcal{J}(\tau, \sigma, x, \omega) \;:=\; \int_\sigma^\tau q(\eta, \xi(\eta), \omega(\eta))\, d\eta \;+\; p(\xi(\tau))$$

to be minimized for any given x, where $\xi = \psi(x, \omega)$.

The conclusions of Lemma 8.1.5 provide infinitesimal information about the variation of V, as follows. Equation (8.14) says that, for each optimal control for (t, x),

$$V(t,x) = \int_t^s q(\eta, \xi(\eta), \omega(\eta))\, d\eta \; + \; V(s, \xi(s))$$

for all s. Thus, the Bellman function evaluated along an optimal trajectory,

$$V(s, \xi(s)),$$

is absolutely continuous and has derivative

$$\frac{dV(s,\xi(s))}{ds} = -q(s, \xi(s), \omega(s))$$

almost everywhere. That is, *along an optimal trajectory the instantaneous decrease in the Bellman function is equal to (minus) the instantaneous cost.*

Because $dV(s,\xi(s))/ds = V_s + V_x \dot{\xi}$, we can also write this equation (using the argument t instead of s):

$$V_t(t,x) = -q(t,x,\omega(t)) - V_x(t,x) f(t,x,\omega(t))\,, \tag{8.20}$$

provided that the partial derivatives make sense. (In interpreting the equation, note that the gradient V_x of V with respect to x is a row vector.) Along arbitrary trajectories, equation (8.13) suggests that the Bellman function will decrease less fast than the instantaneous cost, giving the inequality

$$V_t(t,x) \geq -q(t,x,\nu(t)) - V_x(t,x) f(t,x,\nu(t)) \tag{8.21}$$

for every other (not necessarily optimal) control ν. (Both sides of (8.13) are equal at $s = t$, so at least for small enough $s - t$ one expects that the derivatives with respect to s will satisfy the same inequality, implying (8.21); this is made precise in the proof of the Proposition below.) Note that only the value of the control at the given time t appears in this inequality and in the equality (8.20). If the Bellman function is achieved and if there is enough smoothness, both of (8.20) and (8.21), together with the natural boundary condition on V, can be summarized by the equation:

$$\boxed{V_t(t,x) = -\min_{u \in \mathcal{U}}\{q(t,x,u) + V_x(t,x) f(t,x,u)\},\, V(\tau,x) = p(x) \quad \text{(HJB)}}$$

called the **Hamilton-Jacobi-Bellman** equation associated to the optimal control data.

The significance of (HJB) lies in its reducing the problem to an optimization at each stage (find u that minimizes, for each fixed (t, x), and solve the partial differential equation for V), in a manner totally analogous to the discrete-time case.

Note that in the discrete-time case one has instead the following *discrete HJB equation* or **dynamic programming equation**:

$$\boxed{V(t,x) = \min_{u\in\mathcal{U}}\{q(t,x,u) + V(t+1,\mathcal{P}(t,x,u))\}, V(\tau,x) = p(x) \quad \text{(DPE)}}$$

as a specialization of (8.15). Observe that (DPE) is an equation that evolves backward in time; for differential equations this time reversal is reflected in the negative sign in Equation (HJB).

Definition 8.1.7 *Let* (OCD) *be optimal control data for a continuous-time system. We say that* **there is a smooth solution** *(of the optimal control problem associated to this data) if the following properties are satisfied:*

- *The Bellman function V is of class $\mathcal{C}^1$ for $(t,x) \in [\sigma,\tau] \times \mathcal{X}$, and*
- *for each event $(t,x) \in [\sigma,\tau] \times \mathcal{X}$ there is a continuous control ω such that*

$$V(t,x) = \mathcal{J}(\tau,t,x,\omega)\,.$$

In particular, the Bellman function is achieved. □

The main facts are summarized by the next two results.

Proposition 8.1.8 Assume that there is a smooth solution of the problem (OCD) on $[\sigma,\tau]$. Then,

(i) The Bellman function V satisfies the (HJB) equation for all $(t,x) \in (\sigma,\tau)\times \mathcal{X}$; and

(ii) if (ξ,ω) is any optimal trajectory and ω is continuous, then at all pairs $(t,x) = (t,\xi(t))$ along this trajectory, the minimum in (HJB) is attained at $\omega(t)$.

Proof. The final condition $V(\tau,x) = p(x)$ holds by definition of V. Equation (8.20) holds at each (t,x) along any optimal trajectory corresponding to a continuous control, because of the assumed differentiability of V and the preceding arguments. Therefore, in order to establish (ii) it is only necessary to show that

$$V_t(t,x) + V_x(t,x)f(t,x,u) + q(t,x,u) \geq 0 \tag{8.22}$$

for every control value u, at any given (t,x). Consider any such u and let ω_ε be the control with constant value u on the interval $[t,t+\varepsilon]$. For small enough ε this is admissible for x, and by (8.13),

$$g(\varepsilon) := V(t+\varepsilon,\phi(t+\varepsilon,t,x,\omega_\varepsilon)) - V(t,x) + \int_t^{t+\varepsilon} q(\eta,\xi(\eta),u)\,d\eta \geq 0$$

along $\xi = \psi(x,\omega_\varepsilon)$ for every $\varepsilon > 0$. It follows that $g'(0) \geq 0$, which implies (8.22).

To prove (i), it is only needed to show that every possible $(t,x) \in (\sigma,\tau) \times \mathcal{X}$ appears as part of an optimal trajectory. But this fact is an immediate consequence of the assumption that the problem (OCD) has a smooth solution, when applied to the initial event (t,x). ∎

The best possible situation is that in which the minimization problem in the (HJB) equation has a unique solution as a function of V_x and the rest of the data, at each (t,x). In order to consider this situation, we introduce the **Hamiltonian** associated to the optimal control problem,

$$\boxed{\mathcal{H}(t,x,u,\lambda) := q(t,x,u) + \lambda' f(t,x,u)}$$

seen as a function of $(t,x,u) \in [\sigma,\tau] \times \mathcal{X} \times \mathcal{U}$ and $\lambda \in \mathbb{R}^n$. Thus, the above situation arises when

$$\min_u \mathcal{H}(t,x,u,\lambda)$$

exists and is achieved at a unique u for each fixed (t,x,λ). Actually, we will be interested in the case where this unique minimum is continuous as a function of (t,x,λ), that is, when there is a continuous function

$$\alpha : [\sigma,\tau] \times \mathcal{X} \times \mathbb{R}^n \to \mathcal{U} \tag{8.23}$$

satisfying

$$\min_u \mathcal{H}(t,x,u,\lambda) = \mathcal{H}(t,x,\alpha(t,x,\lambda),\lambda)\,,$$

and this is the unique minimum:

$$q(t,x,\alpha(t,x,\lambda)) + \lambda' f(t,x,\alpha(t,x,\lambda)) < q(t,x,v) + \lambda' f(t,x,v) \tag{8.24}$$

for all

$$(t,x,\lambda) \in [\sigma,\tau] \times \mathcal{X} \times \mathbb{R}^n$$

and all

$$v \in \mathcal{U} \ \text{ so that } \ v \neq \alpha(t,x,\lambda)\,.$$

But even then, this solution when substituted into the system may not result in an admissible control. We need for that one more definition:

Definition 8.1.9 *Let Σ be a system and let $\mathcal{I}$ be an interval in $\mathcal{T}$. A function*

$$k : \mathcal{I} \times \mathcal{X} \to \mathcal{U}$$

is an **admissible feedback law on $\mathcal{I}$** *if for each $\mu \in \mathcal{I}$ and each $x \in \mathcal{X}$ there exists a unique trajectory (ξ,ω) on $J = \{t \,|\, t \in \mathcal{I}, t \geq \mu\}$ such that*

$$\xi(\mu) = x$$

and

$$\omega(t) = k(t,\xi(t))$$

for almost all $t \in J$. In that case ω is called the **closed-loop control starting at (μ,x)**. □

As a trivial example, $k(t,x) := x$ is admissible on $[0,\infty)$ for the linear system $\dot{x} = u$ since there exist solutions of $\dot{x} = x$ defined for all initial states and times, but the feedback $k(t,x) := x^2$ is *not* admissible for this system on the interval $[0,2]$, since solutions do not exist on that interval for $\dot{x} = x^2$ and $x(0) = 1$.

We summarize next the conclusions of the infinitesimal version of the verification result, Lemma 8.1.6, together with the converse provided by the above Proposition:

Theorem 36 *Let* (OCD) *be optimal control data, and assume that*

- *there is some continuous function α as in (8.23) satisfying (8.24);*
- *there is given a continuously differentiable function V on $[\sigma,\tau]\times\mathcal{X}$ so that*

$$k(t,x) := \alpha(t,x,V_x'(t,x))$$

is an admissible feedback on $[\sigma,\tau]$.

Then, the following two statements are equivalent:

1. *V satisfies the HJB equation on $(\sigma,\tau)\times\mathcal{X}$, which can now be written as*

$$\begin{aligned} q(t,x,k(t,x)) + V_x(t,x)f(t,x,k(t,x)) + V_t(t,x) &= 0\,, \\ V(\tau,x) = p(x)\,. \end{aligned}$$

2. *There is a smooth solution of the optimization problem, V is the Bellman function, and for each $(\mu,x)\in[\sigma,\tau]\times\mathcal{X}$ the closed-loop control starting at (μ,x) is the unique optimal control for (μ,x).*

Proof. That (2) implies (1) follows from Proposition 8.1.8. We now prove the converse, using the verification Lemma 8.1.6.

Pick any $\sigma < t < s < \tau$ and any x, and let ν be the closed-loop control starting at (t,x) up to time s. Let ω be any control on $[t,s]$ that is admissible for x. Denote $\xi := \psi(x,\omega)$ and consider the function $B(\eta) := V(\eta,\xi(\eta))$ on $[t,s]$. This is absolutely continuous, and its derivative is

$$V_t(\eta,\xi(\eta)) \;+\; V_x(\eta,\xi(\eta))\,f(\eta,\xi(\eta),\omega(\eta))\,,$$

which, since V satisfies the HJB equation, is the same as

$$V_x(\eta,\xi(\eta)).\{f(\eta,\xi(\eta),\omega(\eta)) - f(\eta,\xi(\eta),k(\eta,\xi(\eta)))\} - q(\eta,\xi(\eta),k(\eta,\xi(\eta)))\,.$$

This last expression is, because of property (8.24), greater than or equal to

$$-q(\eta,\xi(\eta),\omega(\eta))\,,$$

and equal to this when $\omega=\nu$. So

$$V(s,\xi(s)) - V(t,x) \geq -\int_t^s q(\eta,\xi(\eta),\omega(\eta))\,d\eta$$

with equality when $\omega = \nu$. By continuity on t and s, the same inequality holds even if $t = \sigma$ or $s = \tau$. So the hypotheses in Lemma 8.1.6 hold. We conclude that ν is indeed optimal for (μ, x), and it is continuous because α is.

Finally, we prove the uniqueness statement. Assume that ω differs from ν on a set of positive measure. Then $dB/d\eta$ is strictly greater than $-q(\eta, \xi(\eta), \omega(\eta))$ on this set, and hence it follows that

$$V(t, x) < \mathcal{J}(\tau, t, x, \omega)$$

and thus ω cannot be optimal. ∎

Remark 8.1.10 A particular situation in which a function α such as required in the hypothesis of Theorem 36 will exist is as follows. Assume that the system Σ is affine in the control, meaning that

$$f(t, x, u) = A(t, x) + B(t, x)u\,,$$

where A is an n-vector, and B is an $n \times m$ matrix, of C^1 functions of (t, x). This happens for instance if we are dealing with continuous-time linear systems (with $A(t), B(t)$ continuously differentiable), in which case $A(t, x) = A(t)x$ and $B(t, x) = B(t)$.

Assume also that the instantaneous cost is quadratic in u, meaning that we can write

$$q(t, x, u) = u'R(t, x)u + Q(t, x)\,,$$

where R and Q are continuously differentiable functions of (t, x), Q is nonnegative for all (t, x), and R is an $m \times m$ symmetric positive definite matrix for all (t, x). We then may "complete the square" and write (omitting arguments (t, x))

$$u'Ru + \lambda'Bu \;=\; \left(u + \frac{1}{2}R^{-1}B'\lambda\right)' R \left(u + \frac{1}{2}R^{-1}B'\lambda\right) - \frac{1}{4}\lambda'BR^{-1}B'\lambda$$

for each $(t, x, \lambda) \in [\sigma, \tau] \times \mathcal{X} \times \mathbb{R}^n$. It follows that the unique minimum of the Hamiltonian

$$q(t, x, u) + \lambda' f(t, x, u)$$

with respect to u is given by

$$\alpha(t, x, \lambda) := -\frac{1}{2}R^{-1}B'\lambda\,.$$

Note that this defines a continuous (in fact, differentiable) function α. □

A special case of the situation discussed in Remark 8.1.10 is that of continuous-time linear systems $\dot{x} = A(t)x + B(t)u$ over $\mathbb{K} = \mathbb{R}$, when we want to minimize a quadratic cost with $q(t, x, u) := u'R(t)u + x'Q(t)x$, and $p(x) := x'Sx$. (Here R, Q, and S are matrix functions, S constant, of appropriate sizes.) We should expect the Bellman function, if it exists, to be quadratic in x. (Intuitively, we argue as follows. For given initial (μ, x), and any given control ω, we get the state evolution $\xi(t) = \Phi(t,\mu)x + \int_\mu^t \Phi(t,s)B(s)\omega(s)ds$, and when substituted back into the cost $\mathcal{J}(\mu,\tau,x,\omega)$, this cost becomes quadratic on ω and on the initial state x. As a function of ω, we then expect that there will be a unique minimum for each x, and that this minimum will be achieved at an ω which depends linearly on x. When this minimum is substituted in the cost function, we get finally a (homogeneous) quadratic function of x.) Thus we expect to be able to write $V(x) = x'Px$, for some symmetric matrix P, from which it follows that the gradient of V is linear on x: $V_x(x) = 2x'P$. But then, the feedback law in Theorem 36 (p. 361) becomes $k(t,x) = \alpha(t,x,V_x'(t,x)) = -R(t)^{-1}B(t)'P(t)x$, which is linear in x. Substituting in the dynamics of Σ, we get the closed-loop equations (omitting all t arguments): $\dot{x} = (A+BF)x$, $F = -R^{-1}B'P$. This is again a linear system, so it is complete. Thus k is an admissible feedback law. In summary, if V is quadratic, then the optimal controls are induced by linear feedback. Moreover, it turns out, the HJB equation for $V(t,x) = x'P(t)x$ is equivalent to an *ordinary* (not partial) differential equation on the matrix function P, the Riccati Differential Equation.

In order to allow the study of the linear case by itself, we next discuss it in a self-contained manner; however, the previous discussion explains the motivation for the steps to be taken. We return to the general continuous-time nonlinear case in Section 8.5.

8.2 Linear Systems with Quadratic Cost

We consider continuous-time linear systems (with $\mathbb{K} = \mathbb{R}$):

$$\dot{x}(t) = A(t)x(t) + B(t)u(t). \tag{8.25}$$

We assume that A and B are measurable essentially bounded (m.e.b.). Also given are three matrices of functions R, Q, S such that:

- R is an $m \times m$ symmetric matrix of m.e.b. functions, Q is an $n \times n$ symmetric matrix of m.e.b. functions, and S is a constant symmetric $n \times n$ matrix;
- For each t, R is positive definite;
- For each t, Q is positive semidefinite; and
- S is positive semidefinite.

For each pair of real $\sigma \le \tau$ numbers, each state $x \in \mathbb{R}^n$, and each measurable essentially bounded input $\omega : [\sigma, \tau] \to \mathbb{R}^m$, we define:

$$\mathcal{J}(\tau,\sigma,x,\omega) := \int_\sigma^\tau \omega(s)'R(s)\omega(s) + \xi(s)'Q(s)\xi(s)\,ds \;+\; \xi(\tau)'S\xi(\tau),$$

where $\xi(t) = \phi(t, \sigma, x, \omega)$ for all $t \in [\sigma, \tau]$. The problem that we wish to study is this one:

> *For each pair of times $\sigma < \tau$ and state x^0, find a control ω which minimizes $\mathcal{J}(\tau, \sigma, x^0, \omega)$.*

Any control ω with this property will be said to be *optimal for x^0 on the interval* $[\sigma, \tau]$.

Assume $\sigma < \tau$ have been fixed. We denote

$$V(t, x) := \inf\{\mathcal{J}(\tau, t, x, \omega) \mid \omega : [\sigma, \tau] \to \mathbb{R}^m \text{ m.e.b. }\} \tag{8.26}$$

for each $t \in [\sigma, \tau]$ and $x \in \mathbb{R}^n$. This is called the *value* or *Bellman* function associated to the optimization problem. If a control ω^0 which minimizes $\mathcal{J}(\tau, t, x^0, \omega)$ exists, then $V(t, x^0) = \mathcal{J}(\tau, t, x^0, \omega^0)$.

Let $F : [\sigma, \tau] \to \mathbb{R}^{m\times n}$ be m.e.b.. For each $t^0 \in [\sigma, \tau]$ and $x^0 \in \mathbb{R}^n$, consider the solution $x(t)$ of

$$\dot{x}(t) = [A(t) + B(t)F(t)]x(t)\,, \quad x(t^0) = x^0$$

which is obtained when using the feedback law $u(t) = F(t)x(t)$ with the system (8.25), starting from state x^0 at time t^0. The control $\omega(t) = F(t)x(t)$ is called the *F-closed-loop control starting at* (t^0, x^0).

The main result will be as follows.

Theorem 37 *Pick any $\sigma < \tau$. Then there exists an absolutely continuous symmetric $n \times n$ matrix function $P(t)$, defined for $t \in [\sigma, \tau]$, which satisfies the* **matrix Riccati Differential Equation** *on* $[\sigma, \tau]$*:*

$$\boxed{\dot{P} = PBR^{-1}B'P - PA - A'P - Q\,, \quad P(\tau) = S \qquad \text{(RDE)}}\;.$$

Let

$$\boxed{F(t) := -R(t)^{-1}B(t)'P(t)}\;. \tag{8.27}$$

Then the F-closed-loop control starting at (σ, x^0) is the unique optimal control for x^0 on the interval $[\sigma, \tau]$. Moreover,

$$V(t, x) = x'P(t)x$$

for all $t \in [\sigma, \tau]$ and $x \in \mathbb{R}^n$.

The main technical step is to establish the conclusions under the assumption that P exists:

Lemma 8.2.1 Pick any $\sigma < \tau$. Suppose that there exists a matrix function $P(t)$ as in the statement of Theorem 37. Then, with $F(t) = -R(t)^{-1}B(t)'P(t)$, the F-closed-loop control starting at (σ, x^0) is the unique optimal control for x^0 on the interval $[\sigma, \tau]$, and $V(t, x) = x'P(t)x$ for all $t \in [\sigma, \tau]$ and $x \in \mathbb{R}^n$.

We will provide two independent proofs of this Lemma. The first proof is short and uses a simple algebraic computation. The second proof is much longer, but it has the merit of being conceptual and to generalize to large classes of nonlinear systems (it is a particular case of the approach discussed in Section 8.1). Before providing these proofs, we show how the Theorem follows from the Lemma.

We must show for that purpose that the final-value problem (RDE) has a solution on all of the interval $[\sigma, \tau]$. Then it will follow that the resulting P must be symmetric, since its transpose $P(t)'$ solves the same final value problem and solutions of (locally Lipschitz) ODE's are unique.

Consider the final-value problem $\dot{P} = PBR^{-1}B'P - PA - A'P - Q$, $P(\tau) = S$, on the interval $\mathcal{I} = [\sigma, \tau]$. Let $P(\cdot)$ be its maximal solution, defined on some interval $J \subseteq \mathcal{I}$ which is open relative to $\mathcal{I}$. We wish to prove that $J = \mathcal{I}$. Pick any $\sigma' \in J$. Since there does exist a solution of (RDE) on the interval $[\sigma', \tau]$, we may apply Lemma 8.2.1 to the optimal control problem on the restricted interval $[\sigma', \tau]$, and conclude that $V(t, x) = x'P(t)x$ is the Bellman function for the corresponding problem on that interval. Pick any $x \in \mathbb{R}^n$ and any $t \in [\sigma', \tau]$. Then,

$$\begin{aligned} 0 \leq V(t,x) &\leq \mathcal{J}(\tau, t, x, \mathbf{0}) \\ &= x'\Phi(\tau,t)'\left\{\int_t^\tau \Phi(s,\tau)'Q(s)\Phi(s,\tau)ds + S\right\}\Phi(\tau,t)x\,. \end{aligned}$$

In particular, this applies when $t = \sigma'$. As functions of $t \in [\sigma, \tau]$, both $\Phi(\tau, t)$ and the expression in braces $\{\ldots\}$ are bounded. Thus, there is some constant $K > 0$ such that

$$0 \leq x'P(\sigma')x \leq K\,\|x\|^2$$

for all $x \in \mathbb{R}^n$. Since the norm of the symmetric and nonnegative matrix $P(\sigma')$ equals the largest possible value of $x'P(\sigma')x$, $\|x\| = 1$, we conclude that

$$\|P(\sigma')\| \leq K\,.$$

Therefore, the maximal solution $P(\cdot)$ stays in some compact subset of the set of all real $n \times n$ matrices. By Proposition C.3.6 (applied to the time-reversed equation), it follows that P is defined everywhere, that is, $J = \mathcal{I}$, as we wanted to show. This completes the proof of Theorem 37, under the assumption that Lemma 8.2.1 holds. ∎

First Proof of Lemma 8.2.1

Suppose that P satisfies (RDE). Pick any initial state x^0 and any control $\omega : [\sigma, \tau] \to \mathbb{R}^m$, let $\xi(t) = \phi(t, \sigma, x^0, \omega)$, and consider the time derivative of the function $\xi(t)'P(t)\xi(t)$ (defined almost everywhere). Using (RDE) and $\dot{\xi} = A\xi + B\omega$, we have:

$$(\xi'P\xi)^{\cdot} = (\omega + R^{-1}B'P\xi)'R(\omega + R^{-1}B'P\xi) - \omega'R\omega - \xi'Q\xi\,.$$

Integrating, we conclude:

$$\mathcal{J}(\tau,\sigma,x^0,\omega) = x^{0\prime}P(\sigma)x^0 +$$
$$\int_\sigma^\tau \Big(\omega(s)+R^{-1}(s)B(s)'P(s)\xi(s)\Big)' R\Big(\omega(s)+R^{-1}(s)B(s)'P(s)\xi(s)\Big)\,ds\,.$$

From this equality, it is clear that the unique minimum-cost control is that one making $\omega(s) + R^{-1}(s)B(s)'P(s)\xi(s) \equiv 0$, and that the minimum cost is $x^{0\prime}P(\sigma)x^0$. ∎

Second Proof of Lemma 8.2.1

Assume that P satisfies (RDE). Denote, for each t, x, $V(t,x) := x'P(t)x$. (Of course, we do not know yet that V is indeed the value function, but the notation should not cause confusion.) We also write:

$$q(t,x,u) := u'R(t)u + x'Q(t)x\,,\ p(x) := x'Sx\,,$$

and

$$k(t,x) := F(t)x\,,\ f(t,x,u) := A(t)x + B(t)u\,.$$

Since $V_x(t,x) = 2x'P(t)$ and

$$x'[PBR^{-1}B'P - 2PA - Q]x = x'\left[PBR^{-1}B'P - PA - A'P - Q\right]x\,,$$

the property that P satisfies (RDE) translates into the fact that V satisfies the following *Hamilton-Jacobi-Bellman* partial differential equation:

$$V_t(t,x) + q(t,x,k(t,x)) + V_x(t,x)f(t,x,k(t,x)) = 0\,,\quad V(\tau,x) = p(x)\,.$$

We need next this observation:

Claim: Let R be any $m \times m$ symmetric positive definite matrix, $B \in \mathbb{R}^{n\times m}$, and $\lambda \in \mathbb{R}^n$. Then the minimum value of $u'Ru + \lambda'Bu$ over all $u \in \mathbb{R}^m$ is achieved at the unique point $u = -\frac{1}{2}R^{-1}B'\lambda$.

Proof of claim: We just "complete the square":

$$u'Ru + \lambda'Bu = \left(u + \frac{1}{2}R^{-1}B'\lambda\right)' R\left(u + \frac{1}{2}R^{-1}B'\lambda\right) - \frac{1}{4}\lambda'BR^{-1}B'\lambda\,.$$

In particular, when $\lambda' = V_x(t,x) = 2x'P(t)$, this implies that

$$u = -\frac{1}{2}R(t)^{-1}B(t)'\lambda = F(t)x = k(t,x)$$

minimizes $q(t,x,u)+V_x(t,x)f(t,x,u)$. Thus the Hamilton-Jacobi-Bellman equation translates into:

$$V_t(t,x) + \min_{u\in\mathbb{R}^m}\{q(t,x,u) + V_x(t,x)f(t,x,u)\} = 0 \tag{8.28}$$

with the boundary condition $V(\tau,x) = p(x)$. (This is the form of the usual HJB equation associated to the minimization of a cost criterion $\int_\sigma^\tau q(t,x,u)dt+$

$p(x(\tau))$ for a general system $\dot{x} = f(t,x,u)$, cf. Section 8.1. For linear systems with quadratic cost, saying that a quadratic-in-x function $V = x'Px$ solves the HJB equation is exactly the same as saying that P satisfies the RDE.) Yet another way of stating these facts is by means of this pair of equations:

$$V_t(t,x) + V_x(t,x)f(t,x,k(t,x)) = -q(t,x,k(t,x)) \tag{8.29}$$

$$V_t(t,x) + V_x(t,x)f(t,x,u) > -q(t,x,u) \quad \forall u \neq k(t,x), \tag{8.30}$$

understood as being true for almost all $t \in [\sigma,\tau]$ (namely those t for which $\dot{P}(t)$ exists) and all $x \in \mathbb{R}^n$.

Pick any $x^0 \in \mathbb{R}^n$. We must prove that the F-closed-loop control starting at (σ, x^0) is optimal. Call this control $\overline{\omega}$ and let $\overline{\xi}$ be the corresponding state trajectory. Let $\alpha(t) := V(t,\overline{\xi}(t))$, and observe that

$$\alpha(\tau) = p(\overline{\xi}(\tau)) \tag{8.31}$$

because V satisfies the boundary condition $V(\tau,x) = p(x)$. Equation (8.29) implies that $\dot{\alpha}(t) = -q(t,\overline{\xi}(t),\overline{\omega}(t))$, from which, integrating and using Equation (8.31) we conclude that

$$p(\overline{\xi}(\tau)) = \alpha(\sigma) - \int_\sigma^\tau q(t,\overline{\xi}(t),\overline{\omega}(t))\,dt = V(\sigma,x^0) - \int_\sigma^\tau q(t,\overline{\xi}(t),\overline{\omega}(t))\,dt$$

and therefore $V(\sigma,x^0) = \mathcal{J}(\tau,\sigma,x^0,\overline{\omega})$.

To show that V is the value function and that $\overline{\omega}$ is the unique optimal control for x^0 on the interval $[\sigma,\tau]$, we must verify that $V(\sigma,x^0) < \mathcal{J}(\tau,\sigma,x^0,\omega)$ for all other controls ω. So pick any other $\omega : [\sigma,\tau] \to \mathbb{R}^m$, and let $\xi(t) = \phi(t,\sigma,x^0,\omega)$ be the solution of $\dot{\xi}(t) = f(t,\xi(t),\omega(t))$ with $\xi(\sigma) = x^0$. If it were the case that $\omega(t) = k(t,\xi(t))$ for almost all $t \in [\sigma,\tau]$, then ξ would satisfy $\dot{\xi} = f(t,\xi,k(t,\xi))$, so by uniqueness of solutions $\xi = \overline{\xi}$, and thus $\omega = \overline{\omega}$, which is a contradiction. Therefore we know that $\omega(t) \neq k(t,\xi(t))$ for all t in some subset $I \subseteq [\sigma,\tau]$ of positive Lebesgue measure. Write $\beta(t) := V(t,\xi(t))$. We have that

$$\beta(\tau) = p(\xi(\tau)) \tag{8.32}$$

because V satisfies the boundary condition $V(\tau,x) = p(x)$. Equation (8.30) implies that $\dot{\beta}(t) > -q(t,\xi(t),\omega(t))$ for all $t \in I$ and $\dot{\beta}(t) = -q(t,\xi(t),\omega(t))$ for all other t, so integrating and using Equation (8.32) we conclude that

$$p(\xi(\tau)) > \beta(\sigma) - \int_\sigma^\tau q(t,\xi(t),\omega(t))\,dt = V(\sigma,x^0) - \int_\sigma^\tau q(t,\xi(t),\omega(t))\,dt$$

and therefore $V(\sigma,x^0) < \mathcal{J}(\tau,\sigma,x^0,\omega)$, as desired. ∎

The linear quadratic problem solved above is often called the *standard regulator problem.* The solution of the Riccati equation is computed numerically. Note that, by symmetry, one must integrate a system of $n(n+1)/2$ simultaneous nonlinear differential equations. Typically, only the limiting solution, described later, is computed.

Remark 8.2.2 Note that the matrices $P(t)$, though symmetric, are *not* necessarily positive definite. For instance, if

$$S = 0 \quad \text{and} \quad Q \equiv 0,$$

then $\mathcal{J}$ is always minimized by $\omega \equiv 0$, and the optimal cost is therefore

$$V(t,x) \equiv 0$$

so that $P \equiv 0$. Equivalently, the matrix $P \equiv 0$ solves the Riccati equation (RDE) when $S = Q \equiv 0$. □

Exercise 8.2.3 Consider the regulator problem for the system (with $n = m = 1$) on $[0,1]$, $\dot{x} = u$, with $Q = R = 1, S = 0$. Find explicitly the form of the optimal feedback $k(t,x)$. □

Exercise 8.2.4 (*Optimal servo problem.*) Many problems of interest can be reduced to the standard regulator problem. One of these is the *servo problem* defined as follows. The objective is to have the output $y(t) = C(t)x(t)$ of a linear system with outputs Σ closely follow a desired reference signal $r(t)$. Mathematically, we wish to minimize, for given x and $r(\cdot)$, a cost $\mathcal{J}(\sigma,\tau,x,\omega)$ which here equals

$$\int_\sigma^\tau \left\{\omega(t)'R(t)\omega(t) + (C(t)\xi(t) - r(t))'\,Q(t)\,(C(t)\xi(t) - r(t))\right\} ds$$
$$+ (C(\tau)\xi(\tau) - r(\tau))'\,S\,(C(\tau)\xi(\tau) - r(\tau))\,,$$

where (ξ,ω) is the trajectory with $\xi(\sigma) = x$. In one version of this problem, one assumes that $r(t)$ is produced by a linear control generator, that is, r is itself the output of a continuous-time linear system with outputs (and no controls), say

$$\begin{aligned} \dot{z}(t) &= A_0 z(t) \\ r(t) &= C_0 z(t)\,. \end{aligned}$$

Assuming that $R(t)$ is positive definite and $Q(t)$ is positive semidefinite for all t, show how to find a solution for this problem, and show that the solution is unique. (*Hint:* Introduce a large linear system, whose state space consists of all pairs $\binom{x}{z}$. Apply the regulator results to this extended system. The optimal feedback will be a function of t, x, and z.) □

Remark 8.2.5 It may appear that restricting to signals which are outputs of linear systems is not very interesting. On the contrary, many signals of interest can be so represented. For instance, any polynomial of degree d can be obtained as the output (for a suitable initial state) of a system with $y^{(d+1)} \equiv 0$, for instance the output of the system of dimension $d+1$ and

$$A = \begin{pmatrix} 0 & 1 & 0 & \dots & 0 & 0 \\ 0 & 0 & 1 & \dots & 0 & 0 \\ 0 & 0 & 0 & \dots & 0 & 0 \\ \vdots & \vdots & \vdots & \ddots & \vdots & \vdots \\ 0 & 0 & 0 & \dots & 0 & 1 \\ 0 & 0 & 0 & \dots & 0 & 0 \end{pmatrix}, \quad C = (1, 0, \dots, 0).$$

In practice, $z(t)$ itself is not directly available to the controller, so it must be estimated in order to apply the solution obtained in the above problem. For instance, in the case of polynomial signals as in this example, estimation of $z(t)$ could be done by derivation of the reference signal $r(t)$, or by algorithms using observer theory (see later) that do not involve differentiation. □

Exercise 8.2.6 It is sometimes of interest to consider costs that penalize correlations between input and state variables, that is, to include a cross-term of the type $\omega(t)'L(t)\xi(t)$ in q. (For instance, it may be physically undesirable to apply a negative u_1 when x_2 is positive, in which case we can add a term such as $-\rho\omega_1(t)\xi_2(t)$ for some large $\rho > 0$.) Specifically, assume that

$$q(t, x, u) = (u' \; x') \begin{pmatrix} R & L \\ L' & Q \end{pmatrix} \begin{pmatrix} u \\ x \end{pmatrix},$$

where $R(\cdot)$ is positive definite and where the composite matrix (possibly time-varying)

$$\begin{pmatrix} R & L \\ L' & Q \end{pmatrix}$$

is positive semidefinite. *Show* that Theorem 37 (p. 364) remains valid in this case, with a slightly modified feedback law and Riccati equation. (*Hint:* Reduce to the case $L = 0$ by completing the square.) □

Exercise 8.2.7 Obtain an analogue of Theorem 37 (p. 364) for linear discrete-time systems with costs of the type

$$\sum_{k=\sigma}^{\tau-1} \{\omega(k)'R(k)\omega(k) + \xi(k)'Q(k)\xi(k)\} + \xi(\tau)'S\xi(\tau).$$

(The Riccati equation is now of course a *difference* equation; its right-hand side is a bit more complicated than before.) □

Hamiltonian Formulation

In general, a *Hamiltonian matrix* (of size $2n$) is a $2n \times 2n$ real matrix H that satisfies

$$H'J + JH = 0,$$

where

$$J = \begin{pmatrix} 0 & -I \\ I & 0 \end{pmatrix} \tag{8.33}$$

and I is the n-dimensional identity matrix. Equivalently, H has the form

$$H = \begin{pmatrix} M & L \\ N & -M' \end{pmatrix}$$

for some symmetric N and L and arbitrary M.

In particular, we associate to the matrix functions A, B, R, Q the matrix function

$$H := \begin{pmatrix} A & -BR^{-1}B' \\ -Q & -A' \end{pmatrix},$$

which is a Hamiltonian-matrix valued function $H(t), t \in [\sigma, \tau]$. The following fact is proved by a straightforward computation:

Lemma 8.2.8 Let ξ, ω be a pair consisting of an optimal trajectory, starting from a state $\xi(\sigma) = x^0$, and the corresponding optimal control $\omega(t) = -R(t)^{-1}B(t)'P(t)\xi(t)$ on the interval $[\sigma, \tau]$. Then:

$$\frac{d}{dt}\begin{pmatrix} \xi \\ \lambda \end{pmatrix} = H\begin{pmatrix} \xi \\ \lambda \end{pmatrix}, \qquad \xi(\sigma) = x^0, \quad \lambda(\tau) = S\xi(\tau) \tag{8.34}$$

where $\lambda(t) := P(t)\xi(t)$. □

Thus, along optimal trajectories, the vector $(\xi(t)', \lambda(t)')'$ solves the two-point boundary value problem (8.34). This is the form of the "maximum principle" statement of optimality; we show next that in this way one obtains a unique characterization of the optimal trajectory:

Lemma 8.2.9 For each A, B, Q, R, S (m.e.b. entries), there is a unique solution of the two-point boundary value problem (8.34).

Proof. Existence has been proved (use the optimal trajectory and $\lambda \equiv P\xi$). We must show uniqueness. For that, it will be sufficient to show that *every solution of the two-point boundary value problem (8.34) is a solution of the following initial value problem:*

$$\dot{\mu} = H\mu, \quad \mu(\sigma) = \begin{pmatrix} x^0 \\ P(\sigma)x^0 \end{pmatrix}. \tag{8.35}$$

Let $P(\cdot)$ be the solution of (RDE), and consider the unique solution on the interval $[\sigma, \tau]$ of the following matrix linear differential equation:

$$\dot{X} = (A - BR^{-1}B'P)X, \quad X(\tau) = I. \tag{8.36}$$

We let $\Lambda(t) := P(t)X(t)$ and

$$\Psi(t) := \begin{pmatrix} X(t) \\ \Lambda(t) \end{pmatrix}, \quad \text{so that} \quad \Psi(\tau) = \begin{pmatrix} I \\ S \end{pmatrix}.$$

A simple calculation using that P satisfies (RDE) and X satisfies (8.36) shows that this $2n$ by n matrix function solves the corresponding matrix Hamiltonian differential equation:

$$\dot{\Psi} = H\Psi.$$

Now pick any solution $\mu(t) = (\xi(t)', \lambda(t)')'$ of the two-point boundary value problem (8.34). Observe that $\mu(\tau)$ is in the kernel of the n by $2n$ matrix $[S, -I]$ and that the columns of $\Psi(\tau)$ are n linearly independent vectors in this (n dimensional) kernel. Thus, there is some vector $q \in \mathbb{R}^n$ so that $\mu(\tau) = \Psi(\tau)q$. Since $p = \mu - \Psi q$ satisfies $\dot{p} = Hp$, $p(\tau) = 0$, it follows that $\mu \equiv \Psi q$. Thus, in particular, $\xi(\sigma) = X(\sigma)q$, and also

$$\lambda(\sigma) = \Lambda(\sigma)q = P(\sigma)X(\sigma)q = P(\sigma)\xi(\sigma),$$

which is the claim that had to be proved. ∎

Remark 8.2.10 Observe that the first n rows of $\Psi(t)$, that is, $X(t)$, form a nonsingular matrix for each t, since $X(t)$ is a fundamental matrix associated to $\dot{x} = (A - BR^{-1}B'P)x$. Since, by definition, $\Lambda(t) = P(t)X(t)$ for all t, we conclude that

$$P(t) = \Lambda(t)X(t)^{-1}$$

for all t. This provides a way to solve the nonlinear differential equation (RDE): first solve the final value problem $\dot{\Psi} = H\Psi$, $\Psi(\tau) = [I, S']'$, partition as $\Psi = [X', \Lambda']'$, and then take $P = \Lambda X^{-1}$. □

8.3 Tracking and Kalman Filtering*

We now study a problem of output tracking, and, as an application, a deterministic version of Kalman filtering.

Tracking Under Disturbances

A problem which arises often is that of finding controls so as to force the output of a given system to *track* (follow) a desired reference signal $r(\cdot)$. We also allow a disturbance φ to act on the system.

Let Σ be a linear system (over $\mathbb{K} = \mathbb{R}$) with outputs

$$\dot{x} = A(t)x + B(t)u + \varphi(t), \quad y = C(t)x,$$

* This section can be skipped with no loss of continuity.

where φ is an $\mathbb{R}^n$-valued fixed m.e.b. function. Assume given an $\mathbb{R}^p$-valued fixed m.e.b. function r. We consider a cost criterion:

$$\mathcal{J}(\tau,\sigma,x^0,\omega,\varphi,r) := \int_\sigma^\tau \omega(t)'R(t)\omega(t) + e(t)'Q(t)e(t)\,dt \; + \; \xi(\tau)'S\xi(\tau)\,,$$

where $\dot{\xi} = A\xi + B\omega, \xi(\sigma) = x^0$, and $e := C\xi - r$ is the *tracking error.* As in the quadratic optimal control problem, it is assumed that:

- R is an $m \times m$ symmetric matrix of m.e.b. functions, Q is a $p \times p$ symmetric matrix of m.e.b. functions, and S is a constant symmetric $n \times n$ matrix;
- for each t, R is positive definite;
- for each t, Q is positive semidefinite; and
- S is positive semidefinite.

The problem is, then, that of minimizing $\mathcal{J}$, for a given disturbance φ, reference signal r, and initial state, by an appropriate choice of controls. Observe that in the special case when $\varphi \equiv r \equiv 0$ this is a standard linear-quadratic problem, with "Q" matrix equal to $C'QC$ (cost imposed on outputs). For $\varphi \equiv 0$ and r arbitrary, we have a tracking problem; for $r \equiv 0$ and φ arbitrary a problem of regulation under disturbances.

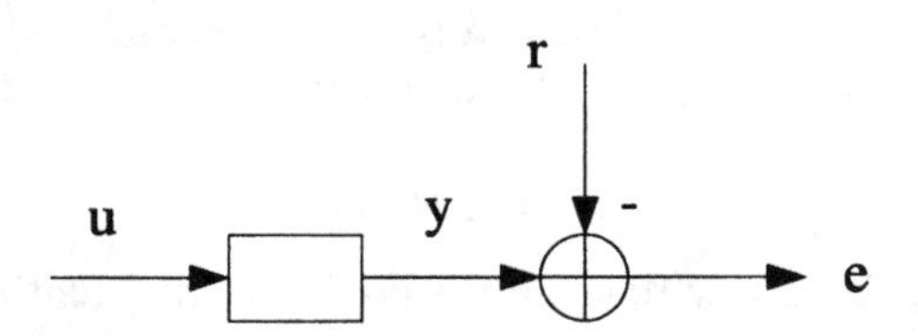

Figure 8.1: *Tracking problem.*

This problem of tracking under disturbances can be reduced to an optimal control problem for a linear system with quadratic costs, as follows. The idea is very simple: add a constant state $x_{n+1} \equiv 1$, allowing the incorporation of φ into A and of r into Q. We do that next.

On the interval $[\sigma, \tau]$ we consider the system (with no outputs)

$$\dot{x} = \widetilde{A}x + \widetilde{B}u\,,$$

where

$$\widetilde{A} := \begin{pmatrix} A & \varphi \\ 0 & 0 \end{pmatrix} \quad \text{and} \quad \widetilde{B} := \begin{pmatrix} B \\ 0 \end{pmatrix}$$

have sizes $(n+1) \times (n+1)$ and $(n+1) \times m$, respectively. We also introduce the associated cost

$$\widetilde{\mathcal{J}}(\tau,\sigma,\widetilde{x^0},\varphi,\omega) := \int_\sigma^\tau \omega(t)'R(t)\omega(t) + \widetilde{\xi}(t)'\widetilde{Q}(t)\widetilde{\xi}(t)\,dt \; + \; \widetilde{\xi}(t)'\widetilde{S}\widetilde{\xi}(t)\,,$$

where tildes on x^0, ξ are used to indicate vectors and vector functions of dimension $n+1$. The new matrices are defined as

$$\widetilde{Q} := \begin{pmatrix} C'QC & -C'Qr \\ -r'QC & r'Qr \end{pmatrix} \quad \widetilde{S} := \begin{pmatrix} S & 0 \\ 0 & 0 \end{pmatrix}$$

(from now on, we drop arguments t whenever clear from the context). In terms of these data,

$$\widetilde{\mathcal{J}}(\tau, \sigma, \widetilde{x^0}, \omega) = \mathcal{J}(\tau, \sigma, x^0, \omega, \varphi, r)$$

for each x^0, ω, where $\widetilde{x^0} := (x^0, 1)'$. Note that the resulting problem will have time-dependent costs even if the original system was time-invariant and the matrices Q, R were constant. The matrices $\widetilde{Q}$ and $\widetilde{S}$ are still semidefinite; in the case of $\widetilde{Q}$ this is a consequence of the formula

$$(z' \ \ a)\,\widetilde{Q}(t)\begin{pmatrix} z \\ a \end{pmatrix} = (Cz - ar(t))'Q(Cz - ar(t)),$$

which holds for each $z \in \mathbb{R}^n$ and each $a \in \mathbb{R}$.

Minimizing $\widetilde{\mathcal{J}}$ with initial states of the form $\widetilde{x^0} = (x^0, 1)'$, for the system $(\widetilde{A}, \widetilde{B})$, provides a solution to the tracking problem. This is achieved using the results earlier in the Chapter. Let $\widetilde{P}$ be the solution of the Riccati Differential Equation (RDE) on page 364,

$$\dot{\widetilde{P}} = \widetilde{P}\widetilde{B}R^{-1}\widetilde{B}'\widetilde{P} - \widetilde{P}\widetilde{A} - \widetilde{A}'\widetilde{P} - \widetilde{Q}, \quad \widetilde{P}(\tau) = \widetilde{S},$$

which one knows exists on all of $[\sigma, \tau]$, by Theorem 37 (p. 364). This solution can be partitioned as

$$\widetilde{P} = \begin{pmatrix} P & \beta \\ \beta' & \alpha \end{pmatrix}$$

(all functions of t), where P is $n \times n$ and α is scalar. In terms of these data, the (RDE) takes the form of the set of equations

$$\dot{P} = PBR^{-1}B'P - PA - A'P - C'QC, \quad P(\tau) = S \tag{8.37}$$

which is itself the Riccati equation for the optimization problem that results when $\varphi \equiv r \equiv 0$ and can be computed independently of the particular r and φ of interest, together with

$$\dot{\beta} = -(A + BF)'\beta + C'Qr - P\varphi, \quad \beta(\tau) = 0 \tag{8.38}$$

and

$$\dot{\alpha} = \beta'BR^{-1}B'\beta - r'Qr - 2\varphi'\beta, \quad \alpha(\tau) = 0, \tag{8.39}$$

where we denoted

$$F := -R^{-1}B'P$$

just as in the standard optimal control problem. For an initial state of the form $\widetilde{x^0} = (x^0, 1)'$ the optimal path satisfies $\dot{\widetilde{x}} = (\widetilde{A} + \widetilde{B}\widetilde{F})\widetilde{x}$, which gives

$$\dot{x} = (A + BF)x - BR^{-1}B'\beta\,, \quad x(\sigma) = x^0\,, \tag{8.40}$$

and the optimal control $u = -R^{-1}\widetilde{B}'\widetilde{P}\widetilde{x}$ can be written in the form

$$u = Fx - R^{-1}B'\beta \tag{8.41}$$

while the optimal cost is

$$(\widetilde{x^0})'\widetilde{P}(\sigma)'\widetilde{x^0} = (x^0)'P(\sigma)'x^0 + 2(x^0)'\beta(\sigma) + \alpha(\sigma)$$

(this is the only place where α appears). Note the form of the solution: The tracking signal and the disturbance drive the system (8.38), which is a system that is formally adjoint to the closed-loop system obtained for no tracking, and the state β of this adjoint system in turn appears as an extra term in the closed-loop system. When $r \equiv 0$ and $\varphi \equiv 0$, also $\alpha \equiv \beta \equiv 0$, and the solution to the standard regulator problem is recovered.

The discussion can be summarized as follows.

Theorem 38 *Assume that A, B, C, Q, R, S are matrix functions as above, and let $\sigma < \tau$. Then the solution of the final-value problem* (8.37) *exists on $[\sigma, \tau]$. Given any m.e.b. functions φ and r on this interval, and any x^0, solve* (8.38) *and* (8.40) *to obtain $x(\cdot)$ and $\beta(\cdot)$ on $[\sigma, \tau]$. Then the control u defined by* (8.41) *has the property that it minimizes*

$$\int_\sigma^\tau \{u'Ru + (Cx - r)'Q(Cx - r)\}\, dt \; + \; x(\tau)'Sx(\tau)$$

subject to $\dot{x} = Ax + Bu + \varphi$, $x(\sigma) = x^0$, among all measurable essentially bounded controls, and is the unique such control. Moreover, the minimum value of this integral is given by

$$x^0P(\sigma)x^0 + 2(x^0)'\beta(\sigma) + a\,,$$

where $a = -\int_\sigma^\tau (\beta'BR^{-1}B'\beta - r'Qr - 2\varphi'\beta)dt$. □

Remark 8.3.1 If S is positive definite, then the matrix $P(\sigma)$ —which depends on the data A, B, R, Q, C but not on the particular x^0, φ, and r— is also positive definite. This can be proved as follows. Assume that

$$(x^0)'P(\sigma)x^0 = 0$$

for some nonzero x^0, and consider the tracking problem with $\varphi \equiv r \equiv 0$. The optimal cost is zero, since $\beta = \alpha \equiv 0$, so the cost along the optimal trajectory (ξ, ω) vanishes:

$$\int_\sigma^\tau \{\omega'R\omega + \xi'C'QC\xi\}\, dt \; + \; \xi(\tau)'S\xi(\tau) \; \equiv \; 0\,,$$

which implies $\omega \equiv 0$ and $\xi(\tau) = 0$ by positive definiteness of R and S. In conclusion, one has a solution of

$$\dot{\xi} = A\xi$$

for which $\xi(\tau) = 0$, which implies $\xi \equiv 0$, so $x^0 = \xi(0) = 0$, as wanted. □

Exercise 8.3.2 Let $\sigma < \tau$ and assume that Q is positive definite and (A, C) is observable on $[\sigma, \tau]$ (but S is merely semidefinite). Show that one can again conclude that $P(\sigma)$ is positive definite. □

(Deterministic) Kalman Filtering

Let (A, B, C) be a time-varying continuous-time linear system as in the previous section, and let Q, R, S be m.e.b. matrix functions of sizes $n \times n$, $m \times m$, and $n \times n$, respectively. Now we assume that both S and $R(\cdot)$ are positive definite.

The problem to be considered in this section is as follows: For any given m.e.b. function $\bar{y}$ on an interval $[\sigma, \tau]$, *minimize the expression*

$$\int_\sigma^\tau \left\{\omega(t)'R(t)\omega(t) + \big(C(t)\xi(t) - \bar{y}(t)\big)'Q(t)\big(C(t)\xi(t) - \bar{y}(t)\big)\right\} dt \; + \; \xi(\sigma)'S\xi(\sigma) \tag{8.42}$$

over the set of all possible trajectories (ξ, ω) *of* $\dot{\xi} = A\xi + B\omega$ *on* $[\sigma, \tau]$.

Remark 8.3.3 This problem is a minor variation of the tracking problem (with no disturbances, that is, with $\varphi \equiv 0$) considered in Theorem 38 (p. 374). The only differences are in the imposition of an initial as opposed to a final state cost and —more importantly— the fact that the minimization is to be carried out over all possible trajectories, not merely among those that start at a fixed initial state. The motivation for studying this question is as follows. Consider the system

$$\begin{aligned} \dot{x} &= Ax + Bu, \quad x(\sigma) = x^0 \\ y &= Cx + v\,, \end{aligned}$$

where the functions u and v are thought of as independent and unknown perturbations affecting the system, and the initial state x^0 is unknown. The function $\bar{y}$ represents an observation of y on the interval $[\sigma, \tau]$, and it is desired to obtain an estimate of the state trajectory, and in particular of the final state $x(\tau)$, based on this observation. Moreover, one wishes to obtain the "best possible" estimate, where "best" is taken to mean the one for which the disturbances u and v, as well as the initial state x^0, would have been as small as possible. The matrices S, Q, and R are used to quantify the relative weighting to be given to each component of the uncertainty. For instance, if it is known that the measurement noise v on some component of the output is very large compared to other components and to the uncertainty in the initial state and the state noise, then one should tend to assign less value to these observations, that is, the corresponding entries of Q should be taken small.

The same question can be interpreted in statistical terms, and then it is more properly called the *Kalman filtering* problem. In that version, $u(\cdot)$ and $v(\cdot)$ are independent white noise processes and x^0 is a random vector, all assumed to be Gaussian distributed and zero mean. With R,Q, and S being, respectively, the *inverses* of the covariance matrices of u,v, and x^0, the minimization problem that we posed is equivalent to that of finding the minimum variance estimate of x given the observations $\overline{y}$ (a conditional mean). The cost function (8.42) is the likelihood function of the problem. The terminology "filtering" reflects the goal of filtering-out the noise in order to recover the state $x(\cdot)$, interpreted as the signal of interest, from the measured signal $\overline{y}$. (Actually, we solve what is called the "smoothing" problem first.) □

Reduction to Tracking Problem

The filtering problem is reduced to the tracking problem (with $\varphi \equiv 0$) by first reversing time, so that the cost is imposed on the final state, and then minimizing over all possible solutions to the tracking problem, over all possible initial states. Finally, and most importantly, a recursive form of the solution is derived, which allows updates of the state estimate as new measurements are taken. Let

$$\begin{aligned} \widetilde{R}(t) &:= R(\tau+\sigma-t)\,, & \widetilde{Q}(t) &:= Q(\tau+\sigma-t)\,, \\ \widetilde{C}(t) &:= C(\tau+\sigma-t)\,, & r(t) &:= \overline{y}(\tau+\sigma-t) \end{aligned}$$

and

$$\widetilde{A}(t) := -A(\tau+\sigma-t)\,, \quad \widetilde{B}(t) := -B(\tau+\sigma-t)$$

for all $t \in [\sigma,\tau]$, for any given $\overline{y}$. For each trajectory (ξ,ω) of (A,B),

$$\widetilde{\xi}(t) := \xi(\tau+\sigma-t)\,, \quad \widetilde{\omega}(t) := \omega(\tau+\sigma-t)$$

provide a trajectory of $(\widetilde{A},\widetilde{B})$ and conversely.

In terms of these data, one can write (8.42) as

$$\int_\sigma^\tau \left\{\widetilde{\omega}(t)'\widetilde{R}(t)\widetilde{\omega}(t) + (\widetilde{C}(t)\widetilde{\xi}(t)-r(t))'\widetilde{Q}(t)(\widetilde{C}(t)\widetilde{\xi}(t)-r(t))\right\}dt \;+\; \widetilde{\xi}(\tau)'S\widetilde{\xi}(\tau)\,, \tag{8.43}$$

and the problem becomes that of minimizing (8.43) over all trajectories of $(\widetilde{A},\widetilde{B})$. For each fixed $\widetilde{x^0}$ this minimum is provided by Theorem 38 (p. 374). We use tildes for the solution of the corresponding Riccati equation:

$$\dot{\widetilde{P}} = \widetilde{P}\widetilde{B}\widetilde{R}^{-1}\widetilde{B}'\widetilde{P} - \widetilde{P}\widetilde{A} - \widetilde{A}'\widetilde{P} - \widetilde{C}'\widetilde{Q}\widetilde{C}\,, \quad \widetilde{P}(\tau) = \widetilde{S}\,. \tag{8.44}$$

The other relevant equations are, then:

$$\dot{\widetilde{\beta}} = \widetilde{P}\widetilde{B}\widetilde{R}^{-1}\widetilde{B}'\widetilde{\beta} - \widetilde{A}'\widetilde{\beta} + \widetilde{C}'\widetilde{Q}r\,, \quad \widetilde{\beta}(\tau) = 0 \tag{8.45}$$

and, with

$$\widetilde{F} = -\widetilde{R}^{-1}\widetilde{B}'\widetilde{P},$$

then

$$\dot{\widetilde{x}} = (\widetilde{A} + \widetilde{B}\widetilde{F})\widetilde{x} - \widetilde{B}\widetilde{R}^{-1}\widetilde{B}'\widetilde{\beta}, \quad \widetilde{x}(\sigma) = \widetilde{x^0} \tag{8.46}$$

is the optimal path. Also from the Theorem it follows that the cost along this trajectory is

$$(\widetilde{x^0})'\widetilde{P}(\sigma)\widetilde{x^0} + 2(\widetilde{x^0})'\widetilde{\beta}(\sigma) + a, \tag{8.47}$$

where a is a constant that does not depend on x^0. Minimizing the expression for x^0 will provide the minimum for the desired problem. By Remark 8.3.1, $\widetilde{P}(\sigma)$ is invertible, so (8.47) equals

$$\left(\widetilde{x^0} + \widetilde{P}(\sigma)^{-1}\widetilde{\beta}(\sigma)\right)' \widetilde{P}(\sigma)\left(\widetilde{x^0} + \widetilde{P}(\sigma)^{-1}\widetilde{\beta}(\sigma)\right) - \widetilde{\beta}(\sigma)'\widetilde{P}(\sigma)^{-1}\widetilde{\beta}(\sigma) + a,$$

which admits the unique minimum

$$\widetilde{x^0} = -\widetilde{P}(\sigma)^{-1}\widetilde{\beta}(\sigma). \tag{8.48}$$

This gives a solution to the observation problem: First solve for $\widetilde{P}$ and $\widetilde{\beta}$, then for $\widetilde{x}$ using the initial condition (8.48), and finally, reversing time, obtain P, β, x. The details are as follows.

Consider the equations:

$$\dot{P} = -PBR^{-1}B'P - PA - A'P + C'QC. \quad P(\sigma) = S \tag{8.49}$$

$$\dot{\beta} = -PBR^{-1}B'\beta - A'\beta - C'Q\bar{y}, \quad \beta(\sigma) = 0 \tag{8.50}$$

as well as

$$F = -R^{-1}B'P$$

and

$$\dot{x} = (A - BF)x + BR^{-1}B'\beta, \quad x(\tau) = -P(\tau)^{-1}\beta(\tau). \tag{8.51}$$

Theorem 39 *Assume that (A, B, C) is a continuous-time linear system with outputs, R and S are positive definite as above, and all entries of A, B, C, R, Q are m.e.b.. Then, the solutions of (8.49) and of (8.50) exist on each interval $[\sigma, \tau]$.*

Given any m.e.b. function $\bar{y}$ on any interval $[\sigma, \tau]$, let $\xi(\cdot)$ be the solution of (8.51) on this interval. Then, there exists some $\omega(\cdot)$ so that (ξ, ω) is the unique trajectory minimizing (8.42). □

Observe that the final state $x(\tau)$ for the optimal solution can be obtained by solving the differential equations for P and β and using simply $x(\tau) = -P(\tau)^{-1}\beta(\tau)$. Thus this value $x(\tau)$, the "filtered" estimate of the state, can be computed "online" on increasing time intervals (increasing τ). One can even avoid the matrix inversion, resulting in a recursive estimate, as follows.

Recursive Formulas for State Estimate

With the previous notations, we let

$$\Pi(t) := P^{-1}(t)$$

for all $t \in [\sigma, \tau]$. Consider the vector

$$z(t) = -\Pi(t)\beta(t)$$

for $t \in [\sigma, \tau]$. It follows from the above discussion that $z(t)$ is the same as the final state $x(t)$ for the optimization problem restricted to $[0, t]$.

From the fact that $\dot{\Pi} = -\Pi\dot{P}\Pi$, substituting $\dot{P}$ from (8.49), and using that $\Pi P = I$, we conclude that Π satisfies the *Filtering Riccati Differential Equation*, or "dual RDE",

$$\boxed{\dot{\Pi} = \Pi A' + A\Pi - \Pi C'QC\Pi + BR^{-1}B', \quad \Pi(\sigma) = S^{-1}. \quad \text{(FRDE)}}$$

Furthermore, from (FRDE) and (8.50) we conclude that

$$\dot{z} = (A - \Pi C'QC)z + \Pi C'Q\overline{y}, \quad z(\sigma) = 0.$$

With the notation

$$L := -\Pi C'Q,$$

this can also be written as

$$\dot{z} = Az + L[Cz - \overline{y}], \quad z(\sigma) = 0 \tag{8.52}$$

which is the final form of the equation for z. We can sumarize as follows:

Theorem 40 *Assume that (A, B, C) is a continuous-time linear system with outputs, R and S are positive definite as above, and all entries of A, B, C, R, Q are m.e.b.. Then, the solution of* (FRDE) *exists on each interval $[\sigma, \tau]$.*

Given any m.e.b. function $\overline{y}$ on any interval $[\sigma, \tau]$, let $\zeta(\cdot)$ be the solution of (8.52) *on this interval. Let (ξ, ω) be the unique trajectory minimizing* (8.42). *Then, $\zeta(\tau) = \xi(\tau)$.* □

Remark 8.3.4 Equation (8.52) has the form of the observer derived earlier, in Chapter 7. (The control does not appear, but see Exercise 8.3.9 below.) Thus, the Kalman filter is an observer where the matrix L has been chosen using an optimization criterion (see Remark 7.1.7). Note also that (FRDE) can be solved independently of the particular observations; in applications one often precomputes the form of Π. Alternatively, one may solve the differential equation online —since (FRDE) is an initial-value rather than a final-value problem, it is possible to solve for $\Pi(\tau)$, for increasing values of τ, as observations are received. Similarly, the estimate $z(\tau)$ can be computed recursively for increasing values of τ, using (8.52), and at each τ it provides the best estimate of $x(\tau)$ based on the observations $\{\overline{y}(t), t \in [\sigma, \tau]\}$. □

Exercise 8.3.5 Derive explicit formulas for the control ω and the corresponding optimal cost in the filtering Theorem. □

Example 8.3.6 Consider the problem of estimating an unknown constant x^0 subject to additive noise, that is, estimating

$$y = x^0 + v$$

from

$$y(t)\,, \quad t \geq 0\,.$$

We take the system

$$\dot{x} = 0\,, \quad y = x + v$$

and choose $S = 1$, $Q = q > 0$, and R arbitrary (since $B = 0$, we may take $m = 1$ and $R = 1$; all formulas including R have B as a factor, so they vanish anyhow). A large q should be used if v is expected to be small compared to the uncertainty in x^0, and vice versa (recall Remark 8.3.3). There results the equation

$$\dot{\Pi} = -\Pi^2 q\,, \quad \Pi(0) = 1\,,$$

so

$$\Pi(t) = \frac{1}{1 + qt}\,,$$

and from here the estimator

$$\dot{z}(t) = \frac{1}{t + 1/q}(y(t) - z(t))\,.$$

Note that for q large (small noise) and t small, the solutions approach y very quickly, consistent with the intuition that the measurements are very reliable. On the other hand, for q small (large noise), z changes slowly, again as expected: More measurements are needed before an accurate estimate is obtained. □

Exercise 8.3.7 Compute explicitly the optimal filter for the case of

$$\dot{x} = u\,, \quad y = x + v$$

and

$$R = S = 1\,, \quad Q = q > 0\,.$$

Analyze the behavior of the filter for q large or small. □

Exercise 8.3.8 Consider the case when $B = 0$, $Q = I$, and S approaches zero. Show that the formulas for least-squares observation in Section 6.3 can be recovered from the results in this section. (*Hint:* The equation for $\widetilde{P}$ can be solved with final condition zero, and its solution at the initial time can be expressed directly in terms of the Wronskian W.) □

Exercise 8.3.9 Consider the problem of minimizing the same cost (8.42) over the set of all trajectories of the system

$$\dot{\xi} = A\xi + G\overline{u} + B\omega\,,$$

where $\overline{u}$ is a known control on $[\sigma, \tau]$. Show that the minimum is achieved through the solution of

$$\dot{z} = Az + L[Cz - \overline{y}] + G\overline{u}, \quad z(\sigma) = 0\,,$$

where L is the same as before. (*Hint:* Simply convert the original problem into an estimation problem for $\dot{\widetilde{\xi}} = A\xi + B\omega$, where $\widetilde{\xi} := \xi - \overline{x}$ and $\overline{x}$ satisfies $\dot{\overline{x}} = A\overline{x} + G\overline{u}$.) □

8.4 Infinite-Time (Steady-State) Problem

We next consider the case of optimal controls for costs functions as earlier, but now for the problem on an infinite interval. To simplify, and because this is the most interesting case when $\tau = \infty$ anyway, we assume from now on that the systems being considered are time-invariant and that the matrices R, Q are also independent of t. Thus, without loss of generality, we can restrict attention to the case of optimization on the interval $[0, \infty)$. The problem to be studied is, then:

> *For the system*
>
> $$\dot{x} = Ax + Bu\,,$$
>
> *and any given initial state x at time $t = 0$, find a control ω on $[0, \infty)$ such that the cost (possibly infinite)*
>
> $$\mathcal{J}_\infty(x, \omega) := \int_0^\infty \{\omega(t)'R\omega(t) + \xi(t)'Q\xi(t)\}\,dt \tag{8.53}$$
>
> *is minimized among all such controls.*

It is again assumed that $R > 0$ and $Q \geq 0$. We define the Bellman function as in the finite interval case, and similarly the notions of optimal control and optimal trajectory.

It will be shown that, under controllability assumptions guaranteeing that the cost can be made finite, a solution again exists and can be computed via an algebraic equation obtained as a limit of the Riccati differential equation. The assumption that $R > 0$ insures that the problem is well-posed, in that no coordinate of the control is "free" of cost. Under the extra assumption that Q is positive definite rather than merely semidefinite, finiteness of the integral along an optimal trajectory will imply that $\xi(t)$ must converge to zero, that is, the optimum system will be stabilized by the optimal control law.

Consider the matrix Riccati equation (RDE) on the interval

$$(-\infty, 0]$$

but with the final condition

$$P(0) = 0 .$$

A solution $P(\cdot)$ exists on this interval. This is because a solution exists on each interval $[\sigma, 0]$ as follows from Theorem 37 (p. 364), and by the ODE Uniqueness Theorem these solutions must coincide on their common domain.

Let $\Pi(t) := P(-t)$. Thus, $\Pi(t)$ satisfies for $t \geq 0$ the equation

$$\dot{\Pi} = -\Pi B R^{-1} B' \Pi + \Pi A + A' \Pi + Q , \quad \Pi(0) = 0 . \tag{8.54}$$

Equivalently, we may by time invariance consider the Riccati equation (RDE) on each interval $[0, t]$, with final condition $P(t) = 0$; then $\Pi(t)$ is the same as the value $P(0)$ for the corresponding solution. It follows that for each $x \in \mathbb{R}^n$ and each $\tau > 0$,

$$x' \Pi(\tau) x = V_\tau(0, x) ,$$

where V_τ is the Bellman function for the optimization problem of minimizing (8.53) but only over the finite interval $[0, \tau]$.

Take any pair $\mu > \tau \geq 0$, and let ν be optimal for the interval $[0, \mu]$ and initial state x. Then,

$$\begin{aligned} x'\Pi(\mu)x &= V_\mu(0,x) = \mathcal{J}(\mu, 0, x, \nu) \\ &= \mathcal{J}(\tau, 0, x, \nu|_{[0,\tau)}) + \int_\tau^\mu \{\nu(t)' R \nu(t) + \xi(t)' Q \xi(t)\} dt \\ &\geq V_\tau(0, x) = x' \Pi(\tau) x . \end{aligned}$$

This holds for all x, so

$$\Pi(\mu) \geq \Pi(\tau) \tag{8.55}$$

in the sense of the partial ordering of symmetric matrices ("$P \geq Q$" means that $x'Px \geq x'Qx$ for all x).

We prove next that, if the system is controllable, then there is a limit Π for $\Pi(t)$ as $t \to \infty$, and will use this fact in order to prove that there are optimal controls for the infinite time problem and that the optimal cost is given by the quadratic form $x'\Pi x$.

Lemma 8.4.1 Assume that Σ is controllable and let R and Q be as above. Then the limit

$$\Pi := \lim_{t \to \infty} \Pi(t)$$

exists. Moreover, Π satisfies the *algebraic Riccati equation*

$$\boxed{\Pi B R^{-1} B' \Pi - \Pi A - A' \Pi - Q = 0 \qquad \text{(ARE)}}$$

Proof. Fix any $x \in \mathbb{R}^n$. We first remark that controllability implies that there is some ω such that

$$\mathcal{J}_\infty(x, \omega) < \infty .$$

Indeed, let ω_1 be any control with

$$\phi(1,0,x,\omega_1)=0$$

and consider the control ω on $[0,\infty)$ that is equal to ω_1 for $t\in[0,1]$ and is identically zero for $t>1$. Then,

$$\mathcal{J}_\infty(x,\omega)=\int_0^1\{\omega_1(t)'R\omega_1(t)+\xi(t)'Q\xi(t)\}dt<\infty\,.$$

It follows that, for each $\tau>0$,

$$x'\Pi(\tau)x=V_\tau(0,x)\leq\mathcal{J}(\tau,0,x,\omega|_{[0,\tau)})\leq\mathcal{J}_\infty(x,\omega)\,. \tag{8.56}$$

Thus,

$$\{x'\Pi(\tau)x\}$$

is not only nondecreasing in τ (equation (8.55)) but it is also bounded above. Thus,

$$\lim_{\tau\to\infty}x'\Pi(\tau)x$$

exists, for each fixed x. Let l_{ij} be the limit obtained when $x=e_i+e_j$, and l_i the limit for $x=e_i$, where e_i is the ith canonical basis element in $\mathbb{R}^n$, $i=1,\ldots,n$. Then, for each pair i,j, using that $\Pi(t)$ is symmetric for each t:

$$\Pi(t)_{ij}=e_i'\Pi(t)e_j=\frac{1}{2}[(e_i+e_j)'\Pi(t)(e_i+e_j)-e_i'\Pi(t)e_i-e_j'\Pi(t)e_j]$$

converges to

$$\frac{1}{2}[l_{ij}-l_i-l_j]\,.$$

So a limit Π exists, as desired. Since $\Pi(t)$ satisfies (8.54),

$$\lim_{t\to\infty}(d/dt)(x'\Pi(t)x)$$

also exists for each x, and hence this limit must be zero. It follows that

$$x'\left[\Pi BR^{-1}B'\Pi-\Pi A-A'\Pi-Q\right]x=0$$

for all x. Symmetry of the matrix in brackets $[\ldots]$ implies that (ARE) must hold. ∎

From equation (8.56), which holds for any control ω, we conclude that

$$x'\Pi x\leq\mathcal{J}_\infty(x,\omega)$$

for all ω. When ν is the closed-loop control on the interval $[0,\infty)$ corresponding to the feedback law

$$k(t,x):=-Fx$$

with

$$F := R^{-1}B'\Pi$$

and starting at $\xi(0) = x$, we have, substituting $\omega = \nu = -R^{-1}B'\Pi\xi$ into (8.53), that

$$\mathcal{J}_\infty(x,\nu) = \int_0^\infty \xi(t)'(\Pi BR^{-1}B'\Pi + Q)\xi(t)\,dt\,.$$

On the other hand,

$$(d/dt)(\xi(t)'\Pi\xi(t)) = -\xi(t)'(\Pi BR^{-1}B'\Pi + Q)\xi(t)\,,$$

because Π satisfies (ARE). It follows that

$$\mathcal{J}_\infty(x,\nu) = x'\Pi x - \lim_{t\to\infty} \xi(t)'\Pi\xi(t) \le x'\Pi x \le \mathcal{J}_\infty(x,\widetilde{\omega})$$

for all $\widetilde{\omega}$ (and, in particular, for ν). Thus,

- ν is optimal;
- $x'\Pi x$ is the Bellman function for this problem;
- $\mathcal{J}_\infty(x,\nu) < \infty$; and
- $\lim_{t\to\infty} \xi(t)'\Pi\xi(t) = 0$.

We now prove that ν is the *unique* optimal control. Let

$$\alpha(t) := \xi(t)'\Pi\xi(t)$$

computed along any arbitrary trajectory (ξ,ω) on $[0,\infty)$ such that $\xi(0) = x$ for which $\omega \neq \nu$. Completing the square, one verifies just as in the finite-time case that the minimum of

$$v'Rv + 2x'\Pi[Ax + Bv]$$

is achieved only at

$$v = -R^{-1}B'\Pi x\,.$$

That is,

$$\omega'R\omega + \dot{\alpha} \ge \left(R^{-1}B'\Pi\xi\right)' R\left(R^{-1}B'\Pi\xi\right) + 2\xi'\Pi\left(A - BR^{-1}B'\Pi\right)\xi$$

for all t, and equality cannot hold almost everywhere since that would mean that $\omega = -R^{-1}B'\Pi\xi$, which we assumed was not the case.

Using this as well as the fact that Π satisfies (ARE), we conclude that

$$\dot{\alpha}(t) \ge -\omega(t)'R\omega(t) - \xi(t)'Q\xi(t)\,,$$

with strict inequality on a set of nonzero measure. Integrating, there exists some $\tau > 0$ such that

$$x'\Pi x < \int_0^\tau \{\omega(t)'R\omega(t) + \xi(t)'Q\xi(t)\}dt + \xi(\tau)'\Pi\xi(\tau)\,. \tag{8.57}$$

Consider $z := \xi(\tau)$ and the control

$$\bar{\omega}(t) := \omega(t+\tau)\,, t \geq 0\,.$$

Since $z'\Pi z$ is the value of the Bellman function at z, the last term in (8.57) is majorized by $\mathcal{J}_\infty(z,\bar{\omega})$. By time invariance,

$$\mathcal{J}(\tau, 0, x, \omega) + \mathcal{J}_\infty(z,\bar{\omega}) = \mathcal{J}_\infty(x,\omega)\,,$$

which is therefore strictly larger than $x'\Pi x$. This shows that the control ω cannot be optimal, and the uniqueness statement follows.

Theorem 41 *Consider the infinite time problem defined by Σ, R, Q, and assume that Σ is controllable. Then there exists for each $x \in \mathbb{R}^n$ a unique optimal control, given by the closed-loop control corresponding to the (constant) feedback*

$$F := -R^{-1}B'\Pi\,,$$

where Π is the solution of the ARE obtained by the above limiting process. Further, if in addition

$$Q \text{ is positive definite,}$$

then

$$\Pi \text{ is also positive definite,}$$

and for any solution of the closed-loop system

$$\dot{\xi} = (A + BF)\xi$$

necessarily

$$\lim_{t\to\infty} \xi(t) = 0\,.$$

Proof. We proved the first part above. Assume now that $Q > 0$. Once we prove that $\Pi > 0$, as we proved already that

$$\xi(t)'\Pi\xi(t) \to 0$$

for all trajectories, it will follow also that $\xi(t) \to 0$, as desired.

Since Π is the increasing limit of the matrices $\Pi(t)$, it will be sufficient to prove that $\Pi(t)$ is positive definite for all $t > 0$. Consider the solution $\Pi(\cdot)$ of the differential equation (8.54). Fix any $x \neq 0$. Let $\beta(t) := x'\Pi(t)x$. Then

$$\dot{\beta}(0) = x'Qx > 0$$

and $\beta(0) = 0$, from which it follows that

$$\beta(t) > 0$$

for all small $t > 0$. Since β is nondecreasing, it follows that $x'\Pi(t)x > 0$ for all t. ∎

Exercise 8.4.2 Show that Theorem 41 (p. 384) is valid if one replaces "Σ is controllable" by "Σ is asycontrollable." Point out exactly where in the proof the weaker hypothesis can be used. □

Given any controllable system Σ, we may always consider the optimal control problem with

$$R = I\,, \quad Q = I\,.$$

Then Theorem 41 can be applied in this case, and we conclude in particular, using Proposition C.5.1:

Corollary 8.4.3 If Σ is a controllable time-invariant linear continuous-time system over $\mathbb{K} = \mathbb{R}$, then there exists an $m \times n$ real matrix F such that $A + BF$ is a Hurwitz matrix. □

The above Corollary is, of course, also a consequence of the Pole-Shifting Theorem, which establishes a far stronger result. Next we consider a discrete-time analogue; its proof follows the same steps as in the continuous-time case.

Lemma/Exercise 8.4.4 Consider the infinite time problem of minimizing

$$\mathcal{J}_\infty(x,\omega) := \sum_{t=0}^{\infty}\{\omega(t)'R\omega(t) + \xi(t)'Q\xi(t)\} \quad \in \mathbb{R}_+ \bigcup \{+\infty\}$$

for a discrete-time linear time-invariant system Σ. Assume that

- Σ is controllable;
- R is positive definite;
- Q is positive semidefinite.

Then there exists for each $x \in \mathbb{R}^n$ a unique optimal control, given by the closed-loop control corresponding to the (constant) feedback

$$F := -(R + B'\Pi B)^{-1}B'\Pi A\,,$$

where Π is the solution of the following discrete-time algebraic Riccati equation obtained as a limit of solutions of appropriate finite-time problems (cf. Exercise 8.2.7):

$$\boxed{\Pi = A'[\Pi - \Pi B(R + B'\Pi B)^{-1}B'\Pi]A + Q \qquad \text{(DARE)}}$$

Further, if in addition

$$Q \text{ is positive definite,}$$

then

$$\Pi \text{ is also positive definite,}$$

and for any solution of the closed-loop system

$$\xi^+ = (A + BF)\xi$$

necessarily

$$\lim_{t \to \infty} \xi(t) = 0 .$$

holds. □

Exercise 8.4.5 Prove, without using the Pole-Shifting Theorem: If Σ is a controllable time-invariant linear discrete-time system over $\mathbb{K} = \mathbb{R}$, then there exists an $m \times n$ real matrix F such that $A + BF$ is a convergent matrix. □

Numerical Computation of Solutions of ARE

Solutions of the ARE can be obtained as limits of solutions of the differential equation or by other numerical techniques. One of the latter is based on the following considerations.

As done earlier, we consider the Hamiltonian matrix:

$$H := \begin{pmatrix} A & -BR^{-1}B' \\ -Q & -A' \end{pmatrix} . \tag{8.58}$$

The next problem shows that finding solutions of the Riccati equation (ARE) is equivalent to finding certain invariant subspaces of this matrix. In general, associated to any matrices A, N, L we may consider the following equation generalizing (ARE):

$$-\Pi L \Pi - \Pi A - A'\Pi + N = 0 \tag{8.59}$$

to be solved for an $n \times n$ matrix Π.

Exercise 8.4.6 Consider the map

$$\alpha : \mathbb{R}^{n \times n} \to G_n(R^{2n}) , \quad \Pi \mapsto \text{col} \begin{pmatrix} I \\ \Pi \end{pmatrix} ,$$

where $G_n(R^{2n})$ is the set of all n-dimensional subspaces of $\mathbb{R}^{2n}$ (the "G" stands for *Grassman manifold*) and where "col" denotes column space. Prove the following facts:

1. α is one-to-one.
2. The image of α restricted to solutions of (8.59) is exactly the set of n-dimensional H-invariant subspaces of the form col $\begin{pmatrix} I \\ P \end{pmatrix}$ for some P.
3. If Π is a solution of Equation (8.59), then the restriction of H to $\alpha(\Pi)$ has matrix $A + L\Pi$ in the basis given by the columns of $\begin{pmatrix} I \\ \Pi \end{pmatrix}$.
4. The matrix Π is symmetric iff $\alpha(\Pi)$ is a Lagrangian subspace. (A Lagrangian subspace V of R^{2n} is one for which $x'Jy = 0$ for all $x, y \in V$, where J is the matrix in Equation (8.33).)

Conclude that, if H is as in (8.58), the symmetric real solutions of the Riccati equation are in one-to-one correspondence with the Lagrangian H-invariant subspaces of $\mathbb{R}^{2n}$ which have an onto projection into the first n coordinates. □

It follows from this exercise that in principle one can compute solutions of (ARE) by spectral analysis of H. In particular, if H has n eigenvectors $v_1, \ldots, v_n$ with the property that, if we partition the $2n \times n$ matrix $(v_1, \ldots, v_n)$ as

$$\begin{pmatrix} X \\ Y \end{pmatrix},$$

with each of X and Y being an $n \times n$ matrix, then X is nonsingular. It follows that YX^{-1} is a (not necessarily symmetric) solution of the ARE.

In general, there is more than one solution of the ARE, even for positive definite Q. However, as we shall prove below, there is only one positive definite solution Π, which is then necessarily (for $Q > 0$) the one needed for the solution of the optimal control problem.

We proved above that $A+BF$ is Hurwitz when Π is the solution of the ARE obtained as the limit of the $\Pi(t)$ and $F = -R^{-1}B'\Pi$. This property holds in fact for *any* positive definite solution of the ARE, as proved next.

Lemma 8.4.7 Assume that Π is a positive definite solution of the ARE and that Q is also positive definite. Then, defining $F := -R^{-1}B'\Pi$, the closed-loop matrix $A_{cl} = A + BF$ is Hurwitz.

Proof. Manipulation of the ARE shows that

$$A_{cl}'\Pi + \Pi A_{cl} = -Q - \Pi B R^{-1} B' \Pi .$$

Thus, Π is a positive definite solution of a *Lyapunov matrix equation*

$$A_{cl}'\Pi + \Pi A_{cl} = Q_1$$

and Q_1 is negative definite, since it is of the form $-Q - Q_2$ with Q positive definite and Q_2 semidefinite. We proved earlier that *the existence of such a solution implies that A_{cl} must indeed be Hurwitz.* It is always instructive to provide alternative proofs, so we establish this directly as follows.

Let λ be any eigenvalue of A_{cl}, and let v be any corresponding eigenvector. Then,

$$A_{cl}v = \lambda v \,, \quad v^* A_{cl}' = \bar{\lambda} v^* .$$

Multiplying the Lyapunov equation on the left by v^* and on the right by v,

$$\operatorname{Re}\lambda = \frac{1}{2}(\lambda + \bar{\lambda}) = (\frac{1}{2})\frac{v^* Q_1 v}{v^* \Pi v} < 0$$

from which we conclude that A is Hurwitz. ∎

Exercise 8.4.8 (Infinite gain margin of LQ feedback.) As above, suppose that Π is a positive definite solution of the ARE and that Q is also positive definite. Pick any $\rho \in [1/2, \infty)$ and let $F := -\rho R^{-1}B'\Pi$. Show that the closed-loop matrix $A_{cl} = A + BF$ is Hurwitz. □

The result in Lemma 5.7.18 is needed in the next proof. This states that the operator

$$\mathcal{L} : \mathbb{R}^{n\times n} \to \mathbb{R}^{n\times n}, \quad \mathcal{L}(X) := MX + XN.$$

is invertible if both M and N are Hurwitz.

Corollary 8.4.9 If Σ is controllable and R, Q are both positive definite, then there is exactly one positive definite solution Π of the ARE.

Proof. Assume that two such solutions Π_1, Π_2 exist. Equating the two corresponding equations, there results the equality $MX + XN = 0$, where

$$\begin{aligned} M &= (A - BR^{-1}B'\Pi_2)' \\ N &= A - BR^{-1}B'\Pi_1 \\ X &= \Pi_1 - \Pi_2 . \end{aligned}$$

Thus, X is in the kernel of the operator $\mathcal{L}$ introduced above. By Lemma 8.4.7, both M and N are Hurwitz. It follows from Lemma 5.7.18 that $\mathcal{L}$ is one-to-one, so $X = 0$, as wanted. ■

Example 8.4.10 Take the system

$$\dot{x} = 3x + 4u$$

and the problem of minimizing

$$\int_0^\infty x^2 + u^2 \, dt .$$

Then the (ARE) is simply

$$16\pi^2 - 6\pi - 1 = 0 ,$$

and it has the two roots $\pi = 1/2$ and $\pi = -1/8$. The unique positive one is the first, and this results then in the optimal feedback law

$$u = -2x$$

and the closed-loop system

$$\dot{x} = -5x ,$$

which clearly has a Hurwitz matrix. Alternatively, one may form the Hamiltonian matrix

$$\begin{pmatrix} 3 & -16 \\ -1 & -3 \end{pmatrix} ,$$

which has eigenvalues ± 5 and corresponding eigenvectors

$$\begin{pmatrix} -8 \\ 1 \end{pmatrix} \text{ and } \begin{pmatrix} 2 \\ 1 \end{pmatrix},$$

which, when taking the quotient of the second coordinate over the first, give again the two solutions $1/2, -1/8$. □

Exercise 8.4.11 For the same system $\dot{x} = 3x + 4u$ consider instead the cost

$$\int_0^\infty u^2 \, dt \, .$$

Show that here there is also a unique positive definite solution of the Riccati equation, but this is *not* the one giving rise to the optimal feedback law. Explain why none of the results given earlier are contradicted by this example. □

The proof of Lemma 8.4.7 involved the fact that the Riccati equation can be rewritten as a Lyapunov equation for the closed-loop matrix. Another, basically equivalent, relation between the two equations is that Lyapunov equations appear as partial differentials of the Riccati equation with respect to Π. More precisely, if we think of the left-hand side of the (ARE) as defining a map

$$\rho(A, B, Q, R, P) := PBR^{-1}B'P - PA - A'P - Q$$

whose domain is the set of all controllable pairs (A, B) and positive definite symmetric Q, R, P (as an open subset of a suitable Euclidean space), then, for any Π_0 for which

$$\rho(\Pi_0) = 0$$

it follows that

$$\rho(\Pi_0 + P) = -[PA_{cl} + A_{cl}'P] + o(\|P\|) \tag{8.60}$$

for all P. By Corollary 8.4.9 there is, for each (A, B, Q, R) as above, a unique solution of (ARE), that is, there is a function α so that

$$\rho(A, B, Q, R, \alpha(A, B, Q, R)) = 0$$

identically.

Exercise 8.4.12 (*Delchamp's Lemma*) Prove that α is differentiable (in fact, real-analytic) by showing that equation (8.60) holds and applying the Implicit Function Theorem. Conclude that there exists a C^1 (and even analytic) function

$$\beta : S_{n,m}^c \to \mathbb{R}^{n \times m}$$

that assigns to each controllable system (A, B) a feedback matrix $F = \beta(A, B)$ such that $A + BF$ is Hurwitz. □

Finally, for the filtering problem too, when the system is time-invariant and R, Q are also independent of t, it is useful to consider the limit of the solution Π of (FRDE) as $t \to \infty$. The gain matrix $L = -\Pi C'Q$ is often called the *steady-state Kalman gain* and the filter equation (8.52) with this gain (now itself a time-invariant observer) the *steady-state Kalman filter.*

Exercise 8.4.13 Show that, if (A, C) is observable, then $\Pi := \lim_{t\to\infty} \Pi(t)$ exists and that it satisfies the dual algebraic Riccati equation

$$A\Pi + \Pi A' - \Pi C'QC\Pi + BR^{-1}B' = 0\,.$$

(*Hint:* Use duality.) □

8.5 Nonlinear Stabilizing Optimal Controls

Theorem 41 shows that the solution of the steady-state linear-quadratic optimal control problem can be expressed in the feedback form $u = -R^{-1}B'\Pi x$, where Π solves the Algebraic Riccati Equation, and, furthermore, this feedback law stabilizes the system. We discuss briefly a partial generalization, to nonlinear continuous-time time-invariant systems, of these facts. Just as the material in Section 8.4 represents an infinite-time version of the finite-time optimal control problem in Section 8.2, the results discussed here are closely related to those in Section 8.1. However, we derive the results in a direct and self-contained manner.

In this section, we work with a continuous-time time-invariant system

$$\dot{x} = f(x,u)$$

(over $\mathbb{K} = \mathbb{R}$). We assume that $0 \in \mathcal{X}$. For each state $x \in \mathcal{X}$, we let Ω_x denote the class of all "stabilizing" controls, meaning by definition the set of locally essentially bounded

$$\omega : [0,\infty) \to \mathcal{U}$$

with the property that the solution $\xi(t) = \phi(t,x,\omega)$ is defined for all $t \geq 0$ (that is, ω is admissible for x), and

$$\lim_{t\to\infty} \xi(t) = 0\,.$$

The objective is to minimize, over all $\omega \in \Omega_x$, the cost

$$\mathcal{J}_\infty(x,\omega) := \int_0^\infty q(\xi(s),\omega(s))\,ds\,,$$

where (ξ,ω) is the trajectory with $\xi(0) = x$. The function

$$q : \mathcal{X} \times \mathcal{U} \to \mathbb{R}_+$$

is assumed to be continuous.

In all the results to be described, we assume given a continuously differentiable function

$$V : \mathcal{X} \to \mathbb{R}, \quad V(0) = 0.$$

This function will play the role of a value or Bellman function, as well as a Lyapunov function for a closed-loop system. Given such a function V, for each $(x,u) \in \mathcal{X} \times \mathcal{U}$ we denote

$$\dot{V}(x,u) := \nabla V(x) \cdot f(x,u).$$

Proposition 8.5.1 Suppose that

$$\dot{V}(x,u) + q(x,u) \geq 0 \quad \text{for all } (x,u) \in \mathcal{X} \times \mathcal{U} \tag{8.61}$$

and that a path (ξ, ω), with $\xi(0) = x^0$ and $\omega \in \Omega_{x^0}$, is so that

$$\dot{V}(\xi(t), \omega(t)) + q(\xi(t), \omega(t)) = 0 \quad \text{for almost all } t \geq 0. \tag{8.62}$$

Then, $V(x^0)$ is the optimal cost from x^0, and ω is an optimal control, i.e.:

$$V(x^0) = \mathcal{J}_\infty(x^0, \omega) = \min_{\nu \in \Omega_{x^0}} \mathcal{J}_\infty(x^0, \nu). \tag{8.63}$$

Proof. Consider any $\nu \in \Omega_{x^0}$, and let $\zeta(t) := \phi(t, x^0, \nu)$. Since $V(\zeta(t))$ is absolutely continuous, we can write, using (8.61):

$$V(\zeta(t)) - V(x^0) = \int_0^t \frac{d}{ds} V(\zeta(s)) ds = \int_0^t \dot{V}(\zeta(s), \nu(s)) ds \geq -\int_0^t q(\zeta(s), \nu(s)) ds$$

for all $t \geq 0$, from which we conclude that

$$V(x^0) \leq V(\zeta(t)) + \int_0^t q(\zeta(s), \nu(s))\, ds.$$

Taking limits as $t \to \infty$, and using $\zeta(t) \to 0$ and continuity of V at zero, we obtain

$$V(x^0) \leq \int_0^\infty q(\zeta(s), \nu(s))\, ds = \mathcal{J}_\infty(x^0, \nu).$$

The same arguments, for $\nu = \omega$ and using (8.62), give $V(x^0) = \mathcal{J}_\infty(x^0, \omega)$. ∎

From this, we conclude immediately that the (infinite-time) Hamilton-Jacobi equation is sufficient for optimality:

Corollary 8.5.2 Suppose that $k : \mathcal{X} \to \mathcal{U}$ is so that

$$\forall x, u \quad \dot{V}(x, k(x)) + q(x, k(x)) = \min_{u \in \mathcal{U}} \left\{ \dot{V}(x,u) + q(x,u) \right\} = 0 \tag{HJB}$$

and that a path (ξ, ω), with $\xi(0) = x^0$ and $\omega \in \Omega_{x^0}$, is so that $\omega(t) = k(\xi(t))$ for all t. Then, $V(x^0)$ is the optimal cost, and ω is an optimal control, from x^0. □

Theorem 42 *Let $k : \mathcal{X} \to \mathcal{U}$ be locally Lipschitz and so that (HJB) holds. Assume that $\mathcal{X} = \mathbb{R}^n$, V is proper and positive definite, and $q(x,u) > 0$ whenever $x \neq 0$ (for all u). Then, for each state x^0, the solution of $\dot{\xi} = f(\xi, k(\xi))$, $\xi(0) = x^0$ exists for all $t \geq 0$, the control $\omega(t) = k(\xi(t))$ is optimal, and $V(x^0)$ is the optimal value from x^0. Moreover, V is a global Lyapunov function for the closed-loop system.*

Proof. Solutions exist locally, since k is locally Lipschitz. Since $\dot{V}(x, k(x)) = -q(x, k(x)) < 0$ whenever $x \neq 0$, V is a Lyapunov function for the closed-loop system. In particular, a solution exists as stated, and its limit is zero as $t \to \infty$, so $\omega \in \Omega_{x^0}$. The claim now follows from Corollary 8.5.2. ∎

We now specialize to C^1 control-affine systems

$$\dot{x} = f(x) + \sum_{i=1}^{m} u_i g_i(x) = f(x) + G(x)u$$

as in Equation (2.34) (but writing here f instead of g_0)) with $\mathcal{X} = \mathbb{R}^n$, $\mathcal{U} = \mathbb{R}^m$, and $f(0) = 0$. We suppose from now on that V is twice continuously differentiable, in addition to being positive definite and proper, and that

$$q(x,u) = u'R(x)u + Q(x)$$

where Q is a continuous function, positive for all $x \neq 0$, and R is a symmetric $n \times n$ matrix of C^1 functions of x, positive definite for each x. We denote

$$L_G V(x) := \nabla V(x)\, G(x) = (L_{g_1} V(x), \ldots, L_{g_m} V(x))$$

and introduce the following algebraic equation:

$$\forall x \quad Q(x) + L_f V(x) - \frac{1}{4} L_G V(x)\, R(x)^{-1} \left(L_G V(x)\right)' = 0\,. \tag{8.64}$$

Corollary 8.5.3 Assume that (8.64) holds, and define

$$k(x) := -\frac{1}{2} R(x)^{-1} \left(L_G V(x)\right)'\,. \tag{8.65}$$

Then, the conclusions of Theorem 42 hold.

Proof. Just as in Remark 8.1.10, we may complete squares to show that (8.65) minimizes the Hamilton-Jacobi expression, i.e., in this special case:

$$k(x) = \operatorname*{argmin}_u \left\{L_f V(x) + L_G V(x) u + u'R(x)u + Q(x)\right\}\,.$$

Note that k is locally Lipschitz, in fact C^1, because V was assumed to be C^2. We are only left to verify the "$= 0$" part of (HJB), i.e.

$$k(x)'R(x)k(x) + Q(x) + L_f V(x) + L_G V(x) k(x) = 0\,.$$

But this is simply (8.64). ∎

Note that an even more special case is that when $R(x) = I$ for all x. Then

$$k(x) := -\frac{1}{2}\left(L_G V(x)\right)',$$

which happens to be (up to a factor 1/2) the stabilizing feedback used in damping control (cf. Lemma 5.9.1). Equation (8.64) reduces, in this case, to:

$$\forall x \quad Q(x) + L_f V(x) - \frac{1}{4}\sum_{i=1}^{m}\left[L_{g_i} V(x)\right]^2 = 0.$$

Exercise 8.5.4 Show that for linear systems and quadratic V, Equation (8.64) reduces to the Algebraic Riccati Equation (ARE) given in Lemma 8.4.1. □

Exercise 8.5.5 Suppose that we drop the requirement that Q be positive definite, asking merely semidefiniteness, but we add, instead, the assumption that k, defined by formula (8.65), globally stabilizes the system. Show that we can still conclude that V is the value function and that the feedback $u = k(x)$ gives the optimal controls. □

Exercise 8.5.6 Consider the following system (with $n = m = 1$): $\dot{x} = xu$, and take the problem of minimizing

$$\int_0^\infty \tfrac{1}{8}x^4 + \tfrac{1}{2}u^2\, dt$$

among all controls making $x(t) \to 0$ as $t \to \infty$. Show that there is a solution, given in feedback form by a quadratic feedback $k(x) = -cx^2$, for some $c > 0$. (*Hint:* Use as a guess a quadratic $V(x) = px^2$. Find the appropriate p by solving the HJB equation.) □

Exercise 8.5.7 Generalize Exercise (8.5.6) to $\dot{x} = uAx$, a system of dimension n and scalar inputs, with A having all eigenvalues with positive real part, and the cost $q(x,u) = \frac{1}{8}(x'Qx)^2 + \frac{1}{2}u^2$, where Q is a positive definite constant symmetric matrix. Show that also here one obtains a quadratic feedback solution. □

Exercise 8.5.8 (Infinite gain margin of LQ feedback; see also Exercise 8.4.8.) Suppose that V is a $\mathcal{C}^2$ function which satisfies (8.64). Pick any $\rho \in [1/2, \infty)$ and let $k(x) := -\rho R(x)^{-1}\left(L_G V(x)\right)'$. Show that $\dot{x} = f(x) + G(x)k(x)$ is globally asymptotically stable with respect to the origin. □

Exercise 8.5.9 Provide an example of a smooth system $\dot{x} = f(x,u)$ and a smooth cost $q(x,u)$ so that (1) for each x^0 there is some $\omega \in \Omega_{x^0}$ so that $\mathcal{J}_\infty(x^0,\omega) < \infty$, but (2) the conclusions of Theorem 42 are false. In fact, show that *any* no-drift system $\dot{x} = G(x)u$ with $\operatorname{rank} G(0) = m < n$ which satisfies the Lie Algebra Accessibility Rank Condition, together with any q, gives rise to such a counterexample. (*Hint:* Use Exercise 5.9.20, noting that k given by (8.65) must be smooth.) □

Exercise 8.5.10 Find the value function V and the optimal feedback solution for the problem of minimizing $\int_0^\infty x^2 + u^2 dt$ for the scalar system $\dot{x} = x^2 + u$. (*Hint:* The HJB equation is now an ordinary differential equation, in fact, a quadratic equation on the derivative of V.) □

Exercise 8.5.11 Suppose that $L_f V(x) \leq 0$ for all x and that $\dot{x} = f(x) + G(x)u$ is globally stabilized by $u = -(\nabla V(x) \cdot G(x))'$, as in Proposition 5.9.1. Show that $u = k(x)$ is an optimal feedback, and V is the value function, for some suitably chosen cost. (*Hint:* Let $Q(x) := -L_f V(x) + \frac{1}{2} L_G V(x) (L_G V(x))'$, which gives (8.64) for which R? Use Exercise 8.5.5.) □

8.6 Notes and Comments

Dynamic Programming

The dynamic programming approach gained wide popularity after the work of Bellman; see, for instance, [41]. Almost every optimization or optimal control textbook contains numerous references to papers on the topic.

The Continuous-Time Case

A central problem in optimal control is the study of existence and smoothness properties. The question of existence often can be settled easily on the basis of general theorems; what is needed is the continuous dependence of $\mathcal{J}$ on ω, with respect to a topology on the space of controls which insures compactness. For bounded inputs and systems linear on controls, Theorem 1 (p. 57), part 2(ii), together with compactness in the weak topology, achieves this purpose. For more general systems, analogous results hold provided that one generalizes the notion of control to include "relaxed controls." See, for instance, [134], [188], [266], or any other optimal control textbook; here we only proved results for a very special class of systems and cost functions, and for these existence can be established in an ad hoc manner.

Continuity of controls, and smoothness of Bellman functions, are a much more delicate matter. It turns out that many problems of interest result in a nonsmooth V. The *regular synthesis* problem in optimal control theory studies generalizations of the results that assumed smoothness to cases where V is *piecewise smooth* in appropriate manners. See, for instance, [134] for more on this topic, as well as [50] and [386]. An alternative is to generalize the concept of solution of a partial differential equation: The notion of *viscosity solution* provides such a possibility, cf. [102]; or the related approach based on *proximal subgradients* can be used, cf. [93].

(Deterministic) Kalman Filtering

A purely deterministic treatment was pursued here so as not to require a large amount of preliminary material on stochastic processes. On the other hand, a probabilistic study is more satisfactory, as the cost matrices have a natural interpretation and formulas can be obtained that quantify the accuracy of the estimator (its covariance is given in fact by the solution to the Riccati equation).

There are excellent textbooks covering Kalman filtering, such as [108] or, for more of an engineering flavor, [16] and [263]. A good reference for practical implementation issues, as well as a detailed derivation in the discrete-time case, is provided by [153]. A very elegant discussion of the discrete-time problem is given by [86], from the point of view of recursive least-squares.

Much has been done regarding nonlinear filtering problems; for some of the theoretical issues involved see, for instance, [209], [293], [312], and the references therein.

Historically, the Kalman filter, introduced for discrete-time in [214] and for continuous-time in [229], appeared in the context of the general problem of "filtering" a signal corrupted by noise. As compared to older Wiener filtering ideas, the Kalman filter modeled the signal of interest as the state trajectory of a linear system driven by white noise. This allowed a highly computable method of solution as well as extensions to nonstationary processes (time-varying state models) and nonlinear systems.

The optimal estimation and optimal control problems can be combined into the stochastic optimal control problem of minimizing a cost criterion for a system such as the one in Exercise 8.3.9, by suitable choice of controls $\overline{u}$, on the basis of noisy observations; the solution to this "LQG" problem —"linear quadratic Gaussian problem," when all noises are assumed to be Gaussian processes— can be obtained from the solutions of the two separate deterministic linear quadratic and linear filtering problems, in much the same fashion as output stabilizers were obtained by combining observers and optimal state feedback laws; see [16] and [263], for instance.

Linear Systems with Quadratic Cost

The treatment of the Linear Quadratic Problem started with Kalman's seminal paper [215]. The literature generated by this problem is immense, as many variants and numerical approaches have been tried. Two excellent texts that deal with such issues are [16] and [263]. The paper [423] discusses many important issues about Riccati equations, including the indefinite case, of interest in game theory as well as in modern H_∞ optimization.

The infinite-dimensional linear case also has been studied in detail; see, for instance, [113] and [133].

Infinite-Time Problems

The computation of solutions to the ARE via Hamiltonian matrices was first suggested by [288] and [322] for the case of distinct eigenvalues; see [212], Section 3.4, for a discussion and many references to numerical techniques associated to this approach.

The result in Exercise 8.4.12 is from [112] (see also [64]).

It is interesting but not very surprising that quadratic problems for linear time-invariant systems give rise to linear solutions (a linear feedback law). When other criteria are used, or when "robust" design is desired, nonlinear or time-varying controllers may be superior, even for linear time-invariant systems. This type of question is explored in [244], [245], and related papers.

Substantial activity has taken place recently on the topic of H_∞ *optimization*, dealing with a different optimization criterion than the linear-quadratic problem. This criterion, which corresponds to the minimization of an operator norm from external "disturbances" to outputs, is of great engineering interest. A book reference is [145], and recent research, resulting in a reduction to a Riccati equation problem, is surveyed in [124]. That reference also shows how the linear-quadratic problem can be viewed as a minimization problem in an input/output context. The norm itself can be computed numerically; see, for instance, [59].

Nonlinear Stabilizing Optimal Controls

The main sense in which the nonlinear results given in Section 8.5 differ from the linear case is in the need to assume that V exists. For linear systems, the ARE always has a solution (under the appropriate controllability conditions). In order to obtain necessary and sufficient results, one must introduce generalized solutions of various types (viscosity, proximal); see for instance [93] and the many references provided therein.

Chapter 9

Optimality: Multipliers

As described in the introduction to Chapter 8, an alternative approach to optimal control relies upon Lagrange multipliers in order to link static optimization problems. In this chapter, we provide a brief introduction to several selected topics in variational, or multiplier-based, optimal control, namely: minimization of Lagrangians (and the associated Hamiltonian formalism) for open input-value sets, the basic result in the classical Calculus of Variations seen as a special case, some remarks on numerical techniques, and the Pontryagin Minimum (or Maximum, depending on conventions) Principle for arbitrary control-value sets but free final state. The area of nonlinear optimal control is very broad, and technically subtle, and, for a more in-depth study, the reader should consult the extensive literature that exists on the subject.

9.1 Review of Smooth Dependence

It is worthwhile at this point to review some of the main conclusions of Theorem 1 (p. 57). Consider the system

$$\dot{x} = f(x,u),$$

under the standard assumption that $f : \mathcal{X} \times \mathcal{U} \to \mathcal{X}$ is C^1 in x, and that f, as well as its Jacobian matrix of partial derivatives f_x with respect to x, are continuous on x, u (cf. Equations (2.22 and (2.23)). The state-space is an open subset $\mathcal{X}$ of $\mathbb{R}^n$, and $\mathcal{U}$ is a metric space. These conclusions characterize the partial derivatives of the final value of the solution, that is $x(\tau)$, with respect to the initial state $x(\sigma)$ and the input $u(\cdot)$ applied during an interval $[\sigma, \tau]$. Here we restate the relevant facts in terms of Jacobians with respect to $x(\sigma)$ and directional derivatives with respect to $u(\cdot)$; Theorem 1 (p. 57) provides a more general statement concerning joint continuous differentiability.

Recall that $\mathcal{L}_{\mathcal{U}}(\sigma, \tau)$ (or $\mathcal{L}^{\infty}_{m}(\sigma, \tau)$, in the special case $\mathcal{U} = \mathbb{R}^m$) denotes the set of measurable essentially bounded maps $\omega : [\sigma, \tau] \to \mathcal{U}$, where "essentially bounded" means that there is some compact subset $C = C_\omega \subseteq \mathcal{U}$ such that $\omega(t) \in C$ for almost all $t \in [\sigma, \tau]$, and we say that an input $\omega \in \mathcal{L}_\mathcal{U}(\sigma, \tau)$ is *admissible for the state* x^0 if the solution of the initial value problem

$$\dot{\xi}(t) = f(\xi(t), \omega(t)), \quad \xi(\sigma) = x^0$$

is well-defined on the entire interval $[\sigma, \tau]$, and we denote, in that case,

$$\phi(\cdot, \sigma, x^0, \omega) = \xi(\cdot).$$

For future reference, we now summarize the main needed facts.

Corollary 9.1.1 (of Theorem 1 (p. 57)). Let $\widetilde{\omega} \in \mathcal{L}_\mathcal{U}(\sigma, \tau)$ be admissible for the state x^0, and write $\widetilde{x}(t) = \phi(t, \sigma, x^0, \widetilde{\omega})$. Denote

$$A(t) = f_x(\widetilde{x}(t), \widetilde{\omega}(t)) \tag{9.1}$$

for each $t \in [\sigma, \tau]$, and let $\Phi(\cdot, \cdot)$ be the fundamental solution matrix associated to $A(\cdot)$:

$$\frac{\partial \Phi(t, r)}{\partial t} = A(t)\Phi(t, r), \quad \Phi(r, r) = I \tag{9.2}$$

for each $t, r \in [\sigma, \tau]$.

Consider the map

$$\Theta : x \mapsto \phi(\tau, \sigma, x, \widetilde{\omega}). \tag{9.3}$$

Then, Θ is well-defined and continuously differentiable in a neighborhood of x^0, and its Jacobian evaluated at x^0 is

$$\Theta_*(x^0) = \Phi(\tau, \sigma). \tag{9.4}$$

Suppose that, in addition, the control-value set $\mathcal{U}$ is an open subset of $\mathbb{R}^m$ and that $f : \mathcal{X} \times \mathcal{U} \to \mathbb{R}^n$ is continuously differentiable, and let

$$B(t) = f_u(\widetilde{x}(t), \widetilde{\omega}(t)). \tag{9.5}$$

Take any fixed input $\mu \in \mathcal{L}_\mathcal{U}(\sigma, \tau)$, and consider the map:

$$\beta : (-h_0, h_0) \to \mathbb{R}^n : h \mapsto \phi(\tau, \sigma, x^0, \widetilde{\omega} + h\mu). \tag{9.6}$$

This map is well-defined for some $h_0 > 0$, is continuously differentiable, and

$$\frac{d\beta}{dh}(0) = \int_\sigma^\tau \Phi(\tau, s)\, B(s)\, \mu(s)\, ds. \tag{9.7}$$

More generally, if $\mu_1, \ldots, \mu_r \in \mathcal{L}_\mathcal{U}(\sigma, \tau)$, the r-parameter map

$$\beta(h_1, \ldots, h_r) := \phi(\tau, \sigma, x^0, \widetilde{\omega} + h_1\mu_1 + \ldots + h_r\mu_r).$$

has Jacobian

$$\beta_*(0) = \left(\int_\sigma^\tau \Phi(\tau, s) B(s) \mu_1(s)\, ds, \ldots, \int_\sigma^\tau \Phi(\tau, s) B(s) \mu_r(s)\, ds \right). \qquad \Box$$

9.2 Unconstrained Controls

When control-value sets are open, various facts about optimal control can be easily derived by finding critical points of appropriate functions, with the use, if necessary, of Lagrange multipliers. In this section, we pursue this line of thought. So we take here $\mathcal{U} \subseteq \mathbb{R}^m$ *open and* $f : \mathcal{X} \times \mathcal{U} \to \mathbb{R}^n$ *continuously differentiable.*

Assume given three C^1 functions

$$q : \mathcal{X} \times \mathcal{U} \to \mathbb{R}\,, \quad p : \mathcal{X} \to \mathbb{R}\,, \quad K : \mathcal{X} \to \mathbb{R}^r\,,$$

to be used, respectively, to denote the instantaneous cost along trajectories, a cost on final states, and a set of r constraints on final states. Also fix an initial state $x^0 \in \mathcal{X}$ and a time interval $[\sigma, \tau]$ on which the optimization is to be carried out. The cost function associated to this problem is as follows:

$$\boxed{\mathcal{J}(\omega) \;:=\; \int_\sigma^\tau q(x_\omega(s), \omega(s))\,ds \;+\; p(x_\omega(\tau))}\,,$$

where $x_\omega(\cdot) := \phi(\cdot, \sigma, x^0, \omega)$ and we define $\mathcal{J}(\omega) := +\infty$ if ω is not admissible (for the initial state x^0). The optimization problem of that of finding

$$\boxed{\min\left\{\mathcal{J}(\omega) \mid \omega \in \mathcal{L}_\mathcal{U}(\sigma, \tau) \text{ admissible, } K(x_\omega(\tau)) = 0\right\}}\,.$$

For example, a quadratic cost along trajectories, as in the classical linear-quadratic problem, would be one of the form $q(x, u) = x'Qx + u'Ru$. A problem in which we want to impose that the final state be a desired target x^{f} is included by letting $K(x) = x - x^{\mathrm{f}}$; on the other hand, $K \equiv 0$ covers the problem of minimizing with no final state constraints. Other special cases of interest, to be discussed separately, are those in which there is no final cost (i.e, $p \equiv 0$) or there is no "Lagrangian" ($q \equiv 0$).

Definition 9.2.1 *The control* $\widetilde{\omega}$ **is optimal for** x^0 *if it is admissible for* x^0 *and has the property that* $\mathcal{J}(\widetilde{\omega}) \le \mathcal{J}(\omega)$ *for all those admissible* ω *for which* $K(x_\omega(\tau)) = 0$. □

Since the domain of $\mathcal{J}$ is open, when $K \equiv 0$ first-order necessary conditions for optimality are obtained by asking that any $\widetilde{\omega}$ minimizing $\mathcal{J}$ must be a critical point of $\mathcal{J}$, i.e. that we have a zero differential at that input: $\mathcal{J}_*[\widetilde{\omega}] = 0$. In the case in which constraints are present, the Lagrange multiplier rule can be employed. This program can be carried out in elementary terms, with no recourse to the Lagrange multiplier rule in infinite-dimensional spaces; we do so next.

Recall the classical Lagrange multiplier rule, applied to a differentiable function $f : \mathcal{O} \to \mathbb{R}$, where $\mathcal{O}$ is an an open subset of Euclidean space: if f achieves a local minimum (or maximum) on $\mathcal{O}$ subject to the differentiable constraints $g_1(y) = \ldots = g_r(y) = 0$ at the point $\widetilde{y}$, and if the Jacobian $G_*(\widetilde{y})$ of $G = (g_1, \ldots, g_r)'$ at $\widetilde{y}$ has rank r, then there exists some r-vector ν so that the function $f + \nu' G$ has a critical point at $\widetilde{y}$. If, instead, the Jacobian has rank less than r, then the rows of $G_*(\widetilde{y})$ are linearly dependent, and therefore there is some nonzero r-vector ν so that $\nu' G_*(\widetilde{y}) = 0$, which means that $0.f + \nu' G$ has a critical point at $\widetilde{y}$. One can summarize both cases in one statement, namely, that there are a scalar $\nu_0 \in \{0, 1\}$ and a vector $\nu \in \mathbb{R}^n$, *not both zero*, such that the function $\nu_0 f + \nu' G$ has a critical point at $\widetilde{y}$. (Example: The minimum at $y = 0$ of $f(y) = y$ under the constraint $y^2 = 0$; here one must take $\nu_0 = 0$.) We next introduce, for our control problem, a condition corresponding to the Jacobian $G_*(\widetilde{y})$ having rank r.

Let $\widetilde{\omega} \in \mathcal{L}_m^\infty(\sigma, \tau)$ be admissible for x^0, write $\widetilde{x}(t) = x_{\widetilde{\omega}}(t) = \phi(t, \sigma, x^0, \widetilde{\omega})$, let $A(\cdot)$ and $B(\cdot)$ be as in Equations (9.1) and (9.5), and let Φ be as in Equation (9.2).

With respect to the given function $K : \mathcal{X} \to \mathbb{R}^r$, we will say that the *linearization along $\widetilde{\omega}$ is output controllable* if

$$\left\{ K_*(\widetilde{x}(\tau)) \int_\sigma^\tau \Phi(\tau, s)\, B(s)\, \mu(s)\, ds \;\middle|\; \mu \in \mathcal{L}_m^\infty(\sigma, \tau) \right\} = \mathbb{R}^r .$$

Equivalently, there must exist r inputs $\mu_1, \ldots, \mu_r$ so that the r vectors

$$\left\{ K_*(\widetilde{x}(\tau)) \int_\sigma^\tau \Phi(\tau, s)\, B(s)\, \mu_1(s)\, ds\,, \;\ldots\;, K_*(\widetilde{x}(\tau)) \int_\sigma^\tau \Phi(\tau, s)\, B(s)\, \mu_r(s)\, ds \right\} \tag{9.8}$$

form a basis in $\mathbb{R}^r$. Observe that, in particular, $K_*(\widetilde{x}(\tau)) \in \mathbb{R}^{r \times n}$ must have rank r, and the vectors in $\mathbb{R}^n$

$$\left\{ \int_\sigma^\tau \Phi(\tau, s)\, B(s)\, \mu_1(s)\, ds\,, \;\ldots\;, \int_\sigma^\tau \Phi(\tau, s)\, B(s)\, \mu_r(s)\, ds \right\} \tag{9.9}$$

must be linearly independent. *If there are no final-state constraints ($r = 0$, $K \equiv 0$), linearized output controllability holds by definition.* (The condition will only play a role when using Lagrange multipliers.)

Remark 9.2.2 One may interpret linearized output controllability in terms of the time-varying linear system that has the above $A(\cdot)$ and $B(\cdot)$ in its state equations, as well as an output function given by the evaluation of the Jacobian of K along the given trajectory: $C(t) := K_*(\widetilde{x}(t))$. The condition says that every possible output value is achievable at time τ. □

We now derive a necessary condition for an extremum. To make the proof more transparent, it is convenient to take first the special case when $q \equiv 0$, and later to extend the result to the general case.

Lemma 9.2.3 Suppose $q \equiv 0$. Assume that $\widetilde{\omega} \in \mathcal{L}_\mathcal{U}(\sigma, \tau)$ is optimal for x^0. Then, there exist a scalar $\nu_0 \in \{0, 1\}$ and a vector $\nu \in \mathbb{R}^r$, not both zero, so that the solution $\lambda : [\sigma,\tau] \to \mathbb{R}^n$ of the adjoint equation $\dot{\lambda}(t) = -A(t)'\lambda(t)$ with final value

$$\lambda(\tau) \;=\; \left(\nu_0\, p_*(\widetilde{x}(\tau)) + \nu' K_*(\widetilde{x}(\tau))\right)'$$

satisfies

$$\lambda(t)'\, B(t) \;=\; 0$$

for almost all $t \in [\sigma,\tau]$. If, in addition, the linearization along $\widetilde{\omega}$ is output controllable, then one may take $\nu_0 = 1$.

Proof. We first treat the case of no final-state constraints ($r = 0$, no "K", and output controllability is by definition satisfied). Take any fixed $\mu \in \mathcal{L}_{\mathcal{U}}(\sigma,\tau)$, and consider the map $\beta(h) := \phi(\tau,\sigma,x^0,\widetilde{\omega} + h\mu)$, as in Equation (9.6). As

$$p(\beta(0)) \;=\; p(\widetilde{x}(\tau)) \;=\; \mathcal{J}(\widetilde{\omega}) \;\leq\; \mathcal{J}(\widetilde{\omega} + h\mu) \;=\; p(\beta(h))$$

for all h near zero, and $p \circ \beta$ is differentiable, it follows that

$$0 \;=\; \frac{d(p\circ\beta)}{dh}(0) \;=\; p_*(\widetilde{x}(\tau))\frac{d\beta}{dh}(0) \;=\; \int_\sigma^\tau p_*(\widetilde{x}(\tau))\, \Phi(\tau,s)\, B(s)\, \mu(s)\, ds\,.$$

Since μ was arbitrary, this means that

$$p_*(\widetilde{x}(\tau))\, \Phi(\tau,t)\, B(t) \;=\; 0$$

for almost all t. Since

$$\frac{\partial\Phi(\tau,t)}{\partial t} \;=\; -\Phi(\tau,t)A(t)\,, \tag{9.10}$$

$\lambda(t) = (p_*(\widetilde{x}(\tau))\Phi(\tau,t))'$ solves $\dot{\lambda}(t) = -A(t)'\lambda(t)$ and has $\lambda(\tau) = p_*(\widetilde{x}(\tau))'$.

Now we treat the case when there are constraints ($r > 0$). There is some integer $1 \leq \ell \leq n$ and inputs $\mu_1,\ldots,\mu_\ell$ so that the following property holds: for every $\mu \in \mathcal{L}_{\mathcal{U}}(\sigma,\tau)$, the vector $\int_\sigma^\tau \Phi(\tau,s)B(s)\mu(s)\,ds$ is a linear combination of

$$\left\{\int_\sigma^\tau \Phi(\tau,s)\, B(s)\, \mu_1(s)\, ds,\; \ldots,\; \int_\sigma^\tau \Phi(\tau,s)\, B(s)\, \mu_\ell(s)\, ds\right\}. \tag{9.11}$$

To see this, we consider the subspace

$$\left\{\int_\sigma^\tau \Phi(\tau,s)B(s)\mu(s)\, ds \;\middle|\; \mu \in \mathcal{L}_{\mathcal{U}}(\sigma,\tau)\right\} \subseteq \mathbb{R}^n$$

(which is the reachable set from the origin for the linearized system along $(\widetilde{x},\widetilde{\omega})$): if this space is $\{0\}$, we pick $\ell := 1$ and take any input as μ_1; otherwise, we let ℓ be the dimension of the space and pick a set of inputs so that the states in (9.11) form a basis.

Consider now the map

$$\beta(h_1,\ldots,h_\ell) \;:=\; \phi(\tau,\sigma,x^0,\widetilde{\omega} + h_1\mu_1 + \ldots + h_\ell\mu_\ell)\,,$$

which is defined on some open neighborhood $\mathcal{O}$ of 0 in $\mathbb{R}^\ell$. We consider the compositions $p \circ \beta$ and $K \circ \beta$ as maps $\mathcal{O} \to \mathbb{R}$. Using h to denote vectors $(h_1, \ldots, h_\ell)'$, we have that

$$(p \circ \beta)(0) = p(\widetilde{x}(\tau)) = \mathcal{J}(\widetilde{\omega}) \leq \mathcal{J}(\widetilde{\omega} + h_1\mu_1 + \ldots + h_\ell\mu_\ell) = (p \circ \beta)(h)$$

for all those $h \in \mathcal{O}$ for which $(K \circ \beta)(h) = 0$. Furthermore, the Jacobian of $K \circ \beta$ at $h = 0$ is $K_*(\widetilde{x}(\tau))\beta_*(0)$, and a formula for the Jacobian $\beta_*(0)$ is given, in turn, in the last part of the statement of Corollary 9.1.1. It follows that the columns of the $r \times \ell$ Jacobian matrix $(K \circ \beta)_*(0)$ are given by the vectors

$$\left\{K_*(\widetilde{x}(\tau)) \int_\sigma^\tau \Phi(\tau, s)\, B(s)\, \mu_1(s)\, ds\,,\ \ldots\,,\ K_*(\widetilde{x}(\tau)) \int_\sigma^\tau \Phi(\tau, s)\, B(s)\, \mu_\ell(s)\, ds\right\}. \tag{9.12}$$

The Lagrange multiplier rule (in the form discussed earlier) provides a necessary condition for a local extremum of $p \circ \beta$ at $h = 0$ subject to the constraint $(K \circ \beta)(h) = 0$: There exist a scalar $\nu_0 \in \{0, 1\}$ and a vector $\nu \in \mathbb{R}^r$, not simultaneously zero, such that

$$\frac{\partial\,(\nu_0\, p \circ \beta\, +\, \nu' K \circ \beta)}{\partial h_i}(0) = 0\,, \quad i = 1, \ldots, \ell\,.$$

That is,

$$0 = (\nu_0\, p + \nu' K)_*(\widetilde{x}(\tau)) \int_\sigma^\tau \Phi(\tau, s)\, B(s)\, \mu_i(s)\, ds\,, \quad i = 1, \ldots, \ell\,.$$

Pick any $\mu \in \mathcal{L}_\mathcal{U}(\sigma, \tau)$. Since, for each μ, $\int_\sigma^\tau \Phi(\tau, s)B(s)\mu(s)\, ds$ can be expressed as a linear combination of the vectors $\int_\sigma^\tau \Phi(\tau, s)B(s)\mu_i(s)\, ds$, we conclude that $\int_\sigma^\tau (\nu_0 p + \nu' K)_*(\widetilde{x}(\tau))\Phi(\tau, s)B(s)\mu(s)\, ds = 0$ for all $u \in \mathcal{L}_\mathcal{U}(\sigma, \tau)$. So

$$(\nu_0\, p + \nu' K)_*\, (\widetilde{x}(\tau))\, \Phi(\tau, t)\, B(t) = 0$$

for almost all t. As with the case $r = 0$, Equation (9.10) says that $\lambda(t) = ((\nu_0 p + \nu' K)_*(\widetilde{x}(\tau))\Phi(\tau, t))'$ solves $\dot{\lambda}(t) = -A(t)'\lambda(t)$ and has the desired final value.

Finally, assume that $r > 0$ and that the output controllability condition holds. Then, there exist inputs $\mu_1, \ldots, \mu_r$ so that the vectors in Equation (9.8) span $\mathbb{R}^r$. Thus, also the vectors in Equation (9.9) are linearly independent, so we can add if necessary inputs $\mu_{r+1}, \ldots, \mu_\ell$ (for some integer $r \leq \ell \leq n$) so that $\mu_1, \ldots, \mu_\ell$ form a generating set as above, and we use these inputs when defining β. Since the first r columns of the $r \times \ell$ matrix $(K \circ \beta)_*(0)$ are given by the vectors in Equation (9.8), $(K \circ \beta)_*(0)$ has full rank r. Then, the Lagrange rule applies in the form usually stated, for nonsingular constraints, and one may indeed pick $\nu_0 = 1$. ∎

Remark 9.2.4 One interpretation of the Lemma is as follows. Let $\mathcal{R}$ denote the set of states reachable from the origin in the interval $[\sigma, \tau]$, for the linear system defined by $A(t), B(t)$ (the linearization along $(\widetilde{\omega}, \widetilde{x})$); that is, $\mathcal{R} = \{\int_\sigma^\tau \Phi(\tau, s)B(s)\mu(s)\,ds, \mu \in \mathcal{L}_m^\infty(\sigma, \tau)\}$. Then, the vector $(\nu_0 p + \nu' K)_*(\widetilde{x}(\tau))$ must be orthogonal to the subspace $\mathcal{R} \subseteq \mathbb{R}^n$. $\square$

Now we treat the general case, when there is a nonzero Lagrangian term q in the cost function. At this point, it is useful to introduce the *Hamiltonian* function associated to the optimal control problem, defined by

$$\boxed{\mathcal{H}(x, u, \eta_0, \eta) := \eta_0 q(x, u) + \eta' f(x, u)}$$

and seen as a function of $\mathcal{X} \times \mathcal{U} \times \mathbb{R} \times \mathbb{R}^n \to \mathbb{R}$.

Theorem 43 *Assume that $\widetilde{\omega} \in \mathcal{L}_\mathcal{U}(\sigma, \tau)$ is optimal for x^0. Then, there exist a scalar $\nu_0 \in \{0, 1\}$ and a vector $\nu \in \mathbb{R}^r$,*

$$(\nu_0, \nu) \neq (0, 0),$$

so that the solution $\lambda : [\sigma, \tau] \to \mathbb{R}^n$ of the final-value problem

$$\dot{\lambda}(t) = -\nu_0 \, q_x(\widetilde{x}(t), \widetilde{\omega}(t))' - A(t)'\lambda(t), \quad \lambda(\tau) = \left(\nu_0 \, p_*(\widetilde{x}(\tau)) + \nu' K_*(\widetilde{x}(\tau))\right)' \tag{9.13}$$

is so that

$$\frac{\partial \mathcal{H}}{\partial u}(\widetilde{x}(t), \widetilde{\omega}(t), \nu_0, \lambda(t)) = 0 \tag{9.14}$$

for almost all $t \in [\sigma, \tau]$. If, in addition, the linearization along $\widetilde{\omega}$ is output controllable, then the same conclusion holds with $\nu_0 = 1$.

Proof. We reduce this problem to the one treated in Lemma 9.2.3. For that purpose, we introduce the system $\dot{x}^\# = f^\#(x^\#, u)$ with state-space $\mathbb{R} \times \mathcal{X} \subseteq \mathbb{R}^{n+1}$ which, when writing $x^\# = \begin{pmatrix} x_0 \\ x \end{pmatrix}$, has equations

$$\begin{aligned} \dot{x}_0 &= q(x, u) \\ \dot{x} &= f(x, u) \end{aligned}$$

(with same input value space $\mathcal{U}$). We also think of K as a function on $\mathbb{R} \times \mathcal{X}$ that depends only on the last n coordinates, that is, we let

$$K^\#(x^\#) := K(x).$$

It is clear that the output controllability condition holds for the extended system if it holds for the original one.

To define a minimization problem for the extended system, we let

$$p^\# \begin{pmatrix} x_0 \\ x \end{pmatrix} := x_0 + p(x) \quad \text{and} \quad q^\# \equiv 0,$$

and consider the initial state

$$x^{0\#} := \begin{pmatrix} 0 \\ x^0 \end{pmatrix}.$$

We write $\mathcal{J}^\#$ for the associated cost function. Observe that an input ω is admissible for x^0 if and only if it is admissible for $x^{0\#}$ because, once that the equation for x has been solved, $x_0(\tau)$ can be obtained by integration: $x_0(\tau) = \int_\sigma^\tau q(x(s), \omega(s))\, ds$. Moreover, the definitions imply that

$$\mathcal{J}^\#(\omega) = \mathcal{J}(\omega)$$

for all ω. Therefore, $\widetilde{\omega}$ is a minimizing input for $\mathcal{J}^\#$, and we can thus apply Lemma 9.2.3 to it.

Let $\nu_0 \in \mathbb{R}$, $\nu \in \mathbb{R}^r$, and $\lambda^\# : [\sigma, \tau] \to \mathbb{R}^{n+1}$ be as in the conclusions of the Lemma, for the extended system. Observe that

$$A^\#(t)' = \begin{pmatrix} 0 & 0 \\ q_x(\widetilde{x}(t), \widetilde{\omega}(t))' & A(t)' \end{pmatrix} \quad \text{and} \quad B^\#(t) = \begin{pmatrix} q_u(\widetilde{x}(t), \widetilde{\omega}(t)) \\ B(t) \end{pmatrix}.$$

As $p_*^\# = (1\,, p_*)$ and $K_*^\# = (0\,, K_*)$, the final condition on $\lambda^\#$ is:

$$\lambda^\#(\tau) = (\nu_0 p_*^\#(\widetilde{x}(\tau)) + \nu' K_*^\#(\widetilde{x}(\tau)))' = \begin{pmatrix} \nu_0 \\ (\nu_0 p_*(\widetilde{x}(\tau)) + \nu' K_*(\widetilde{x}(\tau)))' \end{pmatrix}.$$

We partition

$$\lambda^\#(t) = \begin{pmatrix} \lambda_0(t) \\ \lambda(t) \end{pmatrix}.$$

Comparing the last n coordinates provides the final value $\lambda(\tau)$ shown in (9.13). Moreover, the differential equation $\dot{\lambda}^\#(t) = -A^\#(t)\lambda^\#(t)$ gives that $\lambda_0 \equiv \nu_0$ and that λ satisfies the differential equation in (9.13). Finally, $\lambda^\#(t)'B^\#(t) = 0$ implies that

$$0 = \lambda^\#(t)'B^\#(t) = \nu_0 q_u(\widetilde{x}(t), \widetilde{\omega}(t)) + \lambda(t)'B(t) = \mathcal{H}_u(\widetilde{x}(t), \widetilde{\omega}(t), \nu_0, \lambda(t))$$

for almost all t. ∎

Theorem 43 provides a necessary condition for arbitrary local extrema of $\mathcal{J}$. As in elementary calculus, it cannot distinguish between (local) minima and maxima of $\mathcal{J}$; higher-order derivatives may be used for that purpose (see also the maximum principle, below).

Exercise 9.2.5 Show that, along any optimal trajectory for which $\widetilde{\omega}$ is differentiable, the function $t \mapsto \mathcal{H}(\widetilde{x}(t), \widetilde{\omega}(t), \nu_0, \lambda(t))$ must be constant. (The constancy of $\mathcal{H}(\widetilde{x}(t), \widetilde{\omega}(t), \nu_0, \lambda(t))$ holds, in fact, even if $\widetilde{\omega}$ is not necessarily differentiable, but the proof is a bit harder.) □

Exercise 9.2.6 Provide a version of the Theorem that applies to time-varying systems (and time-dependent costs), that is, $\dot{x} = f(t,x,u)$, and $q = q(t,x,u)$. You may assume that f and q are continuously differentiable. (*Hint:* Easy from the Theorem; you only need to extend the system in a certain way, and then "unpack" the resulting equations.) □

Exercise 9.2.7 A problem for which $q \equiv 0$ is often called a *Meyer problem* of optimal control (this follows similar terminology used in the calculus of variations). A *Lagrange problem*, in contrast, is one for which $p \equiv 0$. (The general problem which we considered, incidentally, is sometimes called a "Bolza" problem.) The main idea which was used in deriving Theorem 43 from Lemma 9.2.3 was the observation that every Lagrange problem can be recast as a Meyer problem. Show that, conversely, every Meyer problem can be reformulated as a Lagrange problem, in such a manner that optimal controls are the same, for every possible initial state. (You may assume that $p \in \mathcal{C}^2$.) □

Exercise 9.2.8 An optimal control $\widetilde{\omega}$ is *abnormal* if it is impossible to obtain the conclusions of Theorem 43 with $\nu_0 = 1$. That is, for each $\lambda(\cdot)$ satisfying (9.13) and (9.14), necessarily $\nu_0 = 0$. Show that $\dot{x} = u^2$, $x(0) = x(1) = 0$, $\mathcal{U} = [-1,1]$, $q(x,u) = u$, $p = 0$, leads to an abnormal optimal control. □

One particular case of interest for the Theorem is as follows. The notations are as earlier.

Corollary 9.2.9 Assume that $\widetilde{\omega}$ is optimal for x^0, that the linearization along $\widetilde{\omega}$ is controllable, and that $r = n$ and the map $K : \mathcal{X} \to \mathbb{R}^n$ has a nonsingular Jacobian $K_*(\widetilde{x}(\tau))$ at the final state. Then, there exists a solution $\lambda : [\sigma,\tau] \to \mathbb{R}^n$ of the adjoint equation

$$\dot{\lambda}(t) \;=\; -q_x(\widetilde{x}(t),\widetilde{\omega}(t))' \;-\; A(t)'\lambda(t) \tag{9.15}$$

so that $\mathcal{H}_u(\widetilde{x}(t),\widetilde{\omega}(t),1,\lambda(t)) = 0$ for almost all $t \in [\sigma,\tau]$.

Proof. We need only to check the output controllability condition. But

$$K_*(\widetilde{x}(\tau)) \left\{ \int_\sigma^\tau \Phi(\tau,s)\, B(s)\, \mu(s)\, ds \;\middle|\; \mu \in \mathcal{L}_m^\infty(\sigma,\tau) \right\} \;=\; \mathbb{R}^n$$

holds because the first matrix is nonsingular and the space $\{\ldots\}$ is, by assumption, equal to $\mathbb{R}^n$. ■

This corollary applies, in particular, when the final state is specified to be a given $x = x^{\mathrm{f}}$. We let in that case $K(x) = x - x^{\mathrm{f}}$, so K has full rank at every state (Jacobian is the identity). Observe that, in contrast with Theorem 43, we did not specify a condition on $\lambda(\tau)$. There is no information to be gained by asking that $\lambda(\tau) = (p_*(\widetilde{x}(\tau)) + \nu' K_*(\widetilde{x}(\tau)))'$ for some ν, because of nonsingularity of $K_*(\widetilde{x})$.

Writing as Hamiltonian Equations

An elegant way of packaging the conclusions of Theorem 43 is by means of Hamiltonian equations, as follows: if $\widetilde{\omega}$ is optimal for x^0, then there is a solution $(x(\cdot), \lambda(\cdot))$ of the following two-point boundary-value problem:

$$\begin{aligned} \dot{x} &= \mathcal{H}_\lambda(x, \widetilde{\omega}, \nu_0, \lambda)', & x(\sigma) &= x^0, \ K(x(\tau)) = 0 \\ \dot{\lambda} &= -\mathcal{H}_x(x, \widetilde{\omega}, \nu_0, \lambda)', & \lambda(\tau) &= \left(\nu_0 p_*(x(\tau)) + \nu' K_*(x(\tau))\right)' \end{aligned}$$

and, along this solution, it must hold that $\mathcal{H}_u(x(t), \widetilde{\omega}(t), \nu_0, \lambda(t)) \equiv 0$. (And, if the linearization along $\widetilde{\omega}$ is output controllable, $\nu_0 = 1$.)

Two special cases merit separate statements. The first is the one when no terminal constraints are imposed. Then the above equations reduce to:

$$\begin{aligned} \dot{x} &= \mathcal{H}_\lambda(x, \widetilde{\omega}, 1, \lambda)', & x(\sigma) &= x^0 \\ \dot{\lambda} &= -\mathcal{H}_x(x, \widetilde{\omega}, 1, \lambda)', & \lambda(\tau) &= \left(p_*(x(\tau))\right)' \end{aligned}$$

and $\mathcal{H}_u(x(t), \widetilde{\omega}(t), 1, \lambda(t)) \equiv 0$. (Linearized output controllability holds by definition, if $r = 0$.)

The other special case is the one when the final state is specified to be a given $x = x^{\mathrm{f}}$ (i.e., $K(x) = x - x^{\mathrm{f}}$). Then, any optimal control $\widetilde{\omega}$ for which the linearization along $\widetilde{\omega}$ is controllable must be such that there is a solution $(x(\cdot), \lambda(\cdot))$ of

$$\begin{aligned} \dot{x} &= \mathcal{H}_\lambda(x, \widetilde{\omega}, 1, \lambda)', & x(\sigma) &= x^0, \ x(\tau) = x^{\mathrm{f}} \\ \dot{\lambda} &= -\mathcal{H}_x(x, \widetilde{\omega}, 1, \lambda)' \end{aligned}$$

and, again, $\mathcal{H}_u((x(t), \widetilde{\omega}(t), 1, \lambda(t)) \equiv 0$. (Linearized output controllability condition holds; cf. Corollary 9.2.9.)

Systems Affine in Controls, Quadratic Costs

An interesting class of systems consists of those that are affine in controls, that is,

$$f(x, u) = F(x) + G(x)u$$

for some n-vector function F and some $n \times m$ matrix function G. (This is as in Equations (4.18), (4.22), and (5.10), but we write here F instead of g_0; note that linear systems have this form, with F a linear function and G constant.) For such systems, it is natural to consider costs along trajectories that are quadratic in u:

$$q(x, u) = u'R(x)u + Q(x)$$

for some $m \times m$ symmetric matrix function $R(x)$ and a scalar function $Q(x)$. Because of the general assumptions made on f and q, we have that F, G, R, and Q are (at least) continuously differentiable. In this context, we will say that

q is *regular* if it is quadratic in u and $R(x)$ is nonsingular for each $x \in \mathcal{X}$, and, in that case, we introduce the map

$$M : \mathcal{X} \times \mathbb{R}^n \to \mathbb{R}^n : (x,\eta) \mapsto q_x(x,\rho(x,\eta))' + f_x(x,\rho(x,\eta))'\eta$$

where $\rho(x,\eta) := -\frac{1}{2}R(x)^{-1}G(x)'\eta$. (For example, if R and G are independent of x then $M(x,\eta) = -Q_*(x) - F_*(x)\eta$.)

Proposition 9.2.10 Assume that $\widetilde{\omega} \in \mathcal{L}_{\mathcal{U}}(\sigma,\tau)$ is optimal for x^0, with respect to the system $\dot{x} = F(x)+G(x)u$, where q is quadratic in u and regular, and that the linearization along $\widetilde{\omega}$ is output controllable. Then, there are a vector $\nu \in \mathbb{R}^r$ and an absolutely continuous function $\lambda : [\sigma,\tau] \to \mathbb{R}^n$, such that $(x,\lambda) = (\widetilde{x},\lambda)$ is a solution of the following system of $2n$ differential equations:

$$\dot{x} = F(x) - \frac{1}{2}G(x)R(x)^{-1}G(x)'\lambda\,, \quad x(\sigma) = x^0 \tag{9.16}$$

$$\dot{\lambda} = -M(x,\lambda)\,, \quad \lambda(\tau) = \big(p_*(x(\tau)) + \nu' K_*(x(\tau))\big)'\,. \tag{9.17}$$

Moreover, if F, G, R, and Q are of class C^k, $k = 1,2,\ldots,\infty$, or (real-)analytic, then $\widetilde{\omega}$ must be of class C^k, or analytic, respectively.*

Proof. As $\mathcal{H}_u(x,u,1,\eta) = 2u'R(x)+\eta'G(x)$ for all vectors u,x,η, the equation $\mathcal{H}_u = 0$ can be solved for u, to give $u = -(1/2)R(x)^{-1}G(x)'\eta$. Therefore,

$$\widetilde{\omega}(t) = -\frac{1}{2}\,R(\widetilde{x}(t))^{-1}\,G(\widetilde{x}(t))'\,\lambda(t)\,. \tag{9.18}$$

This expression for $\widetilde{\omega}$ may then be substituted into the differential equations for x and λ, giving the two-point boundary-value problem in Equations (9.16) and (9.17). The function M has one less degree of differentiability than the data defining the problem. Thus, the set of $2n$ differential equations has a right-hand side of class C^{k-1} on x and λ. It follows that $(\widetilde{x},\lambda)$ is of class C^k in t (and is analytic, in the analytic case); cf. Theorem 1 (p. 57). Thus, from Equation (9.18), we obtain the same differentiability for $\widetilde{\omega}$. ∎

Remark 9.2.11 Recall that the proof of kth order differentiability in the C^k case is very easy: suppose that z is an absolutely continuous function that solves $\dot{z} = W(z)$. Assume that z has been shown, by induction, to be of class C^k in t. Then, provided that W is of class C^k, $W(z(t))$ is of class C^k in t, from which it follows that $\dot{z}$ is of class C^k, from which we conclude that z is in fact of class C^{k+1}. □

The next two problems show that Theorem 43 (p. 403) provides the same characterizations as, respectively, the Linear-Quadratic problem treated in Section 8.2, and the least-squares problem covered in Theorem 5 (p. 109), when costs and constraints are appropriately defined.

---

*That is to say, $\widetilde{\omega}$ is almost everywhere equal to a function that has those properties.

Exercise 9.2.12 Consider the optimal control problem for a linear time-invariant system $\dot{x} = Ax + Bu$ with quadratic costs, that is, $q(x,u) = u'Ru + x'Qx$ and $p(x) = x'Sx$, where R, S, Q are symmetric matrices, Q and S are both positive semidefinite, and R is positive definite. We assume that there are no final-state constraints, $K \equiv 0$. Let $P(\cdot)$ be the solution of the matrix Riccati differential equation $\dot{P} = PBR^{-1}B'P - PA - A'P - Q$ with final condition $P(\tau) = S$. Consider the solution $(x(\cdot), \lambda(\cdot))$ of the two-point boundary value problem obtained in Theorem 43 (p. 403). Show that $\lambda(t) = 2P(t)x(t)$ for all $t \in [\sigma, \tau]$. (*Hint:* You may want to make use of Lemmas 8.2.8 and 8.2.9 — with these, the proof takes only a few lines.) □

Exercise 9.2.13 Consider the optimal control problem for a linear time-invariant controllable system $\dot{x} = Ax + Bu$ with quadratic cost $q(x,u) = \|u\|^2$ and $p(x) \equiv 0$. Assume now that, in addition to the initial state x^0, a final state x^{f} is also specified. Write $\lambda(t) = \Phi(\tau, t)'\nu$, for some vector $\nu \in \mathbb{R}^n$, and solve $\mathcal{H}_u = 0$ for $\tilde{\omega}(t)$ in terms of $\lambda(t)$. Next substitute this expression into the differential equation for $\tilde{x}$ and use the form of the solution to solve for ν in terms of x^0 and x^{f}. At this point, you will have an explicit expression for $\tilde{\omega}$. Compare with Theorem 5 (p. 109). □

Example 9.2.14 Consider the system with $\mathcal{X} = \mathbb{R}^3$, $\mathcal{U} = \mathbb{R}^2$, and equations:

$$\begin{aligned} \dot{x}_1 &= u_1 \\ \dot{x}_2 &= u_2 \\ \dot{x}_3 &= x_1 u_2 \,. \end{aligned}$$

(This is an example of a completely controllable system for which no continuous feedback stabilization is possible, and is a small simplification of Example 5.9.16; see also Exercises 4.3.14 and 4.3.16. The unsimplified equations could be used too.) Consider the problem of minimizing the control energy $\mathcal{J}(u) = \int_\sigma^\tau \frac{1}{2} \|u(s))\|^2 \, ds$, with initial and final states fixed ($K(x) = x - x^{\mathrm{f}}$). We claim that *every optimal control must be of the form*

$$u(t) = a \begin{pmatrix} \cos(\alpha t + b) \\ -\sin(\alpha t + b) \end{pmatrix} \tag{9.19}$$

for some $a, b, \alpha \in \mathbb{R}$. (We will not address the converse question of when such a control is optimal.) To prove this, take any optimal control $u(\cdot)$. If $u \equiv 0$, then u is of the form (9.19), with $a = 0$ and b, α arbitrary. (Observe that the control $u \equiv 0$ is, obviously, the unique optimal control in the interval $[\sigma, \tau]$ that takes x^0 to itself.) So we assume from now on that $u(\cdot) = (u_1(\cdot).u_2(\cdot))'$ is optimal and $u \not\equiv 0$.

We first show that the linearization along the corresponding trajectory $x(\cdot) = \phi(\cdot, \sigma, x^0, u)$ is controllable. We have, along this trajectory, the linearized system matrices:

$$A(t) = \begin{pmatrix} 0 & 0 & 0 \\ 0 & 0 & 0 \\ u_2(t) & 0 & 0 \end{pmatrix} \quad \text{and} \quad B(t) = \begin{pmatrix} 1 & 0 \\ 0 & 1 \\ 0 & x_1(t) \end{pmatrix}.$$

Observe that $B(t)$ happens to be absolutely continuous, in this example, because x_1 is. Let $B_0 := B$ and $B_1 := AB_0 - \dot{B}_0$; then,

$$(B_0\,,\,B_1) \;=\; \begin{pmatrix} 1 & 0 & 0 & 0 \\ 0 & 1 & 0 & 0 \\ 0 & x_1(t) & u_2(t) & u_1(t) \end{pmatrix}$$

This matrix cannot have rank < 3 for almost all $t \in [\sigma, \tau]$, because $u \not\equiv 0$. Therefore, the linearized system described by $(A(t), B(t))$ is controllable, by Corollary 3.5.18. (The statement of that result requires that A and B be smooth, but smoothness is not required when only first derivatives are evaluated. The argument in that case is in fact easy enough that it is worth giving here: if the linearized system would not be controllable, then there would be some vector $p \neq 0$ such that $p'\Phi(\tau, t)B(t) = 0$ for almost all $t \in [\sigma, \tau]$. Thus, since $\Phi(\tau, t)B(t)$ is absolutely continuous, also $-\frac{d}{dt}p'\Phi(\tau, t)B(t) = p'\Phi(\tau, t)A(t)B(t) - p'\Phi(\tau, t)\dot{B}(t) \equiv 0$, from which it would follow that $q(t)'(B_0(t), B_1(t)) \equiv 0$ for $q(t) = \Phi(\tau, t)'p \neq 0$, so $(B_0(t), B_1(t))$ would have rank < 3 for almost all t.)

Since we are dealing with the constraint $K(x) = x - x^{\mathrm{f}}$ and the linearization is controllable, we have linearized output controllability (arguing as in Corollary 9.2.9), and we are in the situation of Proposition 9.2.10. Thus we know that $u(t)$ is analytic in t, and we have that, for some solution λ of $\dot{\lambda} = -A(t)'\lambda$,

$$u_1(t) = -\lambda_1(t) \quad \text{and} \quad u_2(t) = -\lambda_2(t) - \lambda_3(t)x_1(t)\,.$$

Moreover, λ_2 and λ_3 are constant; we let $\lambda_3(t) \equiv \alpha$. Using that $\dot{\lambda}_1(t) = -\alpha u_2(t)$, we have then that

$$\dot{u}_1(t) = \alpha u_2(t) \quad \text{and} \quad \dot{u}_2(t) = -\alpha u_1(t)\,.$$

Therefore u must have the form (9.19). □

9.3 Excursion into the Calculus of Variations

Let $\mathcal{X}$ be an open subset of $\mathbb{R}^n$. By a *curve* we will mean a *Lipschitz continuous* map $x : [\sigma, \tau] \to \mathcal{X}$. Let

$$q \,:\, \mathcal{X} \times \mathbb{R}^n \to \mathbb{R}$$

be a given C^2 function. As a Lipschitz continuous x is absolutely continuous and has an essentially bounded derivative $\dot{x}$, the integral $\int_\sigma^\tau q(x, \dot{x})\,dt$ is well-defined for any curve. Suppose also given an interval $[\sigma, \tau]$ and a point $x^0 \in \mathcal{X}$. The following is the classical problem in the calculus of variations:

minimize the integral $\int_\sigma^\tau q(x, \dot{x})\,dt$ *over all curves* $x : [\sigma, \tau] \to \mathcal{X}$ *with* $x(\sigma) = x^0$.

This is the "endpoint-unconstrained" problem. In addition, one often considers also the problem in which a terminal constraint

$$x(\tau) = x^{\mathrm{f}},$$

where x^{f} is prespecified, is imposed; we will call this the "endpoint-constrained" problem.

Example 9.3.1 Consider the problem of finding the (Lipschitz continuous) $y : [0,1] \to \mathbb{R}_{>0}$ that joins the points $(0,1)$ and $(1,y_1)$ and has the property that, when its graph is revolved around the x-axis, there results a surface of minimal area. From elementary calculus, the surface in question has area $A = \int_0^1 2\pi y\sqrt{1+(dy/dx)^2}dx$. Thus, the minimization problem is the endpoint-constrained problem for which $\sigma = 0$, $\tau = 1$, $x^0 = 1$, $x^{\mathrm{f}} = y_1$, and $q(x,\dot{x}) := x\sqrt{1+\dot{x}^2}$ (the constant factor 2π is irrelevant). □

We call a curve $\widetilde{x} : [\sigma,\tau] \to \mathcal{X}$, with $\widetilde{x}(\sigma) = x^0$, *optimal* if $\int_\sigma^\tau q(\widetilde{\omega},\dot{\widetilde{x}})\,dt \leq \int_\sigma^\tau q(x,\dot{x})\,dt$ for every $x : [a,b] \to \mathcal{X}$ with $x(\sigma) = \xi$. For the endpoint-constrained problem, we ask that $\widetilde{x}(\tau) = x^{\mathrm{f}}$ and that $\int_\sigma^\tau q(\widetilde{\omega},\dot{\widetilde{x}})\,dt \leq \int_\sigma^\tau q(x,\dot{x})\,dt$ for every $x : [\sigma,\tau] \to \mathcal{X}$ with $x(\tau) = x^{\mathrm{f}}$.

Both versions (x^{f} free, or specified,) are (very) special cases of the optimal control problem treated earlier. Indeed, we consider the system $\dot{x} = u$, evolving on the state space $\mathcal{X}$ and with $\mathcal{U} = \mathbb{R}^m$. Note that there is a one-to-one correspondence between curves and controls, since $\dot{x} = u$ (and an absolutely continuous function has an essentially bounded derivative if and only if it is Lipschitz). We introduce the same cost q, thought of as a function $q(x,u)$, and $p \equiv 0$. For the endpoint-constrained problem, we let $K(x) = x - x^{\mathrm{f}}$. Admissibility of a control ω amounts to asking that the solution x_ω remains in $\mathcal{X}$, and a control $\widetilde{\omega}$ is optimal for x^0 in the sense of optimal control if and only if the corresponding trajectory $\widetilde{x}$ is optimal in the variational sense.

From Theorem 43 (p. 403), then, we may conclude as follows.

Corollary 9.3.2 If a curve $\widetilde{x}$ is optimal for the endpoint-unconstrained problem, then $q_u(\widetilde{x}(t),\dot{\widetilde{x}}(t))$ is absolutely continuous,

$$\frac{d}{dt}q_u(\widetilde{x}(t),\dot{\widetilde{x}}(t)) \;=\; q_x(\widetilde{x}(t),\dot{\widetilde{x}}(t)) \tag{9.20}$$

for almost all $t \in [\sigma,\tau]$, and

$$q_u(\widetilde{x}(\tau),\dot{\widetilde{x}}(\tau)) \;=\; 0\,. \tag{9.21}$$

If the curve $\widetilde{x}$ is optimal for the endpoint-constrained problem, then the function $q_u(\widetilde{x}(t),\dot{\widetilde{x}}(t))$ is absolutely continuous, and Equation (9.20) holds for almost all t.

Proof. Take first the unconstrained case ($K \equiv 0$, so we can take $\nu_0 = 1$). Theorem 43 (p. 403) says that there is an absolutely continuous solution of $\dot{\lambda} = (-q_x)'$ (note that $A(t) \equiv 0$), with boundary condition $\lambda(\tau) = 0$, such that $\mathcal{H}_u(\widetilde{x}(t),\dot{\widetilde{x}}(t),1,\lambda(t)) \equiv 0$. Here we have $\mathcal{H}(x,u,1,\eta) := q(x,u) + \eta' u$, so $\mathcal{H}_u \equiv 0$ means that $\lambda = (-q_u)'$ along $\widetilde{x}$. Therefore, $(d/dt)(q_u) = -\dot{\lambda}' = q_x$, as claimed, and the boundary condition $\lambda(\tau) = 0$ gives $q_u = 0$ at the endpoint.

Now consider the endpoint-constrained problem, that is, $K(x) = x - x^{\mathrm{f}}$. The linearization of $\dot{x} = u$ along any trajectory is again $\dot{x} = u$ and is therefore controllable, so that Corollary 9.2.9 applies. The same equation then results. ∎

The differential equation (9.20)

$$\frac{d}{dt} q_u = q_x \,,$$

satisfied along optimal curves, is called the Euler or *Euler-Lagrange* equation associated to the variational problem. If $\widetilde{x}$ is of class C^2, then it solves the equivalent second-order differential equation:

$$q_{uu}\ddot{x} + q_{ux}\dot{x} - q_x = 0 \,. \tag{9.22}$$

Remark 9.3.3 Observe that in the endpoint-unconstrained case one has the boundary conditions $x(\sigma) = x^0$ and $(q_u)|_{t=\tau} = 0$. In the endpoint-constrained case, instead, the boundary conditions are $x(\sigma) = x^0$ and $x(\tau) = x^{\mathrm{f}}$. Thus the "right" number of boundary conditions are imposed, and one may expect, under appropriate technical assumptions, to have only a discrete number of solutions. The Euler-Lagrange equation is merely a necessary condition that must be satisfied by an optimal curve. If an optimal curve is known to exist, and if the Euler-Lagrange equation (and boundary conditions) lead to a unique solution, then this solution must be the optimal one. We do not treat here existence questions; there is a vast literature on the subject. □

Example 9.3.4 Consider the problem of minimizing $\int_0^1 \dot{x}^2 + x^2 \, dt$, $x(0) = 0$, $x(1) = 1$. That is, $q(x,u) = u^2 + x^2$. We have $q_u = 2u$, $q_x = 2x$. So the Euler-Lagrange equation, which must be satisfied for any possible minimizing $\widetilde{x}$, is $d\dot{x}/dt = x$. This means that $\dot{\widetilde{x}}$ must be absolutely continuous and (a.e.) $\ddot{x} = x$. So $\widetilde{x}(t) = ae^t + be^{-t}$, for some constants a, b. The boundary conditions give that $\widetilde{x}(t) = \frac{1}{\sinh 1} \sinh t$. □

Exercise 9.3.5 For each of the following problems, solve the Euler-Lagrange equation (with appropriate boundary conditions).

1. $\int_0^\pi x^2 - \dot{x}^2 \, dt$, $x(0) = x(\pi) = 0$.

2. $\int_0^\pi 4\dot{x}^2 + 2x\dot{x} - x^2 \, dt$, $x(0) = 2, x(\pi) = 0$.

3. $\int_0^1 \cos \dot{x} \, dt$, $x(0) = 0$. (Find only solutions $x(\cdot)$ that are continuously differentiable.)

4. $\int_0^1 \sqrt{1 + \dot{x}^2} \, dt$, $x(0) = a$, $x(1) = b$, where a and b are any two constants.

In the last problem, after you have solved it, explain why the solution should have been obvious. □

Note that, in Example 9.3.4, the equation $\ddot{x} = x$ tells us that $\ddot{\widetilde{x}}$ must be absolutely continuous, so a third derivative exists almost everywhere, and, with an inductive argument, we obtain that all derivatives must exist. Of course, differentiability is also clear once the solution in terms of sinh is obtained, and this solution makes it clear that $\widetilde{x}$ is even a real-analytic function of time. This is a fairly general fact, which we discuss next.

We restrict our attention, for simplicity of exposition, to the one-dimensional case, $n = 1$.

The variational problem is said to be *regular* if

$$q_{uu}(x, u) > 0 \quad \text{for all } (x, u) \in \mathcal{X} \times \mathbb{R}.$$

For instance, in Example 9.3.4, $q(x, u) = u^2 + x^2$, so $q_{uu} = 2$, and regularity holds.

Lemma 9.3.6 Suppose that the problem is regular and that $q \in C^k$, for some $k \in \{2, ..., \infty\}$. Then every solution of the Euler-Lagrange equation, and hence every optimal curve, is (at least) of class C^k on $t \in (\sigma, \tau)$. If q is real-analytic, then every such solution is analytic as well.

Proof. Consider the following map:

$$\alpha : \mathcal{X} \times \mathbb{R} \to \mathcal{X} \times \mathbb{R} : (x, u) \mapsto (x, q_u(x, u)).$$

At each point, the Jacobian of α is

$$\begin{pmatrix} 1 & 0 \\ q_{ux} & q_{uu} \end{pmatrix},$$

which has full rank $= 2$. Therefore, for each two pairs $(x_0, u_0) \in \mathcal{X} \times \mathbb{R}$ and $(x_0, p_0) \in \mathcal{X} \times \mathbb{R}$ such that $\alpha(x_0, u_0) = (x_0, p_0)$ there is some mapping $\gamma : \mathcal{D} \to \mathcal{X} \times \mathbb{R}$, defined on some open neighborhood $\mathcal{D}$ of (x_0, p_0), with the properties that $\gamma(x_0, p_0) = (x_0, u_0)$ and

$$\alpha(\gamma(x, p)) = (x, p)$$

for all $(x, p) \in \mathcal{D}$ (Inverse Function Theorem). Moreover, γ has the same degree of regularity as q_u, so γ is of class C^{k-1} (and is analytic if q is analytic). If we write $\gamma = (\gamma_1, \gamma_2)$, the map $\gamma_2 : \mathcal{D} \to \mathbb{R}$ is so that

$$q_u(x, \gamma_2(x, p)) = p$$

for each $(x, p) \in \mathcal{D}$. Furthermore, since for each fixed $x \in \mathcal{X}$, the map $u \mapsto q_u(x, u)$ is one-to-one (because its derivative is everywhere positive), we have that, for each $(x, p) \in \mathcal{D}$, $u = \gamma_2(x, p)$ is the only solution in $\mathbb{R}$ (not merely in a neighborhood of u_0) of the equation $q_u(x, u) = p$.

Now let $x = \widetilde{x}$ be any solution of the Euler-Lagrange equation. Pick any $t_0 \in (\sigma, \tau)$. Write

$$p(t) = q_u(x(t), \dot{x}(t)) \tag{9.23}$$

for all t, and let $x_0 := x(t_0)$, $u_0 := \dot{x}(t_0)$, and $p_0 := p(t_0)$. Let γ be a function as above, defined on some neighborhood $\mathcal{D}$ of $(x(t_0), p(t_0))$. Since both x and p are (absolutely) continuous (in the case of p, this is asserted in Corollary 9.3.2), there is some neighborhood $\mathcal{I}$ of t_0 so that $(x(t), p(t)) \in \mathcal{D}$ whenever $t \in \mathcal{I}$. Therefore, for all $t \in \mathcal{I}$, we can solve Equation (9.23) for $\dot{x}(t)$ using γ_2:

$$\dot{x}(t) = \gamma_2(x(t), p(t)) . \tag{9.24}$$

So we have that $(x(t), p(t))$ is a solution, on the interval $\mathcal{I}$, of the system of differential equations

$$\begin{aligned} \dot{x} &= \gamma_2(x, p) \\ \dot{p} &= q_x(x, \gamma_2(x, p)) , \end{aligned}$$

(the second one is the Euler-Lagrange equation). Thus, (x, p) has one more degree of regularity than γ_2 and q_u, namely k continuous derivatives (and is analytic, in the analytic case); cf. Propositions C.3.11 and C.3.12. ∎

Corollary 9.3.7 For a regular problem, every solution of the Euler-Lagrange equation, and hence every optimal curve, satisfies

$$q(x(t), \dot{x}(t)) - \dot{x}(t)\, q_u(x(t), \dot{x}(t)) \equiv c \tag{9.25}$$

for some constant c.

Proof. We simply compute $\frac{d}{dt}(q - \dot{x}q_u) = -\dot{x}\,(q_{uu}\ddot{x} + q_{ux}\dot{x} - q_x)$, and remark that this vanishes, because of Equation (9.22). ∎

Remark 9.3.8 We introduced two notions of regularity, one for scalar variational problems and another one for optimal control problems with quadratic costs (cf. Proposition 9.2.10). It is possible to provide a common generalization for both results, in terms of the Hessian with respect to u of the cost function q and a one-to-one condition on its gradient. However, the two cases that we presented are the ones most commonly used, and are the easiest to check. □

Example 9.3.9 We now study the Euler-Lagrange equation for the problem of minimal surfaces of revolution (cf. Example 9.3.1). That is, we consider the problem, on $\mathcal{X} = \mathbb{R}_{>0}$,

$$\int_0^1 x\sqrt{1 + \dot{x}^2}\, dt \;, \quad x(0) = 1 \,, \; x(1) = y_1 \,.$$

Let x be a solution of the Euler-Lagrange equation. Since

$$q_{uu} = \frac{x}{(1 + u^2)^{3/2}}$$

is real-analytic and positive, we have a regular problem, and $x(\cdot)$ is real-analytic. Equation (9.25) gives

$$x = c\sqrt{1+\dot{x}^2},$$

for some constant c which, because $x(0) = 1$, must satisfy $0 < c \leq 1$. From

$$\dot{x}^2 = \left(\frac{x}{c}\right)^2 - 1 \tag{9.26}$$

it follows that

$$\dot{x}\ddot{x} = \frac{1}{c^2}x\dot{x}.$$

We will assume that it is not the case that $\dot{x} \equiv 0$ (which can only happen if $x \equiv c$, and thus $y_1 = 1$). So, by analyticity, there is some nonempty open interval where

$$\ddot{x} = \frac{1}{c^2}x$$

and hence there are constants α_1, β_1 so that

$$x(t) = \alpha_1 \cosh(t/c) + \beta_1 \sinh(t/c)$$

on that interval. By the principle of analytic continuation, x must have this form on the entire interval. The initial condition $x(0) = 1$ implies that $\alpha_1 = 1$, and (9.26) evaluated at $t = 0$ gives $(\beta_1/c)^2 = (1/c)^2 - 1$ and hence $\beta_1 = \pm\sqrt{1-c^2} \in (-1,1)$. Pick $d \in \mathbb{R}$ so that

$$\tanh d = \beta_1,$$

which implies that $\cosh d = 1/c$. Then (using $\cosh(z+y) = \cosh z \cosh y + \sinh z \sinh y$),

$$x(t) = \frac{\cosh(t\cosh d + d)}{\cosh d}. \tag{9.27}$$

Every minimizing curve must be of this form.

We must, however, meet the second boundary condition: $x(1) = y_1$. This can be done if and only if one can solve

$$y_1 = \frac{\cosh(\cosh d + d)}{\cosh d}$$

for d, which requires y_1 to be sufficiently large (approximately > 0.587); in general, there may be none, one, or two solutions. (For instance, if $y_1 = \cosh 1$, the solutions are $d = 0$ and $d \approx -2.3$. The integrals $\int_0^1 x\sqrt{1+\dot{x}^2}dt$ are, respectively, approximately 1.407 and 1.764. So if a minimizer exists, it must be $x(t) = \cosh t$. It turns out, but this is harder to prove, that this function is indeed the unique global minimizer. Incidentally, the truncated cone obtained for the straight line $x(t) = 1-(1-\cosh 1)t$, would give an integral of approximately 1.447.)

Establishing minimality is nontrivial, but is well-studied in the literature. For small y_1, or more generally, for minimization on a large interval (instead of $[0, 1]$), it is physically intuitive that the true minimal-area surface will consist of two disks (at the endpoints), connected by a line. This is not representable as a graph of a function, which means that there are no minimizing extremals of the type we have considered. (In control-theoretic terms, the derivatives are "impulses" and one is led into the area of "impulsive controls".) □

9.4 Gradient-Based Numerical Methods

We now present some remarks concerning numerical approaches to the minimization of $\mathcal{J}(\omega)$. In order to concentrate on the basic ideas, we assume that there are no endpoint constraints (no K) and also that $q \equiv 0$ (though this last simplification can easily be overcome by a system extension, as done in the proof of Theorem 43). Thus $\mathcal{J}(\omega) = p(x_\omega(\tau))$, and we assume that $p : \mathcal{X} \to \mathbb{R}$ is C^1. We still assume that $\mathcal{U}$ is open and f is differentiable.

Classical iterative techniques start with a control $\omega_1 : [\sigma, \tau] \to \mathcal{U}$, and attempt to improve recursively on this initial control, obtaining a sequence $\omega_1, \omega_2, \ldots$ so that $\mathcal{J}(\omega_1) > \mathcal{J}(\omega_2) > \ldots$. Of course, just as with any numerical procedure of this type, it may well be impossible to improve after a certain ω_k, and, even if improvements are always possible, there is no guarantee that the sequence $\mathcal{J}(\omega_k)$ will converge to the infimum value of $\mathcal{J}$. Here, we concentrate on the individual recursion step. (In practice, reinitialization is used in order to restart from another ω_1 an iteration which fails to decrease fast enough.)

Suppose, then, that at a certain stage we have a candidate control $\widetilde{\omega} \in \mathcal{L}_{\mathcal{U}}(\sigma, \tau)$, which is admissible for the initial state x^0. We attempt to compute a perturbation $\mu \in \mathcal{L}_m^\infty(\sigma, \tau)$ with the property that

$$\mathcal{J}(\widetilde{\omega} + h\mu) \;<\; \mathcal{J}(\widetilde{\omega}) \tag{9.28}$$

for a sufficiently small "step size" $h > 0$. Let

$$\beta \,:\, (-h_0, h_0) \to \mathbb{R}^n \;:\; h \mapsto \phi(\tau, \sigma, x^0, \widetilde{\omega} + h\mu) \tag{9.29}$$

be the map considered in Corollary 9.1.1. Then

$$\mathcal{J}(\widetilde{\omega} + h\mu) - \mathcal{J}(\widetilde{\omega}) \;=\; p(\beta(h)) - p(\beta(0)) \;=\; (p \circ \beta)_*(0)\, h \,+\, o(h)\,,$$

will be negative for all small enough $h > 0$, provided that μ is chosen such that $(p \circ \beta)_*(0) < 0$. By the chain rule and Equation (9.7), this last derivative is:

$$p_*(\widetilde{x}(\tau)) \int_\sigma^\tau \Phi(\tau, s)\, B(s)\, \mu(s)\, ds\,, \tag{9.30}$$

where A, B, and Φ are as in Corollary 9.1.1. Thus the objective has been reduced to:

find an input μ such that (9.30) is negative.

If this goal can be met, then, for all small $h > 0$, $\widetilde{\omega} + h\mu$ will be an admissible control in $\mathcal{L}_\mathcal{U}(\sigma, \tau)$ for the same initial state x^0, and thus we will have obtained a new control which leads to a smaller cost. The process can then be repeated, after redefining $\widetilde{\omega} := \widetilde{\omega} + h\mu$. (Note that, however, in practice, once that such a μ has been found, one first performs a "line search" over h, a scalar optimization over the parameter h, in order to minimize $\mathcal{J}(\widetilde{\omega} + h\mu)$, instead of simply taking any small enough h.)

Of course, it may be impossible to make (9.30) negative, for instance because $p_*(\widetilde{x}(\tau)) = 0$. So we will assume that

$$p_*(\widetilde{x}(\tau)) \neq 0.$$

This is often a reasonable assumption: for instance, if $p(x) = \|x\|^2$ then this property will only fail if $x = 0$, but then, the minimum has already been found. Furthermore, we assume from now on that *the linearization along the trajectory $(\widetilde{x}, \widetilde{\omega})$ is controllable as a time-varying linear system*, that is, that the reachability map for this linearized system,

$$L : \mathcal{L}_m^\infty(\sigma, \tau) \to \mathbb{R}^n : \mu \mapsto \int_\sigma^\tau \Phi(\tau, s)\, B(s)\, \mu(s)\, ds$$

is onto. Equivalently (cf. Theorem 5 (p. 109)),

$$W(\sigma, \tau) = \int_\sigma^\tau \Phi(\tau, s)\, B(s)\, B(s)'\, \Phi(\tau, s)'\, ds$$

is a positive definite matrix. When this happens, we say that *the input $\widetilde{\omega}$ is nonsingular for the state x^0.*

Because of controllability, there exist inputs μ such that

$$L\mu = -p_*(\widetilde{x}(\tau))' \tag{9.31}$$

and any such input helps meet the goal of making (9.30) negative (namely, the value becomes $-\|p_*(\widetilde{x}(\tau))\|^2$). There are infinitely many solutions μ to Equation (9.31). One natural choice is the least squares solution, that is, the unique solution of minimum L^2 norm,

$$\mu := -L^\# p_*(\widetilde{x}(\tau))' \tag{9.32}$$

where $L^\#$ denotes the pseudoinverse operator discussed in Theorem 5 (p. 109). This gives the formula

$$\mu(t) = -B(t)'\, \Phi(\tau, t)'\, W(\sigma, \tau)^{-1}\, p_*(\widetilde{x}(\tau))'. \tag{9.33}$$

When $p(x) = \|x\|^2$, this choice of μ amounts to a Newton-method iteration for solving $p(x) = 0$.

Another natural choice for the perturbation μ is:

$$\mu := -L^* p_*(\widetilde{x}(\tau))' \tag{9.34}$$

where L^* is the adjoint of L. This gives the formula

$$\mu(t) = -B(t)' \, \Phi(\tau,t)' \, p_*(\widetilde{x}(\tau))' \tag{9.35}$$

which has the advantage that no matrix inversion is required. With this choice, we have that $L\mu = -LL^* p_*(\widetilde{x}(\tau))' = -W(\sigma,\tau)p_*(\widetilde{x}(\tau))'$. Thus, the expression in (9.30) becomes

$$p_*(\widetilde{x}(\tau)) \int_\sigma^\tau \Phi(\tau,s)\, B(s)\, \mu(s)\, ds = -p_*(\widetilde{x}(\tau))\, W(\sigma,\tau)\, p_*(\widetilde{x}(\tau))' ,$$

which is negative (because $W(\sigma,\tau)$ is positive definite), so indeed we made a good choice of perturbation μ. This is, basically, a steepest descent method for minimizing $\mathcal{J}$.

Exercise 9.4.1 The matrix $W(\sigma,\tau) = \int_\sigma^\tau \Phi(\tau,s)B(s)B(s)'\Phi(\tau,s)'\, ds$ appears in the computation of the perturbation given by formula (9.33). It would seem that in order to compute $W(\sigma,\tau)$, one must first solve the differential equation $\dot{x} = f(x,\widetilde{\omega})$, then solve for the fundamental solution Φ of the variational equation, and, finally, integrate numerically using the formula for $W(\sigma,\tau)$. Such an approach involves large amounts of storage (values of $\widetilde{x}$ to be used in the fundamental equation, as well as in $B(t)$, values of Φ, etc). There is an alternative way of computing $W(\sigma,\tau)$, in "one pass", however, as follows. Let

$$A(x,u) = f_x(x,u) \quad \text{and} \quad B(x,u) = f_u(x,u) ,$$

seen as functions from $\mathcal{X} \times \mathcal{U}$ into $\mathbb{R}^{n\times n}$ and $\mathbb{R}^{n\times m}$ respectively. Now consider the following system of $n+n^2$ differential equations for $x : [\sigma,\tau] \to \mathcal{X}$ and $Q : [\sigma,\tau] \to \mathbb{R}^{n\times n}$:

$$\begin{aligned} \dot{x} &= f(x,\widetilde{\omega}) \\ \dot{Q} &= A(x,\widetilde{\omega})Q + QA(x,\widetilde{\omega})' + B(x,\widetilde{\omega})B(x,\widetilde{\omega})' . \end{aligned}$$

Show that the solution of this equation with initial conditions $x(\sigma) = x^0$ and $Q(\sigma) = 0$ is so that $Q(\tau) = W(\sigma,\tau)$. □

Exercise 9.4.2 Consider the system with $\mathcal{X} = \mathbb{R}^3$, $\mathcal{U} = \mathbb{R}^2$, discussed in Example 9.2.14. Take the control $\widetilde{\omega} \in \mathcal{L}_{\mathcal{U}}(0,2\pi)$ whose coordinates are

$$u_1(t) \equiv 0 , \quad u_2(t) = \sin(t) .$$

Since $\widetilde{\omega} \not\equiv 0$, it is nonsingular for any initial state x^0.
(a) Compute a formula for the pseudoinverse operator $L^\#$, when the control $\widetilde{\omega}$ is used, and x^0 is arbitrary.
(b) Provide a formula for the basic recursive "Newton" step in (9.33), when $p(x) = \|x\|^2$, and compute two iterates, starting from any nonzero state of your choice. (You may use $h=1$.) □

9.5 Constrained Controls: Minimum Principle

Theorem 43 (p. 403) deals exclusively with problems in which the optimal control takes values in the interior of the constraint set. When the input-value set is not open, however, it may provide no useful information.

As an illustration, consider the problem of minimizing $x(1)$, for the system $\dot{x} = u$ with $\mathcal{U} = [-1, 1]$, when the initial state is $x(0) = 0$. Obviously, the optimum is attained with $\widetilde{\omega} \equiv -1$, a control which takes boundary values. If we apply the Theorem, we obtain, letting $\mathcal{H}(x, u, 1, \eta) = \eta u$ (note that we may take $\nu_0 = 1$, since there are no constraints on final states), that $\mathcal{H}_u(\widetilde{x}(t), \widetilde{\omega}(t), 1, \lambda(t)) = \lambda(t) \equiv 1$ (since $\dot{\lambda} = 0$ and $\lambda(1) = 1$); thus it is impossible to satisfy Equation (9.14). Of course, the Theorem could not be applied to begin with, since $\mathcal{U}$ is not open in this example. However, this example already serves to indicate a possible alternative. The condition $\mathcal{H}_u = 0$ is necessary for $\mathcal{H}$ to have a local extremum with respect to u. If we look at $\mathcal{H}$ itself, not its derivative, we have that, along trajectories, $\mathcal{H}(\widetilde{x}(t), \widetilde{\omega}(t), 1, \lambda(t)) \equiv \widetilde{\omega}(t)$. Thus the true solution $\widetilde{\omega} \equiv -1$ is obtained if, instead of setting derivatives to zero, we ask that $\mathcal{H}(\widetilde{x}(t), u, 1, \lambda(t))$ should be minimal, among all $u \in \mathcal{U}$. This is what the Minimum (or "Maximum", depending on sign conventions) Principle asserts.

In this section, we do not need to assume that $f : \mathcal{X} \times \mathcal{U} \to \mathbb{R}^n$ is continuously differentiable nor that the cost functions are differentiable in u. We assume only that $f : \mathcal{X} \times \mathcal{U} \to \mathcal{X}$ is C^1 in x, and that f, as well as its Jacobian matrix of partial derivatives f_x, are continuous on x, u (here $\mathcal{U}$ is a metric space and $\mathcal{X}$ an open subset of some Euclidean space). Similarly, the three functions

$$q : \mathcal{X} \times \mathcal{U} \to \mathbb{R}\,, \quad p : \mathcal{X} \to \mathbb{R}\,, \quad K : \mathcal{X} \to \mathbb{R}^r$$

are assumed to be C^1 on x and continuous on u. The Hamiltonian is defined just as before,

$$\mathcal{H} \,:\, \mathcal{X} \times \mathcal{U} \times \mathbb{R} \times \mathbb{R}^n \to \mathbb{R} \,:\, (x, u, \eta_0, \eta) \mapsto \eta_0 q(x, u) \,+\, \eta' f(x, u)\,.$$

We only prove a result in the (much) simpler case in which the final state is free (no final state constraints given by "K"), in which case η_0 will evaluate to 1, so one can also write $\mathcal{H} = q(x, u) + \eta' f(x, u)$.

Theorem 44 (Minimum Principle) *Assume that $\widetilde{\omega} \in \mathcal{L}_{\mathcal{U}}(\sigma, \tau)$ is optimal for x^0. Then, the solution $\lambda : [\sigma, \tau] \to \mathbb{R}^n$ of the final-value problem*

$$\dot{\lambda}(t) \;=\; -q_x(\widetilde{x}(t), \widetilde{\omega}(t))' \,-\, A(t)'\lambda(t)\,, \qquad \lambda(\tau) \;=\; \left(p_*(\widetilde{x}(\tau))\right)' \tag{9.36}$$

is so that

$$\mathcal{H}(\widetilde{x}(t), \widetilde{\omega}(t), 1, \lambda(t)) \;=\; \min_{u \in \mathcal{U}} \mathcal{H}(\widetilde{x}(t), u, 1, \lambda(t)) \tag{9.37}$$

for almost all $t \in [\sigma, \tau]$.

Proof. Suppose that $\widetilde{\omega}$ is optimal, and let $\widetilde{x}(\cdot)$, $A(\cdot)$, and Φ be as in the proof of Theorem 43. Just as in that proof, we first establish the case $q \equiv 0$ and then derive the general case from it by considering an extended system.

Pick any instant $t_0 \in (\sigma, \tau)$ such that $\dot{\widetilde{x}}(t_0)$ exists and $\dot{\widetilde{x}}(t_0) = f(\widetilde{x}(t_0), \widetilde{\omega}(t_0))$; the set of such t_0's has a complement of zero measure. We will show that Equation (9.37) holds at t_0. Take any element $u \in \mathcal{U}$; we must show that

$$\mathcal{H}(\widetilde{x}(t_0), \widetilde{\omega}(t_0), 1, \lambda(t_0)) \leq \mathcal{H}(\widetilde{x}(t_0), u, 1, \lambda(t_0)). \tag{9.38}$$

For each $\varepsilon > 0$ small enough, we may consider the control ω_ε defined as follows:

$$\omega_\varepsilon := \begin{cases} \widetilde{\omega}(t) & \text{if } t \notin [t_0 - \varepsilon, t_0] \\ u & \text{if } t \in [t_0 - \varepsilon, t_0] \end{cases}$$

(this is often called a "needle variation" of $\widetilde{\omega}$ around $t = t_0$). The control ω_ε is admissible for x^0 provided that ε is small enough. Indeed, by continuity of ϕ, there are two neighborhoods $\mathcal{O}_0$ and $\mathcal{O}$ of $\widetilde{x}(t_0)$, and some $\varepsilon_0 > 0$, with the following properties: (a) if $x \in \mathcal{O}$ then the restriction of $\widetilde{\omega}$ to $[t_0, \tau]$ is admissible for x, and (b) if $x \in \mathcal{O}_0$ then the constant control $\omega \equiv u$ on any interval $[r, r+\varepsilon_0]$ of length ε_0 is admissible for x, and $\phi(t, r, x, \omega) \in \mathcal{O}$ for every $t \in [r, r+\varepsilon_0]$. We may, of course, pick $\mathcal{O}$ to have compact closure, so that $f(x, u)$ and $f(x, \widetilde{\omega}(t))$ are bounded whenever $x \in \mathcal{O}$ and $t \in [\sigma, \tau]$.

Now take any $\varepsilon \leq \varepsilon_0$ such that $\widetilde{x}(t_0 - \varepsilon) \in \mathcal{O}_0$. It follows that one may solve with constant control $\omega \equiv u$ starting from $\widetilde{x}(t_0 - \varepsilon)$ at time $t = t_0 - \varepsilon$, and at time $t = t_0$ the resulting state is in $\mathcal{O}$, so the "tail" of $\widetilde{\omega}$ is admissible after that instant. The complete trajectory obtained by concatenating these solutions is then a solution on the entire interval $[\sigma, \tau]$.

We denote $x_\varepsilon := x_{\omega_\varepsilon}$, that is, $x_\varepsilon(t) = \phi(t, \sigma, x^0, \omega_\varepsilon)$ for all $t \in [\sigma, \tau]$. Next we compare $x_\varepsilon(t_0)$ and $\widetilde{x}(t_0)$. We have

$$\begin{aligned} x_\varepsilon(t_0) - \widetilde{x}(t_0 - \varepsilon) &= \int_{t_0-\varepsilon}^{t_0} f(x_\varepsilon(s), u)\, ds \\ &= \int_{t_0-\varepsilon}^{t_0} [f(x_\varepsilon(s), u) - f(\widetilde{x}(t_0), u)]\, ds + \int_{t_0-\varepsilon}^{t_0} f(\widetilde{x}(t_0), u)\, ds \\ &= \delta_1(\varepsilon) + \varepsilon f(\widetilde{x}(t_0), u) \end{aligned}$$

for some function δ_1 which is $o(\varepsilon)$ as $\varepsilon \searrow 0$. (Such a δ_1 can be obtained as follows. Let M be an upper bound on the values of $\|f(x, u)\|$ and $\|f(x, \omega(t))\|$ for $x \in \mathcal{O}$. Then, for each $s \in [t_0 - \varepsilon, t_0]$, $\|x_\varepsilon(s) - \widetilde{x}(t_0 - \varepsilon)\| = \|x_\varepsilon(s) - x_\varepsilon(t_0 - \varepsilon)\| \leq \varepsilon M$, and also $\|\widetilde{x}(t_0) - \widetilde{x}(t_0 - \varepsilon)\| \leq \varepsilon M$, so $\|x_\varepsilon(s) - \widetilde{x}(t_0)\| \leq 2\varepsilon M$. Now let c be a Lipschitz constant for $f(\cdot, u)$ on $\mathcal{O}$; it follows that we have a bound as above with $\delta(\varepsilon) := 2c\varepsilon^2 M$.)

On the other hand, because we took a point t_0 where the differential equation holds (actually, only differentiability from the left is required), we have that

$$\widetilde{x}(t_0) - \widetilde{x}(t_0 - \varepsilon) = \varepsilon f(\widetilde{x}(t_0), \widetilde{\omega}(t_0)) + \delta_2(\varepsilon)$$

where δ_2 is also $o(\varepsilon)$ as $\varepsilon\searrow 0$. Thus

$$x_\varepsilon(t_0) - \widetilde{x}(t_0) = \varepsilon\,[f(\widetilde{x}(t_0),u) - f(\widetilde{x}(t_0),\widetilde{\omega}(t_0))] + \delta(\varepsilon) \tag{9.39}$$

where $\delta = \delta_1 - \delta_2$ is $o(\varepsilon)$.

Consider the map $\Theta : x \mapsto \phi(\tau, t_0, x, \nu)$, where ν is the control $\widetilde{\omega}$ restricted to the interval $[t_0, \tau]$; Θ is defined on a neighborhood of $x_0^0 := \widetilde{x}(t_0)$, and hence is defined for $x_\varepsilon(t_0)$ when ε is small enough. The first part of Corollary 9.1.1, applied on the interval $[t_0, \tau]$, with respect to the input ν, gives that

$$\begin{aligned}\Theta(x) &= \Theta(x_0^0) + \Theta_*(x_0^0)\,(x - x_0^0) + g(\|x - x_0^0\|) \\ &= \Theta(x_0^0) + \Phi(\tau, t_0)\,(x - x_0^0) + g(\|x - x_0^0\|),\end{aligned}$$

for some function g which is $o(r)$ as $r \to 0$. We substitute $x = x_\varepsilon(t_0)$ into this expression, and use Equation (9.39) as well as the equalities $\Theta(x_\varepsilon(t_0)) = x_\varepsilon(\tau)$ and $\Theta(x_0^0) = \widetilde{x}(\tau)$, to obtain as follows:

$$x_\varepsilon(\tau) = \widetilde{x}(\tau) + \varepsilon\,\Phi(\tau, t_0)\,[f(\widetilde{x}(t_0),u) - f(\widetilde{x}(t_0),\widetilde{\omega}(t_0))] + o(\varepsilon).$$

Finally, since p is differentiable, we may also write

$$\begin{aligned}p(x_\varepsilon(\tau)) &= p(\widetilde{x}(\tau)) + p_*(\widetilde{x}(\tau))\,(x_\varepsilon(\tau) - \widetilde{x}(\tau)) + o(\varepsilon) \\ &= p(\widetilde{x}(\tau)) + \varepsilon\, p_*(\widetilde{x}(\tau))\,\Phi(\tau, t_0)\,[f(\widetilde{x}(t_0),u) - f(\widetilde{x}(t_0),\widetilde{\omega}(t_0))] + o(\varepsilon).\end{aligned}$$

Since, by optimality, $p(\widetilde{x}(\tau)) = \mathcal{J}(\widetilde{\omega}) \leq \mathcal{J}(\omega_\varepsilon) = p(x_\varepsilon(\tau))$, we have that

$$\varepsilon\, p_*(\widetilde{x}(\tau))\,\Phi(\tau, t_0)\,[f(\widetilde{x}(t_0),u) - f(\widetilde{x}(t_0),\widetilde{\omega}(t_0))] + o(\varepsilon) \geq 0$$

for all $\varepsilon > 0$ sufficiently small, so taking $\varepsilon\searrow 0$ we obtain that

$$p_*(\widetilde{x}(\tau))\,\Phi(\tau, t_0)\,[f(\widetilde{x}(t_0),u) - f(\widetilde{x}(t_0),\widetilde{\omega}(t_0))] \geq 0.$$

Let λ be the solution of $\dot{\lambda}(t) = -A(t)'\lambda(t)$ with final value $\lambda(\tau) = (p_*(\widetilde{x}(\tau)))'$. We have proved that

$$\lambda(t_0)'\,[f(\widetilde{x}(t_0),u) - f(\widetilde{x}(t_0),\widetilde{\omega}(t_0))] \geq 0.$$

Recall that $\mathcal{H}(x, v, 1, \eta) = \eta' f(x, v)$ for all $v \in \mathcal{U}$. Thus inequality (9.38) indeed holds.

Now the case of arbitrary q is obtained exactly as in the proof of Theorem 43. There are no changes in the proof until the last sentence, which is now replaced by the equality (with the obvious notations):

$$\mathcal{H}^\#(\widetilde{x}^\#(t), u, 1, \lambda^\#(t)) = \mathcal{H}(\widetilde{x}(t), u, 1, \lambda(t))$$

for all t; the fact that $\widetilde{\omega}(t)$ minimizes $\mathcal{H}^\#(\widetilde{x}^\#(t), u, 1, \lambda^\#(t))$ over $u \in \mathcal{U}$ gives the desired conclusion. ∎

9.6 Notes and Comments

The material on optimization and the Calculus of Variations is classical; see, for instance, [48], [134], [151], [189], [247], [248], [402], [442]. Path minimization problems were posed and solved at least since classical times (Dido's problem, to find a curve of a given length which encloses the largest possible area, appears in Virgil's writings). However, the origins of the field may be traced to the challenge issued "to all mathematicians," in 1696, by Johann Bernoulli, to solve the brachystochrone problem: "If in a vertical plane two points A and B are given, then it is required to specify the orbit AMB of the movable point M, along which it, starting from A, and under the influence of its own weight, arrives at B in the shortest possible time." This problem, which was soon afterwards solved by Newton, Leibniz, and many other mathematicians of the time (including Johann and his brother Jakob), gave rise to the systematic study of the Calculus of Variations. The article [393] explains the historical development, taking the more general point of view afforded by optimal control theory.

A good reference for the material in Section 9.4 is [78]. Under mild conditions, it is possible to prove that generic controls are nonsingular for every state x^0; see [373] and [374], and the related papers [99] and [100].

Section 9.5 merely skims the surface of the subject of the Minimum (or Maximum) Principle. In particular, we did not provide a treatment of the more general case in which final-state constraints $K(x) = 0$ are present. However, the form of the result is easy to guess, and we state it precisely here; see [134] for a proof. Suppose that $K : \mathcal{X} \to \mathbb{R}^r$ is also of class C^1. Assume that $\widetilde{\omega} \in \mathcal{L}_{\mathcal{U}}(\sigma, \tau)$ is optimal for x^0. Then, there exist a scalar $\nu_0 \in \{0, 1\}$ and a vector $\nu \in \mathbb{R}^r$, not both zero, so that the solution $\lambda : [\sigma, \tau] \to \mathbb{R}^n$ of the final-value problem $\dot{\lambda}(t) = -\nu_0 q_x(\widetilde{x}(t), \widetilde{\omega}(t))' - A(t)'\lambda(t)$ with $\lambda(\tau) = (\nu_0 p_*(\widetilde{x}(\tau)) + \nu' K_*(\widetilde{x}(\tau)))'$ (cf. (9.13)) is so that $\mathcal{H}(\widetilde{x}(t), \widetilde{\omega}(t), \nu_0, \lambda(t)) = \min_{u \in \mathcal{U}} \mathcal{H}(\widetilde{x}(t), u, \nu_0, \lambda(t))$ for almost all $t \in [\sigma, \tau]$.

It is also possible to give even more general results, for instance, to deal with mixed initial/final state constraints, as well as non-fixed terminal time and time-optimal problems. For the latter, one must consider variations in which a piece of the control is deleted: such shorter controls cannot be optimal, and the directions generated in this manner lead to an appropriate Hamiltonian. (The only time-optimal problem considered in this text, cf. Chapter 10, is for linear systems, and can be developed in a simple and self-contained fashion using elementary techniques from convex analysis.)

Several extensions of the Maximum Principle have been developed during the past few years. These provide "high order" tests for optimality, and in addition permit the study of far more general classes of systems, including those in which the dynamics does not depend in a Lipschitz continuous manner (or is even discontinuous) on states. A promising direction, in [390], develops an approach to generalized differentials ("multidifferentials"), and proposes their use as a basis for a general nonsmooth version of the maximum principle; references to the extensive literature on the subject are also found there.

Chapter 10

Optimality: Minimum-Time for Linear Systems

We consider time-invariant continuous-time linear systems

$$\dot{x} = Ax + Bu \tag{10.1}$$

with the control-value set $\mathcal{U}$ being a *compact convex* subset of $\mathbb{R}^m$. As usual, a *control* is a measurable map $\omega : [0,T] \to \mathbb{R}^m$ so that $\omega(t) \in \mathcal{U}$ for almost all $t \in [0,T]$. We denote by $\mathcal{L}_m^\infty(0,T)$ the set consisting of measurable essentially bounded maps from $[0,T]$ into $\mathbb{R}^m$ (when $m=1$, just $\mathcal{L}^\infty(0,T)$) and view the set of all controls as a subset $\mathcal{L}_\mathcal{U}(0,T) \subseteq \mathcal{L}_m^\infty(0,T)$. In this chapter, we write simply $\mathcal{L}_\mathcal{U}$ instead of $\mathcal{L}_\mathcal{U}^\infty$, because, $\mathcal{U}$ being compact, all maps into $\mathcal{U}$ are essentially bounded.

For each fixed $x^0 \in \mathbb{R}^n$ and $T > 0$,

$$\alpha_{x^0,T} \; : \; \mathcal{L}_m^\infty(0,T) \to \mathbb{R}^n \; : \; \omega \mapsto e^{TA}x^0 + \int_0^T e^{(T-s)A}B\omega(s)\,ds$$

is the *end-point map* or *reachability map* that sends an input ω into the state reached from x^0 at time T if this input is applied. That is, $\alpha_{x^0,T}(x^0) = x(T)$, where $x(\cdot)$ is the solution of the initial value problem $\dot{x}(t) = Ax(t) + B\omega(t)$, $x(0) = x^0$. We denote the reachable set in time exactly T from state x^0 by

$$\mathcal{R}^T(x^0) \;=\; \alpha_{x^0,T}\left(\mathcal{L}_\mathcal{U}(0,T)\right)$$

and let $\mathcal{R}(x^0) := \bigcup_{T\geq 0} \mathcal{R}^T(x^0)$. For any two fixed x^0 and $x^\mathrm{f} \in \mathcal{R}(x^0)$, we wish to study this problem:

Find a control which takes state x^0 to state x^f in minimal time.

The study of time-optimal problems for linear systems can be pursued as a simple application of basic facts from convex analysis, and we do so here.

We restrict attention to the time-invariant case merely in order to make the presentation as clear as possible. However, it is not difficult to generalize the material here, with similar proofs, to time-varying linear systems.

10.1 Existence Results

Existence of optimal controls will be easy once we establish the following fact.

Theorem 45 *For each $x^0 \in \mathbb{R}^n$ and $T \geq 0$, the set $\mathcal{R}^T(x^0)$ is compact and convex.*

Since $\alpha_{x^0,T}$ is an affine map, and $\mathcal{L}_{\mathcal{U}}(0,T)$ is convex (because $\mathcal{U}$ is), it follows that $\mathcal{R}^T(x^0)$ is convex. The compactness part of Theorem 45 can be proved by first defining a topology that makes $\mathcal{L}_{\mathcal{U}}(0,T)$ compact and then using that $\alpha_{x^0,T}$ is continuous when this topology is used for $\mathcal{L}_{\mathcal{U}}(0,T)$. Since the continuous image of a compact set is compact, this will show compactness of $\mathcal{R}^T(x^0)$. We will follow this idea, but, in order to make the presentation as elementary as possible, develop everything in terms of convergence of sequences.

Weak Convergence

Definition 10.1.1 *The sequence $\{\omega_k\}$ of elements in $\mathcal{L}^\infty_m(0,T)$ is said to* **converge weakly** *to $\omega \in \mathcal{L}^\infty_m(0,T)$, denoted $\omega_k \xrightarrow{w} \omega$, if:*

$$\int_0^T \varphi(s)'\omega_k(s)\,ds \to \int_0^T \varphi(s)'\omega(s)\,ds \quad \text{as } k \to \infty\,,$$

for each integrable function $\varphi : [0,T] \to \mathbb{R}^m$.[†] □

We write

$$\|\omega\|_\infty := \sup_{t\in[0,T]} \|\omega(t)\|$$

(where $\|\omega(t)\|$ means Euclidean norm in $\mathbb{R}^m$, and the supremum is interpreted as essential supremum); this is the infimum of the numbers η with the property that $\|\omega(t)\| \leq \eta$ for almost all $t \in [0,T]$. When we say that a sequence $\{\omega_k\}$ is bounded, we mean that there is some η so that $\|\omega_k\|_\infty \leq \eta$ for all k. We prove later the following result.

Lemma 10.1.2 Every bounded sequence $\{\omega_k\}$ in $\mathcal{L}^\infty_m(0,T)$ has some weakly convergent subsequence.

Lemma 10.1.3 Assume that $\omega \in \mathcal{L}^\infty_m(0,T)$ and the sequence $\{\omega_k\}$ in $\mathcal{L}_{\mathcal{U}}(0,T)$ are such that, for each interval $J \subseteq [0,T]$,

$$\int_J \omega_k(s)\,ds \to \int_J \omega(s)\,ds \quad \text{as } k \to \infty\,.$$

Then, $\omega \in \mathcal{L}_{\mathcal{U}}(0,T)$.

[†] Prime indicates transpose.

Proof. Let L be the intersection of the sets of Lebesgue points in $(0,T)$ for the coordinates of ω. Then $\frac{1}{h}\int_t^{t+h} \|\omega(s)-\omega(t)\|\,ds \to 0$ as $h\searrow 0$, for each $t\in L$, so

$$\lim_{h\searrow 0} \frac{1}{h}\int_t^{t+h} \omega(s)\,ds = \omega(t) \quad \forall t\in L\,. \tag{10.2}$$

Pick any $t\in L$. We will show that $\omega(t)\in\mathcal{U}$. Since the complement $[0,T]\setminus L$ has measure zero, then $\omega(t)\in\mathcal{U}$ for a.a. t.

Choose $h>0$. Applying the hypothesis to the interval $[t,t+h]$, and dividing by the constant h, we have that

$$q_{k,h} := \frac{1}{h}\int_t^{t+h} \omega_k(s)\,ds \;\to\; \frac{1}{h}\int_t^{t+h} \omega(s)\,ds$$

as $k\to\infty$. If we show that

$$q_{k,h}\in\mathcal{U}\,, \tag{10.3}$$

then this limit shows (because $\mathcal{U}$ is closed) that

$$\frac{1}{h}\int_t^{t+h} \omega(s)\,ds \in \mathcal{U}\,.$$

As this holds for any h, taking limit as $h\searrow 0$ and using Equation (10.2), gives $\omega(t)\in\mathcal{U}$, as wanted.

We now prove (10.3). Since $\mathcal{U}$ is a closed convex set, we can represent $\mathcal{U}$ as an intersection of closed half-spaces, $\mathcal{U}=\bigcap_{r\in R}\{v\in\mathbb{R}^m \mid \lambda_r{}'v\le\rho_r\}$, where R is some (in general, infinite) index set. Pick any $r\in R$. Then

$$\lambda_r{}'q_{k,h} = \lambda_r{}'\left(\frac{1}{h}\int_t^{t+h}\omega_k(s)ds\right) = \frac{1}{h}\int_t^{t+h}\lambda_r{}'\omega_k(s)ds \le \frac{1}{h}\int_t^{t+h}\rho_r ds = \rho_r\,.$$

Therefore $q_{k,h}\in\mathcal{U}$. ∎

Corollary 10.1.4 If $\omega_k \xrightarrow{\text{w}} \omega$ and $\omega_k\in\mathcal{L}_{\mathcal{U}}(0,T)$ for all k, then $\omega\in\mathcal{L}_{\mathcal{U}}(0,T)$.

Proof. It is enough to show that the hypothesis of Lemma 10.1.3 holds, that is, that

$$\int_J \omega_{k,i}(s)\,ds \;\to\; \int_J \omega_i(s)\,ds \tag{10.4}$$

for each interval J and for each coordinate $i=1,\ldots,m$. Fix any such J and i. Consider the function $\varphi(s)=I_J(s)e_i$, where I_J is the characteristic function of J and e_i is the ith canonical basis vector in $\mathbb{R}^m$. Since $\int_0^T \varphi_{i,J}(s)'\omega_k ds = \int_J \omega_{k,i}(s)ds$, and similarly for ω, the convergence (10.4) holds. ∎

Together with Lemma 10.1.2, we will then have:

Proposition 10.1.5 Every sequence $\{\omega_k\}$ of elements in $\mathcal{L}_{\mathcal{U}}(0,T)$ has a subsequence which is weakly convergent to some $\omega \in \mathcal{L}_{\mathcal{U}}(0,T)$. □

In order to prove Lemma 10.1.2, we first provide a simple characterization of weak convergence of bounded sequences.

Lemma 10.1.6 Let $\{\omega_k\}$ be a bounded sequence in $\mathcal{L}^\infty(0,T)$, and let $\omega \in \mathcal{L}^\infty(0,T)$. Suppose that, for each interval $J \subseteq [0,T]$,

$$\int_J \omega_k(s)\,ds \to \int_J \omega(s)\,ds \quad \text{as } k \to \infty\,.$$

Then, $\omega_k \xrightarrow{\text{w}} \omega$.

Proof. Let ω and $\{\omega_k\}$ be as in the statement. Pick a number η so that $\|\omega_k\|_\infty \leq \eta$ for all k. Applying Lemma 10.1.3 with $\mathcal{U} = [-\eta,\eta] \subseteq \mathbb{R}$, we have that also $\|\omega\|_\infty \leq \eta$. We claim that

$$\int_S \omega_k(s)\,ds \to \int_S \omega(s)\,ds$$

for every measurable subset $S \subseteq [0,T]$. To see this, pick any $\varepsilon > 0$, and for this ε pick a finite set of disjoint intervals $I_1,\ldots,I_\ell$ so that the set $Q = I_1 \bigcup \ldots \bigcup I_\ell$ satisfies that the symmetric difference $Q\Delta S$ has measure

$$\text{meas}\,(Q\Delta S) < \frac{\varepsilon}{3\eta}\,.$$

(One can always find such intervals: first cover S by an open set Q_0 so that $\text{meas}\,(Q_0 \backslash S) < \varepsilon/2$; since Q is a union of countably many disjoint open intervals, it has a subset Q as above so that $\text{meas}\,(Q_0 \setminus Q) < \varepsilon/2$.) Pick now a k_0 so that

$$k > k_0 \Rightarrow \left|\int_{I_j} \omega_k(s)\,ds - \int_{I_j} \omega(s)\,ds\right| < \frac{\varepsilon}{3\ell}\,, \quad j = 1,\ldots,\ell\,.$$

Then, for each $k > k_0$, $\left|\int_S \omega_k(s)\,ds - \int_S \omega(s)\,ds\right|$ is bounded by

$$\left|\int_Q \omega_k(s)\,ds - \int_S \omega_k(s)\,ds\right| + \left|\int_Q \omega(s)\,ds - \int_S \omega(s)\,ds\right| + \sum_{i=1}^{\ell} \left|\int_{I_j} \omega_k(s)\,ds - \int_{I_j} \omega(s)\,ds\right|,$$

which in turn bounded by $\eta\frac{\varepsilon}{3\eta} + \eta\frac{\varepsilon}{3\eta} + \ell\frac{\varepsilon}{3\ell} = \varepsilon$. As ε was arbitrary, $\int_S \omega_k(s)\,ds \to \int_S \omega(s)\,ds$.

Next, we claim that

$$\int_0^T \varphi(s)\omega_k(s)\,ds \to \int_0^T \varphi(s)\omega(s)\,ds \tag{10.5}$$

for every simple function $\varphi = \sum_{i=1}^{p} c_i I_{S_i}$. Indeed,

$$\int_0^T \varphi(s)\omega_k(s)\,ds = \sum_{i=1}^{p} c_i \int_{S_i} \omega_k(s)\,ds \to \sum_{i=1}^{p} c_i \int_{S_i} \omega(s)\,ds = \int_0^T \varphi(s)\omega(s)\,ds\,.$$

Finally, we claim that Equation (10.5) holds for every integrable function $\varphi : [0,T] \to \mathbb{R}$. To see this, pick any $\varepsilon > 0$; there is then a simple function θ so that $\int_0^T |\theta(s) - \varphi(s)|\,ds < \varepsilon/(4\eta)$. (First find a bounded $\bar{\varphi}$ that approximates φ in $\mathcal{L}^1$, and then approximate $\bar{\varphi}$ uniformly by a simple function.) Pick k_0 so that, for all $k > k_0$, $\left|\int_0^T (\theta(s)\omega_k(s) - \theta(s)\omega(s))\,ds\right| < \varepsilon/2$. Then $\left|\int_0^T \varphi(s)\omega_k(s) - \int_0^T \varphi(s)\omega(s)\,ds\right|$ is bounded by

$$\left|\int_0^T (\varphi(s) - \theta(s))\omega_k(s)\,ds\right| + \left|\int_0^T (\varphi(s) - \theta(s))\omega(s)\,ds\right| + \left|\int_0^T (\theta(s)\omega_k(s) - \theta(s)\omega(s))\,ds\right|$$

$$\leq 2\eta \int_0^T |\theta(s) - \varphi(s)|\,ds + \left|\int_0^T (\theta(s)\omega_k(s) - \theta(s)\omega(s)ds)\right| \leq \varepsilon$$

for all such k. ∎

Example 10.1.7 Let $T > 0$, and pick any number $\rho \in [0,1]$. Define $\omega_k : [0,T] \to \mathbb{R}$, $k = 1,2,\ldots$, by partitioning $[0,T]$ into k equal intervals, and, in each such interval, making its average be ρ:

$$\omega_k(t) = \begin{cases} 1 & \text{if } t \in \left[i\frac{T}{k}, (i+\rho)\frac{T}{k}\right]\,, \quad i = 0,\ldots,k-1 \\ 0 & \text{otherwise}\,. \end{cases}$$

We claim that

$$\omega_k \xrightarrow{\text{w}} \rho$$

(function constantly equal to ρ). If J_0 is an interval of the special form

$$\left[i\frac{T}{k}, (i+1)\frac{T}{k}\right],$$

with $i \in \{0,\ldots,k-1\}$, then $\int_{J_0} \omega(s)ds = \rho T/k$. So, more generally, if $J_0 = \left[i\frac{T}{k}, (i+p)\frac{T}{k}\right]$, with $i+p$ also in $\{0,\ldots,k-1\}$, $\int_{J_0} \omega(s)ds = p\rho T/k = \int_{J_0} \rho ds$. For any interval J, there is some interval $J_0 \subseteq J$ of the form $\left[i\frac{T}{k}, (i+p)\frac{T}{k}\right]$ so that $J \setminus J_0$ has measure less than $2T/k$. Therefore, $\left|\int_J \omega_k(s)ds - \int_J \rho ds\right| \leq \int_{J\setminus J_0} |\omega_k(s) - \rho|\,ds \to 0$. The claim then follows from Lemma 10.1.6. □

Corollary 10.1.8 Assume that $\omega \in \mathcal{L}_m^\infty(0,T)$, and that the sequence $\{\omega_k\}$ in $\mathcal{L}_\mathcal{U}(0,T)$, are such that, for each interval $J \subseteq [0,T]$,

$$\int_J \omega_k(s)\,ds \to \int_J \omega(s)\,ds \quad \text{as } k \to \infty\,.$$

Then, $\omega \in \mathcal{L}_\mathcal{U}(0,T)$ and $\omega_k \xrightarrow{\text{w}} \omega$.

Proof. The first conclusion was given in Lemma 10.1.3. The hypothesis means that $\int_J \omega_{k,i}(s)\,ds \to \int_J \omega_i(s)ds$ for every interval J and each coordinate $i = 1,\ldots,m$. Furthermore, since $\mathcal{U}$ is bounded, the coordinate sequences $\{\omega_{k,i}\}$ are bounded. Pick any $\varphi \in \mathcal{L}_m^1(0,T)$, so each of its coordinates $\varphi_i[0,T] \to \mathbb{R}$ is integrable. Then

$$\int_0^T \varphi(s)'\omega_k(s)\,ds = \sum_{i=1}^m \int_0^T \varphi_i(s)'\omega_{k,i}(s)\,ds$$

converges to

$$\sum_{i=1}^m \int_0^T \varphi_i(s)'\omega_i(s)\,ds = \int_0^T \varphi(s)'\omega(s)\,ds\,,$$

by Lemma 10.1.6. ∎

Exercise 10.1.9 Let $\mathcal{V} \subseteq \mathcal{U}$ be so that the convex hull $\operatorname{co}(\mathcal{V}) = \mathcal{U}$. Show that for each $\omega \in \mathcal{L}_\mathcal{U}(0,T)$ there is some sequence $\nu_k \xrightarrow{\text{w}} \omega$ so that $\nu_k(t) \in \mathcal{V}$ for all $t \in [0,T]$. (Suggestion: you may want to argue as follows. First, show that ω can be weakly approximated by piecewise constant controls, i.e., of the form $\sum_{\text{finite}} I_{J_i} u_i$, for intervals $J_i \subseteq [0,T]$ and elements $u_i \in \mathcal{U}$. Next argue that, on each interval J, a constant control with value $u = \sum_{i=1}^r \rho_i v_i$, with $\sum \rho_i = 1$, $v_i \in \mathcal{V}$, $\rho_i \geq 0$ for all i, can be in turn weakly approximated by controls with values in $\mathcal{V}$. For this last approximation, you may think first of the special case $r = 2$: in that case, the sequence $\omega_k(v_1 - v_2) + v_2$ converges to $\rho v_1 + (1-\rho)v_2$, if ω_k is the sequence constructed in Example 10.1.7.) □

Proof of Lemma 10.1.2

Assume that $\|\omega_k\|_\infty \leq \eta$ for all k. Define $\nu_k(t) := \int_0^t \omega_k(s)ds$. Each element of the sequence $\{\nu_k\}$ is Lipschitz continuous with bound η:

$$\|\nu_k(t_1) - \nu_k(t_2)\| \leq \int_{t_1}^{t_2} \|\omega_k(s)\|\,ds \leq \eta\,|t_1 - t_2|\,, \tag{10.6}$$

and bounded by ηT. Therefore the sequence $\{\nu_k\}$ is equibounded and equicontinuous, and the Arzela-Ascoli Theorem can be applied, from which one concludes that it has a convergent subsequence. Without loss of generality, we assume that $\nu_k \to \nu$ as $k \to \infty$, for some continuous $\nu : [0,T] \to \mathbb{R}^m$. From

Equation (10.6), taking limits as $k \to \infty$, we have that ν is also Lipschitz continuous with bound η. Thus ν is absolutely continuous, so there exists an integrable function $\omega : [0,T] \to \mathbb{R}^m$, namely $\omega = d\nu/dt$, so that $\nu(t) = \int_0^t \omega(s)ds$ for all t. Since ω is the (a.e.) derivative of ν, and ν is Lipschitz with constant η, it follows that ω is almost everywhere bounded by η. Thus $\omega \in \mathcal{L}_m^\infty(0,T)$, and we are only left to establish that $\omega_k \xrightarrow{\text{w}} \omega$. Because of Corollary 10.1.8, it is enough to show that

$$\int_a^b \omega_k(s)ds \to \int_a^b \omega(s)ds$$

for all subintervals $[a,b]$ of $[0,T]$. Take any such $a < b$. As

$$\int_a^b \omega_k(s)ds = \nu_k(b) - \nu_k(a)$$

and $\nu_k \to \nu$ pointwise (even uniformly), we have that $\int_a^b \omega_k(s)ds \to \nu(b) - \nu(a)$. Since $\omega = d\nu/dt$, the last expression is the same as $\int_a^b \omega(s)ds$, so the proof is completed. ∎

Proof of Theorem 45

Convexity is clear, as remarked earlier, because $\alpha_{x^0,T}$ is an affine map, and $\mathcal{L}_\mathcal{U}(0,T)$ is convex.

We now prove that $\mathcal{R}^T(x^0)$ is compact. Let $\{x_k^{\mathrm{f}}\}$ be a sequence of elements of $\mathcal{R}^T(x^0)$, and let the controls $\omega_k \in \mathcal{L}_\mathcal{U}(0,T)$ be so that $\alpha_{x^0,T}(\omega_k) = x_k^{\mathrm{f}}$ for each k. Picking a subsequence, we assume without loss of generality that $\omega_k \xrightarrow{\text{w}} \omega$ for some control $\omega \in \mathcal{L}_\mathcal{U}(0,T)$. Let $x^{\mathrm{f}} := \alpha_{x^0,T}(\omega) \in \mathcal{R}^T(x^0)$. We claim that $x_k^{\mathrm{f}} \to x^{\mathrm{f}}$, which will prove that the original sequence has a convergent subsequence. Indeed, the ith coordinate of x_k^{f}, $i = 1, \ldots, n$, has the form

$$\int_0^T \varphi_i(s)'\omega_k(s)\,ds\,,$$

where $\varphi_i(t)$ is the ith row of $e^{(T-t)A}B$. By weak convergence,

$$\int_0^T \varphi_i(s)'\omega_k(s)\,ds \;\to\; \int_0^T \varphi_i(s)'\omega(s)\,ds$$

as $k \to \infty$, and this last is the ith coordinate of x^{f}. ∎

Remark 10.1.10 The reachable sets $\mathcal{R}^T(x^0)$ can be proved to be closed even for a fairly wide class of nonlinear systems. Consider systems of the type $\dot{x} = f(x) + \sum_{i=1}^m u_i g(x)$ (that is, affine in controls), where $f, g_1, \ldots, g_m$ are locally Lipschitz vector fields. We use $\alpha_{x^0,T}(\omega)$ to denote the solution $x(T)$ of the differential equation, when the input is $u = \omega \in \mathcal{L}_\mathcal{U}(0,T)$ and the initial condition is $x(0) = x^0$, provided that the solution is defined on the entire interval $[\sigma, \tau]$. Thus, α_ω is a map $\mathcal{D}_{x^0,T} \to \mathbb{R}^n$, where $\mathcal{D}_{x^0,T}$ is a subset of $\mathcal{L}_\mathcal{U}(\sigma,\tau)$ (an open subset in the supremum norm, cf. Theorem 1 (p. 57). Assume that $\mathcal{D}_{x^0,T} = \mathcal{L}_\mathcal{U}(\sigma,\tau)$. Then the set $\mathcal{R}^T(x^0)$ is compact. Indeed, the same proof works as in the linear case; the only technical fact needed is that $\alpha_{x^0,T}(\omega_k) \to \alpha_{x^0,T}(\omega)$ whenever $\omega_k \xrightarrow{w} \omega$, and this follows from Theorem 1.

Observe also that Exercise 10.1.9 allows us to conclude, for any $\mathcal{V} \subseteq \mathcal{U}$ for which $\operatorname{co}(\mathcal{V}) = \mathcal{U}$, that $\mathcal{R}^T(x^0) = \operatorname{clos}(\mathcal{R}^T_\mathcal{V}(x^0))$ for all x^0 and T, where $\mathcal{R}^T_\mathcal{V}(x^0)$ denotes the set of states reachable from x^0 using controls with values in $\mathcal{V}$. □

Exercise 10.1.11 Consider the nonlinear, affine in controls, system $\dot{x} = x^2 + u$, with $\mathcal{U} = [-1,1]$ (any other compact convex set $\mathcal{U}$ could be used). Show that there are states x^0 and times T for which $\mathcal{R}^T(x^0)$ is not compact. Why does this not contradict the discussion in Remark 10.1.10? □

Existence of Time-Optimal Controls

Theorem 46 *If $x^f \in \mathcal{R}(x^0)$ then there is a control which takes state x^0 to state x^f in minimal time.*

Proof. Let T be the infimum of the set of nonnegative real numbers

$$\{t \mid x^f \in \mathcal{R}^t(x^0)\}.$$

We need to show that $x^f \in \mathcal{R}^T(x^0)$. Assume that $t_k \searrow T$ are such that $x^f \in \mathcal{R}^{t_k}(x^0)$ for all k. For each k, let ω_k be a control steering x^0 to x^f in time t_k.

Let x^f_k be the state reached from x^0 by restricting ω_k to the interval $[0,T]$. Each $x^f_k \in \mathcal{R}^T(x^0)$, and the sequence $\{x^f_k\}$ converges to x^f. Indeed, we have

$$x^f - x^f_k = \left(e^{\varepsilon_k A} - I\right) x^f_k + \int_0^{\varepsilon_k} e^{(\varepsilon_k - s)A} B\omega_k(T+s)\, ds\,,$$

where we are denoting $\varepsilon_k := t_k - T$, so

$$\left\|x^f - x^f_k\right\| \leq \left\|e^{\varepsilon_k A} - I\right\| \left\|x^f_k\right\| + c\varepsilon_k\,,$$

where c is an upper bound on the possible values of $\left\|e^{tA}Bu\right\|$, for $t \in [0,\tau]$ and $u \in \mathcal{U}$, and τ is any number majorizing t_k for all k. On the other hand,

$$x^f_k = e^{t_k A} x^f + \int_0^{t_k} e^{(t_k - s)A} B\omega_k(s)\, ds\,,$$

implies

$$\left\|x^f_k\right\| \leq e^{\tau\|A\|} \left\|x^f\right\| + c\tau\,,$$

so the sequence $\left\|x^f_k\right\|$ is bounded. Since $\left\|e^{\varepsilon_k A} - I\right\| \to 0$ as $k \to \infty$, we have that $x^f_k \to x^f$.

Since $R^T(x^0)$ is closed (by the previous lemma), it follows that $x^f \in \mathcal{R}^T(x^0)$, as desired. ∎

We call a control that takes x^0 to x^f in minimal time a *time-optimal* control.

Remark 10.1.12 The assumption that $\mathcal{U}$ is compact cannot be dropped from Theorem 46 (p. 430). To see this, consider the one-dimensional system $\dot{x} = u$. If $\mathcal{U} = [0, 1)$ then the infimum time for reaching $x^{\mathrm{f}} = 1$ from $x^0 = 0$ is $T = 1$. However, $1 \notin \mathcal{R}^1(0)$ (because $1 = \int_0^1 \omega(s)ds$ implies $\int_0^1 (1 - \omega(s))ds = 0$, but $1-\omega(s) \geq 0$ for all s, so $\omega(s) = 1$ for a.a. s, a contradiction). Thus boundedness by itself is not enough. Neither is being closed, as illustrated with the same system and $\mathcal{U} = \mathbb{R}$, since, for the same x^{f} and x^0, one has $T = 0$ but 1 is not in $\mathcal{R}^0(0) = \{0\}$. □

10.2 Maximum Principle for Time-Optimality

We review first an elementary fact from convex analysis. Let $C \subseteq \mathbb{R}^n$ be a convex set and let $x^{\mathrm{f}} \in \mathbb{R}^n$. The hyperplane

$$H = \{x \mid \eta' x = \theta\}$$

(where $\eta \in \mathbb{R}^n$ is nonzero) is said to *separate* x^{f} and C if $\eta' z \leq \theta$ for all $z \in C$ and $\eta' x^{\mathrm{f}} \geq \theta$. In particular, $\eta' \left(z - x^{\mathrm{f}}\right) \leq 0$ for all $z \in C$. If there is some such H, then x^{f} cannot be in the interior of C: otherwise, there is some $\varepsilon > 0$ so that $z := x^{\mathrm{f}} + \varepsilon\eta \in C$, and hence

$$\eta' \left(z - x^{\mathrm{f}}\right) = \varepsilon \|\eta\|^2 > 0,$$

a contradiction. Conversely, assume that x^{f} is not in the interior of a convex set C. Then there exists a hyperplane separating x^{f} and C; this is a standard result, see for instance [330], Theorem III.11.6. or [188], Theorem I.9.1 and Corollary I.9.2. (When $x^{\mathrm{f}} \notin C$, this result is a consequence of the Hahn-Banach Theorem, and is valid for arbitrary locally convex topological vector spaces instead of merely for subsets of $\mathbb{R}^n$; see e.g. [348], Theorem II.9.2.) We summarize for future reference:

Lemma 10.2.1 Let $C \subseteq \mathbb{R}^n$ be a convex set and $x^{\mathrm{f}} \in \mathbb{R}^n$. Then, there exists a hyperplane separating x^{f} and C if and only if $x^{\mathrm{f}} \notin \operatorname{int} C$. □

Assume now that $H = \{x \mid \eta' x = \theta\}$ separates x^{f} and C and that x^{f} belongs to C. Any such H is said to be a *support hyperplane* to C at x^{f}. Since $x^{\mathrm{f}} \in C$, the inequality $\eta' x^{\mathrm{f}} \leq \theta$ holds, so $\eta' x^{\mathrm{f}} = \theta$. In other words, H is a support hyperplane to C at x^{f} if and only if there is some nonzero vector $\eta \in \mathbb{R}^n$ so that $H = \{x \mid \eta' \left(x - x^{\mathrm{f}}\right) = 0\}$ and

$$\eta' \left(z - x^{\mathrm{f}}\right) \leq 0 \quad \forall z \in C. \tag{10.7}$$

Any such $\eta \in \mathbb{R}^n$ is said to be *normal to C at x^{f}*. So, from the above Lemma, we have the following fact, which holds in particular for the sets $C = \mathcal{R}^T(x^0)$, for every $T \geq 0$ and $x^0 \in \mathbb{R}^n$.

Lemma 10.2.2 Let $C \subseteq \mathbb{R}^n$ be a closed convex set and $x^{\mathrm{f}} \in C$. Then $x^{\mathrm{f}} \in \partial C$ if and only if there is a normal to C at x^{f}. □

After these preliminaries, we return to the time-optimal problem.

Lemma 10.2.3 Assume that x^0 and x^{f} are two states, and $T > 0$ is such that, for some sequence $t_k \nearrow T$,

$$x^{\mathrm{f}} \notin \mathcal{R}^{t_k}(x^0) \tag{10.8}$$

for all k. Then $x^{\mathrm{f}} \notin \operatorname{int} \mathcal{R}^T(x^0)$.

Proof. Consider any fixed t_k. Since the point x^{f} does not belong to the convex set $\mathcal{R}^{t_k}(x^0)$, in particular it is not in its interior, so there is some hyperplane $\{x \mid \eta_n{}'x = \theta_n\}$ separating x^{f} and $\mathcal{R}^{t_k}(x^0)$. Thus, for each k,

$$\eta_k{}'\left(z - x^{\mathrm{f}}\right) \le 0 \quad \text{for all } z \in \mathcal{R}^{t_k}(x^0). \tag{10.9}$$

As one can multiply each η_k by a positive scalar, we may, and will, assume without loss of generality that η_k has unit norm. By compactness of the unit ball in $\mathbb{R}^n$, a subsequence of $\{\eta_k\}$, which we will take to simplify notations to be again the same sequence, converges to some (unit) vector η. Pick any $z \in \mathcal{R}^T(x^0)$. Let $z_k \to z$ with $z_k \in \mathcal{R}^{t_k}(x^0)$ (for instance, truncate to the intervals $[0, t_k]$ any given control that steers x^0 to z in time T, and argue that $z_k \to z$ as in the proof of Theorem 46 (p. 430)). Then Equation (10.9) gives that $\eta_k{}'\left(z_k - x^{\mathrm{f}}\right) \le 0$ for all such k. Taking limits as $k \to \infty$, Equation (10.7), with $C = \mathcal{R}^T(x^0)$, results, so by Lemma 10.2.2 the conclusion follows. ∎

States reached with time-optimal controls are in the boundary of the respective reachable sets:

Corollary 10.2.4 If $x^{\mathrm{f}} \in \mathcal{R}^T(x^0)$ and T is minimal with this property, then $x^{\mathrm{f}} \in \partial\mathcal{R}^T(x^0)$.

Proof. If $T = 0$ then $x^{\mathrm{f}} = x^0$ and $\mathcal{R}^0(x^0) = \{x^0\}$, and the result is clear. So take $T > 0$. Minimality means that property (10.8) holds, for some (in fact, any) sequence $t_n \nearrow T$. Lemma 10.2.3 then implies that $x^{\mathrm{f}} \notin \operatorname{int} \mathcal{R}^T(x^0)$; since by assumption $x^{\mathrm{f}} \in \mathcal{R}^T(x^0)$, the conclusion follows. ∎

A (very) partial converse to the above result, valid for trajectories starting at $x^0 = 0$, is as follows. The system (10.1), with control-value set $\mathcal{U}$, satisfies the *small-time local controllability (STLC)* condition (at $x^0 = 0$) if

$$0 \in \operatorname{int} \mathcal{R}^{\varepsilon}(0) \quad \forall \varepsilon > 0\,.$$

It was shown in the proof of Proposition 3.6.3 that a sufficient condition for STLC is that the following two properties hold:

- the matrix pair (A, B) is controllable;

- $0 \in \operatorname{int} \mathcal{U}$.

Lemma 10.2.5 Assume that the STCL holds and $0 \le S < T$. Then $\mathcal{R}^S(T) \subseteq \operatorname{int} \mathcal{R}^T(0)$.

Proof. Pick any $x^{\mathrm{f}} = \int_0^S e^{(S-s)A} B\omega(s)\, ds \in \mathcal{R}^S(T)$. Let V be an open subset of $\mathcal{R}^{T-S}(0)$ which contains 0 (STLC assumption). Consider the open set

$$Q := \{e^{SA}v + x^{\mathrm{f}} \mid v \in V\}.$$

Since e^{SA} is invertible and V is open, this is an open set as well, and $x^{\mathrm{f}} \in Q$ (obtained when $v = 0$). We claim that $Q \subseteq \mathcal{R}^T(0)$. Indeed, pick any $v \in V$ and consider the state

$$z = e^{SA}v + x^{\mathrm{f}} = e^{SA}v + \int_0^S e^{(S-s)A} B\omega(s)\, ds.$$

This is the state reached from v in time S, when applying control ω, so $z \in \mathcal{R}^S(v)$. Since $v \in \mathcal{R}^{T-S}(0)$, transitivity of reachability gives that $z \in \mathcal{R}^T(0)$, as claimed. ∎

This Lemma has the following obvious corollary.

Proposition 10.2.6 Assume that $x^{\mathrm{f}} \in \partial\mathcal{R}^T(0)$ and that the STLC is satisfied. Then T is minimal with the property that $x^{\mathrm{f}} \in \mathcal{R}^T(0)$. □

Remark 10.2.7 For *nonzero* initial states x^0, it may be the case that $x^{\mathrm{f}} \in \partial\mathcal{R}^T(x^0)$ yet T is not minimal for steering x^0 to x^{f}, even if the STLC holds. For example, consider the system $\dot{x} = x + u$ with $\mathcal{U} = [-1, 1]$. Here $\mathcal{R}^1(1) = [1, 2e - 1]$, so $1 \in \partial\mathcal{R}^1(1)$, but the minimal time for going from 1 to 1 is $T = 0$, not $T = 1$. And even for initial state $x^0 = 0$ the result may fail, if the STLC does not hold. For an example, take $\dot{x} = u$, $\mathcal{U} = [1, 2]$, and note that $\mathcal{R}^1(0) = [1, 2]$, so $1 \in \partial\mathcal{R}^1(0)$, but 1 can be reached in time 1/2 from 0. □

We have the following characterization of boundary points of reachable sets.

Theorem 47 *(Maximum Principle for boundary of reachable sets.) Assume that $x^{\mathrm{f}} \in \partial\mathcal{R}^T(x^0)$, and that $\widetilde{\omega}$ is any control steering x^0 to x^{f} in time T. Then, for every normal vector η to $\mathcal{R}^T(x^0)$ at x^{f}, the (necessarily everywhere nonzero) solution λ of the adjoint linear differential equation*

$$\dot{\lambda} = -A'\lambda, \quad \lambda(T) = \eta \tag{10.10}$$

satisfies

$$\lambda(t)' B\widetilde{\omega}(t) = \max_{u \in \mathcal{U}} \lambda(t)' Bu \tag{10.11}$$

for almost all $t \in [0, T]$.

Conversely, suppose that $\widetilde{\omega}$ is any control steering x^0 to x^{f} in time T. Assume that there is some nonzero solution λ of Equation (10.10) so that Equation (10.11) holds for almost all t. Then, $x^{\mathrm{f}} \in \partial\mathcal{R}^T(x^0)$.

Proof. Let η be normal to $\mathcal{R}^T(x^0)$ at x^{f}, i.e.,

$$\eta'\left(z-x^{\mathrm{f}}\right) \le 0 \quad \text{for all } z \in \mathcal{R}^T(x^0). \tag{10.12}$$

So

$$\eta' \int_0^T e^{(T-s)A} B\left(\omega(s)-\widetilde{\omega}(s)\right) ds \le 0 \tag{10.13}$$

for all controls $\omega \in \mathcal{L}_{\mathcal{U}}(0,T)$. That is, it must hold that

$$\int_0^T \lambda(s)' B\left(\omega(s)-\widetilde{\omega}(s)\right) ds \le 0 \tag{10.14}$$

for all ω, where $\lambda(s) := e^{(T-s)A'}\eta$ is the solution of Equation (10.10).

Let $J \subseteq [0,T]$ be the set of points where Equation (10.11) fails to hold. Pick any subset $U_0 \subset \mathcal{U}$ that is dense and countable. From the definition of J, the continuity of $\lambda(t)'Bu$ on u, and the density of U_0, it follows that for each $t \in J$ there is some $u_t \in U_0$ so that

$$\lambda(t)'B\widetilde{\omega}(t) < \lambda(t)'Bu_t .$$

We claim that J has measure zero. Otherwise, by countability of U_0, there must exist some fixed element $u_0 \in U_0$, and some subset J_0 of J with nonzero measure, so that $\lambda(t)'B\widetilde{\omega}(t) < \lambda(t)'Bu_0$ for all $t \in J_0$. Let $\omega(t)$ be the control which equals u_0 whenever $t \in J_0$ and equals $\widetilde{\omega}(t)$ elsewhere. Then the integrand in Equation (10.14) is everywhere nonnegative, and is positive on J_0, contradicting the inequality.

We now prove the converse statement. Let $\widetilde{\omega}$ steer x^0 to x^{f} in time T, and let λ be a nonzero solution of Equation (10.10) so that Equation (10.11) holds for almost all t. Consider any other control $\omega \in \mathcal{L}_{\mathcal{U}}(0,T)$. From Equation (10.11), we know that

$$\lambda(t)'B\widetilde{\omega}(t) \ge \lambda(t)'B\omega(t)$$

for almost all $t \in [0,T]$. Therefore,

$$\eta' \int_0^T e^{(T-s)A} B\left(\omega(s)-\widetilde{\omega}(s)\right) ds \le 0 .$$

As this is true for any ω, Equation (10.12) holds, that is, η is a normal vector to $\mathcal{R}^T(x^0)$ at x^{f}. ∎

Corollary 10.2.4 asserts that, if T is the optimal time for reaching x^{f} from x^0, then $x^{\mathrm{f}} \in \partial\mathcal{R}^T(x^0)$. Thus, we have the following *necessary* condition for optimality as a Corollary of Theorem 47.

Theorem 48 *(Maximum Principle for time-optimal control of linear systems.) Assume that $\widetilde{\omega}$ is a control steering x^0 to x^{f} in minimal time $T > 0$. Then, $x^{\mathrm{f}} \in \partial\mathcal{R}^T(x^0)$, and for every normal vector η to $\mathcal{R}^T(x^0)$ at x^{f}, the (necessarily everywhere nonzero) solution λ of the adjoint linear differential equation (10.10) satisfies Equation (10.11) for almost all $t \in [0,T]$.* □

Definition 10.2.8 *The input $\omega \in \mathcal{L}_{\mathcal{U}}(\sigma, \tau)$* **satisfies the maximum principle** *(for the time-optimal control problem) if there is some nonzero solution λ of Equation (10.10) so that Equation (10.11) holds for almost all t.* □

Another terminology often used is "ω is an extremal". Observe that the property means that there exists some nonzero vector $\eta \in \mathbb{R}^n$ so that

$$\eta' e^{(T-t)A} B\omega(t) = 0 \text{ for almost all } t\,.$$

An equivalent statement is that there must exist some nonzero vector $\gamma \in \mathbb{R}^n$ so that $\gamma' e^{-tA} B\omega(t) = 0$ for almost all t (just use $\gamma = e^{TA'}\eta$).

We have proved that a time-optimal control must satisfy the maximum principle, and that a control which satisfies the maximum principle must lead to boundary points. Under additional assumptions, such a control is also time-optimal, thus having a partial converse to Theorem 48, as follows.

Proposition 10.2.9 Assume that the STLC holds and that either $x^0 = 0$ or $x^{\mathrm{f}} = 0$. Then, if ω steers x^0 to x^{f} in time T, while satisfying the maximum principle, then ω steers x^0 to x^{f} in minimal time T.

Proof. The case $x^0 = 0$ follows from the previous discussion. This is because Theorem 47 implies that $x^{\mathrm{f}} \in \partial\mathcal{R}^T(0)$, and, if the STCL holds, this in turn implies optimality (cf. Proposition 10.2.6).

We will reduce the case $x^{\mathrm{f}} = 0$ to the case $x^0 = 0$, as follows. First we introduce $(\widetilde{A}, \widetilde{B}) := (-A, -B)$, and show that the STLC (with the same control-value set $\mathcal{U}$) holds for $(\widetilde{A}, \widetilde{B})$. Next, we remark that, if ω satisfies the maximum principle and steers x^0 into 0, then the new control $\nu \in \mathcal{L}_{\mathcal{U}}(0, T)$ given by

$$\nu(t) := \omega(T-t) \tag{10.15}$$

satisfies the maximum principle with respect to the system $\dot{x} = \widetilde{A}x + \widetilde{B}u$ and steers 0 into x^0 for that system. Applying the previously established case, we conclude that T is the minimal time for steering x^0 to 0, for the new system, and this is equivalent to T being minimal for the original problem. We now fill-in the simple details.

To see that the STLC holds for $(\widetilde{A}, \widetilde{B})$, fix any $T > 0$ and let V be an open neighborhood of 0 included in $\mathcal{R}^T(0)$. It will be enough to show that $-e^{-TA}V \subseteq \widetilde{\mathcal{R}}^T(0)$ (reachable set for new system), since $-e^{-TA}V$ is an open set containing 0. So pick any $x^0 = -e^{-TA}x^{\mathrm{f}}$, with $x^{\mathrm{f}} \in V$. As $x^{\mathrm{f}} \in \mathcal{R}^T(0)$, there is a control $\omega \in \mathcal{L}_{\mathcal{U}}(0, T)$ so that $x^{\mathrm{f}} = \int_0^T e^{(T-s)A} B\omega(s) ds$. Therefore

$$x^0 = \int_0^T e^{s\widetilde{A}} \widetilde{B}\omega(s)\, ds = \int_0^T e^{(T-r)\widetilde{A}} \widetilde{B}\nu(r)\, dr \in \widetilde{\mathcal{R}}^T(0)\,,$$

where we changed variables by $r = T - s$ and defined ν as in Equation (10.15).

Suppose that ω satisfies the maximum principle, that is, there is some nonzero vector γ so that $\gamma' e^{-tA} B\omega(t) = \max_{u \in \mathcal{U}} \gamma' e^{-tA} Bu$ for all $t \in J$, where

J is a set so that meas $([0,T] \setminus J) = 0$. Let ν be as in Equation (10.15). Take the nonzero vector

$$\widetilde{\gamma} := -e^{T\widetilde{A}'}\gamma = -e^{-TA'}\gamma\,.$$

Then for each $s = T - t \in \{T - t, t \in J\}$, the maximum principle holds for ν with respect to the new system:

$$\begin{aligned}\widetilde{\gamma}' e^{-s\widetilde{A}}\widetilde{B}\nu(s) &= \gamma' e^{(T-s)\widetilde{A}} B\omega(T-s) \;=\; \gamma' e^{-tA} B\omega(t) \\ &= \max_{u\in\mathcal{U}} \gamma' e^{-tA} Bu \;=\; \max_{u\in\mathcal{U}} \gamma' e^{(T-s)\widetilde{A}} Bu \;=\; \max_{u\in\mathcal{U}} \widetilde{\gamma}' e^{-s\widetilde{A}}\widetilde{B}u\,.\end{aligned}$$

Finally, we remark that if ω is any control steering a state x^0 to a state x^{f}, then ν defined as in Equation (10.15) steers x^{f} to x^0 for the new system (and vice versa). This is clear either from the explicit formulas for solution, or simply by noticing that if $\dot{x} = Ax + B\omega$ then $z(t) := x(T-t)$ satisfies $\dot{z} = \widetilde{A}z + \widetilde{B}\nu$.

Applied to $\omega \in \mathcal{L}_{\mathcal{U}}$ satisfying the maximum principle and steering x^0 to 0, we conclude that ν steers 0 to x^0 for the new system, and is therefore (case where initial state is zero) time-optimal. But then T is minimal so that $0 \in \mathcal{R}^T(x^0)$, since if $0 = \alpha_{x^0,S}(\omega_0)$ with some $S < T$ then $\nu_0(t) = \omega_0(S-t)$ contradicts optimality of ν. ∎

10.3 Applications of the Maximum Principle

The maximum principle leads us to study, for each nonzero vector γ, the following function:

$$M_\gamma \,:\, \mathbb{R} \to \mathbb{R}^m \;:\; t \to B' e^{-tA'}\gamma\,.$$

This is an analytic function of t, and its derivatives at $t = 0$ are all zero if and only if $\gamma' A^i B = 0$ for $i = 0, 1, \ldots$, so:

Lemma 10.3.1 The matrix pair (A, B) is controllable if and only if $M_\gamma \not\equiv 0$ for each $\gamma \neq 0$. □

Thus, if (A, B) is controllable, and $\gamma \neq 0$, $M_\gamma(t)$ can be zero only for a discrete set of t's.

The "bang-bang" principle says that time-optimal controls take the most advantage of possible control actions at each instant. The name is motivated by the particular case of an interval $\mathcal{U} = [\underline{u}, \overline{u}]$, where optimal controls must switch between the minimal and maximal values $\underline{u}$ and $\overline{u}$. There are various theorems that make this principle rigorous. The following is the simplest one.

Theorem 49 *(Weak bang-bang.) Assume that the matrix pair (A, B) is controllable. Let $\widetilde{\omega}$ be a control steering x^0 to x^{f} in minimal time $T > 0$. Then, $\widetilde{\omega} \in \partial\mathcal{U}$ for almost all $t \in [0, T]$.*

Proof. For each $\eta \in \mathbb{R}^n$ and each $t \in \mathbb{R}$, we introduce the following function:

$$\mu_{\eta,t} : \mathcal{U} \to \mathbb{R} : u \mapsto M_\eta(t)'u.$$

Thus $\mu_{\eta,t}(u) = \eta' e^{-tA} Bu$ for each u.

Note that by the maximum principle, Theorem 48 (p. 434), $x^{\mathrm{f}} \in \partial\mathcal{R}^T(x^0)$. Fix any normal vector η to $\mathcal{R}^T(x^0)$ at x^{f}. Then the solution of Equation (10.10), $\lambda(t) := e^{(T-t)A'}\eta = e^{-tA'}\gamma$, where $\gamma := e^{TA'}\eta \neq 0$, is so that Equation (10.11) is satisfied (a.e.). That is

$$\mu_{\gamma,t}(\widetilde{\omega}(t)) = \max_{u \in \mathcal{U}} \mu_{\gamma,t}(u) \tag{10.16}$$

for almost all $t \in [0,T]$. In general, let $q \in \mathbb{R}^m$ and consider the function $f : \mathcal{U} \to \mathbb{R}$, $f(u) = q'u$. If f is not identically zero, then every point in $\mathcal{U}$ at which f achieves a maximum relative to $\mathcal{U}$ must belong to the boundary of $\mathcal{U}$ (if $u \in \operatorname{int}\mathcal{U}$, then $u + \varepsilon q \in \mathcal{U}$ for some small $\varepsilon > 0$, but then $f(u+\varepsilon q) = f(u) + \varepsilon \|q\|^2 > f(u)$). In particular, by Lemma 10.3.1, $\mu_{\gamma,t}$ can be the zero function only for t in a finite set $J \subseteq [0,T]$. So $\omega(t) \in \partial\mathcal{U}$ for each $t \notin J$ for which Equation (10.16) holds. ∎

Exercise 10.3.2 Give an example of a pair (A,B) (necessarily, not controllable), a convex compact $\mathcal{U}$, and x^0, x^{f}, T, and $\widetilde{\omega}$, so that $\widetilde{\omega}$ steers x^0 to x^{f} in minimal time $T > 0$ but $\widetilde{\omega}(t) \notin \partial\mathcal{U}$ for any $t \in [0,T]$. □

For a large class of systems, one can obtain uniqueness.

Definition 10.3.3 *The system $\dot{x} = Ax + Bu$ with control-value set $\mathcal{U}$ is* **normal*** *if, for each nonzero vector p, for almost all $t \in \mathbb{R}$ the function $\mu_{p,t}$ achieves its maximum at exactly one point of $\mathcal{U}$.* □

In the normal case, given any nonzero vector $\eta \in \mathbb{R}^n$ and any $T > 0$, we define the control

$$\widetilde{\omega}_\eta(t) = \operatorname{argmax}_{u \in \mathcal{U}} \mu_{\gamma,t}(u), \tag{10.17}$$

where $\gamma := e^{TA'}\eta$. The "argmax" in Equation (10.17) is well-defined for almost all t, and we let $\widetilde{\omega}_\eta(t)$ be any fixed element of $\mathcal{U}$ if t is in the set (of measure zero) where the maximum is not uniquely attained.

The main result for the normal case is as follows.

Theorem 50 *Assume that the system (10.1) with control-value set $\mathcal{U}$ is normal. Then, for each two states x^0 and $x^{\mathrm{f}} \in \mathcal{R}(x^0)$, there is a unique time-optimal control $\widetilde{\omega}$ steering x^0 to x^{f}.*

---

*The term "normal" is used in a different manner than for "normal vectors", but, unfortunately, this is by now standard practice.

Proof. Let T be minimal with $x^{\mathrm{f}} \in \mathcal{R}^T(x^0)$. We fix a normal vector η to $R^T(x^0)$ at x^{f}, and define $\widetilde{\omega}_\eta$ as in Equation (10.17). We claim that this is the unique optimal control. Indeed, assume that $\widetilde{\omega}$ is such that $\alpha_{x^0,T}(\widetilde{\omega}) = x^{\mathrm{f}}$. Then the maximum principle and normality (cf. Equations (10.16) and (10.17)) give $\widetilde{\omega}(t) = \widetilde{\omega}_\eta(t)$ for almost all $t \in [0,T]$. ∎

Two important special cases of normality are as follows.

A *strictly convex* set C is a convex set with the property that every supporting hyperplane H to C intersects C at exactly one point. For instance, any ball in Euclidean space is strictly convex. An equivalent statement for compact convex $C \subseteq \mathbb{R}^l$ is that C is strictly convex iff each nonzero linear function $f : \mathbb{R}^l \to \mathbb{R}$ achieves its maximum on C at a unique point of C (since $H = \{x \in \mathbb{R}^n \mid f(x) = \max_{c \in C} f(c)\}$ is a supporting hyperplane to C, and, conversely, any supporting hyperplane can be represented in this fashion).

We say that $\mathcal{U}$ is a *hypercube* if there are real numbers $\underline{u}_i < \overline{u}_i$, $i = 1, \ldots, m$, so that

$$\mathcal{U} = [\underline{u}_1, \overline{u}_1] \times [\underline{u}_2, \overline{u}_2] \times \ldots \times [\underline{u}_m, \overline{u}_m]$$

Observe that for any function of the form $f(u) = p'u$ so that $p \in \mathbb{R}^m$ has all coordinates $p_i \neq 0$, then f achieves its maximum on $\mathcal{U}$ at the unique point $(\widetilde{u}_1, \ldots, \widetilde{u}_m)'$, where

$$\widetilde{u}_i = \begin{cases} \overline{u}_i & \text{if } p_i > 0 \\ \underline{u}_i & \text{if } p_i < 0 \,. \end{cases} \tag{10.18}$$

Lemma 10.3.4 Assume that either:

1. the pair (A, B) is controllable and $\mathcal{U}$ is strictly convex, or
2. each pair (A, b_i) (b_i is the ith column of B) is controllable and $\mathcal{U}$ is a hypercube.

Then the system (10.1), with control-value set $\mathcal{U}$, is normal.

Proof. By definition of strict convexity, $\mu_{\gamma,t}$ achieves its maximum at exactly one point whenever $\mu_{\gamma,t}$ is nonzero; by controllability, this happens for almost all t. In the hypercube case, we need to guarantee that the coordinates of $p = M_\gamma(t) = B'e^{-tA'}\gamma$ are nonzero; but these coordinates are $b_i'e^{-tA'}\gamma$, and are not identically zero because each (A, b_i) is controllable, so the unique minimizing point is given as in Equation (10.18). ∎

It is worth summarizing the conclusions in the hypercube case. Note that these conclusions apply to all single-input ($m = 1$) systems, since every closed convex subset of $\mathbb{R}$ is of the form $[\underline{u}, \overline{u}]$.

Theorem 51 *Suppose that each pair (A, b_i) is controllable and that $\mathcal{U}$ is a hypercube. Then, for each two states x^0 and $x^{\mathrm{f}} \in \mathcal{R}(x^0)$, there is a unique time-optimal control $\widetilde{\omega}$ steering x^0 to x^{f}. Moreover, there is a nonzero vector $\gamma \in \mathbb{R}^n$ so that the ith coordinate of $\widetilde{\omega}$, $i = 1, \ldots, m$, satisfies*

$$\widetilde{\omega}(t)_i = \begin{cases} \overline{u}_i & \text{if } \gamma' e^{-tA} b_i > 0 \\ \underline{u}_i & \text{if } \gamma' e^{-tA} b_i < 0 \end{cases} \tag{10.19}$$

for almost all $t \in [0,T]$. □

Given the control $\tilde{\omega}$ taking x^0 to x^{f} in optimal time, we can always modify $\tilde{\omega}$ on a set of measure zero, so that Equation (10.19) holds for *all* t in the set

$$J = \{t \mid \gamma' e^{-tA} b_i \neq 0 \text{ for all } i = 1, \ldots, m\}.$$

The complement of this set is the union of the sets of zeros of the analytic functions $\gamma' e^{-tA} b_i$ on the interval $[0,T]$, so is finite. So we can rephrase our conclusion by saying that, for the normal hypercube case, optimal controls are *piecewise constant*, taking values in the vertices of $\mathcal{U}$, and have a finite number of possible switches, in the complement of J.

The following is an interesting remark concerning the geometry of reachable sets.

Proposition 10.3.5 Assume that the system (10.1), with control-value set $\mathcal{U}$, is normal. Then, for each state x^0 and each $T > 0$, the set $\mathcal{R}^T(x^0)$ is strictly convex.

Proof. Suppose that $H = \{x \mid \eta' x = \theta\}$ is a supporting hyperplane to $\mathcal{R}^T(x^0)$. Let $\tilde{\omega}_\eta$ be as in Equation (10.17). Pick any two points x_1 and x_2 in $H \bigcap \mathcal{R}^T(x^0)$; we must prove that $x_1 = x_2$. Choose two controls ω_i so that $x_i = \alpha_{x^0,T}(\omega_i)$, $i = 1, 2$. Then, because x_1 and x_2 both belong to $\partial\mathcal{R}^T(x^0)$, and η is normal to $\mathcal{R}^T(x^0)$ at both x_1 and x_2, Theorem 47 (p. 433) says that

$$\omega_1(t) = \tilde{\omega}_\eta = \omega_2(t),$$

so $x_1 = \alpha_{x^0,T}(\omega_1) = \alpha_{x^0,T}(\omega_2) = x_2$, as desired. ■

Scalar Systems

We specialize now to single input systems ($m = 1$), and write b instead of $B = b_1$. In general $\mathcal{U} = [\underline{u}, \overline{u}]$, but we will take, in order to simplify the exposition, $\underline{u} = -1$ and $\underline{u} = 1$. We assume that the pair (A, b) is controllable.

For each two states x^0 and x^{f}, there is a unique time-optimal control $\tilde{\omega}$ steering x^0 to x^{f}, and there is a nonzero vector $\gamma \in \mathbb{R}^n$ so that

$$\tilde{\omega}(t) = \operatorname{sign}(\gamma' e^{-tA} b) \tag{10.20}$$

for all $t \notin S_{\gamma,T}$, where

$$S_{\gamma,T} = \{t \in [0,T] \mid \gamma' e^{-tA} b = 0\}$$

is a finite set.

Although finite, the sets $S_{\gamma,T}$ may have arbitrary cardinalities. For instance, if we take the harmonic oscillator matrix

$$A = \begin{pmatrix} 0 & 1 \\ -1 & 0 \end{pmatrix},$$

we have

$$(0\ \ 1)\, e^{-tA} \begin{pmatrix} 0 \\ 1 \end{pmatrix} = (0\ \ 1) \begin{pmatrix} \cos t & -\sin t \\ \sin t & \cos t \end{pmatrix} \begin{pmatrix} 0 \\ 1 \end{pmatrix} = \cos t\,.$$

Exercise 10.3.6 Give an example of a controllable system with $\mathcal{U} = [-1, 1]$, having the following property: for each integer k there exists a state x_k^{f} such that the (unique) minimum-time control steering 0 to x_k^{f} switches sign k times. (*Hint:* To prove optimality, use Proposition 10.2.9.) □

However, there is a case of some interest where a very strong conclusion may be drawn.

Proposition 10.3.7 Suppose that the matrix A has only real eigenvalues. Then, for each γ, b, and T, $S_{\gamma,T}$ has at most $n-1$ elements.

Proof. By an *exponential polynomial* we will mean a function of the form

$$P(t) = \sum_{i=1}^{k} p_i(t) e^{\alpha_i t}\,,$$

where $\alpha_1 < \alpha_2 < \ldots < \alpha_k$ are real numbers and each p_i a polynomial. We write $\pi(P) := \sum_{i=1}^{k}(1+\delta_i)$, where δ_i is the degree of p_i. We will prove that any such P, if not identically zero, can have at most $\pi(P) - 1$ (distinct) real zeros. This will give the result, because, using the Jordan form of A, we know that $\gamma' e^{-tA} b$ is an exponential polynomial with $\pi(P) \leq n$.

We proceed by induction on $\pi(P)$. If $\pi(P) = 1$ then $P = ce^{\alpha t}$ has no zeros. Suppose now that the result has been proved for n. Take P with $\pi(P) = n+1$ and not identically zero. Assume that P would have $\geq n+1$ zeros. Then

$$P_0(t) := e^{-\alpha_1 t} P(t) = p_1(t) + \sum_{i=2}^{k} p_i(t) e^{(\alpha_i - \alpha_1)t}$$

has the same $\geq n+1$ zeros. This implies that $P_0' = p_1' + \ldots$ has at least n zeros. But $\pi(P_0') \leq n$, so, by the inductive hypothesis, $P_0' \equiv 0$. This means that P_0 is constant, and thus $P = e^{\alpha t} P_0$ is either a constant (if $\alpha_1 = 0$) or an exponential function. In either case, P could not have had $\geq n+1$ zeros. ■

An Example

The simplest nontrivial example is the double-integrator system $\ddot{x} = u$, written in state-space form: $m = 1$, $\mathcal{U} = [-1, 1]$, $n = 2$, and matrices

$$A = \begin{pmatrix} 0 & 1 \\ 0 & 0 \end{pmatrix}, \quad b = \begin{pmatrix} 0 \\ 1 \end{pmatrix}. \tag{10.21}$$

This models a mass moving in one dimension, with no friction, and subject to a force of magnitude at most one in either direction. We know from the previous results that optimal controls are unique (up to sets of measure zero), and are bang-bang with at most one sign change (since $n-1=1$). That is, any time-optimal control joining a pair of states must necessarily be either constant and equal to 1 or -1, or it must start at one of these values and then switch (once) to the other value. For this system, we will solve the problem of time-optimal control from any state x^0 to $x^{\mathrm{f}}=0$ (that is, we will find the "optimal synthesis" that specifies optimal controls from each initial point to the origin).

We have, explicitly,

$$e^{-tA} = \begin{pmatrix} 1 & -t \\ 0 & 1 \end{pmatrix},$$

so time-optimal controls all have the form

$$\widetilde{\omega}(t) = \operatorname{sign}(a-ct) \tag{10.22}$$

for some real numbers a and c which are not both zero (and $\widetilde{\omega}(T_0)$ is arbitrary when $a-cT_0=0$, which can happen at most at one point $T_0 \in [0,T]$). So there are four possibilities for a time-optimal control $\widetilde{\omega}$ defined on an interval $[0,T]$:

- $\widetilde{\omega} \equiv -1$
- $\widetilde{\omega} \equiv 1$
- for some $T_0 \in (0,T)$, $\widetilde{\omega} \equiv -1$ on $[0,T_0)$ and $\widetilde{\omega}(t) \equiv 1$ on $(T_0,T]$
- for some $T_0 \in (0,T)$, $\widetilde{\omega} \equiv 1$ on $[0,T_0)$ and $\widetilde{\omega}(t) \equiv -1$ on $(T_0,T]$.

Conversely, any control of one of these four types, which steers a state x^0 into 0, must be the unique time-optimal control doing so. This is because, first of all, any such control satisfies the maximum principle (just pick $\gamma=(c,a)'$ with a and c chosen so that $a-ct$ has the right sign, switching at T_0 in the two last cases), and second, because in this case (STLC, control to zero final state), the maximum principle is a sufficient as well as necessary test (cf. Proposition 10.2.9). That is, any bang-bang control with at most one switch is optimal, for this problem.

We let S be the set consisting of all those states which can be steered to $x^{\mathrm{f}}=0$ using constant controls. Thus $S=S_- \bigcup S_+$, the sets of all points that can be controlled to zero using $u\equiv -1$ and $u \equiv 1$ respectively. To find these sets, we solve the differential equations $\dot{x}=Ax-b$ and $\dot{x}=Ax+b$ respectively, both with final value $x(t)=0$, obtaining two functions $(-\infty,0] \to \mathbb{R}^2$, ξ_- and ξ_+ respectively:

$$\xi_-(t) = \begin{pmatrix} -\frac{t^2}{2} \\ -t \end{pmatrix}, \quad \xi_+(t) = \begin{pmatrix} \frac{t^2}{2} \\ t \end{pmatrix}.$$

Hence S_- is the half-parabola $\{(-x_2^2/2, x_2), x_2 \geq 0\}$ (in the second quadrant) and S_+ is the half-parabola $\{(x_2^2/2, x_2), x_2 \leq 0\}$ (fourth quadrant). The set

$$S = S_- \bigcup S_+ = \{2x_1 + x_2|x_2| = 0\}$$

splits $\mathbb{R}^2$ into two parts, one "under" S and the other one "over" S.

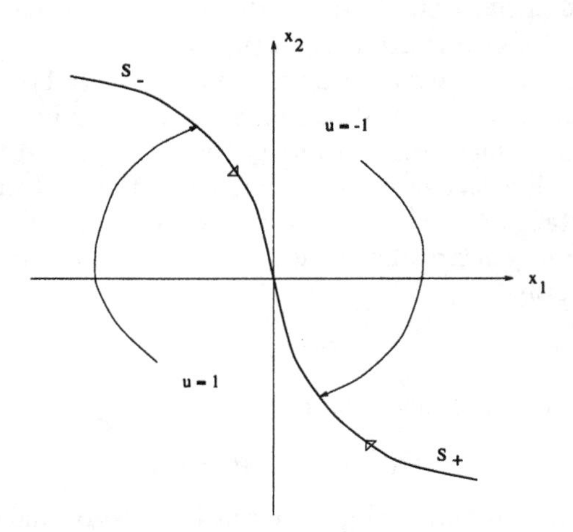

Figure 10.1: *Switching surfaces.*

We claim the following facts:

- If the initial state x^0 is in S_+, then the unique time optimal control to reach 0 is obtained by applying $u \equiv 1$.
- If the initial state x^0 is in S_-, then the unique time optimal control to reach 0 is obtained by applying $u \equiv -1$.
- If the initial state x^0 is "over" the set S, then the unique time optimal control to reach 0 is obtained by applying $u \equiv -1$ until the set S_+ is reached, and then switching to $u \equiv 1$.
- If the initial state x^0 is "under" the set S, then the unique time optimal control to reach 0 is obtained by applying $u \equiv 1$ until the set S_- is reached, and then switching to $u \equiv -1$.

Recall that, for this example, every bang-bang control with at most one switch is optimal, so all we need to do is to show that if we start "over" the set S and apply $u \equiv -1$, the trajectory $x(\cdot)$ that results indeed crosses S_+ (and analogously if we start under S). To establish this, one could compute solutions, but we can also argue as follows: since $\dot{x}_2 = -1 < 0$, eventually $x_2(t)$ becomes negative. (It could not happen that $x(t_0) \in S_-$ for some t_0, since in that case, solving the equation $\dot{x} = Ax - b$ backward from t_0 gives that $x(t) \in S_-$ for all $t \leq t_0$, because of the definition of S_-, contradicting the fact that $x(0) = x^0$ is not in S.) Once that $x_2(t)$ becomes negative, we have that $\dot{x}_1(t) = x_2(t) < 0$, so trajectories move in a "southwest" direction, with $\dot{x}_2 \equiv -1$, and hence must eventually meet S_+. The case where we start under S is proved in the same manner.

Exercise 10.3.8 For the above example, take in particular the initial state $x^0 = (-1, 0)'$. Show that the above synthesis results in the rule "accelerate at maximum power for the first half of the time and decelerate maximally for the second half". Derive this result, for that particular initial state, using elementary arguments not involving the maximum principle. □

10.4 Remarks on the Maximum Principle

For fixed x^0, we may define the Bellman or value function for the problem of time-optimal control from x^0:

$$V(x^{\mathrm{f}}) := \min\{T \geq 0 \mid x^{\mathrm{f}} \in \mathcal{R}^T(x^0)\}.$$

This is a function $\mathcal{R}(x^0) \to \mathbb{R}$. (Reversing time, we could also define an analogous V for the problem of control *to* x.) Being defined by a minimum, this function tends not to be differentiable; but when it is, its gradient is a normal vector as needed for the maximum principle:

Lemma 10.4.1 Assume that V is differentiable at some $x^{\mathrm{f}} \in \operatorname{int} \mathcal{R}(x^0)$. Let $T := V(x^{\mathrm{f}})$. Then $\eta = \nabla V(x^{\mathrm{f}})'$ is normal to $\mathcal{R}^T(x^0)$ at x^{f}.

Proof. We start by establishing the following claim: in general, if $V : \mathcal{O} \to \mathbb{R}$ is a function defined on an open subset of $\mathbb{R}^n$, $C \subseteq \mathbb{R}^n$ is convex, and V achieves a local maximum relative to $C \bigcap \mathcal{O}$ at some point $\zeta \in C \bigcap \mathcal{O}$, then $\nabla V(\zeta)'(z - \zeta) \leq 0$ for all $z \in C$. Pick any $z \in C$, and substitute the vector $h = \varepsilon(z - \zeta)$, for any fixed small enough $\varepsilon \in (0, 1]$, into the expansion $V(\zeta + h) = V(\zeta) + \nabla V(\zeta)'h + o(\|h\|)$. Since $\zeta + h = \varepsilon z + (1 - \varepsilon)\zeta \in C$, if ε is small enough, $V(\zeta + h) \leq V(\zeta)$ gives

$$\nabla V(\zeta)'(z - \zeta) + \frac{o(\varepsilon \|z - \zeta\|)}{\varepsilon} \leq 0.$$

Letting $\varepsilon \to 0$ gives the claim.

Now we apply this general fact with $C = \mathcal{R}^T(x^0)$ and $\mathcal{O} = \operatorname{int} \mathcal{R}(x^0)$; note that V achieves a local maximum at x^{f} because $V(z) \leq T = V(x^0)$ if $z \in \mathcal{R}^T(x^0)$. ■

Remark 10.4.2 When V is not differentiable, it is possible to generalize the interpretation of ∇V by means of "viscosity" or "subgradient" terms. □

Exercise 10.4.3 (a) Consider again the double integrator example in Equation (10.21), with $\mathcal{U} = [-1, 1]$, but now provide a solution to the problem of controllability *from* $x^0 = 0$ *to* an arbitrary state $x^{\mathrm{f}} \in \mathbb{R}^n$. Justify carefully why your solutions are indeed optimal, and make an illustration analogous to Figure 10.1.
(b) For the problem in (a), let $V(x^{\mathrm{f}}) = \min\{T \mid x^{\mathrm{f}} \in \mathcal{R}^T(0)\}$. Provide an explicit formula for the function V. What can you say about differentiability of V? □

Remark 10.4.4 The maximum principle is not merely useful as a technique for finding optimal controls or for characterizing their properties. It is also an extremely useful tool for solving the apparently different problem of simply *finding* a control steering one state to another, because it allows one to restrict the search to a smaller and often more manageable class of controls than all of $\mathcal{L}^\infty$. Suppose that we wish to find a control ω, if one exists, that steers a given initial state x^0 to a desired target state x^{f}. Assuming appropriate technical hypotheses of normality as discussed earlier, if there is a control that works, then there will be a time-optimal one $\widetilde{\omega}$, which satisfies the maximum principle. Define the function $m : \mathbb{R}^n \to \mathcal{U}$ by the formula $m(p) := \operatorname{argmax}_{u \in \mathcal{U}} p'Bu$. Then, we know that there must be some (as yet unknown) nonzero vector $\gamma \in \mathbb{R}^n$ so that, solving the system of $2n$ differential equations

$$
\begin{aligned}
\dot{x} &= Ax + Bm(\lambda)\,, & x(0) &= x^0 \\
\dot{\lambda} &= -A'\lambda\,, & \lambda(0) &= \gamma\,,
\end{aligned}
$$

one obtains that $x(\tau) = x^{\mathrm{f}}$ for some τ. Now, a numerical search can be carried out over the space of possible initial adjoint vectors γ, solving this differential equation, and testing along solutions whether $\|x(t) - x^{\mathrm{f}}\|$ becomes smaller than a certain error tolerance. Thus the search has been reduced from a search in an infinite-dimensional space to searching over a space of dimension n, or, even better, $n-1$ (since, without loss of generality, one could take γ to belong to the unit sphere in $\mathbb{R}^n$). There are technical difficulties to be dealt with, including the fact that the right-hand side of the equation may not depend continuously on λ (because m may not), but the idea is nonetheless very powerful. A related application of the same idea is as a general technique for estimating the boundaries of reachable sets $\mathcal{R}^T(x^0)$, which can proceed in the same manner (since $x(t) \in \mathcal{R}^t(x^0)$ when solutions have been generated in this manner).

A different technique, motivated by bang-bang theorems, is to look for bang-bang controls, parametrized finitely by switching times, assuming an a priori bound is known on the number of switches required (as, for instance, in the cases covered by Proposition 10.3.7). □

10.5 Additional Exercises

In all exercises, we assume given a linear time-invariant continuous time system $\dot{x} = Ax + Bu$ and $\mathcal{U}$ compact convex.

Exercise 10.5.1 Assuming STLC holds, let the minimal time function V be defined as above, for reachability from the initial state $x^0 = 0$. Prove: the level sets $\{V(x) = T\}$ are precisely the boundaries of the reachable sets $\mathcal{R}^T(0)$. □

Exercise 10.5.2 Let $\mathcal{C}^T(x^{\mathrm{f}}) := \{x^0 \mid x^{\mathrm{f}} \in \mathcal{R}^T(x^0)\}$ (set of states that may be steered to x^{f} in time T and let $\mathcal{C}(x^{\mathrm{f}}) = \bigcup_{T \geq 0} \mathcal{C}^T(x^{\mathrm{f}})$. Prove:

- If $x^0 \in \mathcal{C}^T(x^{\mathrm{f}})$ and T is minimal with this property, then $x^0 \in \partial\mathcal{C}^T(x^{\mathrm{f}})$.

- Assume that $x^0 \in \partial\mathcal{C}^T(0)$ and that the STLC is satisfied. Then T is minimal with the property that $x^0 \in \mathcal{C}^T(0)$. □

Exercise 10.5.3 Let $\mathcal{R}^{\leq T}(x^0) := \bigcup_{0 \leq t \leq T} \mathcal{R}^t(x^0)$. Show that, for any x and T, $\mathcal{R}^{\leq T}(x^0)$ is:
- connected,
- compact,
- but not necessarily convex.

(*Hint:* (For compactness.) If $x^{\mathrm{f}}_k \in \mathcal{R}^{t_k}(x^0)$, you may assume that $t_k \searrow T$ or $t_k \searrow T$ for some T (why?). Then, restrict or extend controls ω_k to $[0, T]$. Finally, use a compactness argument as in the proof that $\mathcal{R}^T(x^0)$ is compact.) □

Exercise 10.5.4 For any metric space M, we use $\mathbb{K}(M)$ to denote the family of all nonempty compact subsets of M, and define

$$D(C_1, C_2) := \max\left\{\max_{x\in C_1} d(x, C_2), \max_{x\in C_2} d(x, C_1)\right\}.$$

Show that D defines a metric on $\mathbb{K}$ (usually called the Hausdorff metric). Now consider a linear system $\dot{x} = Ax + Bu$ and a fixed initial state $x^0 \in \mathbb{R}^n$. Show that the mapping $T \mapsto \mathcal{R}^T(x^0)$ is continuous as a map from $\mathbb{R}$ into $\mathbb{K}(\mathbb{R}^n)$. □

Exercise 10.5.5 Consider the undamped harmonic oscillator $\dot{x}_1 = x_2$, $\dot{x}_2 = -x_1 + u$, with control-value set $\mathcal{U} = [-1, 1]$. Show, for each two states x^0, x^{f}:

1. there is a unique control $\widetilde{\omega}$ steering x^0 to x^{f} in minimal time,
2. this control has $|u(t)| = 1$ for almost all t, and
3. the intervals between sign changes have length π.

Finally, provide a complete description of the optimal controls for the problem of transferring any state to $x^{\mathrm{f}} = 0$, including a diagram. (*Hint:* For the controllability part, recall Exercise 3.6.8.) □

Exercise 10.5.6 Consider the system $\dot{x}_1 = x_1 + u$, $\dot{x}_2 = x_2 + u$, with control-value set $\mathcal{U} = [-1, 1]$. Take $x^0 = (1, 0)'$ and $x^{\mathrm{f}} = (2, 0)'$. What is the minimum T so that $x^{\mathrm{f}} \in \mathcal{R}^T(x^0)$? Show that the maximum principle provides no information whatsoever about a time-optimal control steering x^0 to x^{f}. (What property fails?) Show that there are, in fact, infinitely many controls steering x^0 to x^{f} in minimal time. □

Exercise 10.5.7 Consider this nonlinear system, with $\mathcal{X} = \mathbb{R}^2$ and $\mathcal{U} = [-1, 1]$:

$$\begin{aligned} \dot{x}_1 &= (1 - x_2^2)\, u^2 \\ \dot{x}_2 &= u\,. \end{aligned}$$

Show that the system is complete, that is, solutions exist for every initial state and every control. Show, however, that $\mathcal{R}^1(0)$ is not a closed set. Why does this not contradict the discussion in Remark 10.1.10? (*Hint:* Consider the states $x^{\mathrm{f}}_k = \alpha(\omega_k)$, where ω_k is a control that switches fast between $+1$ and -1.) □

Exercise 10.5.8 (a) Consider a controllable single-input system $\dot{x} = Ax + bu$. Show that there is some $\delta > 0$ with the following property: for each $\gamma \neq 0$, the function $t \mapsto \gamma' e^{-tA} b$ has at most $n-1$ zeros in the interval $[0, \delta]$.
(b) Conclude that, for controllable single-input systems, with $\mathcal{U} = [\underline{u}, \overline{u}]$, there is some $\delta > 0$ so that, whenever x^0 and x^{f} are so that x^{f} can be reached in time at most δ from x^0, the time optimal control steering x^0 to x^{f} is piecewise constant, taking values $\underline{u}$ or $\overline{u}$ in at most n intervals.
(c) Conclude further that, if T is the minimum time for steering x^0 to x^{f}, then the time optimal control steering a state x^0 to a state x^{f} is piecewise constant, with at most Tn/δ switches.
(*Hint:* (For (a).) Assume the property is false on $[0, 1/k]$. Note that one may restrict attention to γ's so that $\|\gamma\| = 1$. What can be said about the zeros of derivatives of $\gamma' e^{-tA} b$? Take limits $k \to \infty$.) □

10.6 Notes and Comments

Weak convergence as defined here is a particular case of a concept that is standard in functional analysis. In general, if N is a normed space and N^* is its dual (the set of continuous linear functionals $N \to \mathbb{R}$), the *weak-star* topology on N^* is the topology defined by the property that a net $\{x_k^*\}$ in N^* converges to $x^* \in N^*$ if and only if $x_k^*(x) \to x^*(x)$ for all $x \in N$. In particular, we may take $N = \mathcal{L}^1_m(0, T)$, the set of integrable functions $\varphi : [0, T] \to \mathbb{R}^m$ with the norm $\int_0^T \|\omega(s)\|\, ds$. The space $\mathcal{L}^\infty_m(0, T)$, with the norm $\|\cdot\|_\infty$, can be naturally identified with the dual of N via

$$\omega(\varphi) := \int_0^T \varphi(s)'\omega(s)\, ds\,.$$

Thus, a net ω_k (in particular, a sequence) converges to ω in the weak-star sense if and only if $\int_0^T \varphi(s)'\omega_k(s)\, ds \to \int_0^T \varphi(s)'\omega(s)\, ds$ for all integrable functions $\varphi : [0, T] \to \mathbb{R}^m$. So what we called weak convergence is the same as weak-star convergence. Alaoglu's Theorem (see e.g. [98], Theorem V.3.1). asserts that the closed unit ball in N^* (or equivalently, any closed ball) is compact in the weak-star topology. This proves Lemma 10.1.2. However, we provided a self-contained proof.

In connection with Remark 10.1.12, we note that it is possible to relax the convexity assumption; see [188], which may also be consulted for many more results on the material covered in this chapter.

In connection with Remark 10.4.2, the reader is directed to [93].

Appendix A

Linear Algebra

A.1 Operator Norms

If $x \in \mathbb{C}^l$, we introduce the Euclidean norm

$$\|x\| = \sqrt{|x_1|^2 + \ldots + |x_l|^2}.$$

If A is a complex $n \times m$ matrix, by $\|A\|$ we denote the operator norm

$$\|A\| := \max_{\|\mu\|=1} \|A\mu\| \tag{A.1}$$

with respect to the above norm on $\mathbb{C}^n$ and $\mathbb{C}^m$.

Numerically, $\|A\|$ is just the largest singular value of A, that is, the square root of the largest eigenvalue λ_1 of the Hermitian matrix

$$Q := A^* A$$

(where the star indicates conjugate transpose), as discussed later in this Appendix. If v_1 is an eigenvector corresponding to λ_1, then $\|Av_1\| = \|A\| \, \|v_1\|$. In particular, when A is a real matrix, v_1 can be chosen real, so the same value $\|A\|$ is obtained when maximizing just over *real* vectors μ such that $\|\mu\| = 1$.

It is a basic fact from linear algebra that there are positive constants c_1, c_2, which depend only on n, m, such that

$$c_1 \sqrt{\sum_{i,j} a_{ij}^2} \;\leq\; \|A\| \;\leq\; c_2 \sqrt{\sum_{i,j} a_{ij}^2} \tag{A.2}$$

for all such A.

If $\{A_l\}$ is a sequence of matrices (all of the same size $n \times m$), we define convergence

$$A_l \to B$$

to mean that $\|A_l - B\| \to 0$. Equivalently, because of (A.2), this means that

$$(A_l)_{ij} \to B_{ij}$$

for each i, j.

A.2 Singular Values

The concepts of singular value and singular value decompositions are very useful in linear control theory. We develop here the basic facts about these.

Let A be a complex $n \times l$ matrix of rank r. Using "$*$" to denote conjugate transpose, let $Q := A^*A$. Since Q is Hermitian and positive semidefinite, it admits an orthonormal basis of eigenvectors,

$$Qv_i = \lambda_i v_i\,, \quad i = 1, \ldots, l\,, \tag{A.3}$$

where the eigenvalues λ_i are real,

$$0 \le \lambda_l \le \lambda_{l-1} \le \ldots \le \lambda_1\,.$$

Note that $\operatorname{rank} Q = r$, so $\lambda_{r+1} = \ldots = \lambda_l = 0$.

We define $\sigma_i := \sqrt{\lambda_i}$ for each $i = 1, \ldots, r$. By definition, these are the **singular values** of the matrix A. In particular, one often denotes σ_r as

$$\sigma_{\min}(A)\,,$$

and this is the *smallest singular value* of A. When A has full column rank —that is, $\ker A = 0$, $r = l$, and Q is nonsingular— $\|Ax\|^2 = x^*Qx \ge \sigma_r^2 \|x\|^2$, so

$$\|Ax\| \;\ge\; \sigma_{\min}(A)\;\|x\| \tag{A.4}$$

for all vectors x; so $\sigma_{\min}(A)$ provides a measure of how far A is from being singular, in the sense of having a nonzero kernel; in fact, as proved below, this number measures the precise distance to the set of singular matrices. As $\|Ax\|^2 = x^*Qx \le \sigma_1^2 \|x\|^2$ for all x and $\|Av_1\|^2 = \sigma_1^2 \|v_1\|^2$, the largest singular value σ_1 is the same as the operator norm of A with respect to Euclidean norms on vectors, as discussed above in Section A.1.

Then, we introduce

$$u_i := \frac{1}{\sigma_i} A v_i \in \mathbb{C}^n\,, \quad i = 1, \ldots, r\,,$$

and note that

$$u_i^* u_j = \frac{1}{\sigma_i \sigma_j} v_i^* Q v_j = \frac{\lambda_j}{\sigma_i \sigma_j} v_i^* v_j$$

for each i, j between 1 and r, and therefore that the set

$$\{u_1, \ldots, u_r\}$$

is orthonormal. By the Gram-Schmidt procedure, we complete this to an orthogonal basis of $\mathbb{C}^n$ and let

$$U := (u_1, \ldots, u_n)$$

be the unitary matrix that results from listing all elements of that basis. Note that, for $j \leq r$ and any i,

$$u_i^* A v_j = \sigma_j u_i^* u_j = \begin{cases} 0 & \text{if } i \neq j \\ \sigma_j & \text{otherwise,} \end{cases}$$

and that for every $j = r+1, \ldots, l$ the product is zero since v_j is in the kernel of A. Thus, if we define V to be the unitary matrix whose columns are the vectors v_j, $j = 1, \ldots, l$, then $U^*AV = \Sigma$, where Σ is the matrix

$$\begin{matrix} & r & l-r \\ r & \begin{pmatrix} \begin{pmatrix} \sigma_1 & & \\ & \ddots & \\ & & \sigma_r \end{pmatrix} & 0 \\ n-r \quad 0 & 0 \end{pmatrix} \end{matrix} \tag{A.5}$$

that is, an $n \times l$ matrix whose only nonzero entries are the first r principal diagonal entries Σ_{ii}, and these are precisely the singular values of A. One calls the equality

$$A = U \Sigma V^* \tag{SVD}$$

a **singular value decomposition (SVD)** of A. From this, one can also write A as a sum of rank one matrices

$$A = \sum_{i=1}^{r} \sigma_i u_i v_i^* \,,$$

and this sum is itself sometimes called the SVD of A. Note that for real matrices A, both U and V can be taken to be real, that is, orthogonal matrices, since the constructions can all be done in that case over $\mathbb{R}$. Singular value decompositions can be computed efficiently and reliably using any of the standard numerical packages such as LINPACK.

Exercise A.2.1 Assume that there exist unitary matrices U, V so that $A = U\Sigma V^*$, where Σ is as in (A.5). Prove that the σ_i's must be the singular values of A. □

Lemma/Exercise A.2.2 Prove that for every matrix A: (i) A, its conjugate transpose A^*, and its transpose A' all have the same singular values; and (ii) for every nonzero $\lambda \in \mathbb{C}$ and every i, $\sigma_i(\lambda A) = |\lambda| \, \sigma_i(A)$. □

One simple application of SVDs is in computing the *Moore-Penrose inverse* or *pseudoinverse* of A. (Also called the *generalized inverse.*) This is the matrix

$$A^{\#} := V\Sigma^{\#}U^{*}\,, \tag{A.6}$$

where $\Sigma^{\#}$ is the $l \times n$ matrix

$$\begin{array}{c} \\ r \\ \\ l-r \end{array} \begin{pmatrix} \begin{pmatrix} \sigma_1^{-1} & & \\ & \ddots & \\ & & \sigma_r^{-1} \end{pmatrix} & 0 \\ 0 & 0 \end{pmatrix} \quad (\text{columns: } r,\ n-r)$$

whose only nonzero elements are the inverses of the singular values of A, placed on the main diagonal. The pseudoinverse solves the general least-squares problem of finding the vector x of minimum norm among all those vectors minimizing

$$\|Ax - b\|\,.$$

We claim that $x := A^{\#}b$ is the (unique) such minimal solution. Indeed, given any x and b,

$$\|Ax - b\|^2 = \|\Sigma y - c\|^2\,,$$

where $y := V^{*}x$ and $c := U^{*}b$; since x and y have the same norm, because V is unitary, it is equivalent to minimize as a function of y and then to substitute again x. But this expression is just

$$\sum_{i=1}^{r} |\sigma_i y_i - c_i|^2 + \sum_{i=r+1}^{l} |c_i|^2\,,$$

which is minimized when $y_i := c_i/\sigma_i$ for $i = 1, \ldots, r$. The other coordinates of y have no influence. Among all such vectors, minimum norm is achieved when setting all other coordinates of y to zero, that is,

$$y = \Sigma^{\#}c = \Sigma^{\#}U^{*}b\,,$$

from which it follows that $x = Vy = A^{\#}b$, as claimed.

In the text we deal with pseudoinverses of operators between an infinite dimensional space and a finite dimensional one, under a full rank assumption that results in a simpler formula for the pseudoinverse; the next exercise provides the connection.

Exercise A.2.3 Assume that A has full row rank $r = n$. Show that, then, the formula for the pseudoinverse simplifies to

$$A^{\#} = A^{*}(AA^{*})^{-1}\,,$$

which can be evaluated without computing singular values. (What is the analogous formula when A has full column rank?) In particular, for square matrices of full rank, $A^{\#} = A^{-1}$. □

As remarked above, $\sigma_{\min}(A)$ measures how far the matrix A is from being singular. We now make this precise. As usual, $\|\cdot\|$ denotes operator norm.

Lemma A.2.4 For any matrix A of full column rank $r = l$,

$$\sigma_{\min}(A) = \text{dist}(A, \text{sing}) := \min\{\|\Delta\| \mid \ker(A + \Delta) \neq 0\}.$$

Proof. Note that the minimum in the right-hand side of the claim is achieved, since matrices B with nonzero kernel form a closed set (for instance, defined by the vanishing of the determinant of B^*B). Let Δ and $x \neq 0$ be such that $\|\Delta\| = \text{dist}(A, \text{sing})$ and $(A + \Delta)x = 0$. We write $A = A + \Delta - \Delta$ and apply (A.4) to get:

$$\sigma_{\min}(A)\ \|x\| \ \leq\ \|(A + \Delta)x\|\ +\ \|\Delta\|\ \|x\| = \|\Delta\|\ \|x\|\ ,$$

from which it follows that

$$\sigma_{\min}(A) \leq \|\Delta\| = \text{dist}(A, \text{sing}).$$

Conversely, consider the matrix

$$\Delta := -\sigma_{\min}(A) u_r v_r^*$$

obtained by replacing all other singular values in (SVD) by zero. Since

$$A + \Delta = \sum_{i=1}^{r-1} \sigma_i u_i v_i^*$$

has rank less than l,

$$\text{dist}(A, \text{sing}) \leq \|\Delta\| = \sigma_{\min}(A)$$

and the equality follows. ∎

This result is used in the text in order to obtain characterizations of how uncontrollable a system is. From Lemmas A.2.4 and A.2.2 we also know that:

Corollary A.2.5 For any matrix A of full row rank $r = n$,

$$\sigma_{\min}(A) = \min\{\|\Delta\| \mid (A + \Delta) \text{ is not of full row rank }\}. \qquad \square$$

A result in a latter section of this Appendix establishes that the singular values of a matrix, and in particular $\sigma_{\min}(A)$, are continuous functions of the entries of A, for matrices of any fixed rank.

A.3 Jordan Forms and Matrix Functions

Jordan forms are not very useful numerically but are convenient when establishing various results. We work here with matrices over $\mathbb{C}$, but the existence of Jordan forms can be proved over arbitrary algebraically closed fields.

Recall the basic result, which can be found in any elementary linear algebra book. (We state this in a weaker form than usual; the complete result includes a uniqueness statement, but this is not needed in the text.)

Proposition A.3.1 For any matrix $A \in \mathbb{C}^{n\times n}$ there exists an invertible $n \times n$ complex matrix T and there is some integer $l \geq 1$ such that $T^{-1}AT$ has the block structure

$$\begin{pmatrix} J_1 & 0 & \cdots & 0 & 0 \\ 0 & J_2 & \cdots & 0 & 0 \\ \vdots & \vdots & \ddots & \vdots & \vdots \\ 0 & 0 & \cdots & J_{l-1} & 0 \\ 0 & 0 & \cdots & 0 & J_l \end{pmatrix} \tag{A.7}$$

where each block J_i is a $k_i \times k_i$ matrix of the form

$$\begin{pmatrix} \lambda & 1 & 0 & \cdots & 0 & 0 \\ 0 & \lambda & 1 & \cdots & 0 & 0 \\ \vdots & \vdots & \vdots & \ddots & \vdots & \vdots \\ 0 & 0 & 0 & \cdots & 1 & 0 \\ 0 & 0 & 0 & \cdots & \lambda & 1 \\ 0 & 0 & 0 & \cdots & 0 & \lambda \end{pmatrix} \tag{A.8}$$

for some eigenvalue $\lambda = \lambda_i$ of A. □

A matrix of the form (A.8) is a *Jordan block* (corresponding to the eigenvalue λ).

Note that the integers k_i must add up to n. The same eigenvalue of A may appear in more than one block, and the multiplicity of each eigenvalue λ is the same as the sum of those integers k_i for which $\lambda_i = \lambda$. Note that the only possible eigenvectors of each Jordan block J are those of the form

$$(\alpha, 0, \ldots, 0)' \tag{A.9}$$

(prime indicates transpose), so for each eigenvalue λ of J (equivalently, of the original matrix A) the *geometric multiplicity* of λ, that is, the dimension of the λ-eigenspace

$$\ker(\lambda I - A), \tag{A.10}$$

is equal to the number of blocks corresponding to λ.

Let $f : \mathbb{C} \to \mathbb{C}$ be any entire function, that is, a function analytic on the entire plane, and let

$$f(z) = \sum_{i=0}^{\infty} a_i z^i$$

be its expansion at the origin (which has an infinite radius of convergence). For any square matrix A, we define $f(A)$ by the series

$$f(A) := \sum_{i=0}^{\infty} a_i A^i .$$

This series is convergent, because the inequalities

$$\|A^i\| \leq \|A\|^i$$

imply that the sequence of partial sums is a Cauchy sequence in $\mathbb{C}^{n\times n}$:

$$\left\| \sum_{i=p}^{q} a_i A^i \right\| \leq \sum_{i=p}^{q} |a_i| \, \|A\|^i ,$$

and the right-hand side goes to zero because

$$\sum_{i=0}^{\infty} |a_i| r^i < \infty$$

for all real numbers r.

Assume that f and g are entire and consider the entire function $h := f \cdot g$. Since the Taylor series for h about $z = 0$ can be computed as the product of the series of f and g, it follows by substitution that

$$(fg)(A) = f(A)g(A) \tag{A.11}$$

for all matrices A. Since $fg = gf$, it follows also that

$$f(A)g(A) = g(A)f(A)$$

for all A and all functions. (Using an integral formula, it is possible to define $f(A)$ for functions f that are not entire but merely analytic in a neighborhood of the spectrum of A, in such a way that these properties hold, but we do not need that more general definition here.)

The following is very easy to establish.

Lemma/Exercise A.3.2 Let f be an entire function and assume that J is of size $k \times k$ and has the form (A.8). Then,

$$f(J) = \begin{pmatrix} f(\lambda) & f'(\lambda) & \frac{f''(\lambda)}{2!} & \cdots & \frac{f^{(k-2)}(\lambda)}{(k-2)!} & \frac{f^{(k-1)}(\lambda)}{(k-1)!} \\ 0 & f(\lambda) & f'(\lambda) & \cdots & \frac{f^{(k-3)}(\lambda)}{(k-3)!} & \frac{f^{(k-2)}(\lambda)}{(k-2)!} \\ \vdots & \vdots & \vdots & \ddots & \vdots & \vdots \\ 0 & 0 & 0 & \cdots & f'(\lambda) & \frac{f''(\lambda)}{2!} \\ 0 & 0 & 0 & \cdots & f(\lambda) & f'(\lambda) \\ 0 & 0 & 0 & \cdots & 0 & f(\lambda) \end{pmatrix} . \tag{A.12}$$

As a consequence, the *Spectral Mapping Theorem* is proved: If A is any matrix with eigenvalues

$$\lambda_1, \ldots, \lambda_n$$

(counted with repetitions), then the eigenvalues of $f(A)$ are the numbers

$$f(\lambda_1), \ldots, f(\lambda_n)$$

(also with repetitions). □

If v is an eigenvector of A, $Av = \lambda v$, then $A^i v = \lambda^i v$ for all i, so $f(A)v = f(\lambda)v$ and v is an eigenvector of $f(A)$, too. From this, or from the Jordan form, we conclude that the geometric multiplicity of $f(\lambda)$ with respect to $f(A)$ is greater than or equal to that of λ with respect to A.

Note that, if $B = T^{-1}AT$, then also $B^i = T^{-1}A^iT$ for all i, so

$$B = T^{-1}AT \Rightarrow f(B) = T^{-1}f(A)T \tag{A.13}$$

for all A, all entire f, and all invertible T. Together with Proposition A.3.1 and Lemma A.3.2 this gives a way of computing $f(A)$ if one has obtained the Jordan form (A.7): $f(A)$ is of the form TBT^{-1}, where B is a block matrix each of whose blocks has the form (A.12). In particular, it follows that $f(A) = g(A)$ if it is only known that f and g, and their respective derivatives of high enough order, coincide on the spectrum of A.

A particular case of interest is $f(z) = e^{tz}$ for any fixed t. Let A be any fixed (square) matrix. As a function of t,

$$e^{tA} = \sum_{m=0}^{\infty} \frac{1}{m!} t^m A^m \tag{A.14}$$

converges uniformly on compacts, since

$$\left\| \sum_{m=k}^{\infty} \frac{1}{m!} t^m A^m \right\| \leq \sum_{m=k}^{\infty} \frac{1}{m!} (|t| \; \|A\|)^m .$$

Similarly, taking termwise derivatives of (A.14) results in the series for Ae^{tA}, which is also uniformly convergent on t. It follows that

$$\frac{de^{tA}}{dt} = Ae^{tA} , \tag{A.15}$$

so e^{tA} solves the matrix differential equation $\dot{X} = AX$. From Lemma A.3.2 together with (A.13) it follows that each entry of e^{tA} is a linear combination of terms of the form

$$t^s e^{\lambda t} , \tag{A.16}$$

where s is a nonnegative integer and λ is an eigenvalue of A.

The following fact is of use when discussing controllability and observability under sampling.

Lemma A.3.3 Assume that f is an entire function that is one-to-one on the spectrum of A, and let λ be an eigenvalue of A for which $f'(\lambda) \neq 0$. Then the geometric multiplicity of $f(\lambda)$ with respect to $f(A)$ is the same as that of λ with respect to A. Moreover, the $f(\lambda)$-eigenspace of $f(A)$ is the same as the λ-eigenspace of A.

Proof. We may assume that A has the form (A.7). Since eigenvectors corresponding to $f(\lambda)$ are sums of eigenvectors corresponding to the various blocks $f(J_i)$ for which $\lambda_i = \lambda$ (by the Spectral Mapping Theorem plus the one-to-one hypothesis), it is enough to show that the geometric multiplicity of $f(\lambda)$ is one for the block (A.12), under the hypothesis that $f'(\lambda) \neq 0$. To show this, it is enough to see that there cannot be any eigenvector that is not of the form (A.9). But the equation

$$f(J)\begin{pmatrix} \alpha_1 \\ \alpha_2 \\ \vdots \\ \alpha_k \end{pmatrix} = f(\lambda)\begin{pmatrix} \alpha_1 \\ \alpha_2 \\ \vdots \\ \alpha_k \end{pmatrix}$$

implies (by induction, starting with the $(k-1)$st row) that

$$\alpha_k = \alpha_{k-1} = \ldots = \alpha_2 = 0\,,$$

as desired. ∎

Actually, a stronger result holds. If f is as in the Lemma and f' does not vanish at any eigenvalue of A, then A is a function $g(f(A))$ for some analytic function g. This can be proved using the Implicit Function Theorem, but requires the definition of analytic (not necessarily entire) functions of a matrix.

Recall that a *cyclic* $n \times n$ matrix F is one for which there exists some vector $v \in \mathbb{C}^n$ (a *cyclic vector for* F) so that

$$\operatorname{span}\{v, Fv, \ldots, F^k v, \ldots,\} = \mathbb{C}^n\,. \tag{A.17}$$

If F is cyclic, then translating it by a multiple of the identity does not affect cyclicity, that is, for any real number α the matrix $F + \alpha I$ is again cyclic. This is because, for each k and each vector v,

$$\operatorname{span}\{v, Fv, \ldots, F^k v\} = \operatorname{span}\{v, Fv + \alpha v, \ldots, F^k v + k\alpha F^{k-1} v + \ldots \alpha^k v\}\,. \tag{A.18}$$

Observe that any Jordan block J is cyclic, using the vector $(0, 0, \ldots, 0, 1)'$.

Exercise A.3.4 Let f be an entire function and assume that J is of size $k \times k$ and has the form (A.8). Assume further that $f'(\lambda) \neq 0$ for each eigenvalue λ of A. Prove that then $f(J)$ is cyclic. □

A.4 Continuity of Eigenvalues

The fact that eigenvalues depend continuously on matrix entries is essential in understanding many linear systems results, especially those dealing with robustness, and is useful as well as a technique of proof for matrix identities. In this section we provide a precise statement of this fact. Since eigenvalues of A are zeros of the characteristic polynomial of A, and in turn the coefficients of this characteristic polynomial are continuous functions of the entries of A, it is sufficient to prove the continuous dependence of zeros on the coefficients of a (monic) polynomial.

For each complex number λ and real ε, we let $B_\varepsilon(\lambda)$ (respectively, $S_\varepsilon(\lambda)$) be the open disk (respectively, circle) of radius ε centered at λ.

Lemma A.4.1 Let $f(z) = z^n + a_1 z^{n-1} + \ldots + a_n$ be a polynomial of degree n and complex coefficients having distinct roots

$$\lambda_1, \ldots, \lambda_q ,$$

with multiplicities

$$n_1 + \ldots + n_q = n ,$$

respectively. Given any small enough $\varepsilon > 0$ there exists a $\delta > 0$ so that, if

$$g(z) = z^n + b_1 z^{n-1} + \ldots + b_n , \quad |a_i - b_i| < \delta \text{ for } i = 1, \ldots, n , \tag{A.19}$$

then g has precisely n_i roots in $B_\varepsilon(\lambda_i)$ for each $i = 1, \ldots, q$.

Proof. We assume without loss of generality that ε is smaller than all differences $|\lambda_i - \lambda_j|/2$ for all $i \neq j$, and denote $S_i := S_\varepsilon(\lambda_i)$ for each i. These circles are disjoint and the interiors contain exactly n_i roots of f counted with multiplicities, so from the Residue Theorem

$$n_i = \frac{1}{2\pi i} \int_{S_i} \frac{f'(z)}{f(z)} \, dz$$

(the integral evaluated counterclockwise) for each i.

Consider the polynomial g in (A.19) as a function $g(b, z)$ of $b = (b_1, \ldots, b_n)$ and z. As such it is continuous, and the same holds for $\frac{g'(b,z)}{g(b,z)}$ provided that $g(b, z) \neq 0$. By uniform continuity when z is on the closure of the union of the disks $B_\varepsilon(\lambda_i)$ and b is near $a = (a_1, \ldots, a_n)$, it follows that $g(z) \neq 0$ on every S_i and

$$\left| \frac{f'(z)}{f(z)} - \frac{g'(z)}{g(z)} \right| < \frac{1}{\varepsilon}$$

when g is as in (A.19) and δ is sufficiently small. But then

$$\frac{1}{2\pi i} \int_{S_i} \frac{g'(z)}{g(z)} \, dz \; - \; n_i \; < \; 1 ,$$

which implies that this integral —which is an integer that counts the number of zeros of g in $B_\varepsilon(\lambda_i)$— is also equal to n_i, and the conclusion of the Lemma follows. ∎

We now list a few useful but very simple corollaries. The first is just a restatement of the Lemma.

Corollary A.4.2 Let U be any open subset of $\mathbb{C}$ and fix any two nonnegative integers $k \leq n$. Then the set of monic polynomials of degree n having at least k roots in U is open, when polynomials are topologized by their coefficients as above. □

Corollary A.4.3 Let n be any fixed positive integer. For each $a = (a_1, \ldots, a_n)$, let

$$r_1(a) \leq r_2(a) \leq \ldots \leq r_n(a)$$

be the real parts of the zeros of $P_a(z) = z^n + a_1 z^{n-1} + \ldots + a_n$ arranged in nondecreasing order. Then, as a function $\mathbb{C}^n \to \mathbb{R}$, each r_i is continuous.

Proof. Fix any a and let $r_i := r_i(a)$ for each i. Let $\lambda_1, \ldots, \lambda_q$ be the distinct zeros of P_a, with multiplicities $n_1, \ldots, n_q$, and arranged so that their real parts are nondecreasing. Pick any

$$\varepsilon < \min \left\{ \frac{|\operatorname{Re} \lambda_i - \operatorname{Re} \lambda_j|}{2} \;\middle|\; \operatorname{Re} \lambda_i \neq \operatorname{Re} \lambda_j \right\}$$

and find δ as in Lemma A.4.1. We claim that

$$|r_i(a) - r_i(b)| < \varepsilon$$

for each i whenever all $|a_i - b_i| < \delta$, which will prove continuity.

Fix any $i = 1, \ldots, q$, and observe that

$$r_{n_{i-1}+1}(a) = \ldots = r_{n_i}(a) = \operatorname{Re} \lambda_i \tag{A.20}$$

(denoting $n_0 := 0$). For b close to a, the polynomial P_b has exactly n_i roots at distance less than ε from λ_i, so the real parts of these n_i zeros are at distance less than ε from $\operatorname{Re} \lambda_i$.

Assume that $\operatorname{Re} \lambda_i > \operatorname{Re} \lambda_j$ for $j = 1, \ldots, \alpha$, $\alpha < i$, and that $\operatorname{Re} \lambda_i < \operatorname{Re} \lambda_j$ for $j = \beta, \ldots, q$, for some $\beta > i$. From the choice of ε and the fact that these real parts are at a distance less than ε from $r_1(a), \ldots, r_{n_\alpha}(a)$, it follows that

$$r_1(b), \ldots, r_{n_\alpha}(b)$$

are at distance larger than ε from $r_{n_i}(a)$, and the same is true for $r_j(b)$, $j \geq \beta$. Thus, the zeros within distance ε of $\operatorname{Re} \lambda_i$ are precisely $r_{n_\alpha+1}(b), \ldots, r_{n_\beta-1}(b)$, and in particular

$$|r_j(b) - r_j(a)| < \varepsilon$$

for $j = n_{i-1}+1, \dots, n_i$. This shows that r_j is continuous, for each such j. Since i was arbitrary, the result holds for all r_j's. ∎

The same proof of course is applicable if instead of real parts one looks at magnitudes of zeros.

Corollary A.4.4 For each positive integer n, there exist n continuous functions

$$\lambda_1, \dots, \lambda_n : \mathbb{C}^{n\times n} \to \mathbb{R}$$

such that $\lambda_1(A) \leq \ldots \leq \lambda_n(A)$ for all A, and if A is Hermitian, $\lambda_1(A), \dots, \lambda_n(A)$ are the eigenvalues of A.

Proof. Just apply the previous corollary to the characteristic polynomial and use that the eigenvalues are all real when the matrix is Hermitian. ∎

Corollary A.4.5 For each pair of positive integers l and n, there exist l continuous functions

$$\sigma_1, \dots, \sigma_l : \mathbb{C}^{n\times l} \to \mathbb{R}$$

such that for each A of rank $r \leq l$, $\sigma_r(A) \leq \ldots \leq \sigma_1(A)$ are the singular values of A. In particular, on matrices of any fixed rank r, $\sigma_{\min}(A)$ is a continuous function of A.

Proof. Just let $\sigma_i(A) := \sqrt{\lambda_{l-i+1}(A^*A)}$, where the λ_i are the functions in Corollary A.4.4. ∎

As a different example of the use of Lemma A.4.1, consider the following fact, which is used in the text. It is typical of the use of continuous dependence in order to reduce questions about arbitrary matrices to simpler cases.

Proposition A.4.6 Let A and B be two $n \times n$ complex matrices, and consider the linear operator

$$L_{A,B} : \mathbb{C}^{n\times n} \to \mathbb{C}^{n\times n} : X \mapsto AX - XB' .$$

Then each eigenvalue of $L_{A,B}$ is of the form

$$\lambda - \mu$$

for some eigenvalues λ and μ of A and B, respectively.

Proof. First consider the case of matrices A and B for which all of the possible n^2 differences

$$\lambda_i - \mu_j \tag{A.21}$$

are distinct, where $\lambda_1, \dots, \lambda_n$ and $\mu_1, \dots, \mu_n$ are, respectively, the eigenvalues of A and B. In particular, each of A and B has distinct eigenvalues. In this case there exist two bases of eigenvectors

$$Av_i = \lambda_i v_i\,, \quad Bw_i = \mu_i w_i\,, \quad i = 1, \dots, n\,.$$

Let $X_{ij} := v_i w_j'$ for each pair i, j. We claim that these are eigenvectors of $L_{A,B}$ with corresponding eigenvalues $\lambda_i - \mu_j$. Indeed,

$$AX_{ij} - X_{ij}B' = Av_i w_j' - v_i(Bw_j)' = \lambda_i v_i w_j' - v_i(\mu_j \omega_j)' = (\lambda_i - \mu_j)X_{ij}\,.$$

Since there are n^2 such eigenvalue/eigenvector pairs, every eigenvalue has the required form.

We now use continuity on eigenvalues in order to obtain the general result. Given any two matrices A, B, assume that there is some eigenvalue σ of $L_{A,B}$ that is not of the form $\lambda - \mu$. Let $\varepsilon > 0$ be such that

$$|\sigma - (\lambda - \mu)| > \varepsilon \tag{A.22}$$

for all eigenvalues λ and μ of A and B, respectively. Pick $\delta > 0$ so that, if $\|\widetilde{A} - A\| < \delta$ and $\|\widetilde{B} - B\| < \delta$, then all eigenvalues of $\widetilde{A}$ (respectively, $\widetilde{B}$, $L_{\widetilde{A},\widetilde{B}}$) are at distance less than $\varepsilon/3$ from eigenvalues of A (respectively, B, $L_{A,B}$). The Lemma is used here; note that any fixed matrix representation of $L_{A,B}$ is continuous on A, B.

It is always possible to find a pair $(\widetilde{A}, \widetilde{B})$ so that $\|\widetilde{A} - A\| < \delta$ and $\|\widetilde{B} - B\| < \delta$ and all of the n^2 differences $\widetilde{\lambda}_i - \widetilde{\mu}_j$ are distinct. (Just perturb the Jordan forms for A and B.) Using one such pair and applying the conclusion in the special case proved first, there must exist some eigenvalues $\widetilde{\lambda}$ and $\widetilde{\mu}$ of $\widetilde{A}$ and $\widetilde{B}$, respectively, so that

$$\left|\sigma - (\widetilde{\lambda} - \widetilde{\mu})\right| < \varepsilon/3\,,$$

while at the same time $\left|\widetilde{\lambda} - \lambda\right| < \varepsilon/3$ and $|\widetilde{\mu} - \mu| < \varepsilon/3$ for some eigenvalues of A and B, respectively. This contradicts (A.22). ∎

Corollary A.4.7 The operator $L_{A,B}$ is nonsingular if and only if A and B have no common eigenvalues.

Proof. By the Proposition, nonsingularity is equivalent to asking that $\lambda - \mu \neq 0$ for all eigenvalues λ of A and μ of B. ∎

Remark A.4.8 There is another proof of Corollary A.4.7 that does not require explicitly the result on continuity of eigenvalues. This is based on the observation that the determinant of $L_{A,B}$ is the same as the resultant of the characteristic polynomials of A and B, and hence the determinant is zero if and only if these two polynomials have a common root, that is, if A and B share a common eigenvalue.

The fact about the resultant is in turn proved as follows. For matrices A, B so that all the numbers $\lambda_i - \mu_j$ are distinct, the first part of the proof of Proposition A.4.6 shows that the determinant is the product of all of the terms $(\lambda_i - \mu_j)$, that is, the resultant. Since the resultant is a polynomial —and therefore continuous— function of the entries of A and B, and pairs of matrices

A, B so that all of the $(\lambda_i - \mu_j)$'s are distinct form a dense subset, the result follows for all matrices.

This technique is different, and it has the advantage that it extends to arbitrary fields, and in fact arbitrary commutative rings. That is, over any commutative ring R the determinant of $L_{A,B}$, seen as an operator on $n \times n$ matrices over R, is the resultant of χ_A and χ_B. This is because any polynomial identity on matrices that is proved for complex matrices is necessarily true over arbitrary commutative rings. (Using some algebra: It is only necessary to restrict to the subring generated by the entries of all matrices appearing in the identity; this is a quotient of an extension of finite transcendence degree of $\mathbb{Z}$ and is therefore a subring of $\mathbb{C}$. So identities over $\mathbb{C}$ project into identities over the ring.) □

Exercise A.4.9 Prove the Cayley-Hamilton Theorem for complex matrices by first reducing it to diagonal matrices and using continuity of characteristic polynomials and invariance of the Theorem's conclusions under matrix similarity. (As mentioned in Remark A.4.8, this also automatically establishes the Cayley-Hamilton Theorem over arbitrary commutative rings.) □

Appendix B

Differentials

This Appendix reviews basic notions of differentiability, first for multivariable maps and then for maps between normed spaces. The latter are used in the text in order to characterize linearizations of continuous-time systems.

B.1 Finite Dimensional Mappings

We first recall some basic concepts from multivariable calculus. A mapping

$$T = \begin{pmatrix} T_1 \\ \vdots \\ T_n \end{pmatrix} : \mathcal{U} \to \mathcal{V} \tag{B.1}$$

defined on an open subset $\mathcal{U} \subseteq \mathbb{R}^m$ and with values in an open subset $\mathcal{V} \subseteq \mathbb{R}^n$ is said to be *k-times differentiable* or is *of class C^k* if each of its components

$$T_i : \mathcal{U} \to \mathbb{R}$$

has continuous partial derivatives of order up to k. In the case $k = \infty$, one says that T is *smooth*; when $k = 1$ one just says *differentiable*. When $k = 0$, one means simply that the map is continuous. When the set $\mathcal{U}$ is not necessarily open, we define smoothness to mean that there is a smooth extension to some neighborhood of $\mathcal{U}$.

Given T as in (B.1) but over the complexes, that is, with $\mathcal{U} \subseteq \mathbb{C}^m$ and $\mathcal{V} \subseteq \mathbb{C}^n$, we may also see T as a $2n$-vector of real-valued maps, depending on $2m$ real variables. We shall say that such a T is of class C^k if it is so as a map over the reals. This should not be confused with analyticity, which corresponds to differentiability as a function of m complex variables. Everything in this section will apply to systems over $\mathbb{C}$ by consideration of the real-valued systems of twice the dimension.

The *Jacobian* of the differentiable map T at the element $u^0 \in \mathcal{U}$ is the matrix

$$T_*[u^0] = \left(\left. \frac{\partial T_i(u)}{\partial u_j} \right|_{u=u^0} \right),$$

which is continuous as a function of u^0. Continuity of the matrix means that each entry is continuous, or equivalently continuity of T as a map from $\mathcal{U}$ into $\mathbb{R}^{n\times m}$, when the latter is given the topology that results from listing all coordinates and identifying $\mathbb{R}^{n\times m}$ with $\mathbb{R}^{nm}$. If $n = 1$, the Jacobian is the same as the *gradient*

$$\nabla T[u^0] = \left(\frac{\partial T(u)}{\partial u_1}(u^0), \ldots, \frac{\partial T(u)}{\partial u_m}(u^0) \right)$$

of $T = T_1$.

The chain rule states that the Jacobian of the composition is the composition (matrix product) of the Jacobians:

$$(T \circ S)_*[u^0] = T_*[S(u^0)]S_*[u^0]$$

whenever T and S are differentiable and $S(u^0)$ is in the domain of T.

One may also think of $T_*[u^0]$, for each fixed u^0, as a linear mapping from $\mathbb{R}^m$ to $\mathbb{R}^n$, the **differential** of T at u^0. Thus, the Jacobian is nothing more than the matrix of the differential with respect to standard bases. One advantage of considering the differential is that it can be defined naturally in infinite dimensional spaces, and we do that below. The generalization will follow from the following characterization of Jacobians, which can be found in any elementary analysis book (see, e.g., [121]):

Fact B.1.1 The map T is differentiable if and only if there exists a continuous function

$$J : \mathcal{U} \to \mathbb{R}^{n\times m}$$

such that, for each $u^0 \in \mathcal{U}$,

$$\lim_{u\to u^0} \frac{\|T(u) - T(u^0) - J(u^0)(u - u^0)\|}{\|u - u^0\|} = 0. \tag{B.2}$$

If such a J exists, then necessarily $J(u^0) = T_*[u^0]$. □

We are using the notation $\|x\|$ for the Euclidean norm

$$\|x\| = \sqrt{x_1^2 + \ldots + x_n^2}$$

of a vector in Euclidean space.

More generally, one applies these concepts for maps between infinite dimensional spaces as follows.

B.2 Maps Between Normed Spaces

Recall that a *normed space* N over $\mathbb{K} = \mathbb{R}$ or $\mathbb{C}$ is a vector space over $\mathbb{K}$, together with a map

$$N \to \mathbb{R},\ x \mapsto \|x\|\ ,$$

which satisfies the following properties for all $x, z \in N$ and $k \in \mathbb{K}$:

1. $\|x\| > 0$ if $x \neq 0$ and $\|0\| = 0$.
2. $\|kx\| \;=\; |k|\ \|x\|$.
3. $\|x + y\| \;\leq\; \|x\| + \|y\|$.

A normed space N is a metric space when distance is defined as

$$\mathrm{d}\,(x, z) := \|x - z\|\ .$$

If N is complete under this metric, it is a *Banach space.*

The notion of derivative, and more generally that of a Jacobian of a multivariable map, can be generalized to maps between normed spaces. This generalization is based on the interpretation of a Jacobian of a map from $\mathbb{K}^m$ into $\mathbb{K}^n$ at a point x^0 as itself a map from $\mathbb{K}^m$ into $\mathbb{K}^n$, and it satisfies the basic properties of derivatives, such as the chain rule. We review here the basic definition and the main properties that we shall need. More details can be found in any modern analysis book, e.g., [121].

A continuous mapping

$$F : \mathcal{O} \to N_2$$

from an open subset $\mathcal{O}$ of a normed space N_1 to another normed space N_2 is *(Fréchet) differentiable* at a point $x^0 \in \mathcal{O}$ iff there is a linear mapping

$$F_*[x^0] : N_1 \to N_2$$

such that

$$\left\|F(x) - F(x^0) - F_*[x^0](x - x^0)\right\| \;=\; o\left(\left\|x - x^0\right\|\right)$$

as $x - x^0 \to 0$. If such a map exists, it is unique and is a bounded operator, called the *differential of F at x^0*. For each $v \in N_1$, $F_*[x^0](v)$ is the *directional derivative* of F at x^0 in the direction of v.

If F has a differential at each point of $\mathcal{O}$, the *derivative* of F is the mapping

$$F_* :\ \mathcal{O} \to L(N_1, N_2),\ x \mapsto F_*[x]\,,$$

where $L(N_1, N_2)$ is the set of all bounded linear maps from N_1 to N_2 endowed with the operator norm. One says that F *is of class C^1* if this mapping is continuous. Second derivatives of F are defined via derivatives of F_*, and so on inductively.

Remark B.2.1 There is another notion of differentiability which is often useful. The mapping $F : \mathcal{O} \to N_2$ is *Gateaux differentiable* at a point $x^0 \in \mathcal{O}$ if there is a linear bounded operator $F_*[x^0] : N_1 \to N_2$ such that, for each $v \in N_1$,

$$\left\|F(x^0 + hv) - F(x^0) - F_*[x^0](hv)\right\| = o(h) \tag{B.3}$$

as $h \to 0$. If F is Fréchet differentiable, then it is also Gateaux differentiable, with the same $F_*[x^0]$. But Gateaux differentiability is weaker, because the limit is taken along each direction $x - x^0 = hv$, $h \to 0$, separately. Equation (B.3) says that the mapping

$$\beta : \mathcal{I} \to N_2 : h \mapsto F(x^0 + hv)$$

(defined on some interval $(-h_0, h_0)$ about zero) is differentiable at $h = 0$, and has $\dot{\beta}(0) = F_*[x^0](v)$. □

Provided that $\dim N_1 \geq \dim N_2$, we say that the map F *has full rank at* x^0 if F is a submersion there, i.e., if it is differentiable and $F_*[x^0]$ is onto.

We shall need to apply the Implicit Function Theorem to differentiable mappings. In general, various technical conditions are needed in order for the Theorem to hold for maps among infinite dimensional spaces. However, in the cases of interest to us, one of the spaces is finite dimensional, so there is no difficulty. We start with the Rank Theorem.

Theorem 52 *Assume that $F : \mathcal{O} \to N_2$ is of class C^1 and of full rank at x^0, and let $y^0 := F(x^0)$. Assume further that N_2 is finite dimensional. Then there exists an open neighborhood $\mathcal{V} \subseteq N_2$ of y^0 and a map*

$$G : \mathcal{V} \to \mathcal{O}$$

of class C^1 such that $G(y^0) = x^0$ and so that

$$F(G(y)) = y$$

for each $y \in \mathcal{V}$. □

This is an immediate consequence of the Rank Theorem for finitedimensional mappings, which can be found in any advanced calculus book, because if F has full rank, then there is some finite dimensional subspace

$$\widetilde{N}_1 \subseteq N_1$$

so that the restriction of F to

$$\mathcal{O} \bigcap \widetilde{N}_1$$

still has full rank, and the finite dimensional theorem can be applied to the restricted mapping.

The Implicit Function Theorem follows from this. In general, if

$$F : \mathcal{O} \times \mathcal{Q} \to N_3$$

is differentiable at

$$(x^0, y^0) \in \mathcal{O} \times \mathcal{Q} \subseteq N_1 \times N_2\,,$$

where N_1, N_2, N_3 are normed spaces, the *partial derivative* with respect to N_2 at (x^0, y^0) is the linear bounded operator

$$\frac{\partial}{\partial y} F[x^0, y^0] : N_2 \to N_3$$

given by

$$F_*[x^0, y^0](0, \cdot)\,,$$

which can be identified naturally to the differential of

$$F(x^0, \cdot)$$

as a map from $\mathcal{Q}$ into N_3. If this partial derivative is of full rank, then also the mapping

$$\widetilde{F} : \mathcal{O} \times \mathcal{Q} \to N_1 \times N_3 : (x, y) \to (x, F(x, y))$$

is of full rank. Applying the Rank Theorem to $\widetilde{F}$ we conclude the Implicit Function Theorem:

Theorem 53 *Assume that $F : \mathcal{O} \times \mathcal{Q} \to N_3$ is of class C^1, where $\mathcal{O}, \mathcal{Q}$ are open subsets of normed spaces N_1 and N_2, respectively. Assume further that $\frac{\partial}{\partial y} F$ is of full rank at (x^0, y^0), that $z^0 = F(x^0, y^0)$, and that both N_1 and N_3 are finite dimensional. Then there exist open neighborhoods $\mathcal{V} \subseteq N_3$ of z^0 and $\mathcal{W} \subseteq \mathcal{O}$ of x^0 and a map*

$$G : \mathcal{W} \times \mathcal{V} \to \mathcal{Q}$$

of class C^1 such that $G(x^0, z^0) = y^0$ and so that

$$F(x, G(x, z)) = z$$

for each $(x, z) \in \mathcal{W} \times \mathcal{V}$. □

The following Lemma is particularly useful when applied to mappings sending controls into final states. In those cases one often applies the result to conclude under certain conditions that controllability using arbitrary controls implies controllability with nice controls, such as piecewise constant controls or polynomial ones.

Lemma B.2.2 Assume that $\widetilde{N}_1$ is a dense subspace of N_1, that $f : \mathcal{O} \to N_2$ is of class C^1, and that $\dim N_2 = n < \infty$. Assume further that f has full rank at some $x^0 \in \mathcal{O}$. Then there are elements $\widetilde{x^0}$ in

$$\widetilde{\mathcal{O}} := \mathcal{O} \bigcap \widetilde{N}_1$$

arbitrarily close to x^0 such that the mapping

$$\widetilde{f} := f|_{\widetilde{\mathcal{O}}} : \widetilde{\mathcal{O}} \to N_2$$

is of full rank at $\widetilde{x^0}$. If $x^0 \in \widetilde{\mathcal{O}}$, then $\widetilde{f}$ is of full rank at x^0.

Proof. Note that $\widetilde{f}$ is also of class $\mathcal{C}^1$, directly from the definition of differentiability; its differential at any $x \in \widetilde{\mathcal{O}}$ is the restriction of $f_*[x]$ to $\widetilde{N}_1$. Let

$$v_1, \ldots, v_n$$

be elements of N_1 so that

$$f_*[x^0](v_1), \ldots, f_*[x^0](v_n)$$

form a basis of N_2. Since $f_*[x^0]$ is continuous, we may approximate the v_i's by elements of $\widetilde{N}_1$ while preserving linear independence. So we assume from now on that $v_i \in \widetilde{N}_1$ for all i. If $x^0 \in \widetilde{N}_1$, then this means that $\widetilde{f}$ has full rank there, as wanted. If not, consider a sequence

$$x_j \to x^0 \quad \text{as } t \to \infty$$

of elements of $\widetilde{\mathcal{O}}$. By continuous differentiability,

$$\lim_{j \to \infty} f_*[x_j](v_i) = f_*[x](v_i)$$

for each i. It follows that for all j large enough

$$f_*[x_j](v_1), \ldots, f_*[x_j](v_n)$$

are linearly independent, and they are the same as

$$\widetilde{f}_*[x_j](v_1), \ldots, \widetilde{f}_*[x_j](v_n)$$

because the x_j's are in $\widetilde{N}_2$. ■

Appendix C

Ordinary Differential Equations

As explained in the text, it is convenient to introduce some concepts from Lebesgue integration theory. However, readers who have not seen measure theory before still should be able to follow the rest of this book if they substitute "piecewise continuous function" instead of "measurable function" and they interpret "almost everywhere" as "everywhere except at most for a countable set of points." At some points in a few proofs, results from Lebesgue integration theory will be used, but all statements as well as most proofs will not require such knowledge. Two good references for real-variables concepts are [190] and [264]. Except for some notation introduced there, the next section can be skipped by readers willing to restrict themselves to piecewise continuous controls.

C.1 Review of Lebesgue Measure Theory

The simplest way to introduce measurable functions is by first defining sets of measure zero and then defining measurable functions as the pointwise limits of step (piecewise constant) functions. We do this next.

A subset S of $\mathbb{R}^n$ is said to have *zero measure* if for each $\varepsilon > 0$ there is a countable union of balls $B_1, B_2, \ldots$ of volume ε_i such that

$$\sum_{i=1}^{\infty} \varepsilon_i < \varepsilon$$

and

$$S \subseteq \bigcup_{i=1}^{\infty} B_i \,.$$

Two functions $f, g : O \to T$ from some subset $O \subseteq \mathbb{R}^n$ into some other set T

are said to be *equal almost everywhere (a.e.)* if the set

$$\{x \in O \mid f(x) \neq g(x)\}$$

has zero measure. In general, a property is said to hold almost everywhere provided that it only fails in a set of measure zero.

If $\mathcal{U}$ is a metric space and $\{f_i, i = 1, 2, \ldots\}$ is a sequence of functions from $O \subseteq \mathbb{R}^n$ into $\mathcal{U}$, and also $f : O \to \mathcal{U}$, one says that $\{f_i\}$ *converges to f almost everywhere*, denoted

$$f_i \to f \quad \text{(a.e.)}$$

if the set

$$\{x \in O \mid f_i(x) \not\to f(x)\}$$

has measure zero. In other words, except for a negligible set, the sequence converges at every point. (For instance,

$$f_i(x) : [0,1] \to [0,1] : x \mapsto x^i$$

converges almost everywhere to the function $f(x) \equiv 0$.)

Let $\mathcal{I}$ be an interval in $\mathbb{R}$ and $\mathcal{U}$ a metric space. A *piecewise constant* function $g : \mathcal{I} \to \mathcal{U}$ is one that is constant in each element I_j of a finite partition of $\mathcal{I}$ into subintervals. An $f : O \to \mathcal{U}$ is a *measurable function* if there exists some sequence of piecewise constant functions $\{f_i, i = 1, 2, \ldots\}$ so that $\{f_i\}$ converges to f almost everywhere. Clearly, continuous functions are measurable, and in general $g \circ f$ is measurable if f is measurable and g is continuous.

By abuse of notation, it is common to identify functions that are equal almost everywhere. In particular, given a $\omega \in \mathcal{U}^{[\sigma,\tau)}$ we may, and will, also think of it as a function defined on the closed interval $[\sigma, \tau]$.

The next two Remarks do use some less trivial facts from Lebesgue measure. However, they are not needed for the understanding of the further material in this Appendix, and only play a role in an approximation result in the main text.

Remark C.1.1 Assume that the metric space $\mathcal{U}$ is separable, that is, it has a countable dense subset. Consider the following property for an $\omega : \mathcal{I} \to \mathcal{U}$:

$$\omega^{-1}(V) \text{ is measurable for each open subset } V \subseteq \mathcal{U}\,, \tag{C.1}$$

where "measurable" means Lebesgue measurable. (The measurable sets are those obtained by countable intersections, unions, and complements, starting with open subsets of $\mathcal{I}$ and sets of measure zero.) Then, *the map ω is measurable if and only if it satisfies Property (C.1).* This is a very useful fact, standard when $\mathcal{U}$ is a subset of a Euclidean space. In this generality, it is also known but less accessible, so we give now a proof. The necessity of (C.1) is a consequence of the fact that an almost everywhere limit of functions satisfying (C.1) again satisfies (C.1) (see, for instance, [264], Fact M7 in Chapter X), and the fact that piecewise constant functions obviously satisfy (C.1). The converse can be proved as follows. Assume that ω satisfies the property. By [264], Fact M11, and Remark 3 on page 234 of that reference, ω is known to be the limit of simple functions. Recall that ω is simple if there is a partition of $\mathcal{I}$ into finitely many measurable sets $J_0, \ldots, J_k$ such that $J_1, \ldots, J_k$ have finite measure and ω is constant on each J_i. It will be enough, then, to show that every simple function is an almost everywhere limit of piecewise constant ones. So let ω be simple, and assume that

$$\omega(t) = u_i \quad \text{if } t \text{ is in } J_i$$

and that all the u_i's are distinct.

We claim that it is enough to consider the case in which $k = 1$. Indeed, assume that one has found sequences $\{\omega_n^{(i)}, n = 1, 2, \ldots\}$ and sets Z_i of measure zero for each $i = 1, \ldots, k$, such that for each i and each $t \notin Z_i$ there is some $N = N(t, i)$ such that

$$\begin{aligned} t \in J_i \quad &\Rightarrow \quad \omega_n^{(i)}(t) = u_i \\ t \notin J_i \quad &\Rightarrow \quad \omega_n^{(i)}(t) = u_0 \end{aligned}$$

for all $n > N$ (intuitively, these are approximating the "characteristic functions" of the J_i's, but with values u_i, u_0 instead of $1, 0$). Then, we can define a function ω_n as follows: For any $t \in \mathcal{I}$, if there is exactly one of the $\omega_n^{(i)}(t)$ equal to u_i and the rest equal u_0, then we let $\omega_n(t) := u_i$; otherwise, we let $\omega_n(t) := u_0$. This is again piecewise constant, as it is constant on each interval where all of the $\omega_n^{(i)}$'s are constant. Now, for each

$$t \notin Z := Z_1 \cup \ldots \cup Z_k$$

there is some N so that $\omega_n(t) = \omega(t)$ for all $n > N$: It is enough for this to take $N > N(t, i)$ for this t and $i = 1, \ldots, k$. So we assume that $k = 1$.

The problem is now that of showing that if $J \subseteq \mathcal{I}$ has finite measure and u, v are two elements of $\mathcal{U}$, with $\omega(t) = u$ if $t \in J$ and $\omega(t) = v$ otherwise, then there exists a sequence of piecewise constant functions $\{\omega_n\}$ so that $\omega_n(t) = \omega(t)$ for all $n = n(t)$ large enough, except on a set of measure zero. To see this, start with a sequence of open sets V_n so that

$$\mu(V_n \Delta J) \to 0,$$

where μ indicates Lebesgue measure and Δ indicates symmetric difference. Since each V_n is a countable union of intervals and has finite measure, without loss of generality we can assume that each V_n is a finite union of (open) intervals. Furthermore, we may also assume that

$$\mu(V_n \Delta J) < 2^{-n}$$

for all n. Let ω_n be equal to u on V_n and v otherwise; this is a piecewise constant function. It agrees with ω whenever $t \notin V_n \Delta J$. Letting

$$W_N := \bigcup_{n \geq N} (V_n \Delta J),$$

one has that $\mu(W_N) \to 0$, so also $Z := \cap\{W_N, N \geq 0\}$ has zero measure, and for each $t \notin Z$ one has that $\omega_n(t) = \omega(t)$ for large N. □

A function $\omega : \mathcal{I} \to \mathcal{U}$ is *(essentially) bounded* if it is measurable and there exists a compact subset $K \subseteq \mathcal{U}$ such that

$$\omega(t) \in K \quad \text{for almost all} \quad t \in \mathcal{I},$$

and it is *locally* (essentially) bounded if the restriction of ω to every bounded subinterval of $\mathcal{I}$ is bounded.

Remark C.1.2 Assume, as in Remark C.1.1, that $\mathcal{U}$ is separable. Then: *If ω is essentially bounded, then there is always a sequence of piecewise constant functions ω_n converging almost everywhere to ω and equibounded* (that is, $\omega_n(t) \in K$ for almost all t, for some compact K independent of n). This is because if K is a compact containing the values of ω, we may replace ω by a function almost everywhere equal having image entirely contained in K, and now see K itself as a separable metric space (as a subspace of $\mathcal{U}$). Then Property (C.1) holds for maps into K, so ω is measurable also as a map into K; therefore an approximation can be found with values entirely contained in K.

Often one can obtain approximations by more regular controls. For instance, assume that $\mathcal{U}$ is a convex subset of $\mathbb{R}^n$. Then each piecewise constant control can be approximated almost everywhere by continuous controls (just interpolate linearly on small intervals about the discontinuities), and any bound on the norm of the original control is preserved. Thus, every essentially bounded measurable control can be approximated in this case by an equibounded sequence of continuous controls. If in addition $\mathcal{U}$ is open, then one can approximate, as long as the interval $\mathcal{I}$ of definition is finite, by analytic, and even polynomial, controls. This is because one can first assume that the interval is closed, then approximate by continuous controls, and then approximate these uniformly via the Weierstrass Theorem by polynomials in each coordinate; openness of $\mathcal{U}$ guarantees that the values of the controls eventually lie entirely in $\mathcal{U}$. □

For measurable functions $h : \mathcal{I} \to \mathbb{C}^n$ on a finite interval one may define the Lebesgue integral $\int h(\tau)d\tau$ via the limits of integrals of suitable sequences of approximating piecewise constant functions. (For essentially bounded functions, the only ones that are really used in this text, any bounded sequence of piecewise constant functions will do.) One can also define integrals of functions on infinite intervals in a similar fashion. An *integrable* function is one for which $\int \|h(\tau)\| \, d\tau$ is finite; the function h is *locally integrable* if

$$\int_a^b \|h(\tau)\| \, d\tau < \infty$$

for each $a < b$, $a, b \in \mathcal{I}$.

A function $\xi : \mathcal{I} \to \mathbb{C}^n$ defined on a compact interval $\mathcal{I} = [a, b]$ is said to be *absolutely continuous* if it satisfies the following property: for each $\varepsilon > 0$ there is a $\delta > 0$ such that, for every k and for every sequence of points

$$a \leq a_1 < b_1 < a_2 < b_1 < \ldots < a_k < b_k \leq b$$

so that $\sum_{i=1}^{k}(b_i - a_i) < \delta$, it holds that $\sum_{i=1}^{k} \|\xi(b_i) - \xi(a_i)\| < \varepsilon$. The function ξ is absolutely continuous if and only if it is an indefinite integral, that is, there is some integrable function h such that

$$\xi(t) \;=\; \xi(a) + \int_a^t h(\tau)\, d\tau \tag{C.2}$$

for all $t \in \mathcal{I}$. An absolutely continuous function is differentiable almost everywhere, and

$$\dot{\xi}(t) = h(t)$$

holds for almost all t, if h is as in (C.2). Note that a function is Lipschitz continuous if and only if it is absolutely continuous and has an essentially bounded derivative. If $\mathcal{I}$ is an arbitrary, not necessarily compact, interval, a *locally absolutely continuous* function $\xi : \mathcal{I} \to \mathbb{C}^n$ is one whose restriction to each compact interval is absolutely continuous; equivalently, there is some $\sigma \in \mathcal{I}$ and some locally integrable function h so that $\xi(t) = \xi(\sigma) + \int_\sigma^t h(\tau)\, d\tau$ for all $t \in \mathcal{I}$.

If $\mathcal{I}$ is an interval and $\mathcal{U}$ is a metric space, we denote the sets of all (essentially) bounded measurable maps and all locally bounded maps into $\mathcal{U}$ by

$$\mathcal{L}^\infty_\mathcal{U}(\mathcal{I}) \quad \text{and} \quad \mathcal{L}^{\infty,loc}_\mathcal{U}(\mathcal{I}),$$

respectively. When $\mathcal{U} = \mathbb{R}^n$ (or $\mathbb{C}^n$, depending on context) we write simply

$$\mathcal{L}^\infty_n(\mathcal{I}) \quad \text{and} \quad \mathcal{L}^{\infty,loc}_n(\mathcal{I}),$$

respectively. Elements of such spaces are equivalence classes of functions that coincide almost everywhere, and there is no distinction between

$$\mathcal{L}^\infty_n[\sigma,\tau), \quad \mathcal{L}^\infty_n[\sigma,\tau], \quad \mathcal{L}^\infty_n(\sigma,\tau), \quad \mathcal{L}^\infty_n(\sigma,\tau]$$

since these can be all naturally identified with each other. The space $\mathcal{L}^\infty_n(\mathcal{I})$ is a Banach space under the norm

$$\|\xi\|_\infty := \text{ess.sup.}\,\{\|\xi(t)\|\,, t \in \mathcal{I}\};$$

that is, $\|\xi\|_\infty$ is the infimum of the numbers c such that $\{t \mid \|\xi(t)\| \geq c\}$ has measure zero, and with the distance

$$\mathrm{d}(\xi,\zeta) = \|\xi - \zeta\|_\infty$$

the space is a complete metric space. This is the standard "sup norm" save for the fact that for measurable functions one ignores values that occur only exceptionally.

More generally, for every metric space $\mathcal{U}$ we view $\mathcal{L}^\infty_\mathcal{U}(\mathcal{I})$ as a metric space (itself complete, if $\mathcal{U}$ is complete) when endowed with the distance

$$\mathrm{d}_\infty(\omega,\nu) := \sup_{t\in\mathcal{I}} \mathrm{d}(\omega(t),\nu(t))$$

(uniform convergence), again understood in the "essential" sense that behavior on sets of measure zero is disregarded in computing the supremum.

If $\sigma < \tau$ are finite, the notation $\mathcal{C}^0_n(\sigma,\tau)$ is used for the subspace of $\mathcal{L}^\infty_n(\sigma,\tau)$ consisting of all *continuous* maps

$$\xi : [\sigma,\tau] \to \mathbb{R}^n .$$

(This is of course very different from the space of continuous maps on the *open* interval; it may be identified to the space of such maps whose one-sided limits exist at $t \to \sigma^+$ and $t \to \tau^-$.) This is again a Banach space. It follows that if ρ is any positive real number, the ball centered at any $\xi^0 \in \mathcal{C}^0_n(\mathcal{I})$,

$$\overline{\mathcal{B}}_\rho(\xi^0) := \{\zeta \in \mathcal{C}^0_n(\mathcal{I}) \mid \|\xi-\zeta\|_\infty \le \rho\}, \tag{C.3}$$

is also a complete metric space. We shall use the fact that the Contraction Mapping Theorem holds on each such ball as well as on the whole space. That is, if

$$S : \overline{\mathcal{B}}_\rho(\xi^0) \to \overline{\mathcal{B}}_\rho(\xi^0)$$

is any map such that, for some $\lambda < 1$,

$$\|S\xi - S\zeta\|_\infty \le \lambda \|\xi-\zeta\|_\infty$$

for all ξ,ζ, then S has a (unique) fixed point in $\overline{\mathcal{B}}_\rho(\xi^0)$, and similarly for the entire space ($\rho = \infty$).

Remark C.1.3 Assume that $\mathcal{I}$ is an interval and $\mathcal{U}$ is a metric space. Suppose that we have a convergent sequence $\omega_i \to \omega$ in $\mathcal{L}^\infty_\mathcal{U}(\mathcal{I})$. We prove here that the set of these controls is equibounded, that is, there is some compact subset $\widehat{K}$ of $\mathcal{U}$ so that, for almost all $t \in \mathcal{I}$, $\omega(t) \in \widehat{K}$ and $\omega_i(t) \in \widehat{K}$ for all i. By definition of $\mathcal{L}^\infty_\mathcal{U}(\mathcal{I})$ there are compact subsets K and K_i, $i \ge 1$, so that, except for $t \in \mathcal{I}$ in a set of zero measure, $\omega(t) \in K$ and $\omega_i(t) \in K_i$ for all i. For each $\varepsilon > 0$, let B_ε be the set of points at distance at most ε from K. Since $\omega_i \to \omega$ in uniform norm, for almost all t we have that $\mathrm{d}(\omega_i(t),\omega(t)) < \varepsilon_i$, where $\{\varepsilon_i\}$ is a sequence of real numbers converging to zero, and so $\omega_i(t) \in B_{\varepsilon_i}$ for all i. (Observe that these sets need not be compact unless, for example, one assumes the space $\mathcal{U}$ to be locally compact.) Replacing K_i by its intersection with B_{ε_i}, we may take each K_i as contained in B_{ε_i}. It will be enough to show that the set $\widehat{K}$ defined as the union of K and of all the sets K_i is also compact. We pick a sequence $\{u_i\}$ of elements of $\widehat{K}$ and show that it admits a convergent subsequence in $\widehat{K}$.

If this sequence is contained in a finite union $A = K_1 \bigcup K_2 \bigcup \ldots \bigcup K_r$, then it has a convergent subsequence, since A is compact, and we are done. So assume that the sequence is not contained in any finite union of this form. Then, there are subsequences u_{j_ℓ} and K_{k_ℓ} so that $u_{j_\ell} \in K_{k_\ell}$ for all ℓ. (Such a subsequence can be obtained by induction: if we already have $j_1, \ldots, j_\ell$ and $k_1, \ldots, k_\ell$, let $A = K_1 \bigcup K_2 \bigcup \ldots \bigcup K_{k_\ell}$ and let B be a union of finitely many K_i's which contains $u_1, \ldots u_{j_\ell}$; as the entire sequence cannot be contained in the union of A and B, there are $j > j_\ell$ and $k > k_\ell$ so that $u_j \in K_k$, so we may take $j_{\ell+1} := j$ and $k_{\ell+1} := k$.) We renumber and suppose from now on that $u_i \in K_i$ for each i. Pick for each $i = 1, 2, \ldots$ some $y_i \in K$ so that $\mathrm{d}(u_i, y_i) \leq \varepsilon_i$. Since K is compact, there is a subsequence of $y_1, y_2, \ldots$ which converges to some element $y \in K$, and renumbering again, we assume without loss of generality that $y_i \to y$ as $i \to \infty$. Thus $u_i \to y$ as well, so the original sequence has a convergent subsequence, establishing compactness. □

C.2 Initial-Value Problems

When dealing with continuous-time systems, it is necessary to have a good understanding of the basic facts regarding initial-value problems

$$\dot{\xi} = f(t, \xi(t)), \quad \xi(\sigma^0) = x^0, \tag{IVP}$$

where $\xi(t) \in \mathcal{X} \subseteq \mathbb{R}^n$. Such equations result when a control is substituted in the right-hand side of a control equation

$$\dot{x} = f(x, u).$$

(As a rule, when we are proving results for differential equations we use ξ to denote a solution and reserve the letter "x" for states, that is, values $\xi(t)$. During informal discussions, examples, or exercises, we often revert to the more standard use of x for the solution function ξ itself. Similar remarks apply to the notation ω for control functions and u for control values.)

Since it is essential to allow for the possibility of discontinuous controls, the right hand side of (IVP) cannot be assumed to be continuous on its first argument t, and hence this results in a possibly nondifferentiable solution $\xi(\cdot)$. Thus, the meaning of "solution" of the differential equation must be clarified. The simplest approach is to *define* a solution of (IVP) on an interval $\mathcal{I}$ as a locally absolutely continuous function

$$\xi : \mathcal{I} \to \mathcal{X} \tag{C.4}$$

such that the integral equation

$$\xi(t) = x^0 + \int_{\sigma^0}^{t} f(\tau, \xi(\tau))\, d\tau \tag{INT}$$

holds. In order for this integral to be meaningful, we first remark that $f(t, \xi(t))$ is measurable. More precisely, we assume that

$$f : \mathcal{I} \times \mathcal{X} \to \mathbb{R}^n$$

is a vector function, where $\mathcal{I}$ is an interval in $\mathbb{R}$ and $\mathcal{X}$ is an open subset of $\mathbb{R}^n$, and that the following two properties hold:

$$f(\cdot, x) : \mathcal{I} \to \mathbb{R}^n \text{ is measurable for each fixed } x \tag{H1}$$

and

$$f(t, \cdot) : \mathbb{R}^n \to \mathbb{R}^n \text{ is continuous for each fixed } t\,. \tag{H2}$$

Under these hypotheses, it is also true that

$$f(t, \xi(t))$$

is measurable as a function of t, for any given continuous (C.4). This is proved as follows. If ξ would be piecewise constant on a partition $I_1, \ldots, I_k$ of $\mathcal{I}$, say with $\xi(t) = x_i$ when $t \in I_i$, then

$$f(t, \xi(t)) = \sum_{i=1}^{k} f(t, x_i)\chi_i(t)\,,$$

where χ_i is the characteristic function of I_i. This function is a sum of products of measurable ones, hence measurable itself. If now ξ is continuous, we can always find a sequence of piecewise constant functions $\{\xi_i\}$ such that

$$\xi_i \to \xi$$

(uniformly on compact subsets of $\mathcal{I}$). Since f is continuous on the second coordinate (assumption (H2)) it follows that pointwise

$$f(t, \xi_i(t)) \to f(t, \xi(t))\,,$$

and therefore the latter is also measurable as claimed.

Definition C.2.1 *A* **solution** *of* (IVP) *on an interval* $J \subseteq \mathcal{I}$ *containing* σ^0 *is a locally absolutely continuous function* $\xi : J \to \mathcal{X}$ *such that* (INT) *holds for all* $t \in J$. □

In most cases in this text, the derivative of ξ is (almost everywhere) bounded on each finite interval, because the right-hand side of equations involve controls (which are assumed, by definition, to be essentially bounded) and the continuous function ξ. Thus, solutions will be in fact locally Lipschitz, not merely absolutely continuous.

C.3 Existence and Uniqueness Theorem

The following result is known as the *Bellman-Gronwall* or the *Gronwall Lemma.* It is central to the proof of uniqueness and well-posedness of ODEs. By "interval" we mean either finite or infinite interval.

Lemma C.3.1 Assume given an interval $\mathcal{I} \subseteq \mathbb{R}$, a constant $c \geq 0$, and two functions

$$\alpha, \mu : \mathcal{I} \to \mathbb{R}_+$$

such that α is locally integrable and μ is continuous. Suppose further that for some $\sigma \in \mathcal{I}$ it holds that

$$\mu(t) \leq \nu(t) := c + \int_\sigma^t \alpha(\tau)\mu(\tau)\, d\tau \tag{C.5}$$

for all $t \geq \sigma, t \in \mathcal{I}$. Then, it must hold that

$$\mu(t) \leq c e^{\int_\sigma^t \alpha(\tau)\, d\tau} . \tag{C.6}$$

Proof. Note that $\dot{\nu}(t) = \alpha(t)\mu(t) \leq \alpha(t)\nu(t)$ almost everywhere, so

$$\dot{\nu}(t) - \alpha(t)\nu(t) \leq 0 \tag{C.7}$$

for almost all t. Let

$$\pi(t) := \nu(t) e^{-\int_\sigma^t \alpha(\tau)\, d\tau} .$$

This is a locally absolutely continuous function, and

$$\dot{\pi}(t) = [\dot{\nu}(t) - \alpha(t)\nu(t)]\, e^{-\int_\sigma^t \alpha(\tau)\, d\tau}$$

is ≤ 0 by (C.7). So π is nonincreasing, and thus,

$$\nu(t) e^{-\int_\sigma^t \alpha(\tau)\, d\tau} = \pi(t) \leq \pi(\sigma) = \nu(\sigma) = c ,$$

from which the conclusion follows. ∎

The following remark is very simple, but the resulting inequality will appear repeatedly.

Lemma C.3.2 Let $f, g : L \times \mathcal{X} \to \mathbb{R}^n$ both satisfy the hypotheses (H1), (H2), where L is an interval in $\mathbb{R}$, let $X_0 \subseteq X \subseteq \mathbb{R}^n$ be two subsets, and let

$$\xi : L \to X \quad \text{and} \quad \zeta : L \to X_0$$

be continuous. Suppose that there exist two locally integrable functions

$$\alpha, \beta : L \to \mathbb{R}_+$$

such that

$$\|f(t,x) - f(t,y)\| \leq \alpha(t)\, \|x - y\| \tag{C.8}$$

for all $x, y \in X$ and all $t \in L$, and

$$\|f(t,x) - g(t,x)\| \leq \beta(t) \tag{C.9}$$

for each $x \in X_0$ and all $t \in L$. Then, if x^0, z^0 are arbitrary elements of $\mathbb{R}^n$ and $\sigma^0 \in L$, and we denote

$$\tilde{\xi}(t) := x_0 + \int_{\sigma^0}^{t} f(\tau, \xi(\tau))\, d\tau$$

and

$$\tilde{\zeta}(t) := z_0 + \int_{\sigma^0}^{t} g(\tau, \zeta(\tau))\, d\tau\,,$$

it holds that

$$\left\|\tilde{\xi}(t) - \tilde{\zeta}(t)\right\| \;\leq\; \left\|x^0 - z^0\right\| \;+\; \int_{\sigma^0}^{t} \alpha(\tau)\, \|\xi(\tau) - \zeta(\tau)\|\, d\tau \;+\; \int_{\sigma^0}^{t} \beta(\tau)\, d\tau$$

for all $t \geq \sigma^0$ such that $t \in L$.

Proof. Just write

$$f(\tau, \xi) - g(\tau, \zeta) = f(\tau, \xi) - f(\tau, \zeta) + f(\tau, \zeta) - g(\tau, \zeta)\,,$$

take norms, and use the triangle inequality. ∎

We now state the main Theorem on existence and uniqueness.

Theorem 54 *Assume that $f : \mathcal{I} \times \mathcal{X} \to \mathbb{R}^n$ satisfies the assumptions* (H1) *and* (H2)*, where $\mathcal{X} \subseteq \mathbb{R}^n$ is open and $\mathcal{I} \subseteq \mathbb{R}$ is an interval, and the following two conditions also hold:*

1. *f is locally Lipschitz on x; that is, there are for each $x^0 \in \mathcal{X}$ a real number $\rho > 0$ and a locally integrable function*

$$\alpha : \mathcal{I} \to \mathbb{R}_+$$

such that the ball $B_\rho(x^0)$ of radius ρ centered at x^0 is contained in $\mathcal{X}$ and

$$\|f(t, x) - f(t, y)\| \leq \alpha(t)\, \|x - y\|$$

for each $t \in \mathcal{I}$ and $x, y \in B_\rho(x^0)$.

2. *f is locally integrable on t; that is, for each fixed x^0 there is a locally integrable function $\beta : \mathcal{I} \to \mathbb{R}_+$ such that*

$$\left\|f(t, x^0)\right\| \leq \beta(t)$$

for almost all t.

Then, for each pair $(\sigma^0, x^0) \in \mathcal{I} \times \mathcal{X}$ there is some nonempty subinterval $J \subseteq \mathcal{I}$ open relative to $\mathcal{I}$ and there exists a solution ξ of (IVP) *on J, with the following property: If*

$$\zeta : J' \to \mathcal{X}$$

is any other solution of (IVP), *where* $J' \subseteq \mathcal{I}$, *then necessarily*

$$J' \subseteq J \quad \text{and} \quad \xi = \zeta \quad \text{on } J'.$$

The solution ξ *is called the* **maximal solution** *of the initial-value problem in the interval* $\mathcal{I}$.

Before proving the Theorem, we remark that the assumptions imply also a stronger version of Condition 2, namely that it holds also uniformly on compacts. That is, for any compact $K \subseteq \mathcal{X}$ there is a locally integrable γ such that

$$\|f(t,x)\| \leq \gamma(t) \quad \text{for all} \quad t \in \mathcal{I}, x \in K. \tag{C.10}$$

This is proved as follows. Given any $x^0 \in K$, there are $\rho > 0$ and functions α, β as in the Theorem. Thus, for any element $x \in B_\rho(x^0)$ and each $t \in \mathcal{I}$,

$$\begin{aligned} \|f(t,x)\| &\leq \|f(t,x^0)\| + \|f(t,x) - f(t,x^0)\| \\ &\leq \beta(t) + \rho\alpha(t). \end{aligned} \tag{C.11}$$

Call this last function γ_{x^0} and note that it is locally integrable, too. Consider the open covering of K by the sets of the form $B_\rho(x^0)$; by compactness there is a finite subcover, corresponding to, say, balls centered at $x_1, \ldots, x_l$. Take then

$$\gamma(t) := \max\{\gamma_{x_1}, \ldots, \gamma_{x_l}\}.$$

This is again locally integrable, and (C.11) gives then the desired Property (C.10). □

We now prove the Theorem.

Proof. Assume without loss of generality that $\mathcal{I} \neq \{\sigma^0\}$; otherwise the problem is trivial. We first show that there exists some $\delta > 0$ such that problem (IVP) has a solution on the interval $[\sigma^0, \sigma^0 + \delta] \bigcap \mathcal{I}$. If σ^0 is the right endpoint of $\mathcal{I}$, then this is obvious, so we assume that such is not the case. Pick, for the given x^0, a number ρ and locally integrable functions α and β as in the statement of the Theorem. Note that, because of local integrability, the function

$$a(t) := \int_{\sigma^0}^{t+\sigma^0} \alpha(\tau)\, d\tau \to 0$$

as $t \to 0^+$, and similarly for

$$b(t) := \int_{\sigma^0}^{t+\sigma^0} \beta(\tau)\, d\tau.$$

Note also that both a and b are nonnegative and nondecreasing. Thus, there exists a $\delta > 0$ such that $\sigma^0 + \delta \in \mathcal{I}$ and

(i) $a(t) \leq a(\delta) = \lambda < 1$ for all $t \in [0, \delta]$, and

(ii) $a(t)\rho + b(t) \leq a(\delta)\rho + b(\delta) < \rho$ for all $t \in [0, \delta]$.

Let $\xi_0 \equiv x^0$ on the interval $[\sigma^0, \sigma^0 + \delta]$ and consider the ball of radius ρ centered at this point in the space $\mathcal{C}_n^0(\sigma^0, \sigma^0 + \delta)$,

$$\overline{\mathcal{B}} := \overline{\mathcal{B}}_\rho(\xi_0) .$$

Introduce the operator

$$S : \overline{\mathcal{B}} \to \mathcal{C}_n^0(\sigma^0, \sigma^0 + \delta)$$

defined by

$$(S\xi)(t) := x^0 + \int_{\sigma^0}^t f(\tau, \xi(\tau))\, d\tau .$$

Since the values $\xi(\tau), \tau \in [\sigma^0, \sigma^0 + \delta]$ all belong to some compact set K (by continuity of ξ,) it follows from the discussion before the proof that $f(\cdot, \xi(\cdot))$ is locally integrable, since

$$\int_{\sigma^0}^t \|f(\tau, \xi(\tau))\|\, d\tau \leq \int_{\sigma^0}^t \gamma(\tau)\, d\tau .$$

Thus, S is well defined, and $S\xi$ is absolutely continuous, for each ξ. We claim that $\overline{\mathcal{B}}$ is invariant under S. Indeed, take any $\xi \in \overline{\mathcal{B}}$ and apply Lemma C.3.2 with $L = [\sigma^0, \delta + \sigma^0]$, $g \equiv 0$, same $\alpha, \beta, x^0, \xi, f$ as here, $X = B_\rho(x^0)$, $X_0 = \{x^0\}$, $\zeta := \xi_0$, and $z^0 = x^0$. Since then $\widetilde{\xi} = S\xi$ and $\widetilde{\zeta} = \xi_0$, we conclude that

$$\|S\xi - \xi_0\|_\infty \leq \|\xi - \xi_0\|_\infty\, a(\delta) + b(\delta) < \rho,$$

the last inequality because of Property (ii). Therefore, S can be thought of as a map $\overline{\mathcal{B}} \to \overline{\mathcal{B}}$. Next we prove that it is a contraction there.

For that we again apply Lemma C.3.2, this time with the following choices, given any $\xi, \zeta \in \overline{\mathcal{B}}$: $X = X_0 = B_\rho(x^0)$, α, f, x^0 as here, $g = f$, $\beta \equiv 0$, and $L = [\sigma^0, \delta + \sigma^0]$. Note that now the first and third terms in the conclusion of the Lemma vanish, and we have

$$\|S\xi - S\zeta\|_\infty \leq \lambda\, \|\xi - \zeta\|_\infty ,$$

so S is a contraction.

It follows then from the Contraction Mapping Theorem that there is some fixed point $\xi = S\xi$, and this is, then, a solution of (IVP) in the interval $[\sigma^0, \delta + \sigma^0]$. If σ^0 is not a left endpoint of $\mathcal{I}$, a similar argument proves that there is a solution in some interval of the form $[\sigma^0 - \delta, \sigma^0]$, and therefore by concatenating both solutions, we conclude that there is a solution in an interval $[\sigma^0 - \delta, \sigma^0 + \delta]$. If σ^0 is an endpoint, we have a solution in an interval of the form $[\sigma^0, \sigma^0 + \delta]$ or $[\sigma^0 - \delta, \sigma^0]$, and in any case this gives a solution defined on a neighborhood of σ^0 in $\mathcal{I}$. Moreover, this local existence result holds for every initial pair (σ^0, x^0).

We now prove the following uniqueness statement: *If ξ and ζ are two solutions of the problem* (IVP) *on some interval $J \subseteq \mathcal{I}$, $\sigma^0 \in J$, then $\xi \equiv \zeta$ on J.* We prove that they coincide for $t \in J, t \geq \sigma^0$, the case $t \leq \sigma^0$ being analogous.

We first prove that ξ and ζ must coincide in some (possibly small) interval of the form $[\sigma^0, \sigma^0 + \delta]$, for some $\delta > 0$. (If σ^0 is the right endpoint, there is nothing to prove.) Choose again ρ as in Condition 1, for the given x^0. By continuity of both ξ and ζ, there is a δ small enough that all values $\xi(t)$ and $\zeta(t)$, for $t \in [\sigma^0, \sigma^0 + \delta]$, belong to the ball $B_\rho(x^0)$. Once more we apply Lemma C.3.2, this time taking $X = X_0 = \mathcal{X}$, $f, x^0, \xi, \zeta, \alpha$ as here, $z^0 = x^0$, $g = f$, $L = [\sigma^0, \sigma^0 + \delta]$, and $\beta \equiv 0$. As before, the first and last term vanish, so the Lemma gives that

$$\|\xi(t) - \zeta(t)\| \leq \int_{\sigma^0}^{t} \alpha(\tau) \, \|\xi(\tau) - \zeta(\tau)\| \, d\tau$$

for all $t \in [\sigma^0, \sigma^0 + \delta]$. Gronwall's Lemma then gives that the left-hand side must be zero, since "c" is zero. This proves uniqueness on that (small) interval.

Assume now that there would exist any $t > \sigma^0, t \in J$, such that $\xi(t) \neq \zeta(t)$. Define

$$\sigma^1 := \inf\{t \in J, t > \sigma^0 \mid \xi(t) \neq \zeta(t)\}.$$

So $\xi \equiv \zeta$ on $[\sigma^0, \sigma^1)$, and hence by continuity of both ξ and ζ it also holds that $\xi(\sigma^1) = \zeta(\sigma^1)$. The above local uniqueness proof was equally valid for any initial $\sigma^0 \in \mathcal{I}$ and any $x^0 \in \mathcal{X}$. Thus, we apply it to the initial-value problem with σ^1 as initial time and initial state $x^1 := \xi(\sigma^1)$. It follows that $\xi \equiv \zeta$ on some interval of the form $[\sigma^1, \sigma^1 + \delta]$ for some $\delta > 0$, contradicting the definition of σ^1.

We now show that there is a maximal solution. For this, consider

$$\tau_{\min} := \inf\{t \in \mathcal{I} \mid \text{ there is a solution on } [t, \sigma^0]\}$$

and

$$\tau_{\max} := \sup\{t \in \mathcal{I} \mid \text{ there is a solution on } [\sigma^0, t]\}.$$

(Possibly $\tau_{\max} = +\infty$ or $\tau_{\min} = -\infty$.) From the local existence result we know that $\tau_{\min} < \tau_{\max}$. Consider the open interval $(\tau_{\min}, \tau_{\max})$. There is a solution defined on this interval, because there are two sequences

$$s_n \downarrow \tau_{\min} \quad \text{and} \quad t_n \uparrow \tau_{\max},$$

and there is a solution on each interval (s_n, t_n), by definition of $\tau_{\min}$ and $\tau_{\max}$, and these solutions coincide on their common domains by the uniqueness statement.

We now pick J as follows. If $\tau_{\min}$ and $\tau_{\max}$ are both in the interior of $\mathcal{I}$, then

$$J := (\tau_{\min}, \tau_{\max}).$$

If $\tau_{\min}$ is the left endpoint (that is, $\mathcal{I}$ has the form $[\tau_{\min}, b]$ or $[\tau_{\min}, b)$), then $\tau_{\min}$ is added to J provided that a solution exists on an interval including $\tau_{\min}$. Similarly for $\tau_{\max}$. With this definition, J is open relative to $\mathcal{I}$ and nonempty.

Moreover, if $\tau_{\min}$ is interior to $\mathcal{I}$, then there can be no solution ζ of (IVP) which is defined on $[\tau_{\min}, \sigma^0]$. Otherwise, from the local existence Theorem applied to the initial-value problem with $\xi(\tau_{\min}) = \zeta(\tau_{\min})$, there would result a solution on some interval

$$(\tau_{\min} - \delta, \tau_{\min}]$$

and hence by concatenation also one on $(\tau_{\min} - \delta, \sigma^0]$, contradicting the definition of $\tau_{\min}$. Similarly for $\tau_{\max}$. In conclusion, any solution of (IVP) must have domain included in the above J, and therefore the last part of the Theorem follows again from the uniqueness statement. ■

When $\tau_{\max}$ is in the interior of $\mathcal{I}$, one calls $t = \tau_{\max}$ a *finite escape time* for the problem (IVP).

Exercise C.3.3 Let $\mathcal{X} = \mathcal{I} = \mathbb{R}$. Consider the differential equation

$$\dot{x} = x^2$$

with the following initial values, and find explicitly the maximal solution in each case:

1. $x(0) = 1$.

2. $x(1) = 0$.

3. $x(1) = 1$. □

The local Lipschitz property typically is not hard to verify, due to the fact to be proved next.

Proposition C.3.4 Assume that $f : \mathcal{I} \times \mathcal{X} \to \mathbb{R}^n$ satisfies assumption (H1), where $\mathcal{X}$ is open in $\mathbb{R}^n$ and $\mathcal{I}$ is an interval. Suppose further that for each t the function $f(t, \cdot)$ is of class $\mathcal{C}^1$ and that for each compact set $K \subseteq \mathcal{X}$ there is some locally integrable function $\alpha : \mathcal{I} \to \mathbb{R}_+$ such that

$$\left\| \frac{\partial f}{\partial x}(t, x) \right\| \leq \alpha(t) \quad \text{for all } t \in \mathcal{I}, x \in K\,. \tag{C.12}$$

Then f satisfies the local Lipschitz Property (1) in Theorem 54. In the particular case when f is independent of t, Property (C.12) is automatically satisfied.

Proof. Given any $x^0 \in \mathcal{X}$, pick any ρ such that $\overline{B_\rho(x^0)} \subseteq \mathcal{X}$. Apply the above property with $K := \overline{B_\rho(x^0)}$. Thus, for any two elements x, y in K we have that for each coordinate $f_i, i = 1, \ldots, n$ of f,

$$f_i(t, x) - f_i(t, y) = \sum_{i=1}^{n} \frac{\partial f_i}{\partial x_j}(t, z)(x_j - y_j)$$

for some z in the line joining x and y, by the Mean Value Theorem applied to the function $f_i(\mu x + (1-\mu)y)$ of $\mu \in [0,1]$. It follows that for some constant c,

$$\|f(t,x) - f(t,y)\| \le c\alpha(t)\,\|x-y\| \ ,$$

so $c\alpha(t)$ is as desired. If f (or even if just $\partial f/\partial x$) is independent of t, continuity of $\partial f/\partial x$ insures that this is bounded on compact sets. ∎

The proof of the Theorem shows that solutions exist at least on the interval $[\sigma^0, \sigma^0 + \delta]$, if δ is so that properties (i) and (ii) in the proof hold. This gives (very conservative) lower bounds on the interval of definition. The next problem illustrates how one may use such estimates, for an equation that arises naturally from a control system.

Exercise C.3.5 Consider the system

$$\dot{x}(t) = x^2(t) + u(t)\,. \tag{C.13}$$

Show that there is a solution with $x(0) = 1$, defined on the interval $[0, 0.1]$, if u is any control that satisfies

$$\int_0^{0.1} |u(\tau)|\,d\tau \ < \ 0.1\,.$$

(*Hint:* Try $\rho := 1$ in the Theorem.) □

The following fact is often useful.

Proposition C.3.6 Assume that the hypotheses of the above Theorem hold and that in addition it is known that there is a compact subset $K \subseteq \mathcal{X}$ such that the maximal solution ξ of (IVP) satisfies $\xi(t) \in K$ for all $t \in J$. Then

$$J = [\sigma^0, +\infty) \bigcap \mathcal{I}, \tag{C.14}$$

that is, the solution is defined for all times $t > \sigma^0$, $t \in \mathcal{I}$.

Proof. Assume that (C.14) does not hold. Thus, ξ is defined on $[\sigma^0, \tau_{\max})$, $\tau_{\max} < \infty$, but there is no solution on $[\sigma^0, \tau_{\max}]$, and $\tau_{\max} \in \mathcal{I}$. Find a γ as in the discussion preceding the proof of Theorem 54, for the compact set K in the statement of the Proposition, and consider the function

$$\widetilde{\xi}(t) := \begin{cases} \xi(t) & \text{if } t \in [\sigma^0, \tau_{\max}) \\ x^* & \text{if } t = \tau_{\max} \end{cases}$$

where x^* is any element in K. Thus, $f(t, \widetilde{\xi}(t))$ is bounded by $\gamma(t)$, which is an integrable function on $[\sigma^0, \tau_{\max}]$, and is therefore integrable itself. *Define* now

$$\xi(\tau_{\max}) := x^0 + \int_{\sigma^0}^{\tau_{\max}} f(\tau, \widetilde{\xi}(\tau))\,d\tau\,.$$

Since this is the limit of the integrals $\int_{\sigma^0}^t$ as $t \to \tau_{\max}$ and K is closed, $\xi(\tau_{\max}) \in K$ and in particular $\xi(\tau_{\max}) \in \mathcal{X}$. Also, since $f(\tau, \widetilde{\xi}(\tau)) = f(\tau, \xi(\tau))$ almost everywhere, we conclude that ξ satisfies

$$\begin{aligned} \xi(t) &= x^0 + \int_{\sigma^0}^t f(\tau, \xi(\tau))\, d\tau \\ &= x^0 + \int_{\sigma^0}^t f(\tau, \widetilde{\xi}(\tau))\, d\tau \end{aligned}$$

on the interval $[\sigma^0, \tau_{\max}]$; thus, it is absolutely continuous (from the first equality) and it satisfies (IVP) (from the second). So we defined an extension to $[\sigma^0, \tau_{\max}]$, contradicting maximality. ∎

A related result is as follows:

Proposition C.3.7 Assume that $\mathcal{X} = \mathbb{R}^n$ and that the hypotheses in Theorem 54 hold but with the boundedness Property (2) being global: There is some locally integrable β such that

$$\|f(t,x)\| \leq \beta(t) \quad \text{for all } x \in \mathbb{R}^n\,.$$

Then again (C.14) holds.

Proof. We use the same argument as in the proof of Proposition C.3.6, except that now $f(t, \widetilde{\xi}(t))$ is bounded by β instead of γ. ∎

Note that the assumption that $\mathcal{X} = \mathbb{R}^n$ is crucial in having a well defined $\xi(\tau_{\max})$ in this argument. Otherwise boundedness is not sufficient, as illustrated by

$$\dot{\xi} = 1, \quad x^0 = 0$$

on $\mathcal{X} = (-1, 1)$ and $\mathcal{I} = \mathbb{R}$, for which $J = (-1, 1) \neq \mathcal{I}$.

Finally, another sufficient condition for global solutions to exist is that the Lipschitz condition holds globally:

Proposition C.3.8 Let $\mathcal{X} = \mathbb{R}^n$ and assume that $[\sigma^0, +\infty) \subseteq \mathcal{I}$. Suppose that f satisfies the assumptions of Theorem 54 except that the function α can be chosen independently of x^0, with $\rho = \infty$; that is, that there exists a locally integrable α so that

$$\|f(t,x) - f(t,y)\| \leq \alpha(t)\, \|x - y\|$$

for each $t \in \mathcal{I}$ and $x, y \in \mathbb{R}^n$. Then $J = [\sigma^0, +\infty)$.

Proof. This is proved in a manner very similar to Theorem 54. Let δ be such that (i) in the proof of the Theorem holds, i.e., so that

$$a(t) \leq a(\delta) = \lambda < 1 \text{ for all } t \in [0, \delta]\,.$$

Then, the proof of the Theorem, arguing with all of $\mathcal{C}_n^0(\sigma^0, \sigma^0 + \delta)$ instead of $\overline{\mathcal{B}}_\rho(\xi_0)$, shows that solutions exist for (IVP) at least on the interval $[\sigma^0, \sigma^0 + \delta]$, and that this is true for any such initial-value problem, independently of the particular σ^0 and x^0. Given now a fixed x^0 and σ^0, consider the maximal solution ξ. Assume the Proposition to be false, so in particular $\tau_{\max} < \infty$. Let $\sigma_1 := \tau_{\max} - \delta/2$ and $x_1 := \xi(\sigma_1)$. The initial-value problem has a maximal solution ζ with these initial data and, as remarked before, ζ is defined at least in the interval

$$[\sigma_1, \sigma_1 + \delta] .$$

If we let $\widetilde{\xi}(t)$ be defined as $\xi(t)$ for $t \leq \sigma_1$ and equal to $\zeta(t)$ for $t > \sigma_1$, we obtain a solution of the original (IVP) on

$$[\sigma^0, \tau_{\max} + \delta/2] ,$$

contradicting maximality of $\tau_{\max}$. ∎

Exercise C.3.9 Consider the nonlinear pendulum

$$\begin{aligned} \dot{x}_1 &= x_2 \\ \dot{x}_2 &= -\sin x_1 + u(t) \end{aligned}$$

with $\mathcal{I} = \mathbb{R}$.

1. Prove that if $u \in \mathcal{L}_m^{\infty,loc}(\mathbb{R})$ and σ^0, x^0 are arbitrary, the corresponding initial-value problem has $\tau_{\max} = +\infty$.

2. Prove the same fact for the equation that results when one considers the closed-loop system resulting from the feedback law $u = -2x_1 - x_2$.

3. Find a feedback law (and an initial state) for which the closed-loop system has a finite $\tau_{\max}$. □

Exercise C.3.10 Show that with the feedback $u = -x^3$, system (C.13) in Exercise C.3.5, with $\mathcal{I} = \mathcal{X} = \mathbb{R}$, has no escape times, for any initial condition. □

All of the facts given above hold for complex initial-value problems (IVP) for which the set $\mathcal{X}$ is a subset of a complex Euclidean space $\mathbb{C}^n$ and f takes values in $\mathbb{C}^n$. It is only necessary to consider separately the resulting differential equations for the real and imaginary parts of ξ and f. That is, the resulting equation is equivalent to an equation in $\mathbb{R}^{2n}$, and the hypotheses of the various results must be verified for this real system. Observe that properties such as local integrability, essential boundedness, or continuity, can be defined in terms of complex numbers, using the norm

$$\sqrt{\sum_{i=1}^{n} |x_i|^2}$$

in $\mathbb{C}^n$, where $|\lambda|$ is the magnitude of the complex number λ. Equivalently, one may use the standard Euclidean norm in $\mathbb{R}^{2n}$ and ask that these properties hold for the corresponding real systems; the same concepts result.

It is many times the case that f in (IVP) is smooth in both x and t; in control applications this will correspond to the case in which the dynamics as well as the control being applied are well behaved. In that case one may conclude that solutions are smooth as functions of time:

Proposition C.3.11 Assume that $f : \mathcal{I} \times \mathcal{X} \to \mathbb{R}^n$ is of class $\mathcal{C}^k$, where $k \geq 1$. Then, the conclusions of Theorem 54 hold, and the maximal solution ξ of (IVP) is of class $\mathcal{C}^{k+1}$.

Proof. By Proposition C.3.4, we know that the hypotheses of the Theorem hold, so there is indeed a solution. By induction, assume that ξ is of class $\mathcal{C}^l$, with some $l \leq k$. (The case $l = 0$ is clear by the definition of solution.) Then, in the equation

$$\dot{\xi}(t) = f(t, \xi(t)),$$

the right-hand side is of class $\mathcal{C}^l$. Thus, the derivative of ξ is l times continuously differentiable, from which it follows that ξ is of class $\mathcal{C}^{l+1}$, establishing the induction step. ∎

The result also holds for functions of class $\mathcal{C}^\omega$. Recall that an *analytic* (or more precisely, *real analytic*) *function*

$$f : \mathcal{V} \to \mathbb{R},$$

where $\mathcal{V}$ is an open subset of $\mathbb{R}^m$, is one with the property that for each $v^0 \in \mathcal{V}$ there exists some $\rho > 0$ and a series

$$f(v_1, \ldots, v_m) = \sum_{i_1=0}^{\infty} \cdots \sum_{i_m=0}^{\infty} a_{i_1 \ldots i_m} (v_1 - v_1^0)^{i_1} \ldots (v_m - v_m^0)^{i_m} \tag{C.15}$$

convergent for all $v \in B_\rho(v^0)$. A vector function $f : \mathcal{V} \to \mathbb{R}^n$ is analytic provided that each coordinate is. We also say that f *is of class* $\mathcal{C}^\omega$. When $\mathcal{V}$ is not necessarily open, we define analyticity by the requirement that there exists an extension to an analytic map in a neighborhood of $\mathcal{V}$.

Proposition C.3.12 Assume that $f : \mathcal{I} \times \mathcal{X} \to \mathbb{R}^n$ is of class $\mathcal{C}^\omega$. Then, the conclusions of Theorem 54 hold, and the maximal solution ξ of (IVP) is of class $\mathcal{C}^\omega$.

Proof. Because analyticity is a local property, and using the fact that solutions are unique, it is enough to prove that there is an analytic solution defined in some interval $(\sigma^0 - \delta, \sigma^0 + \delta)$ about σ^0. (Even if σ^0 is an endpoint of $\mathcal{I}$, by definition of analytic function there is an extension of f to some open neighborhood of $\mathcal{I} \times \mathcal{X}$, so we may assume that σ^0 is interior to $\mathcal{I}$.) Adding if necessary a coordinate satisfying

$$\dot{\xi}_{n+1} = 1, \quad \xi_{n+1}(\sigma^0) = \sigma^0 ,$$

we may assume without loss of generality that f is independent of t, that is, $f : \mathcal{X} \to \mathbb{R}^n$. Since the series (C.15) for f is convergent about x^0, it also defines an extension to an analytic function g of complex variables $(z_1, \ldots, z_n)$ defined in a neighborhood $\mathcal{Z}$ of x^0 in complex space $\mathbb{C}^n$. We shall show that there is a continuous function

$$\widetilde{\xi} : \{|\sigma^0 - \sigma| \leq \delta\} \to \mathcal{Z} \subseteq \mathbb{C}^n$$

defined on some complex disk of radius $\delta > 0$ about σ^0, such that

$$\widetilde{\xi}(\sigma) = x^0 + \int_{\sigma^0}^{\sigma} f(\widetilde{\xi}(z))\, dz \quad \text{if } |\sigma^0 - \sigma| < \delta , \tag{C.16}$$

where the integral is interpreted as a line integral along the line joining σ^0 to σ, such that $\widetilde{\xi}$ is analytic on $|\sigma^0 - \sigma| < \delta$ (as a function of a complex variable). Then the restriction of $\widetilde{\xi}$ to the real interval $(\sigma^0 - \delta, \sigma^0 + \delta)$ will be as desired.

In order to prove the existence of $\widetilde{\xi}$ we introduce for each positive real δ the space $\mathcal{D}_n^\omega(\sigma^0, \delta)$ consisting of all those

$$\widetilde{\xi} : \{|\sigma^0 - \sigma| \leq \delta\} \to \mathbb{C}^n$$

that have the property that they are analytic in the interior of the disk and are continuous everywhere, thought of as a normed space with

$$\|\xi\|_\infty := \sup_{|\sigma^0 - \sigma| \leq \delta} \|\xi(\sigma)\| .$$

This space is complete and is hence a Banach space, as proved in most functional analysis texts (see, for instance, [399]). It is naturally identified to a subspace of the Hardy space H^∞ consisting of all functions that are analytic in the interior of the disk and are bounded. (Completeness is an easy consequence of Morera's Theorem: Any Cauchy sequence has a continuous limit, by the completeness of the space of all continuous functions on the disk, and a uniform limit of analytic functions must be analytic, since along each closed path one has that

$$0 = \int_\Gamma f_n\, dz \to \int_\Gamma f\, dz$$

and therefore the integral of the limit must be zero, too.)

We also introduce for each $\rho > 0$ the ball

$$\overline{\mathcal{B}}_\rho(\xi_0) \subseteq \mathcal{D}_n^\omega(\sigma^0, \delta)$$

consisting of all elements with $\|\xi - \xi_0\| \leq \rho$, where ξ_0 is the function constantly equal to x^0, and introduce the operator $S : \overline{\mathcal{B}}_\rho(\xi_0) \to \overline{\mathcal{B}}_\rho(\xi_0)$ defined by

$$(S\xi)(\sigma) := x^0 + \int_{\sigma^0}^{\sigma} f(\xi(z))\, dz .$$

As before, both ρ and δ can be chosen small enough that this is a contraction; the result follows from the Contraction Mapping Theorem. ∎

We now prove that solutions depend continuously on initial conditions as well as on the right-hand side of the equation.

Theorem 55 *Let $\underline{\alpha} : \mathcal{I} \to \mathbb{R}_+$ be an integrable function, where $\mathcal{I} = [\sigma, \tau]$ is a bounded closed interval in $\mathbb{R}$, $\mathcal{X}$ an open subset of $\mathbb{R}^n$, and D a positive real number. Suppose that f and h are two mappings $\mathcal{I} \times \mathcal{X} \to \mathbb{R}^n$ which satisfy the hypotheses of Theorem 54 (namely (H1), (H2), and the local Lipschitz and integrability properties), and that $\xi : \mathcal{I} \to \mathcal{X}$ is a solution of*

$$\dot{\xi}(t) = f(t, \xi(t))$$

such that the D-neighborhood of its range:

$$K = \{x \mid \|x - \xi(t)\| \leq D \text{ for some } t \in [\sigma, \tau]\}$$

is included in $\mathcal{X}$. Let

$$H(t) := \int_\sigma^t h(s, \xi(s))\, ds \quad t \in \mathcal{I}$$

and $\underline{H} := \sup_{t \in \mathcal{I}} \|H(t)\|$. Assume that

$$\max\left\{\underline{H}\,,\, \|\xi(\sigma) - z^0\|\right\} \leq \frac{D}{2} e^{-\int_\sigma^\tau \underline{\alpha}(s)\, ds} \tag{C.17}$$

and, with $g := f + h$,

$$\|g(t,x) - g(t,z)\| \leq \underline{\alpha}(t)\, \|x - z\| \quad \text{for all } x, z \in \mathcal{X} \text{ and } t \in \mathcal{I}. \tag{C.18}$$

Then, the solution ζ of the perturbed equation

$$\dot{\zeta} = g(t, \zeta) = f(t, \zeta) + h(t, \zeta), \quad \zeta(\sigma) = z^0 \tag{C.19}$$

exists on the entire interval $[\sigma, \tau]$, and is uniformly close to the original solution in the following sense:

$$\|\xi - \zeta\|_\infty \leq \left(\|\xi(\sigma) - z^0\| + \underline{H}\right) e^{\int_\sigma^\tau \underline{\alpha}(s)\, ds}. \tag{C.20}$$

Proof. Let $\zeta : J \to \mathbb{R}^n$ be the maximal solution of the initial value problem (C.19), $J \subseteq \mathcal{I}$. This solution exists because f and h, and therefore g, satisfy the hypotheses of the existence theorem. For any $t \in J$ we have:

$$\xi(t) - \zeta(t) = \xi(\sigma) - z^0 + \int_\sigma^t [g(s, \xi(s)) - g(s, \zeta(s))]\, ds - \int_\sigma^t h(s, \xi(s))\, ds.$$

Therefore, for each $t \in J$:

$$\|\xi(t) - \zeta(t)\| \leq \|\xi(\sigma) - z^0\| + \|H(t)\| + \int_\sigma^t \underline{\alpha}(s)\, \|\xi(s) - \zeta(s)\|\, ds$$

and hence from Gronwall's inequality we conclude:

$$\|\xi(t) - \zeta(t)\| \leq \left(\|\xi(\sigma) - z^0\| + \underline{H}\right) e^{\int_\sigma^\tau \underline{\alpha}(s)\, ds}. \tag{C.21}$$

In particular, this implies that $\|\xi(t) - \zeta(t)\| \leq D$, and therefore $\zeta(t) \in K$. Thus the maximal solution ζ is included in a compact subset of $\mathcal{X}$, which by Proposition C.3.6 implies that $J = \mathcal{I}$, and this means that Equation (C.21) holds for all $t \in \mathcal{I}$, which is what Equation (C.20) asserts. ∎

Remark C.3.13 In Theorem 1 (p. 57), and also in Section 4.2 (starting on page 147), we establish several additional properties regarding joint continuity of $\xi(t)$ on t and the initial condition. □

The following characterization of maximal solutions is often very useful.

Exercise C.3.14 Show that if $\xi : [\sigma^0, \tau_{\max}) \to \mathcal{X}$ is a maximal solution of $\dot{x} = f(t, x)$ then

$$\lim_{t \nearrow \tau_{\max}} \|\xi(t)\| = \infty. \tag{C.22}$$

(*Hint:* Suppose that there would exist a sequence $t_n \nearrow \tau_{\max}$ so that $\xi(t_n)$ converges to some state z^0; a solution ζ exists on $[\tau_{\max} - \delta, \tau_{\max} + \delta]$, for some $\delta > 0$, with $\zeta(\tau_{\max}) = z^0$; now use continuity on initial conditions to extend ξ.) □

C.4 Linear Differential Equations

We now review some facts about the solutions of linear differential equations. We take here $\mathbb{K} = \mathbb{R}$ or $\mathbb{C}$.

Assume that $A(\cdot)$ is an $n \times n$ matrix of locally (essentially) bounded (measurable) functions on an interval $\mathcal{I}$ with values in $\mathbb{K}$. Consider the initial-value problem

$$\dot{\xi}(t) = A(t)\xi(t), \quad \xi(\sigma^0) = x^0. \tag{C.23}$$

We may assume that $\mathcal{I} = \mathbb{R}$ simply by defining $A(t) \equiv 0$ outside $\mathcal{I}$. Since the right-hand side is globally Lipschitz, Proposition C.3.8 insures that there is a (unique) solution for all $t \geq \sigma^0$. Solving the equation for $t < \sigma^0$ is equivalent to solving the reversed-time equation $\dot{\xi} = -A\xi$, which is also globally Lipschitz. Thus, solutions are defined on $(-\infty, +\infty)$. Observe that, in the complex case, the resulting system in $\mathbb{R}^{2n}$ is again linear, so the same conclusions apply for both real and complex equations.

A convenient way to study all solutions of a linear equation, for all possible initial values simultaneously, is to introduce the matrix differential equation

$$\dot{X}(t) = A(t)X(t), \quad X(\sigma) = I, \tag{C.24}$$

with $X(t) \in \mathbb{K}^{n \times n}$ for each $t \in \mathbb{R}$. Since this can be seen as a linear differential equation over $\mathbb{K}^{n^2}$, it has a global solution on $(-\infty, \infty)$.

We denote by $\Phi(\tau, \sigma)$ the solution of (C.24) at time τ, that is, $X(\tau)$. This is defined for each real σ and t; Φ is the *fundamental solution* associated to A. Observe that the solution of (C.23) is then given by

$$\xi(t) = \Phi(t, \sigma^0)x^0 .$$

More generally, the solution of a nonhomogeneous equation

$$\dot{\xi} \;=\; A(t)\xi \;+\; \nu, \quad \xi(\sigma^0) = x^0 \tag{C.25}$$

on the interval $[\sigma^0, t]$, where ν is any locally integrable n-vector of functions, is given by the *variation of parameters formula*

$$\xi(t) = \Phi(t, \sigma^0)x^0 \;+\; \int_{\sigma^0}^{t} \Phi(t, s)\nu(s)ds \tag{C.26}$$

also in terms of the fundamental solution. This formula can be verified by derivation; it is well defined because continuity of Φ implies that the integrand is locally integrable.

A number of properties of the matrix function $\Phi(\tau, \sigma)$ can be proved as simple consequences of the uniqueness of solutions of ode's. Note that equation (C.24) says that

$$\frac{\partial \Phi(\tau, \sigma)}{\partial \tau} = A(\tau)\Phi(\tau, \sigma) .$$

Lemma/Exercise C.4.1 Prove that, for all $\sigma, \tau, \mu \in \mathbb{R}$:

(a) $\Phi(\tau, \tau) = I$.

(b) $\Phi(\tau, \sigma) = \Phi(\tau, \mu)\Phi(\mu, \sigma)$.

(c) $\Phi(\tau, \sigma)^{-1} = \Phi(\sigma, \tau)$.

(d) $\frac{\partial \Phi(\sigma, \tau)}{\partial \tau} = -\Phi(\sigma, \tau)A(\tau)$.

(e) If $A(t) \equiv A$ is constant, then $\Phi(\tau, \sigma) = e^{(\tau - \sigma)A}$.

(f) $\det \Phi(\tau, \sigma) = e^{\int_{\sigma}^{\tau} \operatorname{trace} A(\tau)\, d\tau}$. □

Sometimes the term fundamental solution is used for the matrix function of just one variable $\widetilde{\Phi}(t) := \Phi(t, 0)$, instead of Φ. These two objects carry the same information, since Φ can be recovered from $\widetilde{\Phi}$ via the formula $\Phi(t, \sigma) = \widetilde{\Phi}(t)\widetilde{\Phi}(\sigma)^{-1}$.

The exponential e^{tA} that appears in (e) above is defined by the matrix series

$$e^{tA} = I + tA + \frac{t^2}{2}A^2 + \ldots + \frac{t^n}{n!}A^n + \ldots .$$

This series is normally convergent, and

$$\|e^{tA}\| \le \sum_{n=0}^{\infty} \frac{t^n}{n!} \|A\|^n = e^{t\|A\|}.$$

One method of computing the exponential is via Jordan forms —not at all suitable for accurate numerical computation, however— or more generally by finding a similar matrix for which the exponential is easy to compute. When all of the eigenvalues of A are distinct, the matrix can be diagonalized. That is, there is an invertible matrix T and a diagonal matrix

$$\Lambda = \operatorname{diag}(\lambda_1, \ldots, \lambda_n)$$

(the λ_i being the eigenvalues of A) so that

$$T^{-1}AT = \Lambda.$$

Then also $A^l = T\Lambda^l T^{-1}$ for all l, and therefore

$$e^A = Te^{\Lambda}T^{-1} = T\operatorname{diag}(e^{\lambda_1}, \ldots, e^{\lambda_n})\,T^{-1}.$$

Consider as an example the equation

$$\dot{x} = \begin{pmatrix} 0 & 1 \\ -1 & 0 \end{pmatrix} x$$

corresponding to a pendulum linearized about the position $\theta = 0$ (normalizing constants so that $m = g = 1$). Since the eigenvalues $\pm i$ of A are distinct, we can diagonalize; computing eigenvectors gives:

$$A = \begin{pmatrix} 1 & 1 \\ i & -i \end{pmatrix} \begin{pmatrix} i & 0 \\ 0 & -i \end{pmatrix} \begin{pmatrix} 1/2 & -i/2 \\ 1/2 & i/2 \end{pmatrix}.$$

Thus

$$\begin{aligned} e^{tA} &= \begin{pmatrix} 1 & 1 \\ i & -i \end{pmatrix} \begin{pmatrix} e^{it} & 0 \\ 0 & e^{-it} \end{pmatrix} \begin{pmatrix} 1/2 & -i/2 \\ 1/2 & i/2 \end{pmatrix} \\ &= \begin{pmatrix} (e^{it}+e^{-it})/2 & (e^{it}-e^{-it})/2i \\ -(e^{it}-e^{-it})/2i & (e^{it}+e^{-it})/2 \end{pmatrix} = \begin{pmatrix} \cos t & \sin t \\ -\sin t & \cos t \end{pmatrix}. \end{aligned}$$

In the nonconstant case, the exponential is generalized to the use of a power series expansion, for $\Phi(t,\sigma)$, called the *Peano-Baker formula*, for any fixed real numbers σ, t:

$$\begin{aligned} \Phi(t,\sigma) &= I + \int_\sigma^t A(s_1)\,ds_1 + \int_\sigma^t \int_\sigma^{s_1} A(s_1)A(s_2)\,ds_2 ds_1 + \ldots \\ &+ \int_\sigma^t \int_\sigma^{s_1} \cdots \int_\sigma^{s_l - 1} A(s_1)A(s_2)\ldots A(s_l)\,ds_l \ldots ds_2 ds_1 + \ldots \end{aligned} \tag{C.27}$$

This is obtained formally when solving by iteration for $X(\cdot)$ the fixed-point equation

$$X(t) = I + \int_{\sigma}^{t} A(s)X(s)ds$$

corresponding to the initial-value problem (C.24). The series (C.27) converges uniformly on compacts with respect to σ, t. Indeed, since the entries of $A(\cdot)$ are locally bounded, there is for each finite interval $\mathcal{I}$ a constant K such that

$$|A(s)| < K$$

for almost all $s \in I$; therefore the lth term of the series is bounded by

$$\frac{K^l(t-\sigma)^l}{l!}.$$

The partial sums of the series obtained by differentiating each term with respect to t are simply the products of $A(t)$ by the partial sums of the above series. Therefore that termwise differentiated series converges uniformly, too. It follows that the derivative of the sum exists and it equals this termwise derivative, that is, $A(t)\Phi(t,\sigma)$. So $\Phi(t,\sigma)$ is a solution for (C.24) in $[\sigma, t]$ and hence is the fundamental solution matrix introduced earlier.

An advantage of this derivation is that it provides an explicit formula for the solution, in terms of the power series (C.27). This forms the basis of some numerical integration methods for linear systems. For constant matrices, it is the same as the series defining the exponential.

Exercise C.4.2 Substitute the Peano-Baker series for $\Phi(t,\sigma)$ into the variation of parameters formula (C.26), assuming for simplicity the case $x^0 = 0$, and obtain in this manner a series expansion for $\xi(t)$. □

Remark C.4.3 When $A(t)$ is analytic (respectively, k-times continuously differentiable, smooth) on an open interval $\mathcal{I}$ (meaning that each entry of A is, as a function of t), then also $\Phi(t,s)$ is analytic (respectively, $k+1$-times continuously differentiable, smooth), as a function of $(t,s) \in \mathcal{I} \times \mathcal{I}$. This is because, for any fixed $\sigma \in \mathcal{I}$,

$$\Phi(t,s) = \Phi(t,\sigma)\Phi(s,\sigma)^{-1},$$

and each factor has the desired degree of smoothness, by Propositions C.3.11 and C.3.12, being solutions of Equation (C.24). □

Exercise C.4.4 For discrete-time difference equations, one may define

$$\Phi(t,\sigma) := A(t-1)A(t-2)\dots A(\sigma)$$

whenever $t > \sigma$, and

$$\Phi(\sigma,\sigma) := I.$$

Derive then an expansion analogous to that in equation (C.26) for the discrete-time case. Note that in the special situation in which every matrix $A(k)$ is invertible, one can then also define

$$\Phi(t,s) := \Phi(s,t)^{-1}$$

for $t < s$, and Property (b) in Lemma C.4.1 holds. □

C.5 Stability of Linear Equations

We review here some basic facts about the asymptotic behavior of linear differential and difference equations.

Proposition C.5.1 Let A be an $n \times n$ real (respectively, complex) matrix. Then:

1. All real (respectively, complex) solutions of the differential equation

   $$\dot{\xi}(t) = A\xi(t)$$

   satisfy that

   $$\xi(t) \to 0 \text{ as } t \to \infty \tag{C.28}$$

   if and only if all eigenvalues of A have negative real parts.

2. All real (respectively, complex) solutions of the difference equation

   $$\xi(t+1) = A\xi(t)$$

   satisfy (C.28) if and only if all eigenvalues of A have magnitude less than 1.

Proof. From the Jordan form of A it follows that each entry of e^{tA} is a finite linear combination of terms of the form

$$ct^i e^{\lambda t},$$

where λ is an eigenvalue of A, i is an integer, and c is a constant, and that each entry of A^k is a finite linear combination of terms of the form

$$ck^i \lambda^k$$

(see Lemma A.3.2 in Appendix A.3). Thus, $e^{tA} \to 0$ or $A^k \to 0$ if the respective eigenvalue condition holds.

Conversely, assume that there is some eigenvalue λ of A with

$$a := \operatorname{Re} \lambda \geq 0 .$$

Let v be a corresponding eigenvector. Then

$$\left\|e^{tA}v\right\| = e^{at}\left\|v\right\| .$$

This is constant if $a = 0$, and tends to ∞ if $a > 0$. If $\mathbb{K} = \mathbb{C}$, this means that the solution $e^{tA}v$ contradicts the assumption on A that all solutions converge to zero. If $\mathbb{K} = \mathbb{R}$, then the real and imaginary parts of $e^{tA}v$ are solutions of

$$\dot{\xi} = A\xi ,$$

which cannot both converge to zero, again providing a contradiction. If there is some eigenvalue λ with $|\lambda| \geq 1$, an analogous argument applies to $\left\|A^k v\right\|$. ∎

Definition C.5.2 *The complex matrix A is said to be*

- **Hurwitz** *if all its eigenvalues have a negative real part;*
- **convergent** *or discrete-time Hurwitz if all of its eigenvalues have magnitude less than* 1. □

Some authors call a *Schur* matrix what we are calling here a convergent matrix.

It is possible to determine whether a given matrix is Hurwitz (or convergent) without having to compute eigenvalues, directly, by checking certain polynomial inequalities on the coefficients of A. This is typically done through a *Routh-Hurwitz* type of test; see, for instance, [34] for extensive discussions of such issues, and also Remark 5.7.20 in Chapter 5.

Bibliography

In addition to the references cited in the text, listed here are some additional papers and books. These contain many further references, and a few provide introductions to some of the areas that have not been covered, but the list is in no way comprehensive or balanced in terms of the different possible topics.

Other Books

Some books that should be of interest are included in this Bibliography. The three-part book [230] was in a sense the first "modern systems and control theory" treatise. Algebraic topics on realization are covered there, and the close analogies between automata and linear systems results are discussed. Though somewhat out of date, it is still worth reading.

There are many automata theory texts; some examples include: [52], [97] (material on observability), [125] (an excellent treatise of many topics in automata theory, very algebraic in character; it also includes a chapter on linear systems as well as many results on what can be described as "systems over semirings," a class that includes both linear systems and automata), and [156].

For general topics on linear systems, good references are for instance [17], [58] (focuses on performance issues), [65] (a classic, but apparently out of print), [88], [114] (a brief introduction to basic facts about linear systems, including controllability, observability, and minimality), [144] (a very detailed reference for elementary algebraic theory of linear systems and classical control theory; it has many realistic examples), [212] (modern, very algebraic, treatment of linear systems, emphasizing the 'polynomial matrix' approach), [314] (an elementary introduction to control theory, including comparisons to automata and a brief treatment of some 'advanced' topics such as category-theoretic methods), [335] (a rather summarized treatment of algebraic theory of time-invariant linear systems), [337] (introduces linear systems and optimal linear control, including many stochastic topics), and [430] (a very algebraic approach to linear regulator design, based on the author's "geometric" method of solution).

There are literally hundreds of textbooks dealing with frequency-response and basic linear theory from an engineering standpoint, geared to an undergraduate audience but very useful nonetheless for learning the basic concepts. One particularly nice book is [260], as it contains very detailed examples of applications.

In optimal control the choice is also very large. For excellent introductions to quadratic/linear problems, including some study of the effect of nonlinearities as well as numerical methods, see [16] and [263]. More generally, some suggested texts are: [134] (one of the best treatises of optimal control; results proved rigorously and in great generality), [188] (especially for the time-optimal problem for linear systems), and [266] (an excellent "classical" reference, somewhat outdated; exposition of linear and nonlinear optimal control, and of basic properties of nonlinear systems; little algebraic theory of linear systems). Also, [108] provides an excellent brief exposition to filtering and stochastic control. Among other books in optimal control are: [25], [44], [48], [103], [151], [165], [189], [247], [248], [276], [289], [315], [379], and [442].

On the control of nonlinear systems there are several books available, including: [115] (operator approach), [199], [311] (state-space methods, differential-geometric in character), [414] (a very nice exposition of material on ODEs relevant to control theory, including stability, as well as the operator approach to input/output nonlinear systems), and [422] (operator approach).

For infinite-dimensional systems, some books are [107], [133], [147], and [281]. Many others are cited in them.

For applications of control theory see for instance the following books: [20] (economics), [45] (management), [146] (automobiles), [168] (biology), [317] (economics), [325] (oil recovery), [328] (physiology), [346] (economics), [347] (management), [350] (economics and engineering applications), [380] (chemical process control), [394] (biomedicine), [401] (psychology), [420] (robotics), and [427] (signal processing). Also, the *Systems & Control Encyclopedia* [352] and the *The Control Handbook* [274] contain many articles on applications as well as on further theoretical questions.

1. Abed, E.H., and J-H. Fu, "Local stabilization and bifurcation control, I. Hopf bifurcation," *Systems and Control Letters* **7**(1986): 11-17.
2. Abraham, R., and J.E. Marsden, *Foundations of Mechanics, second edition*, Benjamin Cummings, Reading, 1978.
3. Adamjan, V.M., D.Z. Arov, and M.G. Krein, "Analytic properties of Schmidt pairs for a Hankel operator and the generalized Schur-Takagi problem," *Math. USSR Sbornik* **15**(1971): 31-73.
4. Aeyels, D., "Stabilization of a class of nonlinear systems by a smooth feedback control," *Systems and Control Letters* **5**(1985): 289-294.
5. Aeyels, D., "Generic observability of differentiable systems," *SIAM J. Control & Opt.* **19**(1981): 595-603.
6. Aeyels, D., "On the number of samples necessary to achieve observability," *Systems and Control Letters* **1**(1981): 92-94.
7. Aeyels, D., "Global observability of Morse-Smale vector fields," *J. Diff. Eqs.* **45**(1982): 1-15.
8. Aeyels, D., and M. Szafranski, "Comments on the stabilizability of the angular velocity of a rigid body," *Systems and Control Letters* **10**(1988): 35-39.
9. Ahmed, N.U., *Elements of Finite-Dimensional Systems and Control Theory*, Wiley, New York, 1988.
10. Akaike, H., "Stochastic theory of minimal realization," *IEEE Trans. Autom. Control* **AC-19**(1974): 667-674.
11. Akaike, H., "Markovian representation of stochastic processes by canonical variables," *SIAM J. Control & Opt.***13**(1975):162-173.
12. Albertini, F., and P. Dai Pra, "Forward accessibility for recurrent neural networks," *IEEE Trans. Automat. Control* **40** (1995): 1962-1968.
13. Albertini, F., and E.D. Sontag, "For neural networks, function determines form," *Neural Networks* **6**(1993): 975-990.
14. Albertini, F., and E.D. Sontag, "State observability in recurrent neural networks," *Systems & Control Letters* **22**(1994): 235-244.
15. Alur, R., T.A. Henzinger, and E.D. Sontag (eds.), *Hybrid Systems III. Verification and Control*, Springer Verlag, Berlin, 1996.
16. Anderson, B.D.O., and J.B. Moore, *Linear Optimal Control*, Prentice-Hall, Englewood Cliffs, 1971.
17. Antsaklis, P.J., and A.N. Michel, *Linear Systems*, McGraw Hill, New York, 1997.
18. Antoulas, A.C., "On recursiveness and related topics in linear systems," *IEEE Trans. Autom. Control* **AC-31**(1986): 1121-1135.
19. Antoulas, A.C., and B.D.O. Anderson, "On the scalar rational interpolation problem," *IMA J. Math. Control Infor.***3**(1986):61-88.
20. Aoki, M., *Optimal Control and System Theory in Dynamic Economic Analysis*, North Holland, New York, 1976.

21. Arbib, M.A., "A characterization of multilinear systems," *IEEE Trans. Autom. Control* **14**(1969): 699-702.
22. Arbib, M.A., and E. Manes, "Machines in a category: an expository introduction," *SIAM Review* **57**(1974): 163-192.
23. Artstein, Z., "Stabilization with relaxed controls," *Nonl. Anal., TMA* **7**(1983): 1163-1173.
24. Astrom, K.J., *Introduction to Stochastic Control Theory*, Academic Press, New York, 1970.
25. Athans, M., and P.L. Falb, *Optimal Control: An Introduction to the Theory and its Applications*, McGraw-Hill, New York, 1966.
26. Aubin, J.-P., and A. Cellina, *Differential Inclusions: Set-Valued Maps and Viability Theory*, Springer-Verlag, Berlin, 1984.
27. Baccioti, A., and P. Boieri, "Linear stabilizability of planar nonlinear systems," *Math. Control Signals Systems* **3**(1990): 183-193.
28. Baillieul, J., "Controllability and observability of polynomial dynamical systems," *Nonl.Anal.,TMA* **5**(1981): 543-552.
29. Baillieul, J., R.W. Brockett, and R.B. Washburn, "Chaotic motion in nonlinear feedback systems," *IEEE Trans. Circuits and Systems* **CAS-27**(1980): 990-997.
30. Baillieul, J., S.S. Sastry, and H.J. Sussmann (eds.), *Essays on Mathematical Robotics*, Springer-Verlag, New York, 1998.
31. Banks, S.P., *Control Systems Engineering: Modelling and Simulation, Control Theory, and Microprocessor Implementation*, Prentice-Hall, Englewood Cliffs, 1986.
32. Banks, S.P., "Stabilizability of finite- and infinite-dimensional bilinear systems, *IMA J. of Math. Control and Inform.* **3**(1986): 255-271.
33. Barmish, B.R., M.J. Corless, and G. Leitmann, "A new class of stabilizing controllers for uncertain dynamical systems," *SIAM J. Control & Opt.* **21**(1983): 246-255.
34. Barnett, S., *Polynomials and Linear Control Systems*, Marcel Dekker, New York, 1983.
35. Barnett, S., *Matrices in Control Theory*, Krieger, Malabar, 1984.
36. Barnett, S., *Introduction to Mathematical Control Theory*, Oxford University Press, New York, 1985.
37. Bartlett, A.C., C.V. Hollot, and H. Lin, "Root locations of an entire polytope of polynomials: It suffices to check the edges," *Math. Control Signals Systems* **1**(1988): 61-71.
38. Bartosiewicz, Z., "Minimal polynomial realizations," *Math. Control Signals Systems* **1**(1988): 227-231.
39. Basile, G., and G. Marro, "Controlled and conditioned invariant subspaces in linear system theory," *J. Optimiz. Theo. & Appl.* **3**(1969): 306-315.
40. Belevitch, V., *Classical Network Theory*, Holden-Day, San Franscisco, 1968.

41. Bellman, R., *Dynamic Programming*, Princeton University Press, Princeton, 1957.
42. Bellman, R.E., *Introduction to the Mathematical Theory of Control Processes*, Academic Press, New York, 1967.
43. Bellman, R.E., and R. Kalaba (eds.), *Selected Papers on Mathematical Trends in Control Theory*, Dover, New York, 1964.
44. Bellman, R.E., and R. Kalaba, *Dynamic Programming and Modern Control Theory*, Academic Press, New York, 1965.
45. Bensoussan, A., E.G. Hurst, and B. Naslund, *Management Applications of Modern Control Theory*, Elsevier, New York, 1974.
46. Benveniste, A., and P. Le Guernic, "Hybrid dynamical systems theory and the SIGNAL language," *IEEE Trans. Autom. Control* **AC-35**(1990): 535-546.
47. Berger, M.S., and M.S. Berger, *Perspectives in Nonlinearity: An Introduction to Nonlinear Analysis*, Benjamin, New York, 1968.
48. Berkovitz, L.D., *Optimal Control Theory*, Springer-Verlag, New York, 1974.
49. Bhatia, N.P., and G.P. Szegö, *Stability Theory in Dynamical Systems*, Springer-Verlag, Berlin, 1970.
50. Boltyansky, V.G., "Sufficient conditions for optimality and the justification of the dynamic programming method," *SIAM J. Control & Opt.* **4**(1966): 326-361.
51. Bonnard, B., "Controlabilité des systèmes non linéaires," *C.R. Acad. Sci. Paris, Serie I,* **292**(1981): 535-537.
52. Booth, T.L., *Sequential Machines and Automata Theory*, Wiley, New York, 1967.
53. Boothby, W. M., *An Introduction to Differentiable Manifolds and Riemannian Geometry*, Academic Press, New York, 1975.
54. Boothby, W.M., and R. Marino, "Feedback stabilization of planar nonlinear systems," *Systems and Control Letters* **12**(1989): 87-92.
55. Boothby, W.M., and E.N. Wilson, "Determination of the transitivity of bilinear systems," *SIAM J.Control & Opt.* **17**(1979): 212-221.
56. Bose, N.K., "Problems and progress in multidimensional systems theory," *Proceedings IEEE* **65**(1976): 824-840.
57. Bose, N.K., *Multidimensional Systems Theory*, Reidel, Dordrecht, 1985.
58. Boyd, S.P., and C.H. Barratt, *Linear Controller Design*, Prentice Hall, Englewood Cliffs, 1991.
59. Boyd, S., V. Balakrishnan, and P. Kabamba, "On computing the H_∞ norm of a transfer matrix," *Math. Control Signals Systems* **2** (1989): 207-219.
60. Brammer, R., "Controllability in linear autonomous systems with positive controllers," *SIAM J. Control* **10**(1972): 339-353.
61. Brewer, J.W., J.W. Bunce, and F.S.Van Vleck, *Linear Systems over Commutative Rings*, Marcel Dekker, New York, 1986.

62. Brewer, J.W., and L. Klinger, "Pole shifting for families of systems: the dimension one case," *Math. Control Signals Systems* **1**(1988): 285-292.
63. Brockett, R.W., "The status of stability theory for deterministic systems," *IEEE Trans. Autom. Control* **AC-11**(1966): 596-606.
64. Brockett, R.W., "Structural properties of the equilibrium solutions of Riccati equations," in *Lect.Notes in Math., vol. 132* (A.V. Balakrishnan et al., eds.), Springer-Verlag, Heidelberg, 1970,pp. 61-69.
65. Brockett, R.W., *Finite Dimensional Linear Systems*, Wiley, New York, 1970.
66. Brockett, R.W., "On the algebraic structure of bilinear systems," in *Theory and Applications of Variable Structure Systems* (R. Mohler and A. Ruberti, eds.,) Academic Press, New York, 1972,pp. 153-168.
67. Brockett, R.W., "System theory on group manifolds and coset spaces," *SIAM J. Control* **10**(1972): 265-284.
68. Brockett, R.W., "Lie theory and control systems defined on spheres," *SIAM J. Control & Opt.* **25**(1973): 213-225.
69. Brockett, R.W., "Volterra series and geometric control theory," *Automatica* **12**(1976): 167-176.
70. Brockett, R.W., "Nonlinear systems and differential geometry," *Proc. IEEE* **64**(1976): 61-72.
71. Brockett, R.W., "Feedback invariants for nonlinear systems," *Proc. IFAC Congress*, Helsinki, 1978.
72. Brockett, R.W., "Asymptotic stability and feedback stabilization," in *Differential Geometric Control Theory* (R.W. Brockett, R.S. Millman, and H.J. Sussmann, eds.), Birkhauser, Boston, 1983, pp. 181-191.
73. Brockett, R.W., R.S. Millman, and H.J. Sussmann (eds.), *Differential Geometric Control Theory*, Birkhauser, Boston, 1983.
74. Brockett, R.W., and J.L. Willems, "Discretized partial differential equations: Examples of control systems defined on modules," *Automatica* **10**(1974): 507-515.
75. Brockett, R.W., and A.S. Willsky, "Finite group homomorphic sequential systems," *IEEE Trans. Autom. Control* **AC-17**(1972): 483-490.
76. Brogan, W.L., *Modern Control Theory*, Prentice-Hall, Englewood Cliffs, 1985.
77. Brunovsky, P., "A classification of linear controllable systems," *Kybernetika* **6**(3)(1970): 173-188.
78. Bryson, A.E., and Y-C. Ho, *Applied Optimal Control*, Wiley, New York, 1969.
79. Bumby, R., E.D. Sontag, H.J. Sussmann, and W. Vasconcelos, "Remarks on the pole-shifting problem over rings," *J. Pure Appl. Algebra* **20**(1981): 113-127.
80. Byrnes, C.I., and A. Isidori, "New results and counterexamples in nonlinear feedback stabilization," *Systems and Control Letters* **12**(1989): 437-442.

81. Byrnes, C.I., "On the control of certain infinite-dimensional systems by algebro-geometric techniques," *Amer. J. Math.* **100**(1978): 1333-1381.

82. Campbell, S.L., *Singular Systems of Differential Equations, I* and *II*, Pitman, Marshfield, 1980, 1982.

83. Campbell, S.L., "Control problem structure and the numerical solution of linear singular systems," *Math. Control Signals Systems* **1**(1988): 73-87.

84. Canudas de Wit, C.A., *Adaptive Control for Partially Known Systems: Theory and Applications*, Elsevier, New York, 1988.

85. Casdagli, M., "Nonlinear prediction of chaotic time series," *Physica D* **35**(1989): 335-356.

86. Catlin, D.E., *Estimation, Control, and the Discrete Kalman Filter*, Springer-Verlag, New York, 1989.

87. Celle, F., J.P. Gauthier, D. Kazakos, and G. Sallet, "Synthesis of nonlinear observers: A harmonic analysis approach," *Math. Systems Theory (Special Issue on Mathematical Control Theory)* **22**(1989): 291-322.

88. Chen, C.T., *Introduction to Linear System Theory*, Holt, Reinhart and Winston, New York, 1970.

89. Chen, K.T., "Iterated path integrals," *Bulletin AMS* **83**(1977): 831-879.

90. Cho, H., and S.I. Marcus, "On supremal languages of classes of sublanguages that arise in supervisor synthesis problems with partial observation," *Math. Control Signals Systems* **2**(1989): 47-69.

91. Chow, W.L., "Uber systeme von linearen partiellen differentialgleichungen erster ordnung," *Math. Ann.* **117**(1939): 98-105.

92. Clarke, F.H., Yu.S. Ledyaev, E.D. Sontag, and A.I. Subbotin, "Asymptotic controllability implies feedback stabilization," *IEEE Trans. Autom. Control* **42**(1997): 1394-1407.

93. Clarke, F.H., Yu.S. Ledyaev, R.J. Stern, and P. Wolenski, *Nonsmooth Analysis and Control Theory*, Springer-Verlag, New York, 1997.

94. Cobb, J.D., "Feedback and pole placement in descriptor variable systems," *Intern. J. Control* **33**(1981): 1135-1146.

95. Conte, G., and A.M. Perdon, "An algebraic notion of zeros for systems over rings," in *Mathematical Theory of Networks and Systems*, (P.A.Fuhrmann, ed.), Springer-Verlag, Berlin, 1984, pp. 323-330.

96. Constanza, V., B.W. Dickinson, and E.F. Johnson, "Universal approximations of discrete-time control systems over finite time," *IEEE Trans. Autom. Control* **AC-28**(1983): 439-452.

97. Conway, J.B., *Regular Algebra and Finite Machines*, Chapman and Hall, London, 1971.

98. Conway, J.B., *A Course in Functional Analysis, Second Edition*, Springer-Verlag, New York, 1990.

99. Coron, J-M., "Global asymptotic stabilization for controllable systems without drift," *Math of Control, Signals, and Systems* **5**(1992): 295-312.

100. Coron, J-M., "Stabilization in finite time of locally controllable systems by means of continuous time-varying feedback laws," *SIAM J. Control and Opt.* **33**(1995): 804-833.
101. Coron, J-M., L. Praly, and A. Teel, "Feedback stabilization of nonlinear systems: sufficient conditions and Lyapunov and input-output techniques," in *Trends in Control* (A. Isidori, ed.), Springer-Verlag, Berlin, 1995, pp. 293-348.
102. Crandall, M.G., L.C. Evans, and P.L. Lions, "Some properties of viscosity solutions of Hamilton-Jacobi equations," *Trans. AMS* **282**(1984): 487--502.
103. Craven, B.D., *Mathematical Programming and Control Theory*, Wiley, New York, 1978.
104. Crouch, P.E., "Dynamical realizations of finite Volterra series," Ph.D. Thesis, Harvard, 1977.
105. Crouch, P.E., "Spacecraft attitude control and stabilization: applications of geometric control theory," *IEEE Trans. Autom. Control* **29**(1984): 321-333.
106. Crouch, P.E., and A.J. van der Schaft, *Variational and Hamiltonian Control Systems*, Springer-Verlag, New York, 1987.
107. Curtain, R., and A.J. Pritchard, *Infinite Dimensional Linear Systems Theory*, Springer-Verlag, Berlin, 1978.
108. Davis, M.H.A., *Linear Estimation and Stochastic Control*, Chapman and Hall, London, 1977.
109. Dayawansa, W.P., and C.F. Martin, "Asymptotic stabilization of two dimensional real-analytic systems," *Systems Control Lett.* **12**(1989): 205-211.
110. de Jong, L.S., *Numerical Aspects of Realization Algorithms in Linear Systems Theory*, Ph.D. Thesis, Univ. of Eindhoven, Mathematics Department, 1975.
111. Deimling, K., *Multivalued Differential Equations*, de Gruyter, Berlin, 1992.
112. Delchamps, D.F., "Analytic stabilization and the algebraic Riccati equation," in *Proc. IEEE Conf. Dec.and Control, Dec. 1983*, IEEE Publications, pp. 1396-1401.
113. Delfour, M.C., C. McCalla, and S.K. Mitter, "Stability and the infinite-time quadratic cost problem for linear hereditary differential systems," *SIAM J. Control* **13**(1975): 48-88.
114. Desoer, C.A., *Notes for a Second Course on Linear Systems*, Van Nostrand, New York, 1970.
115. Desoer, C.A., and M. Vidyasagar, *Feedback Synthesis: Input-Output Properties*, Academic Press, New York, 1975.
116. Desoer, C.A., and M.G. Kabuli, "Nonlinear plants, factorizations, and stable feedback systems," in *Proc. IEEE Conf. Decision and Control*, Los Angeles, Dec 1987, pp. 155-156.

117. Dickinson, B.W., "On the fundamental theorem of linear state variable feedback," *IEEE Trans. Autom.Cntr.* **AC-19**(1974):577-579.
118. Dickinson, B.W., "A new characterization of canonical convolutional encoders," *IEEE Trans. Inform. Theory* **IT-20**(1976): 352-354.
119. Dickinson, B.W., "The classification of linear state variable control laws," *SIAM J. Control & Opt.* **14**(1976): 467-477.
120. Dickinson, B.W., and E.D. Sontag, "Dynamical realizations of sufficient sequences," *IEEE Trans. Inform.Th.* **IT-31**(1985):670-676.
121. Dieudonné, J., *Foundations of Modern Analysis*, Academic Press, New York, 1960.
122. Dolezal, V., *Monotone Operators and Applications in Control and Network Theory*, Elsevier, New York, 1979.
123. Dorato, P. (ed.), *Robust Control*, IEEE Press, New York, 1987.
124. Doyle, J.C., K. Glover, P.P. Khargonekar, and B.A. Francis, "State-space solutions to the standard H_2 and H_∞ control problems," *IEEE Trans. Autom. Control* **AC-34**(1989): 831-846.
125. Eilenberg, S., *Automata, Languages, and Machines, Volume A*, Academic Press, New York, 1974.
126. El-Assoudi, R., and J.P. Gauthier, "Controllability of right-invariant systems on real simple Lie groups of type F_4, G_2, C_n, and B_n," *Math. Control Signals Systems* **1**(1988): 293-301
127. Elliott, D.L., "A consequence of controllability," *J. Diff. Eqs.* **10** (1971): 364-370.
128. Emre, E., "On necessary and sufficient conditions for regulation of linear systems over rings," *SIAM J.Control & Opt.* **20**(1982): 155-160.
129. Emre, E., and P.P. Khargonekar, "Regulation of split linear systems over rings: coefficient-assignment and observers," *IEEE Trans. on Autom. Control* **AC-27**(1982): 104-113.
130. Farmer, J.D., and J.J. Sidorowich, "Exploiting chaos to predict the future and reduce noise," in *Evolution, Learning, and Cognition* (Y.C. Lee, ed.) World Scientific Press, 1988, p. 277ff.
131. Faurre, P., "Stochastic realization algorithms," in *System Identification - Advances and Case Studies* (R.K. Mehra and D.G. Lainiotis, eds.), Academic Press, New York, 1976, pp. 1-25.
132. Faurre, P., M. Clerget, and F. Germain, *Operateurs Rationnels Positifs*, Dunod, Paris, 1979.
133. Feintuch, A., and R. Saeks, *System Theory. A Hilbert Space Approach*, Academic Press, New York, 1982.
134. Fleming, W.H., and R.W. Rishel, *Deterministic and Stochastic Optimal Control*, Springer-Verlag, New York, 1975.
135. Fleming, W.H., et. al., *Future Directions in Control Theory: A Mathematical Perspective*, SIAM Publications, Philadelphia, 1988.
136. Fliess, M., "Sur certaines familles de séries formelles," Thèse de Doctorat d'Etat, Université Paris VII, 1972.

137. Fliess, M., "Matrices de Hankel," *J. Math Pures et. Appl.* **53** (1974): 197-224.
138. Fliess, M., "Sur la realization des systemes dynamiques bilineaires," *C.R. Acad. Sc. Paris* **A277**(1973): 243-247.
139. Fliess, M., "The unobservability ideal for nonlinear systems," *IEEE Trans. Autom. Control* **AC-26**(1981): 592-593.
140. Fliess, M., "Fonctionnelles causales non linéaires et indéterminées non commutatives," *Bull. Soc. Math. France* **109**(1981): 3-40.
141. Fliess, M., "Quelques définitions de la théorie des systèmes à la lumière des corps diffèrentiels," *C.R.Acad. Sc. Paris* **I**(1987): 91-93.
142. Fliess, M., and I. Kupka, "A finiteness criterion for nonlinear input-output differential systems," *SIAM J.Control & Opt.* **21**(1983): 721-728.
143. Fornasini, E., and G. Marchesini, "Algebraic realization theory of bilinear discrete-time input/output maps," *J.Franklin Inst.* **301**(1976): 143-159.
144. Fortmann,T.E., and K.L.Hitz, *An Introduction to Linear Control Systems*, Marcel Dekker, New York, 1977.
145. Francis, B.A., *A Course in H_∞ Control Theory*, Springer-Verlag, New York, 1987.
146. Fruechte, R.D., *Application of Control Theory in the Automotive Industry*, Interscience Enterprises, Geneve, 1983.
147. Fuhrmann, P.A., *Linear Systems and Operators in Hilbert Space*, McGraw-Hill, New York, 1981.
148. Fujiwara, H., "Computational complexity of controllability/observability problems for combinatorial circuits," in *Proc. 18th IEEE Int. Symp. on Fault Tolerant Computing*, IEEE Publ., 1988, pp. 64-69.
149. Fuller, A.T., "The early development of control theory, parts I and II," *J. Dynamic Systems, Measurement, and Control* **98**(1976): 109-118 and 224-235.
150. Fuller, A.T., "In the large stability of relay and saturated control systems with linear controllers," *Int. J. Control*, **10**(1969): 457-480.
151. Gamkrelidze, R.V., *Principles of Optimal Control Theory*, Plenum Press, New York, 1978.
152. Gantmacher, F.R., *Theory of Matrices* (English transl.), Chelsea, New York, 1959.
153. Gelb, A., et.al., *Optimal Estimation*, MIT Press, Cambridge, MA, 1974.
154. Gibson, J.A., and T.T. Ha, "Further to the preservation of controllability under sampling," *Int. J. Control* **31**(1980): 1013-1026.
155. Gilbert, E., "Controllability and observability in multivariable control systems," *SIAM J. Control* **1**(1963): 128-151.
156. Ginsburg, S., *An Introduction to Mathematical Machine Theory*, Addison-Wesley, Reading, MA, 1962.
157. Glass, M., "Exponential stability revisited," *Int. J. Control* **46**(1987): 1505-1510.

158. Gopal, M., *Modern Control System Theory*, Wiley, New York,1984.
159. Gopinath, B., "On the control of multiple input-output systems," *Bell Systems Tech. J.*, March 1971.
160. Grasse, K.A., "Perturbations of nonlinear controllable systems," *SIAM J. Control & Opt.* **19**(1981): 203-220.
161. Grasse, K.A., "Nonlinear perturbations of control-semilinear control systems," *SIAM J. Control & Opt.* **20**(1982): 311-327.
162. Grasse, K.A., "On accessibility and normal accessibility: the openness of controllability in the fine C^0 topology," *J.Diff. Eqs.* **53**(1984): 387-414.
163. Grasse, K.A., and H.J. Sussmann, "Global controllability by nice controls," in *Nonlinear Controllability and Optimal Control* (H.J. Sussmann, ed.), Dekker, New York, 1990, pp. 33-79.
164. Grasselli, O.M., and A. Isidori, "An existence theorem for observers of bilinear systems," *IEEE Trans. Autom.Control* **26**(1981):1299-1301.
165. Grimble, M.J., and M.A. Johnson, *Optimal Control and Stochastic Estimation: Theory and Applications*, Wiley, New York, 1988.
166. Grizzle, J.W., "Controlled invariance for discrete-time nonlinear systems with an application to the decoupling problem," *IEEE Trans. Autom. Control* **AC-30**(1985): 868-874.
167. Grizzle, J.W., and A. Isidori, "Block noninteracting control with stability via static state feedback," *Math. Control Signals Systems* **2**(1989): 315-341.
168. Grodins, F.S., *Control Theory and Biological Systems*, Columbia University Press, New York, 1963.
169. Hahn, W., *Stability of Motion*, Springer-Verlag, New York, 1967.
170. Hájek, O., *Control Theory in the Plane*, Springer-Verlag, Berlin, 1991.
171. Hammer, J., "Fraction representations of nonlinear systems: a simplified approach," *Int. J. Control* **46**(1987): 455-472.
172. Hammer, J., "On nonlinear systems, additive feedback, and rationality," *Int. J. Control* **40**(1984): 953-969.
173. Harris, C.J., and S.A. Billings, *Self-Tuning and Adaptive Control: Theory and Applications*, P. Peregrinus, London, 1985.
174. Harvey, C.A., and E.B. Lee, "On stability with linear feedback control," Honeywell Corp. Memo Number 8150 (to C.R. Stone), January 8, 1963.
175. Hautus, M.L.J., "Controllability and observability conditions for linear autonomous systems," *Ned. Akad. Wetenschappen, Proc. Ser. A* **72** (1969): 443-448.
176. Hautus, M.L.J., "Stabilization, controllability, and observability for linear autonomous systems," *Ned. Akad. Wetenschappen, Proc. Ser. A* **73**(1970): 448-455.
177. Hautus, M.L.J., "A simple proof of Heymann's Lemma," Memo COSOR 76-17, Department of Mathematics, Eindhoven Univ. of Technology, Nov.76.

178. Hautus, M.L.J., "(A, B)-invariant and stabilizability subspaces, a frequency domain description," *Automatica* **16**(1980): 703-707.

179. Hautus, M.L.J., and E.D. Sontag, "New results on pole-shifting for parametrized families of systems", *J.Pure Appl. Algebra* **40**(1986): 229-244.

180. Hazewinkel, M., and A-M. Perdon, "On families of systems: pointwise-local-global isomorphism problems," *Int.J. Control* **33**(1981): 713-726.

181. Helmke, U., "Linear dynamical systems and instantons in Yang-Mills theory," *IMA J.Math. Control and Information* **3**(1986): 151-166.

182. Hermann, R., "On the accessibility problem in control theory," in *Int. Symp. Diff. Eqs. and Mechanics*, Academic Press, New York, 1973.

183. Hermann, R., *Geometric Structure Theory of Systems-Control, Theory and Physics*, Math. Sci. Press, Brookline, MA, 1974.

184. Hermann, R., *Cartanian Geometry, Nonlinear Waves, and Control Theory*, Math. Sci. Press, Brookline, MA, 1979.

185. Hermann, R., and A.J. Krener, "Nonlinear controllability and observability," *IEEE Trans. Autom. Control* **22**(1977): 728-740.

186. Hermes, H., "On a stabilizing feedback attitude control," *J. Optim. Theory and Appl.* **31**(1980): 373-384.

187. Hermes, H., "On the synthesis of a stabilizing feedback control via Lie algebraic methods," *SIAM J. Control & Opt.* **18**(1980):352-361.

188. Hermes, H., and J.P. LaSalle, *Functional Analysis and Time-Optimal Control*, Academic Press, New York, 1969.

189. Hestenes, M.R., *Calculus of Variations and Optimal Control Theory*, R. E. Kriegerz, Huntington, New York, 1980.

190. Hewitt, E., and K. Stromberg, *Real and Abstract Analysis*, Springer-Verlag, New York, 1965.

191. Heymann, M., "Comments on 'Pole assignment in multi-input controllable linear systems'," *IEEE Trans. Autom. Control* **AC-13** (1968): 748-749.

192. Hinrichsen, D., and J. O'Halloran, "A complete characterization of orbit closures of controllable singular systems under restricted system equivalence," *SIAM J. Control & Opt.* **28**(1990): 602-623.

193. Hirschorn, R.M., "Controllability in nonlinear systems," *J. Diff. Eqs.* **19**(1975): 46-61.

194. Hirschorn, R.M., "Invertibility of nonlinear control systems," *SIAM J. Control & Opt.* **17**(1979): 289-297.

195. Hirschorn, R.M., "(A,B)-invariant distributions and disturbance decoupling in nonlinear systems," *SIAM J. Control & Opt.* **19** (1981): 1-19.

196. Hsu, F.H., *On Reachable Sets*, Ph.D. Thesis, Case Western Reserve University, Cleveland, 1973.

197. Hunt, L.R., R. Su, and G. Meyer, "Design for multi-input nonlinear systems," in *Differential Geometric Control theory* (R.W.Brockett, R.S.Millman, and H.J.Sussmann, eds.), Birkhauser, Boston, 1983.

198. Isidori, A., "Direct construction of minimal bilinear realizations from nonlinear input/output maps," *IEEE Trans. Autom. Control* **18**(1973): 626-631.

199. Isidori, A., *Nonlinear Control Systems, Third Edition*, Springer-Verlag, London, 1995.

200. Isidori, A., A.J. Krener, C. Gori-Giorgi, and S. Monaco, "Nonlinear decoupling via feedback: A differential geometric approach," *IEEE Trans. Autom. Control* **26**(1981): 331-345.

201. Jacobson, C.A., and C.N. Nett, "Linear state-space systems in infinite-dimensional space: The role and characterization of joint stabilizability/detectability," *IEEE Trans. Autom. Control* **33**(1988): 541-549.

202. Jacobson, D.H., *Extensions of Linear-Quadratic Control, Optimization and Matrix Theory*, Academic Press, New York, 1977.

203. Jakubczyk, B., "Invertible realizations of nonlinear discrete time systems," in *Proc. Princeton Conf.Inf. Sc.and Systs.*(1980): 235-239.

204. Jakubczyk, B., "Poisson structures and relations on vector fields and their Hamiltonians," *Bull. Polish Acad Sci* **34**(1986): 713-721.

205. Jakubczyk, B., "Feedback linearization of discrete-time systems," *Systems and Control Letters* **9**(1987): 441-416.

206. Jakubczyk, B., and D. Normand-Cyrot, "Orbites de pseudo-groupes de diffeomorphismes et commandabilité des systemes non linearires en temps discret," *C.R.Acad. Sc. Paris* **298-I**(1984): 257-260.

207. Jakubczyk, B., and W. Respondek, "On linearization of control systems," *Bull. Acad. Pol. Sci., Ser. Sci. Math. Astr. Phys.* **28**(1980): 517-522.

208. Jakubczyk, B., and E.D. Sontag, "Controllability of nonlinear discrete-time systems: A Lie-algebraic approach," *SIAM J. Control & Opt.* **28**(1990): 1-33.

209. James, M.R., "Asymptotic nonlinear filtering and large deviations with application to observer design," Systems Research Center Report TR-88-28, University of Maryland, 1988. (Ph. D. Dissertation.)

210. Jurdjevic, V., and J.P. Quinn, "Controllability and stability," *J. Diff.Eqs.* **28**(1978): 381-389.

211. Jurdjevic, V., and H.J. Sussmann, "Control systems on Lie groups," *J.Diff.Eqs.* **12**(1972): 313-329.

212. Kailath,T., *Linear Systems*, Prentice Hall, Englewood Cliffs, 1980.

213. Kalman, R.E., "Nonlinear aspects of sampled-data control systems," in *Proc. of Symp. on Nonlinear Circuit Analysis* (J. Fox, ed.), Polytechnic Institute of Brooklyn, 1956, pp. 273-313.

214. Kalman, R.E., "A new approach to linear filtering and prediction problems," *ASME Trans.* **82D**(1960): 33-45.

215. Kalman, R.E., "Contributions to the theory of optimal control," *Bol. Soc. Mat. Mex.* **5**(1960): 102-119.

216. Kalman, R.E., "On the general theory of control systems," in *Proc. First IFAC Congress Automatic Control*, Moscow, 1960, Butterworths, London, vol. 1, pp. 481-492.

217. Kalman, R.E., "Canonical structure of linear dynamical systems," *Proc. Natl. Acad. Sci.* **48**(1962): 596-600.

218. Kalman, R.E., "Mathematical description of linear dynamical systems," *SIAM J. Control* **1**(1963): 152-192.

219. Kalman, R.E., "The theory of optimal control and the calculus of variations," in *Mathematical Optimization Techniques* (R. Bellman, ed.), Univ. of California Press, 1963, pp. 309-331.

220. Kalman, R.E., "Algebraic structure of linear dynamical systems, I. The module of Σ," *Proc. Natl. Acad. Sci. (USA)* **54**(1965): 1503-1508.

221. Kalman, R.E., "Irreducible realizations and the degree of a rational matrix," *J. SIAM* **13**(1965): 520-544.

222. Kalman, R.E., "Linear stochastic filtering - Reappraisal and outlook," in *Proc. Symp. on System Theory* (J. Fox, ed.), Polytechnic Press, New York, 1965, pp. 197-205.

223. Kalman, R.E., "Algebraic aspects of the theory of dynamical systems," in *Differential Equations and Dynamical Systems* (J.K. Hale and J.P. LaSalle, eds.), Academic Press, New York, 1967,pp.133-146.

224. Kalman, R.E., "Kronecker invariants and feedback," in *Ordinary Differential Equations* (L. Weiss, ed.), Academic Press, New York, 1972, pp. 459-471.

225. Kalman, R.E., "Algebraic-geometric description of the class of linear systems of constant dimension," in *Proc. 8th Princeton Conf. Inform. Sci. and Systems*, 1974, pp. 189-191.

226. Kalman, R.E., "Optimization, mathematical theory of, control theory" in *Encyclopaedia Britannica, Fifteenth Edition*, 1974, pp. 636-638.

227. Kalman, R.E., "Algebraic aspects of the generalized inverse," in *Generalized Inverses and Applications* (M. Zuhair Nashed, ed.), Academic Press, New York, 1976, pp. 111-124.

228. Kalman, R.E., "Identifiability and modeling in econometrics," *Developments in Statistics* **4**(1983): 97-135.

229. Kalman, R.E., and R.S. Bucy, "New results in linear filtering and prediction theory," *Trans. ASME (J. Basic Eng.)* **83D**(1961): 95-108.

230. Kalman, R.E., P.L. Falb, and M.A. Arbib, *Topics in Mathematical System Theory*, McGraw-Hill, New York, 1969.

231. Kalman, R.E., Y.C. Ho, and K.S. Narendra, "Controllability of linear dynamical systems," *Control Diff. Eqs.* **1**(1963): 189-213.

232. Kalouptsidis, N., and J. Tsinias, "Stability improvement of nonlinear systems by feedback," *IEEE Trans. Autom. Control* **29**(1984): 346-367.

233. Kamen, E.W., "On an algebraic theory of systems defined by convolution operators," *Math. Systems Theory* **9** (1975): 57-74.

234. Kamen, E.W., "Module structure of infinite-dimensional systems with applications to controllability," *SIAM J.Control & Opt.* **14**(1976):389-408.

235. Kamen, E.W., P.K. Khargonekar, and K.R. Polla, "A transfer function approach to linear time-varying systems," *SIAM J.Control & Opt.* **23**(1985): 550-565.

236. Kappel, F., K. Kunisch, and W. Schappacher, *Control Theory for Distributed Parameter Systems and Applications*, Springer-Verlag, New York, 1983.

237. Karny, M., K. Warwick, and V. Kurkova (eds.), *Dealing with Complexity: a Neural Network Approach*, Springer-Verlag, London, 1997.

238. Kawski, M., "Stabilization of nonlinear systems in the plane," *Systems and Control Letters* **12**(1989): 169-176.

239. Kawski, M., "High-order small-time local controllability," in *Nonlinear Controllability and Optimal Control* (H.J. Sussmann, ed.), Dekker, New York, 1990, pp. 431-467.

240. Kawski, M., "The complexity of deciding controllability," in *Systems Control Lett.* **15** (1990): 9-14.

241. Kawski, M., and Sussmann, H.J., "Noncommutative power series and formal Lie-algebraic techniques in nonlinear control theory," in *Operators, Systems and Linear Algebra: Three Decades of Algebraic Systems Theory* (U. Helmke, D. Praetzel-Wolters, and E. Zerz, eds.), B.G. Teubner, Stuttgart, 1997, pp. 111-129.

242. Kenney, C., and A.J. Laub, "Controllability and stability radii for companion form systems," *Math.Control Signals Systems* **1**(1988):239-256.

243. Khargonekar, P.P., "On matrix fraction representations for linear systems over commutative rings," *SIAM J. Control & Opt.* **20**(1982): 172-197.

244. Khargonekar, P.P., T.T. Georgiou, and A.M. Pascoal, "On the robust stabilizability of linear time invariant plants with unstructured uncertainty," *IEEE Trans. Autom. Control* **AC-32**(1987): 201-208.

245. Khargonekar, P.P., K. R. Poolla, and A. Tannenbaum, "Robust control of linear time-invariant plants by periodic compensation," *IEEE Trans. Autom. Control* **AC-30**(1985): 1088-1096.

246. Khargonekar, P.P., and E.D. Sontag, "On the relation between stable matrix fraction decompositions and regulable realizations of systems over rings," *IEEE Trans. Autom. Control* **27**(1982): 627-638.

247. Kirk, D.E., *Optimal Control Theory; an Introduction*, Prentice-Hall, Englewood Cliffs, 1970.

248. Knobloch, H.W., *Higher Order Necessary Conditions in Optimal Control Theory*, Springer-Verlag, New York, 1981.

249. Koditschek, D.E., "Adaptive techniques for mechanical systems," in *Proc.5th. Yale Workshop on Adaptive Systems*, Yale University, New Haven, 1987, pp. 259-265.

250. Kokotovic, P.V., and H.J. Sussmann, "A positive real condition for global stabilization of nonlinear systems," *Systems Control Letters* **13**(1989): 125-134.

251. Kolmanovsky, I., and N.H. McClamroch, "Developments in nonholonomic control problems," *Control Systems Magazine* **15**(1995): 20-36.

252. Koppel, L.B., *Introduction to Control Theory, with Applications to Process Control*, Prentice-Hall, Englewood Cliffs, 1968.

253. Kou, S.R., D.L. Elliott, and T.J. Tarn, "Exponential observers for nonlinear dynamical systems," *Inform.Contr* **29**(1975): 204-216.

254. Krasnoselskii, M.A., and P.P. Zabreiko, *Geometric Methods of Nonlinear Analysis*, Springer-Verlag, Berlin, 1983.

255. Krener, A.J., "A generalization of Chow's theorem and the bang-bang theorem to nonlinear control systems," *SIAM J. Control* **12** (1974): 43-52.

256. Krener, A.J., "The high order maximal principle and its application to singular extremals," *SIAM J.Control & and Opt.* **15**(1977): 256-293.

257. Krener, A.J., "(Adf,g), (adf,g) and locally (adf,g) Invariant and Controllability Distributions," *SIAM J.Control & Opt.* **23**(1985): 523-549.

258. Krener, A.J., "Nonlinear controller design via approximate normal forms," in *Proc. IMA Conf.on Signal Processing, Minneapolis, June-Aug. 1988*, Institute for Mathematics and its Applications, 1989.

259. Krstic, M., I. Kanellakopoulos, and P. Kokotovic, *Nonlinear and Adaptive Control Design*, Wiley, New York, 1995.

260. Kuo, B.C., *Automatic Control Systems*, Seventh Edition, Prentice Hall, Upper Saddle River, NJ, 1995.

261. Kupka, I., and G. Sallet, "A sufficient condition for the transitivity of pseudo-groups: Application to system theory," *J. Diff. Eqs.* **47**(1973): 462-470.

262. Kurzweil, J., "On the inversion of Lyapunov's second theorem on stability of motion," *Amer. Math. Soc. Transl., Ser.2*, **24** (1956): 19-77.

263. Kwakernaak, H., and R. Sivan, *Linear Optimal Control Systems*, Wiley, New York, 1972.

264. Lang, S., *Analysis II*, Addison-Wesley, Reading, MA, 1969.

265. Langenhop, C.E., "On the stabilization of linear systems," *Proc. AMS* **15**(1964): 735-742.

266. Lee, E.B., and L. Markus, *Foundations of Optimal Control Theory*, Wiley, New York, 1968.

267. Lee, K.K., and A. Arapostathis, "On the controllability of piecewise linear hypersurface systems," *Systems & Control Letters* **9**(1987): 89-96.

268. Lee, K.K., and A. Arapostathis, "Remarks on smooth feedback stabilization of nonlinear systems," *Systems & Control Letters* **10**(1988):41-44.

269. Lee, H.G., A. Arapostathis, and S.I. Marcus, "On the linearization of discrete time systems," *Int. J. Control* **45**(1987): 1103-1124.

270. Leigh, J.R., *Functional Analysis and Linear Control Theory*, Academic Press, New York, 1980.

271. Lenz, K., J. Doyle, and P. Khargonekar, "When is a controller H_∞ optimal?" *Math. Control Signals Systems* **1**(1988): 107-122.

272. Leontaritis, I.J., and S.A. Billings, "Input-output parametric models for nonlinear systems," Parts I and II, *Int. J. Control* **41** (1985): 303-344.

273. Levi, M., and F.M.A. Salam (eds.), *Dynamical Systems Approaches to Nonlinear Problems in Systems and Circuits*, SIAM Publications, Philadelphia, 1988.

274. Levine, W.S. (ed.), *The Control Handbook* CRC Press, Boca Raton, 1995, pp. 895-908.

275. Levis, A.H., et. al., "Challenges to control: A collective view. Report from the workshop held at the University of Santa Clara on September 18-19, 1986," *IEEE Trans. Autom. Control* **32**(1987): 275-285.

276. Lewis, F.L., *Optimal Estimation: With an Introduction to Stochastic Control Theory*, Wiley, New York, 1986.

277. Lewis, F.L., "A survey of linear singular systems," *Circuits, Systs., and Signal Proc.* **5**(1986): 3-36.

278. Lin, Y., and E.D. Sontag, "Control-Lyapunov universal formulae for restricted inputs," *Control: Theory and Advanced Technology* **10**(1995): 1981-2004.

279. Lin, Y., E.D. Sontag, and Y. Wang, "A smooth converse Lyapunov theorem for robust stability," *SIAM J. Control and Optimization* **34** (1996): 124-160.

280. Lindquist, A., and G. Picci, "Realization theory for multivariate stationary Gaussian processes," *SIAM J.Control & Opt.* **23**(1985): 809-857.

281. Lions, J.L., *Optimal Control of Systems Governed by Partial Differential Equations*, Springer-Verlag, Berlin, 1971.

282. Ljung, L., and T. Söderström, *Theory and Practice of Recursive Estimation*, MIT Press, Cambridge, 1983.

283. Lo, J.T., "Global bilinearization of systems with controls appearing linearly," *SIAM J.Control* **13**(1975): 879-885.

284. Lobry, C., "Controllabilité des systèmes non linéaires," *SIAM J. Control* **8**(1970): 573-605.

285. Logemann, H., "On the transfer matrix of a neutral system: Characterizations of exponential stability in input-output terms," *Systems and Control Letters* **9**(1987): 393-400.

286. Longchamp, R., "State-feedback control of bilinear systems," *IEEE Trans. Autom. Cntr.* **25**(1980): 302-306.

287. Luenberger, D.G., "An introduction to observers," *IEEE Trans. Autom. Cntr.* **16**(1971): 596-602.

288. MacFarlane, A.G.J., "An eigenvector solution of the optimal linear regulator," *J.Electron. Control* **14**(1963): 643-654.

289. Macki, J., and A. Strauss., *Introduction to Optimal Control Theory*, Springer-Verlag, New York, 1982.
290. Manes, E.G., ed., *Category Theory Applied to Computation and Control, Proceedings of the First International Symposium* (San Francisco, February 25-26, 1974), Springer-Verlag, New York, 1975.
291. Manitius, A., "Necessary and sufficient conditions of approximate controllability for general linear retarded systems," *SIAM J. Control & Opt.* **19**(1981): 516-532.
292. Marcus, S.I., "Optimal nonlinear estimation for a class of discrete time stochastic systems," *IEEE Trans. Autom. Cntr.* **24**(1979): 297-302.
293. Marcus, S.I., "Algebraic and geometric methods in nonlinear filtering," *SIAM J. Control & Opt.* **22**(1984): 817-844.
294. Marino, R., "Feedback stabilization of single-input nonlinear systems," *Systems and Control Letters* **10**(1988): 201-206.
295. Markus, L., and H. Yamabe, "Global stability criteria for differential systems," *Osaka J. Math.* **12**(1960): 305-317.
296. Massera, J.L., "Contributions to stability theory," *Annals of Math* **64** (1956): 182-206. Erratum in *Annals of Math* **68**(1958): 202.
297. Massey, J.L., and M.K. Sain, "Codes, automata and continuous systems: Explicit interconnections," *IEEE Trans. Autom. Control* **12**(1970): 644-650.
298. Mayeda, H., and T. Yamada, "Strong structural controllability," *SIAM J. Control & Opt.* **17**(1979): 123-138.
299. Mayr, O., *The Origins of Feedback Control*, MIT Press, Cambridge, MA, 1970.
300. Mesarovic, M.D., and Y.Takahara, *General Systems Theory: Mathematical Foundations*, Academic Press, New York, 1975.
301. Minorsky, N., *Theory of Nonlinear Control Systems*, McGraw-Hill, New York, 1969.
302. Monaco, S., and D. Normand-Cyrot, "Invariant distributions for nonlinear discrete-time systems," *Systems and Control Letters* **5**(1984): 191-196.
303. Moog, C. H., "Nonlinear decoupling and structure at infinity," *Math. Control Signals Systems* **1**(1988): 257-268
304. Moroney, P., *Issues in the Implementation of Digital Feedback Compensators*, MIT Press, Cambridge, MA, 1983.
305. Morse, A.S., "Ring models for delay differential systems," *Automatica* **12**(1976): 529-531.
306. Morse, A.S., and L.M. Silverman, "Structure of index-invariant systems," *SIAM J. Control* **11**(1972): 215-225.
307. Narendra, K.S., and K. Parthasarathy, "Identification and control of dynamical systems using neural networks," *IEEE Trans. Neural Nets* **1**(1990): 4-27.
308. Nelson, E., *Tensor Analysis*, Princeton University Press, Princeton, NJ, 1967.

309. Nieuwenhuis, J.W., and J.C. Willems, "Continuity of dynamical systems - a system theoretic approach," *Math. Control Signals Systems* **1**(1988): 147-165.

310. Nijmeijer, H., "Observability of autonomous discrete-time nonlinear systems: a geometric approach," *Int. J. Control* **36**(1982): 867-874.

311. Nijmeijer, H., and A.V. Van der Schaft, *Nonlinear Dynamical Control Systems*, Springer-Verlag, New York, 1990.

312. Ocone, O., "Probability densities for conditional statistics in the cubic sensor problem," *Math. Control Signals Systems* **1**(1988): 183-202.

313. Ozveren, C.M., G.C. Verghese, and A.S. Willsky, "Asymptotic orders of reachability in perturbed linear systems," *IEEE Trans. Autom. Control* **33**(1988): 915-923.

314. Padulo, L., and M.A. Arbib, *System Theory. A Unified State-Space Approach to Continuous and Discrete Systems*, Saunders, Philadelphia, 1974. Republished by Hemisphere, Washington, DC.

315. Pallu de La Barriere, R., *Optimal Control Theory; A Course in Automatic Control Theory*, Saunders, Philadelphia, 1967.

316. Pandolfi, L., "Canonical realizations of systems with delays," *SIAM J. Control & Opt.* **21**(1983): 598-613.

317. Pitchford, J.D., and S.J. Turnovsky, *Applications of Control Theory to Economic Analysis*, Elsevier/North-Holland, Amsterdam, 1977.

318. Polderman, J.W., "A state space approach to the problem of adaptive pole placement," *Math. Control Signals Systems* **2**(1989): 71-94

319. Pontryagin, L.S., V.G. Boltyanskii, R.V. Gamkrelidze, and E.F. Mischenko, *The Mathematical Theory of Optimal Processes*, Wiley, New York, 1962.

320. Popov, V.M., "On a new problem of stability for control systems," *Autom. Remote Control* (English translation) **24**(1963): 1-23.

321. Popov, V.M., *Hyperstability of Control Systems*, Springer-Verlag, Berlin, 1973. (Translation of *Hiperstabilitatea Sistemelor Automate*, published in 1966 in Rumanian.)

322. Potter, J.E., "Matrix quadratic solutions," *SIAM J. Appl. Math.* **14** (1966): 496-501.

323. Praly, L., and Y. Wang, "Stabilization in spite of matched unmodelled dynamics and an equivalent definition of input-to-state stability," *Math. Control Signals Systems* **9**(1996): 1-33.

324. Prichard, A.J., and J. Zabczyk, "Stability and stabilizability of infinite dimensional systems," *SIAM Review* **23**(1983): 25-52.

325. Ramirez, W.F., *Application of Optimal Control Theory to Enhanced Oil Recovery*, Elsevier, New York, 1987.

326. Rantala, J., *An Application of Stochastic Control Theory to Insurance Business*, University of Tampere, Tampere, 1984.

327. Reutenauer, C., *Free Lie Algebras*, Oxford University Press, New York, 1993.

328. Riggs, D.S., *Control Theory and Physiological Feedback Mechanisms*, Krieger, Huntington, New York, 1976.

329. Rissanen, J., "Control system synthesis by analogue computer based on the 'generalized linear feedback' concept," in *Proc. Symp. Analoge Comp. Applied to the Study of Chem. Processes*, Presses Académiques Européennes, Brussels, 1961.

330. Rockafellar, R.T., *Convex Analysis*, Princeton University Press, 1970.

331. Rosenbrock, H.H., *State Space and Multivariable Theory*, Nelson, London, 1970.

332. Rouchaleau, Y., B.F. Wyman, and R.E. Kalman, "Algebraic structures of linear dynamical systems III. Realization theory over a commutative ring," *Proc. Nat. Acad. Sci.* **69**(1972): 3404-3406.

333. Rouchaleau, Y., and B.F. Wyman, "Linear dynamical systems over integral domains," *J. Comput. System Sci.* **9**(1974): 129-142.

334. Rouche, N., P. Habets, and M. Laloy, *Stability Theory by Lyapunov's Direct Method*, Springer-Verlag, New York, 1977.

335. Rugh, W.J., *Mathematical Description of Linear Systems*, Marcel Dekker, New York, 1975.

336. Rugh, W.J., *Nonlinear System Theory: the Volterra/Wiener Approach*, The Johns Hopkins University Press, Baltimore, 1981.

337. Russell, D.L., *Mathematics of Finite Dimensional Control Systems*, Marcel Dekker, New York, 1979.

338. Ryan, E.P., "On Brockett's condition for smooth stabilizability and its necessity in a context of nonsmooth feedback," *SIAM J. Control Optim.* **32**(1994): 1597-1604.

339. Ryan, E.P., and N.J. Buckingham, "On asymptotically stabilizing feedback control of bilinear systems," *IEEE Trans. Autom. Cntr.* **28**(1983): 863-864.

340. Sage, A.P., and J.L. Melsa, *Estimation Theory with Applications to Communications and Control*, Krieger, Huntington, 1979.

341. Salam, F.M.A., "Feedback stabilization of the nonlinear pendulum under uncertainty: a robustness issue," *Systems and Control Letters* **7**(1986): 199-206.

342. Saskena, V.R., J.O'Reilly, and P.V. Kokotovic, "Singular perturbations and two-scale methods in control theory: survey 1976-1983," *Automatica* **20**(1984): 273-293.

343. Schmitendorf, W.E., and B.R. Barmish, "Null controllability of linear systems with constrained controls," *SIAM J. Control & Opt.* **18**(1980): 327-345.

344. Schwartz, C.A., B.W. Dickinson, and E.D. Sontag, "Characterizing innovation representations for discrete-time random processes," *Stochastics* **11**(1984): 159-172.

345. Seidman, T.I., "How violent are fast controls?," *Math. Control Signals Systems* **1**(1988): 89-95.

346. Seierstad, A., and K. Sydsater, *Optimal Control Theory with Economic Applications*, North-Holland, New York, 1987.

347. Sethi, S.P., and G.L. Thompson, *Optimal Control Theory: Applications to Management Science*, Kluwer, Boston, 1981.

348. Shaefer, H.H., *Topological Vector Spaces*, Springer-Verlag, New York, 1980.

349. Sharma, P.K., "On pole assignment problem in polynomial rings," *Systems and Control Letters* **5**(1984): 49-54.

350. Siljak, D.D., *Large Scale Dynamic Systems*, North-Holland, New York, 1978.

351. Silva Leite, F., and P.E. Crouch, "Controllability on classical Lie groups," *Math. Control Signals Systems* **1**(1988): 31-42.

352. Singh, M.G. (ed.), *Systems & Control Encyclopedia: Theory, Technology, Applications*, Pergamon Press, Oxford, 1987.

353. Sira-Ramirez, H., "A geometric approach to pulse-width-modulated control in dynamical systems," *IEEE Trans. Autom. Control* **34** (1989): 184-187.

354. Skowronski, J.M., *Control Theory of Robotic Systems*, World Scientific, Teaneck, 1989.

355. Slemrod, M., "Stabilization of bilinear control systems with applications to nonconservative problems in elasticity," *SIAM J.Control & Opt.* **16**(1978): 131-141.

356. Sontag, E.D., "On certain questions of rationality and decidability," *J. Comp. Syst. Sci.* **11**(1975): 375-381.

357. Sontag, E.D., "Linear systems over commutative rings: A survey," *Ricerche di Automatica* **7**(1976): 1-34.

358. Sontag, E.D., "On the observability of polynomial systems," *SIAM J.Control & Opt.* **17**(1979):139-151.

359. Sontag, E.D., "Realization theory of discrete-time nonlinear systems: Part I- The bounded case," *IEEE Trans. Circuits and Syst.* **CAS-26**(1979): 342-356.

360. Sontag, E.D., *Polynomial Response Maps*, Springer-Verlag, New York, 1979.

361. Sontag, E.D., "Conditions for abstract nonlinear regulation," *Information and Control* **51**(1981): 105-127.

362. Sontag, E.D., "Nonlinear regulation: The piecewise linear approach," *IEEE Trans. Autom. Control* **AC-26**(1981): 346-358.

363. Sontag, E.D., "A Lyapunov-like characterization of asymptotic controllability," *SIAM J. Control & Opt.* **21**(1983): 462-471.

364. Sontag, E.D., "An algebraic approach to bounded controllability of nonlinear systems," *Int. J. Control* **39**(1984): 181-188.

365. Sontag, E.D., "A concept of local observability," *Systems and Control Letters* **5**(1984): 41-47.

366. Sontag, E.D., "Controllability is harder to decide than accessibility," *SIAM J.Control & Opt.* **26**(1988): 1106-1118.
367. Sontag, E.D., "Finite dimensional open-loop control generators for nonlinear systems," *Int. J. Control* **47** (1988): 537-556.
368. Sontag, E.D., "Remarks on stabilization and input-to-state stability," *Proc. IEEE Conf. Dec.and Control, Tampa, Dec. 1989*, IEEE Publications, pp. 1376-1378.
369. Sontag, E.D., "A 'universal' construction of Artstein's theorem on nonlinear stabilization," *Systems and Control Letters* **13**(1989): 117-123.
370. Sontag, E.D., "Smooth stabilization implies coprime factorization," *IEEE Trans. Autom. Control* **34**(1989): 435-443.
371. Sontag, E.D., "Integrability of certain distributions associated with actions on manifolds and applications to control problems," in *Nonlinear Controllability and Optimal Control* (H.J. Sussmann, ed.), pp. 81-131, Marcel Dekker, New York, 1990.
372. Sontag, E.D., "Feedback stabilization of nonlinear systems," in *Robust Control of Linear Systems and Nonlinear Control* (M.A. Kaashoek, J.H. van Schuppen, and A.C.M. Ran, eds.), Birkhäuser, Cambridge, MA, 1990, pp. 61-81.
373. Sontag, E.D., "Spaces of observables in nonlinear control," in *Proc. Intern. Congress of Mathematicians 1994*, Volume 2, Birkhäuser Verlag, Basel, 1995, pp. 1532-1545.
374. Sontag, E.D., "Control of systems without drift via generic loops," *IEEE Trans. Autom. Control* **40**(1995): 1210-1219.
375. Sontag, E.D., and H.J. Sussmann, "Complete controllability of continuous-time recurrent neural networks," *Systems and Control Letters* **30**(1997): 177-183.
376. Sontag, E.D., and Y. Wang, "New characterizations of the input to state stability property," *IEEE Trans. Autom. Control* **AC-41**(1996): 1283-1294.
377. Sontag, E.D., and Y. Wang, "Output-to-state stability and detectability of nonlinear systems," *Systems and Control Letters* **29**(1997): 279-290.
378. Spong, M.W., and M. Vidyasagar, *Robot Dynamics and Control*, Wiley, New York, 1989.
379. Stengel, R.F., *Stochastic Optimal Control: Theory and Application*, Wiley, New York, 1986.
380. Stephanopoulos, G., *Chemical Process Control: An Introduction to Theory and Practice*, Prentice-Hall, Englewood Cliffs, 1984.
381. Sussmann, H.J., "Semigroup representations, bilinear approximations of input-output maps, and generalized inputs," in *Mathematical Systems Theory, Udine 1975* (G. Marchesini, ed.), Springer-Verlag, New York, pp. 172-192.
382. Sussmann, H.J., "Minimal realizations and canonical forms for bilinear systems," *J. Franklin Institute* **301**(1976): 593-604.

383. Sussmann, H.J., "Existence and uniqueness of minimal realizations of nonlinear systems," *Math. Sys. Theory* **10** (1977): 263-284.

384. Sussmann, H.J., "Single-input observability of continuous-time systems," *Math. Systems Theory* **12**(1979): 371-393.

385. Sussmann, H.J., "Subanalytic sets and feedback control," *J.Diff. Eqs.* **31**(1979): 31-52.

386. Sussmann, H.J., "Analytic stratifications and control theory," in *Proc. 1978 Internat. Congr. Mathematicians, Helsinki*, pp. 865-871.

387. Sussmann, H.J., "Lie brackets, real analyticity, and geometric control," in *Differential Geometric Control theory* (R.W. Brockett, R.S. Millman, and H.J. Sussmann, eds.), Birkhauser, Boston, 1983.

388. Sussmann, H.J., "A general theorem on local controllability," *SIAM J. Control & Opt.* **25**(1987): 158-194.

389. Sussmann, H.J., "Minimal realizations of nonlinear systems," in *Geometric Methods in Systems Theory* (D.Q. Mayne and R.W. Brockett, eds.), Reidel, Dordrecht, 1973, pp. 243-252.

390. Sussmann, H.J., "Multidifferential calculus: chain rule, open mapping and transversal intersection theorems," in *Optimal Control: Theory, Algorithms, and Applications* (W.W. Hager and P.M. Pardalos, eds.), Kluwer, 1997.

391. Sussmann, H.J., and V. Jurdjevic, "Controllability of nonlinear systems," *J. Diff. Eqs.* **12**(1972): 95-116.

392. Sussmann, H.J., E.D. Sontag, and Y. Yang, "A general result on the stabilization of linear systems using bounded controls," *IEEE Trans. Autom. Control* **39**(1994): 2411-2425.

393. Sussmann, H.J., and J.C. Willems, "300 years of optimal control: From the brachystochrone to the maximum principle," *Control Systems Magazine* **17**(1997): 32-44.

394. Swan, G.W., *Applications of Optimal Control Theory in Biomedicine*, Marcel Dekker, New York, 1984.

395. Tannenbaum, A., *Invariance and System Theory: Algebraic and Geometric Aspects*, Springer-Verlag, New York, 1980.

396. Tannenbaum, A., "On pole assignability over polynomial rings," *Systems and Control Letters* **2**(1982):13-16.

397. Tannenbaum,A., "Polynomial rings over arbitrary fields in two or more variables are not pole assignable," *Systems and Control Letters* **2**(1982): 222-224.

398. Tay, T.T., and J.B. Moore, "Left coprime factorizations and a class of stabilizing controllers for nonlinear systems," in *Proc. IEEE Conf. Decision and Control*, Austin, Dec.1988, pp. 449-454.

399. Taylor, A.E., *Introduction to Functional Analysis*, Wiley, New York, 1958.

400. Thompson, W. (Lord Kelvin), "Mechanical integration of the general linear differential equation of any order with variable coefficients," *Proc. Roc. Soc.* **24**(1876): 271-275.

401. Toates, F.M., *Control Theory in Biology and Experimental Psychology*, Hutchinson, London, 1975.

402. Troutman, J.L., *Variational Calculus and Elementary Convexity*, Springer Verlag, New York, 1983.

403. Tsinias, J., "Sufficient Lyapunovlike conditions for stabilization," *Math. Control Signals Systems* **2**(1989): 343-357.

404. Tsinias, J., and N. Kalouptsidis, "Prolongations and stability analysis via Lyapunov functions of dynamical polysystems," *Math. Systems Theory* **20**(1987): 215-233.

405. Tsitsiklis, J.N., "On the control of discrete-event dynamical systems," *Math. Control Signals Systems* **2**(1989): 95-107.

406. Utkin, V.I., *Sliding Modes and Their Application in Variable Structure Systems*, Mir, Moscow, 1978.

407. Van der Schaft, A.J., "Stabilization of Hamiltonian Systems," *J. Nonlinear Analysis, MTA* **10**(1986): 1021-1035.

408. Van der Schaft, A.J., "On realizations of nonlinear systems described by higher-order differential equations," *Math Systems Theory* **19** (1987): 239-275.

409. Van Schuppen, J.H., "Stochastic realization problems," in *Three Decades of Mathematical System Theory* (Nijmeijer, H., and J.M. Schumacher, eds.), Springer-Verlag, Berlin, 1989.

410. Varaiya, P.P., and R. Liu, "Bounded-input bounded-output stability of nonlinear time-varying differential systems," *SIAM J.Control* **4**(1966): 698-704.

411. Veliov, V.M., and M.I. Krastanov, "Controllability of piecewise linear systems," *Systems and Control Letters* **7**(1986): 335-341.

412. Verma, M.S., "Coprime fractional representations and stability of nonlinear feedback systems," *Int. J. Control* **48**(1988): 897-918.

413. Vidyasagar, M., "Input-output stability of a broad class of linear time-invariant multivariable systems," *SIAM J.Ctr.***10**(1972):203-209.

414. Vidyasagar, M., *Nonlinear Systems Analysis*, Prentice-Hall, Englewood Cliffs, 1978.

415. Vidyasagar, M., "Decomposition techniques for large-scale systems with nonadditive interactions: stability and stabilizability," *IEEE Trans. Autom. Control* **25** (1980): 773-779.

416. Vidyasagar, M., "On the stabilization of nonlinear systems using state detection," *IEEE Trans. Autom. Control* **25**(1980): 504-509.

417. Vidyasagar, M., *Control Systems Synthesis: A Factorization Approach*, MIT Press, Cambridge, MA, 1985.

418. Vinter, R.B., "New results on the relationship between dynamic programming and the maximum principle," *Math. Control Signals Systems* **1**(1988): 97-105.

419. Viswanadham, N., and M. Vidyasagar, "Stabilization of linear and nonlinear dynamical systems using an observer-controller configuration," *Systems and Control Letters* **1** (1981): 87-91.

420. Vukobratovic, M., and D. Stokic, *Control of Manipulation Robots: Theory and Application*, Springer-Verlag, Berlin, 1982.

421. Wang, Y., and E.D. Sontag, "Algebraic differential equations and rational control systems," *SIAM J. Control and Opt.* **30**(1992): 1126-1149.

422. Willems, J.C., *The Analysis of Feedback Systems*, MIT Press, Cambridge, MA, 1971.

423. Willems, J.C., "Least squares stationary optimal control and the algebraic Riccati equation," *IEEE Trans. Autom. Control* **AC-16**(1971): 621-634.

424. Willems, J.C., "From time series to linear systems, Part I: Finite dimensional time-invariant systems," *Automatica* **22**(1986):561-580.

425. Willems, J.C., "From time series to linear systems, Part II: Exact Modelling," *Automatica* **22**(1986): 675-694.

426. Willems, J.C., "From time series to linear systems, Part III: Approximate Modelling," *Automatica* **23**(1987): 87-115.

427. Willsky, A.S., *Digital Signal Processing and Control and Estimation Theory: Points of Tangency, Areas of Intersection, and Parallel Directions*, MIT Press, Cambridge, MA, 1979.

428. Wilson, F.W., "Smoothing derivatives of functions and applications," *Trans. AMS* **139**(1969): 413-428.

429. Windeknecht, T.G., *General Dynamical Processes*, Academic Press, New York, 1971.

430. Wonham, M., *Linear Multivariable Control*, Springer-Verlag, New York, 1974.

431. Wonham, W.M., "On pole assignment in multi-input controllable linear systems," *IEEE Trans. Autom. Control* **AC-12**(1967):660-665.

432. Wonham, W.M., and A.S. Morse, "Decoupling and pole assignment in linear multivariable systems: a geometric approach," *SIAM J. Control* **8**(1970): 1-18.

433. Wonham, W.M., and P.J. Ramadge, "Modular supervisory control of discrete event systems," *Math. Control Signals Systems* **1**(1988): 13-30.

434. Wyman, B.F., "Time-varying linear discrete-time systems - realization theory," in *Advances in Math., Supp. Studies, Vol. I*, Academic Press, New York, 1978, 233-258.

435. Wyman, B.F., "Time-varying linear discrete-time systems: duality," *Pacific J. of Math.* **86**(1980): 361-377.

436. Wyman, B.F., and M.K. Sain, "The zero module and essential inverse systems," *IEEE Trans. Circuits and Systems* **CAS-28**(1981): 112-126.

437. Yamamoto, Y., "Realization theory of infinite dimensional linear systems, Parts I,II," *Math Sys. Theory* **15**(1981): 55-77, 169-190.
438. Yamamoto, Y., "Pseudo-rational input/output maps and their realizations: a fractional representation approach to infinite-dimensional systems," *SIAM J. Control & Opt.* **26-6**(1988): 1415-1430.
439. Yamamoto, Y., "Correspondence of internal and external stability - Realization, transfer functions and complex analysis," in *Realization and Modelling in System Theory* (M.A. Kaashoek, J.H. van Schuppen, and A.C.M. Ran, eds.), Birkhäuser, Cambridge, MA, 1990.
440. Yamamoto, Y., and E.D. Sontag, "On the existence of approximately coprime factorizations for retarded systems," *Systems and Control Letters* **13**(1989): 53-58.
441. Youla, D.C., H.A. Jabr, and J.J. Bongiorno, "Modern Wiener-Hopf design of optimal controllers: part II," *IEEE Trans. Autom. Cntr.* **AC-21**(1976): 319-338.
442. Young, L.C., *Lectures on the Calculus of Variations and Optimal Control Theory*, Chelsea, New York, 1980.
443. Zabczyk, J., "Some comments on stabilizability," *Applied Math Optimiz.* **19**(1989): 1-9.
444. Zadeh, L.A., and C.A. Desoer, *Linear System Theory — The State-Space Approach*, McGraw-Hill, New York, 1963.

List of Symbols

Symbol	Description	Page
$'$	transpose	
$\cdot$	derivative (with respect to time)	
$\rightsquigarrow$	reachability relation	28
$\Vert\cdot\Vert_\infty$	(essential) supremum norm	56
$\sim$	indistinguishability	263
$*$	(subscript) Jacobian, Fréchet differential	141,462,463
$*$	(subscript) linearization (differential) of system	40,51
$*$	(superscript) conjugate transpose; adjoint operator	89,105
$*$	convolution, only in Chapter 7	326
$\#$	pseudoinverse of operator or matrix	107,450
$\diamond$	empty (i.e., length zero) control	26
∇	gradient	141
$[\cdot,\cdot]$	Lie bracket of vector fields	142
ad_f	ad operator (Lie bracket)	142
α	map from initial state and control to final state	57
A	state matrix for linear systems	37,47,92
Ad_{tX}	"Ad" operator on vector fields (conjugation by e^{tX})	165
B	input matrix for linear systems	37,47,92
$\mathcal{B}_\rho(x)$	open ball of radius ρ about x	123
$\overline{\mathcal{B}}_\rho(x)$	closed ball of radius ρ about x	123,472
$\mathbf{B}_{n,m}$	class of B matrices for recurrent nets	129
co	convex hull	428
$\mathbb{C}$	complexes	
C	output matrix for linear systems	37,47
$\mathcal{C}(x)$	states that can be controlled to x	87
$\mathcal{C}^T(x)$	states controllable to x in time exactly T	87
$\mathcal{C}^{\leq T}(x)$	states controllable to x in time $\leq T$	156

$\mathcal{C}_{\mathcal{V}}^{\leq T}(x)$	states controllable to x in time $\leq T$ without leaving $\mathcal{V}$	156
$\mathcal{C}^1$	continuously differentiable functions	463
$\mathcal{C}^\infty$	smooth (infinitely differentiable) functions	461
$\mathcal{C}^k$	functions k-times continuously differentiable	461
$\mathcal{C}_n^0(\sigma, \tau)$	continuous maps on $[\sigma, \tau]$	472
$\mathcal{C}^\omega$	analytic functions	484
χ	characteristic polynomial	89
$\mathcal{D}$	domain of dynamics (continuous-time)	44
$\mathcal{D}_{\sigma,\tau}$	domain of transition map, fixed interval $[\sigma, \tau]$	53
$\mathcal{D}_f$	domain of flow associated to vector field f	147
$\mathcal{D}_\phi$	domain of transition map	26
$\mathcal{D}_\lambda$	domain of response map	30
δ	sampling interval	7,72
Δ	distribution (in the sense of differential geometry)	169
e^{tA}	exponential of matrix tA	488
e^A	exponential of matrix A	488
$e^{tf}x^0$	alternative notation for flow $\phi(t, x^0)$	150
$f(t, x, u)$	right-hand side for continuous-time system	41
$f(x, u)$	rhs for continuous-time time-invariant system	41
f, g	vector fields (typically associated to controls)	141
f_u	vector field associated to input value u	154
ϕ	transition (global dynamics) map	26
$\mathbb{F}$	smooth functions	141
Φ	fundamental matrix, for linear differential equation	488
Γ	trajectory: pair $(\xi(\cdot), \omega(\cdot))$	28
h	output map	27,33
$\mathcal{H}$	Hankel matrix	285
$\operatorname{int}(V)$	interior of set V	
I	identity matrix	
I	in Section 6.2, indistinguishability relation	269
$\mathcal{I}$	interval in $\mathbb{R}$ or relative to $\mathcal{T}$	25
φ	smooth function (nonlinear controllability chapter)	141
$\mathcal{J}$	total cost in optimal control problems	350,363
$k(t)$	kernel of integral operator	105
$k(x)$	feedback	236
κ_i	Kronecker indexes	190
$\mathbb{K}$	field; for continuous-time systems $= \mathbb{R}$ or $\mathbb{C}$	36

$\mathcal{R}(A,B)$	reachability space for pair	92
R	integral cost on controls, in LQ problems	363
$\mathcal{R}(x)$	reachable set from state x	84
$\mathcal{R}^T(x)$	reachable set from x in time exactly T	84
$\mathcal{R}^{\leq T}(x)$	reachable set from x in time $\leq T$	156
$\mathcal{R}_{\mathcal{V}}^{\leq T}(x)$	reachable set from x in time $\leq T$ without leaving $\mathcal{V}$	156
S	cost on final states, in LQ problems	363
$\mathcal{S}_{n,m}$	pairs $A \in \mathbb{K}^{n\times n}, B \in \mathbb{K}^{n\times m}$	96
$\mathcal{S}_{n,m}^c$	controllable pairs in $\mathcal{S}_{n,m}$	96
σ	time, usually initial	
$\sigma_{\min}$	smallest singular value	448
Σ	system	26
$\Sigma_{[\delta]}$	δ-sampled continuous-time system	72
τ	time, usually final	26
$\tau_{\min}, \tau_{\max}$	minimal and maximal time for solution	479
$\mathcal{T}$	time set	26
u	input value	33,41
	also input *functions* in informal discussions (instead of ω)	
$\mathcal{U}$	input-value space	26
$\mathcal{U}^{[\sigma,\tau)}$	controls $[\sigma,\tau) \to \mathcal{U}$	26
	(for continuous time, same as $(\sigma,\tau) \to \mathcal{U}$ or $[\sigma,\tau] \to \mathcal{U}$)	
$\mathbb{V}$	smooth vector fields	141
V	(control-) Lyapunov function	219
$\xrightarrow{\text{w}}$	weak convergence	424
ω	input (control) functions	26
x	state	41,26,33
	also state *functions* in informal discussions (instead of ξ)	
x^0	state, usually initial, or equilibrium	31
x^{f}	final state (in optimal control problems, typically)	423
$\mathcal{X}$	state space	26
X, Y, Z	vector fields (typically in generated Lie Algebra)	143
ξ	state path	28
y	output value	27,33
$\mathcal{Y}$	output-value space	27
$\mathbb{Z}$	integers	

Index

CPSIA information can be obtained
at www.ICGtesting.com
Printed in the USA
LVHW051741280523
748263LV00005B/286